Environmental History and Palaeolimnology

Developments in Hydrobiology 67

Series editor

H. J. Dumont

Environmental History
and Palaeolimnology

Proceedings of the Vth International Symposium on Palaeolimnology,
held in Cumbria, U.K.

Edited by
J. P. Smith, P. G. Appleby, R. W. Battarbee, J. A. Dearing, R. Flower,
E. Y. Haworth, F. Oldfield & P. E. O'Sullivan

Reprinted from Hydrobiologia, vol. 214 (1991)

Springer-Science+Business Media, B.V.

ISBN 978-0-7923-1318-2   ISBN 978-94-011-3592-4 (eBook)
DOI 10.1007/978-94-011-3592-4

Printed on acid-free paper

Dedicated to the memory of
Edward Smith Deevey
1914–1988

Contents

Preface xi

Obituary: Edward Smith Deevey (1914–1988) 1

Palaeolimnology in the English Lakes – some questions and answers over fifty years
    by W. Pennington 9

Accuracy and precision in sediment chronology
    by I. U. Olsson 25

^{241}Am dating of lake sediments
    by P. G. Appleby, N. Richardson & P. J. Nolan 35

Accuracies in Po-210 determination for lead-210 dating
    by F. El-Daoushy, K. Olsson & R. Garcia-Tenorio 43

How long was the Younger Dryas? Preliminary evidence from annually laminated sediments of
Soppensee (Switzerland)
    by A. F. Lotter 53

Absolute dating of late Quaternary Lacustrine sediments by high resolution varve chronology
    by B. Zolitschka 59

The record of deposition of radionuclides in the sediments of Ponsonby Tarn, Cumbria
    by P. J. P. Bonnett & R. S. Cambray 63

Deposition and transport of radionuclides within an upland drainage basin in mid-Wales
    by P. J. P. Bonnett & P. G. Appleby 71

Progress in understanding the chemical stratigraphy of metals in lake sediments in relation to acidic
precipitation
    by S. A. Norton & J. S. Kahl 77

Spheroidal carbonaceous particles as a marker for recent sediment distribution
    by M. Wik & I. Renberg 85

Magnetic spherules in recent lake sediments
    by D. McLean 91

Lake sediment records of erosional processes
    by J. A. Dearing 99

A multidisciplinary study of the lake Bjäresjösjön (S Sweden): land-use history, soil erosion, lake
trophy and lake-level fluctuations during the last 3000 years
    by M.-J. Gaillard, J. A. Dearing, F. El-Daoushy, M. Enell & H. Håkansson 107

Paleolimnology of Slapton Ley, Devon, UK
    by P. E. O'Sullivan, A. L. Heathwaite, P. G. Appleby, D. Brookfield, M. W. Crick,
    C. Moscrop, T. B. Mulder, N. J. Vernon & J. M. Wilmshurst 115

Sequential inorganic chemical analysis of a core from Slapton Ley, Devon, UK
    by A. L. Heathwaite & P. E. O'Sullivan 125

Sediment characteristics in relation to cultivation history in two varved lake sediments from East
Finland
    by E. Grönlund 137

The influence of land use on the sedimentation of the river delta in the Kyrönjoki drainage basin
    by R. Heikkilä 143
Heavy metals (Cu and Zn) in recent sediments of Llangorse Lake, Wales: non-ferrous smelting, Napoleon and the price of wheat – a palaeoecological study
    by R. Jones, F. M. Chambers & K. Benson-Evans 149
A comparative study of heavy metal contamination and pollution in four Reservoirs in the English Midlands, UK
    by I. D. L. Foster, S. M. Charlesworth & D. H. Keen 155
Sedimentary diatom concentrations and accumulation rates as predictors of lake trophic state
    by T. J. Whitmore 163
The sediment column as a record of trophic status: examples from Bosherston Lakes, SW Wales
    by A. W. G. Rees, G. C. F. Hinton, F. G. Johnson & P. E. O'Sullivan 171
Recent changes to upland tarns in the English Lake District
    by E. Y. Haworth & J. P. Lishman 181
Palaeolimnological study of an environmental monitoring area, or, Are there pristine lakes in Finland?
    by H. Simola, P, Huttunen, J. Rönkkö & P. Uimonen-Simola 187
The eutrophication history of Lake Särkinen, Finland and the effects of lake aeration
    by O. Sandman, K. Eskonen & A. Liehu 191
Are we building enough bridges between paleolimnology and aquatic ecology?
    by J. P. Smol 201
Weichselian chironomid and cladoceran assemblages from maar lakes
    by W. Hofmann 207
Stratigraphy of the fossil Chironomidae (Diptera) from Lake Grasmere, South Island, New Zealand, during the last 6000 years
    by B. Schakau 213
Modern assemblages of arctic and alpine Chironomidae as analogues for late-glacial communities
    by I. R. Walker 223
Paleolimnology of Neusiedlersee, Austria: the succession of ostracods
    by H. Löffler 229
Palaeolimnology of Neusiedlersee – II: the distribution of nutrients and trace metals
    by A. Gunatilaka 239
Paleolimnological investigation of three manipulated lakes from Sudbury, Canada
    by S. S. Dixit, A. S. Dixit & J. P. Smol 245
Dominant diatoms in the interglacial lake sediments of the Middle Pleistocene in Central and Eastern Poland
    by B. Marciniak 253
Fossil diatom inferred reconstruction of the pH history of two acidic, clear water lakes from insular Newfoundland, Canada
    by D. A. Scruton, J. K. Elner & S. N. Ray 259
Taphonomy and diagenesis in diatom assemblages; a Late Pleistocene palaeoecological study from Lake Magadi, Kenya
    by P. Barker, F. Gasse, N. Roberts & M. Taieb 267
Palaeolimnological aspects of a Late-Glacial shallow lake in Sandy Flanders, Belgium
    by L. Denys, C. Verbruggen & P. Kiden 273
Palaeolimnological studies of laminated sediments from the Shropshire–Cheshire meres
    by K. M. Farr, D. M. Jones, P. E. O'Sullivan, G. Eglinton, D. H. Tarling & R. E. M. Hedges 279

Paleolimnological studies using sequential lipid extraction from recent lacustrine sediment: recognition of source organisms from biomarkers
  by P. A. Cranwell 293

Vegetation change and pollen recruitment in a lowland lake catchment: Groby Pool, Leics (England)
  by C. David & N. Roberts 305

Seasonal changes in sedimenting material collected by high aspect ratio sediment traps operated in a holomictic eutrophic lake
  by R. J. Flower 311

Paleolimnology of a Polar Oasis, Truelove Lowland, Devon Island, N.W.T., Canada
  by R. H. King 317

An environmental history of two freshwater lakes in the Larsemann Hills, Antarctica
  by D. S. Gillieson 327

Paleolimnology of Qilu Hu, Yunnan Province, China
  by M. Brenner, K. Dorsey, Song Xueliang, Wang Zuguan, Long Ruihua, M. W. Binford, T. J. Whitmore & A. M. Moore 333

Sedimentary features and the evolution of lake Honghu, central China
  by Shuming Cai & Zhaolu Yi 341

Palaeolakes of the south central Sahara – problems of palaeoclimatological interpretation
  by R. Baumhauer 347

Holocene environments in the central Sahara
  by E. Schulz 359

Chronology of the major palaeohydrological events in NW Africa during the late Quaternary: PALHYDAF results
  by J. Ch. Fontes & F. Gasse 367

Paleolimnology, William Morris and *The Magic Flute*
  by P. E. O'Sullivan 373

Preface

The fifth International Symposium on Palaeolimnology was held at Ambleside in the English Lake District from August 31 to September 6, 1989. During the symposium 65 papers were presented at seven sessions and 52 posters displayed. Three late afternoon/evening special lectures were given, one of which was a memorial to the late Ed. Deevey, to whom this volume is dedicated.

Associated with the symposium were five excursions to various parts of the UK and Ireland, and a visit to the laboratories of the Freshwater Biological Association and Institute of Freshwater Ecology. Conference participants were also invited to a buffet party and visit to the Lake District National Park Centre at Brockhole as the guests of the Park Authority.

The local organising committee for the symposium also formed the editorial panel for this volume. They included: Peter Appleby, Rick Battarbee, John Dearing, Roger Flower, Elizabeth Haworth, Frank Oldfield, Paddy O'Sullivan and John Smith.

Support for the conference is gratefully acknowledged from the following organisations;

The Royal Society
Department of the Environment
US Army European Research Office
Barclays Bank
Central Electricity Generating Board
Lake District Special Planning Board
South Lakeland District Council
Charlotte Mason College
Molspin Limited

The conference is also indebted to the many individuals who provided such effective help in the preparation and smooth running of the programme.

J. P. SMITH
May 1991

Hydrobiologia **214:** 1–7, 1991.
J. P. Smith, P. G. Appleby, R. W. Battarbee, J. A. Dearing, R. Flower, E. Y. Haworth, F. Oldfield & P. E. O'Sullivan (eds),
Environmental History and Palaeolimnology.

1

Edward Smith Deevey
1914–1988

I first heard of Ed Deevey in 1950, when I was a beginning Ph.D. student at Yale. I came from a three-person biology department whose minuscule library did not include the American Journal of Science, in which Ed had published his most important work. I had, however, been poking around the lakes and bogs of Nova Scotia for five years and was full of half-formed ideas about paleolimnology, climatic change, and the relevance of past climate to the modern distribution of plants and animals. I was chagrined to find a man I had never heard of, an American at that, who'd had all the same ideas ten years before, formulated them more clearly and carried out the work with an ingenuity and completeness beyond my aspirations. For a few weeks I was afraid to

2

read another Deevey paper, and when the man himself returned from Yucatan I became his first graduate student.

It is difficult in retrospect to remember just how precocious the young Ed Deevey had been. A summa cum laude graduate of Yale in botany at the age of 19, he had fallen under the magnetic influence of G. Evelyn Hutchinson, whose encyclopedic interests and theoretical approach contrasted sharply with contemporary post-Clementsian plant ecology. By the age of 23, when many of today's students are still trying to find themselves, Ed had completed his Ph.D. in zoology. In Hutchinson's laboratory, zoology covered animals, but also plants, the earth, mankind and a very large part of the contemporary world of ideas.

By the age of thirty Ed had a formidable record of research accomplishment. He had shown that the last interglacial period was wet in the Himalayas, applied pollen analysis to the archaeology of Mexico, and published two major monographs establishing the pollen stratigraphy of New England and its vegetational significance, three major monographs on the comparative limnology of New York and New England, and a series of shorter papers on arctic alpine limnology, redox potentials in lakes, harpacticoid copepods, Thoreau's Walden Pond, freshwater jellyfish and the population structure of the black widow spider.

Best of all, he had completed his pioneering biostratonomy of Linsley Pond, which applied quantitative methods to the paleolimnological record to test major theoretical ideas adapted from ecology, physiology and embryology. Without radiocarbon dating, but with a remarkably incisive and imaginative mind, he fearlessly attempted a quantitative reconstruction of such ecosystem properties as nutrient status, primary productivity and trophic structure. The world is not so simple, as the Windermere group realized even then, and in later years Ed was very aware of its complexity. I had a hand in weakening one of the Linsley Pond conclusions, but that gallant attempt to test ecological theory against paleolimnological observation still seems to me the

most splendid thing of its kind in our subject.

By the time I met him, Ed had introduced life tables into field ecology, discovered the late-glacial in North America, demonstrated a 5 degree C Pleistocene lowering of temperature in the Gulf of Mexico, dated the last rise of sea level in southern New England, shown that the end of the last ice age was synchronous in the old and New Worlds and in both polar hemispheres, and produced his masterful review of the Pleistocene biogeography of Europe and North America.

Over the next few decades Ed set up one of the first post-Libby wave of radiocarbon laboratories and published eight long lists of radiocarbon dates and, with Margaret Davis, the first pollen deposition rate diagram. With a series of geochemical collaborators he pioneered the use of stable carbon and sulphur isotopes in paleolimnology. I'd like to remind John Smol gently that the word 'paleolimnology' was already in use at that remote epoch – Ed Deevey used it himself in the title of a paper in 1953. The question of whether we needed more bridges between paleolimnology and modern ecology didn't arise for a paleolimnologist whose daily companions included Evelyn Hutchinson, Gordon Riley & John Brooks. Nor could it, without surgical severing of several critical commissures in Ed's brain, for he was working actively in both areas himself. Those were splendid times to be a paleolimnologist, largely because of Ed Deevey's example.

I'm not going to say anything about Ed's work in China, nor in Florida and Middle America, although a very good case could be made that his paleolimnological study of Mayan culture was his real life's work. By that time I had despaired of matching him on his home ground, and had moved my research to Africa, where you didn't have to be as smart as Ed Deevey to do something worthwhile. By that time, too, my emotional investment in my own students had surpassed that in my old professors. Readers in search of a more balanced treatment of his whole career should consult Brenner (in press), Rice & Brenner (1989) and Livingstone & Brenner (1989). I would, however, like to say a little about Ed Deevey the man during the years that I knew him.

His outstanding traits of personality were shyness and a keen sense of humor. The shyness sometimes kept the humor from full expression, as when he was addressing a large audience. This was the way too many people saw him, and it was only his close friends, those who knew him in a small group, who were able to savor the full flavor of the man. His store of puns and limericks was inexhaustible. He was unfailingly kind and gentle to us as students and not less so with the passing years. He had a deep human sympathy, and an uncanny intuition, that enabled him to reach out, even by post from half the world away, with just exactly the right word of encouragement and understanding when things were going wrong.

I learned from one of my Antipodean colleagues recently that some believe it to be a perquisite of professorial life in America to have echelons of indentured graduate students. Any exploitation in my relationship with Ed Deevey went the other way. When I needed a field assistant to test the Mk I of what now seems to be called the modified Livingstone sampler, Ed Deevey lent a hand. I will not burden you with his appreciative, if scatological, commentary on its performance. He was still writing final reports on the grant that financed my doctoral field-work years after I left Yale, although he and his own research did not profit at all from the support of that grant. Sometimes graduate students cannot see what a professor does for them. If they are misplaced in science, and don't enjoy what they are doing, they may feel that the professor is driving them to it. Ed Deevey drove no one. He played the game of discovery, the greatest game in the world, and welcomed us to play it too.

The easy democracy and intellectual excitement of Deevey's laboratory recruited even Yale undergraduates to science, and attracted to paleoecology many excellent graduate students. His influence was even wider through people who were more nearly his equals in seniority and accomplishment, his postdoctoral fellows, research associates and sabbatical visitors. These people returned to Australia, England, Fennoscandia, France, Ireland, Japan, and the Netherlands, bearing something of the Deevey approach to lakes and their history with them. Anne Tutin has given us an example of how his publications affected an even wider circle.

Ed Deevey followed a deliberate policy of attacking only problems that no one else would tackle if he didn't. This policy produced the enormous scientific contribution that we honor today, but it was less a recipe for professional success than an affirmation that one person could make a difference. In retrospect, the affirmation wasn't needed: his erudition, his self-effacing modesty, his wit, his gentle good humor and his kindness would have made a difference to all who knew him, even without his professional accomplishment.

References

Brenner, Mark, In press. Edward S. Deevey 1914–1988. Palaeogeogr., Palaeoclimat., Palaeoecol.

Livingstone, D. A. & Mark Brenner, 1989. Resolution of Respect. Edward S. Deevey 1914–1988. Bull. Ecol. Soc. Amer. June 1989: 137–140.

Rice, Prudence M. & Mark Brenner, 1989. Edward Smith Deevey, Jr. 1914–1988. SAS Bull., Soc. Archaeol. Sci. 12 : 2.

Bibliography, 1937–Present

1937   Deevey, E. S. Pollen from interglacial beds in the Panggong Valley, and its climatic interpretation. Am. J. Sci. 235: 44–56.

1939   Deevey, E. S. (Discussion) Arctic-Alpine limnology. Am. J. Sci. 237: 830–833.

1939   Deevey, E. S. Studies on Connecticut lake sediments, I. A postglacial climatic chronology for southern New England. Am. J. Sci. 237: 691–724.

1939   Hutchinson, G. E., E. S. Deevey & A. Wollack. The oxidation-reduction potentials of lake waters and their ecological significance. Proc. Nat. Acad. Sci. 25: 87–90.

1940   Deevey, E. S. Limnological studies in Connecticut, V. A contribution to regional limnology. Am. J. Sci. 238: 717–741.

1941   Deevey, E. S. Notes on the encystment of the harpacticoid copepod *Canthocamptus staphylinoides* Pearse. Ecology 22: 197–200.

1941   Deevey, E. S. Limnological studies in Connecticut, VI. The quantity and composition of the bottom fauna of thirty-six Connecticut and New York lakes. Ecol. Mon. 11: 413–455.

1942 Deevey, E. S. & J. S. Bishop. Limnology. pp. 69–121, 296–298, in: A fishery survey of important Connecticut lakes, Lyle M. Thorpe, ed., Conn. Geol. and Nat. Hist. Surv. Bull., 63.

1942 Deevey, E. S. Studies on Connecticut lake sediments, III. The biostratonomy of Linsley Pond. Am. J. Sci. 240: 233–264, 313–338.

1942 Deevey, E. S. A re-examination of Thoreau's 'Walden'. Quat. Rev. Biol. 17: 1–11.

1942 Deevey, E. S. Some geographic aspects of limnology. Sci. Month. 55: 423–434.

1943 Deevey, E. S. Additional pollen analyses from southern New England. Am. J. Sci. 241: 717–752.

1943 Deevey, E. S. & J. L. Brooks. *Craspedacusta* in open water, Lake Quassapaug, Connecticut. Ecology, 24: 266–267.

1944 Deevey, E. S. Pollen analysis and history. Am. Sci. 32: 39–53.

1944 Deevey, E. S. Pollen analysis and Mexican archaeology: an attempt to apply the method. Am. Antiq. 10: 134–149. (Also published as: Intento para datar las Cultures Medias del Valle de Mexico mediante analysis de polen. Ciencia 4: 97–105, 1943).

1945 Deevey, G. B. & E. S. Deevey. A life table for the black widow. Trans. Conn. Acad. Arts and Sci., 36: 115–134.

1946 Deevey, E. S. (Discussion) An absolute pollen chronology in Switzerland. Am. J. Sci 244: 442–447.

1946 Deevey, E. S. Scientific Intelligence, Geology. A review of Zeuner, Frederick E., The Pleistocene Period, Its Climate, Chronology and Faunal Successions. Am. J. Sci. 244: 373–376.

1946 Deevey, E. S. Problems of the spring overturn in stratified lakes. J. N. Engl. Water Works Assoc. 60: 182–188.

1947 Deevey, E. S. Life tables for natural populations of animals. Quart. Rev. Biol. 22: 283–314. (Reprinted in: Readings in population and community ecology, W. E. Hazen, ed., Saunders, Philadelphia, 1964; 1970; 1975).

1948 Deevey, E. S. On the date of the last rise of sea level in southern New England, with remarks on the Grassy Island site. Am. J. Sci. 246: 329–352.

1949 Deevey, E. S. Biogeography of the Pleistocene. Part I. Europe and North America. Bull. Geo. Soc. Am. 60: 1315–1416.

1949 Hutchinson, G. E. & E. S. Deevey. Ecological studies on populations. pp. 325–359, in: A survey of biological progress, G. S. Avery, ed., vol. 1, Academic Press, New York.

1949 Deevey, E. S. Living records of the Ice Age. Sci. Am. 180(5): 48–51.

1950 Deevey, E. S. The probability of death. Sci. Am. 182(4): 58–60

1950 Deevey, E. S. Hydroids from Louisiana and Texas, with remarks on the Pleistocene biogeography of the western Gulf of Mexico. Ecology 31: 334–367.

1950 Deevey, E. S. Ecology, the ice age, and the Gulf of Mexico. February 1950, Yale Sci. Mag. pp. 7–8; 18; 20.

1951 Deevey, E. S. Late-glacial and postglacial pollen diagrams from Maine. Am. J. Sci. 249: 177–207.

1951 Deevey, E. S. & J. E. Potzger. Peat samples for radiocarbon analysis: problems in pollen statistics. Am. J. Sci. 249: 473–511.

1951 Deevey, E. S. Recent textbooks of human ecology. Ecology 32: 347–51.

1951 Deevey, E. S. Life in the depths of the pond. Sci. Am. 185(4): 68–72.

1951 Flint, R. F. & E. S. Deevey. Radiocarbon dating of late-Pleistocene events. Am. J. Sci. 249: 257–300.

1951 Deevey, E. S. A brief discussion of the relation of some radiocarbon dates to the pollen chronology. In: Radiocarbon dating, Frederick Johnson, ed., Mem. Soc. Am. Archaeol. 8: 56–57.

1952 Deevey, E. S. Radiocarbon dating. Sci. Am. 186(2): 24–28.

1953 Deevey, E. S. Utilizzazione del radiocarbonio nella determinazione delle età geologiche. Riv. Sci. Preist. 6: 115–125.

1953 Blau, M., E. S. Deevey & M. S. Gross. Yale natural radiocarbon measurements, I. Pyramid Valley, New Zealand, and its problems. Science 118: 1–6.

1953 Deevey, E. S. Paleolimnology and climate. pp. 273–318, in: Climatic change, H. Shapley, ed., Harvard Univ. Press, Cambridge.

1954 Deevey, E. S., M. S. Gross, G. E. Hutchinson & H. Kraybill. The natural C14 contents of materials from hard-water lakes. Proc. Nat. Acad. Sci. 40: 285–288.

1954 Deevey, E. S. Hydroids of the Gulf of Mexico. In: The Gulf of Mexico, its origin, waters, and marine life, P. S. Galstoff, ed., U.S. Fish and Wildlife Serv., Fishery Bull. Vol. 55, 89: 267–272.

1954 Deevey, E. S. The end of the moas. Sci. Amer. 190(2): 84–90. (Repr. 1955, in: First Book of Animals, Simon & Schuster, xi, 240 pp.).

1955 Deevey, E. S. Some biogeographic implications of paleolimnology. Int. Assoc. Theoret. Appl. Limnol. Proc. 12: 654–659.

1955 Deevey, E. S. Limnological studies in Guatemala and El Salvador. Int. Assoc. Theoret. Appl. Limnol. Proc. 12: 278–283.

1955 Deevey, E. S. Paleolimnology of the upper swamp

| | deposit, Pyramid Valley. Rec. Canterbury Mus. 6: 291–344. |

1955   Deevey, E. S. The obliteration of the hypolimnion. Mem. Ist. Ital. Idrobiol. Suppl. 8: 9–38. (Colloque I.U.B.S., no. 19).

1955   Preston, R. S., E. Person & E. S. Deevey. Yale natural radiocarbon measurements, II. Science 122: 954–960.

1956   Deevey, E. S. The human crop. Sci. Am., 194(4): 105–112.

1956   Hutchinson, G. E., R. Patrick & E. S. Deevey. Sediments of Lake Patzcuaro, Michoacan, Mexico. Bull. Geol. Soc. Amer. 67: 1491–1540.

1957   Deevey, E. S. & R. F. Flint. Postglacial hypsithermal interval. Science 125: 182–184.

1957   Barendsen, G. S., E. S. Deevey & L. J. Gralenski. Yale natural radiocarbon measurements, III. Science, 126: 908–919.

1957   Deevey, E. S. Limnologic studies in Middle America, with a chapter on Aztec limnology. Trans. Conn. Acad. Arts Sci. 39: 213–328.

1957   Deevey, E. S. Archeologic identification. Science, 126: 412–413. (Reprinted as General Summary, in: The identification of non-artifactual archaeological materials, W. W. Taylor, ed., Nat. Acad. Sci. Nat. Res. Coun. Pub. 55).

1957   Deevey, E. S. Department of Amplification. The New Yorker, March 16, 1957, pp. 98–100.

1958   Deevey, E. S. Final report on radiocarbon dating. Office of Naval Research, Contract Nonr 609, 17 pp., processed.

1958   Deevey, E. S. Radiocarbon-dated pollen sequences in eastern North America. Veröffentl. Geobot. Inst. Rübel, Zürich 34: 30–37.

1958   Deevey, E. S. The equilibrium population. pp. 64–86, in: The population ahead, R. G. Francis, ed., Univ. of Minnesota Press, Minneapolis.

1959   Deevey, E. S., L. J. Gralenski & V. Hoffren. Yale natural radiocarbon measurements, IV. Am. J. Sci. Radiocarbon Suppl. 1: 144–172.

1959   Deevey, E. S. The hare and the haruspex. Yale Review, 49: 161–179. (Reprinted in: Am. Sci.; in: Sierra Club Bull; in: Man alone, Dell, New York, 1963; and in: Human ecology, J. B. Bresler ed., Addison-Wesley, 1965).

1960   Oana, S. & E. S. Deevey. Carbon 13 in lake waters, and its possible bearing on paleolimnology. Am. J. Sci., Bradley Vol., 258-A: 253–272.

1960   Stuiver, M., E. S. Deevey & L. J. Gralenski. Yale natural radiocarbon measurements, V. Am. J. Sci., Radiocarbon Suppl. 2: 49–61.

1960   Deevey, E. S. The human population. Sci. Am., 203(3): 194–204.

1961   Deevey, E. S. Recent advances in Pleistocene stratigraphy and biogeography. pp. 594–623, in: Vertebrate speciation, W. Frank Blair, ed., University of Texas Press, Austin.

1961   Stuiver, M. & E. S. Deevey. Yale natural radiocarbon measurements, VI. Radiocarbon 3: 126–140.

1962   Deevey, E. S. Animal populations. pp. 18–26, in: Frontiers of modern biology, G. B. Moment ed., Houghton Mifflin, Boston.

1962   Deevey, E. S. & N. Nakai. Fractionation of sulfur isotopes in lake waters. pp. 169–178, in: Biogeochemistry of sulfur isotopes, M. L. Jensen, ed., (Proc. NSF Symp., April 12–14, 1962.), Yale Univ. Dept. Geol., New Haven.

1962   Stuiver, M. & E. S. Deevey. Yale natural radiocarbon measurements, VII. Radiocarbon 4: 250–262.

1963   Brooks, J. L. & E. S. Deevey. New England. pp. 117–162, in: Limnology in North America, D. G. Frey, ed., Univ. Wisconsin Press, Madison.

1963   Deevey, E. S., N. Nakai & M. Stuiver. Fractionation of sulfur and carbon isotopes in a mesomictic lake. Science 139: 407–408.

1963   Stuiver, M., E. S. Deevey & I. Rouse. Yale natural radiocarbon measurements, VIII. Radiocarbon, 5: 312–341.

1963   Deevey, E. S., M. Stuiver & N. Nakai. Use of light nuclides in limnology. pp. 471–475, in: Radioecology, V. Schultz & A. W. Klement, Jr. eds., Reinhold, New York and AIBS, Washington.

1963   Deevey, E. S. General and urban ecology. pp. 20–32, in: The urban condition, L. J. Duhl, ed., Basic Books, New York.

1964   Deevey, E. S. Preliminary account of fossilization of zooplankton in Rogers Lake. Int. Assoc. Theor. Appl. Limnol. Proc. XV: 981–992.

1964   Deevey, E. S., M. Stuiver & N. Nakai. Isotopes of carbon and sulfur as tracers of lake metabolism. Int. Assoc. Theor. Appl. Limnol. Proc. XV: 284–288.

1964   Deevey, E. S. General and historical ecology. BioScience, 14(7): 33–35.

1964   Deevey, E. S. & M. Stuiver. Distribution of natural isotopes of carbon in Linsley Pond and other New England lakes. Limnol. Oceanogr. 9: 1–11.

1964   Davis, M. B. & E. S. Deevey. Pollen accumulation rates: estimates from late-glacial sediment of Rogers Lake. Science 145: 1293–1295.

1965   Deevey, E. S. Sampling lake sediments by use of the Livingstone sampler. pp. 521–529, in: Handbook of paleontological techniques, B. Kummel & D. Raup, eds., Freeman, San Francisco.

1965   Deevey, E. S. Pleistocene nonmarine environments. pp. 643–952, in: The Quaternary of the

6

United States, H. E. Wright & D. G. Frey, eds., Princeton Univ. Press, Princeton, N.J.

1967 Deevey, E. S. The reply: Letter from Birnam Wood. Yale Review 56: 631–640.

1967 Tsukada, M. & E. S. Deevey. Pollen analyses from four lakes in the southern Maya area of Guatemala and El Salvador. pp. 303–311, in: Quaternary paleoecology, E. Cushing & H. E. Wright, eds., Yale Univ. Press, New Haven.

1967 Deevey, E. S. Introduction. pp. 63–72, in: Pleistocene extinctions: The Search for a cause, P. S. Martin & H. E. Wright, eds., Yale Univ. Press, New Haven.

1969 Deevey, E. S. Review of Margalef, R., Perspectives in ecological theory. Limnol. Oceanogr. 14: 313–315.

1969 Deevey, E. S. Cladoceran populations of Rogers Lake, Connecticut, during late and postglacial time. Mitt. Int. Ver. Limnol. 17: 56–63.

1969 Deevey, E. S. Coaxing history to conduct experiments. BioScience, 19(1): 40–43.

1969 Deevey, E. S. Specific diversity in fossil assemblages. pp. 224–241, in: Diversity and stability in ecological systems, Brookhaven Nat. Lab., Brookhaven Symp. Biol., no. 22.

1970 Deevey, E. S. In defense of mud. Bull. Ecol. Soc. Am. 51: 5–8.

1970 Deevey, E. S. Mineral cycles. Sci. Am. 223(3): 148–158.

1971 Deevey, E. S. The quality of environment. pp. 421–423, in: Topics in the study of life: the Bio source book, Amy Kramer, ed., Harper & Row, New York.

1971 Deevey, E. S. & G. B. Deevey. The American species of *Eubosmina* Seligo (Crustacea, Cladocera). Limnol. Oceanogr. 16: 201–218.

1971 Deevey, E. S. Section 4, Biogeochemistry of lakes; Section 5, sediments and history of lakes. In: Program VIII, Freshwater lakes and their management, D. K. Todd ed., Univ. California Exten., Water Resour Engin. Educ. Series, 21 pp., processed.

1971 Deevey, E. S. The chemistry of wealth (presidential address). Ecol. Soc. Am., Bull. 52(4): 3–8.

1972 Deevey, E. S. Biogeochemistry of lakes: major substances. In: Nutrients and eutrophication, G. E. Likens, ed., Limnol. and Oceanogr., Spec. Symp., 1: 14–20.

1972 Deevey, E. S., ed. Growth by intussusception; ecological essays in honor of G. Evelyn Hutchinson. Conn. Acad. Arts. & Sci., Trans. 44: 1–443.

1973 Deevey, E. S. Sulfur, nitrogen, and carbon in the biosphere. In: Carbon and the biosphere, G. M. Woodwell & E. V. Pecan, eds., Brookhaven Nat. Lab., Brookhaven Symp. Biol. no. 24: 182–190.

1976 Deevey, E. S., ed. Air, water, and land: the global ecosystem. A Scientific American Offprint Reader (Introduction by E. S. Deevey, pp. 3–10), Freeman, San Francisco.

1976 Deevey, E. S., ed. Social Ecology. A Scientific American Reader (Introduction by E. S. Deevey), Freeman, San Francisco.

1976 Deevey, E. S. Time-worn highlands and coastal plain. pp. 202–233, in: Our continent: a natural history of North America, S. L. Fishbein, ed., Nat. Geo. Soc., Washington.

1977 Deevey, E. S., H. Vaughan & G. B. Deevey. Lakes Yaxha and Sacnab, Peten, Guatemala: planktonic fossils and sediment focusing. pp. 189–196, in: Interactions between sediments and fresh water, H. L. Golterman, ed., Junk, The Hague and PUDOC, Wageningen.

1978 Deevey, E. S. Holocene forests and Maya disturbance near Quexil Lake, Peten, Guatemala. Polsk. Arch. Hydrobiol. 25: 117–129.

1978 Deevey, E. S. & Mark Brenner. Sedimentary history of Spanish Pond. Appendix E, 10 pp., in: El Pantano de los Espanioles, L. D. Harris and others, eds., U.S. Nat. Park. Serv., Contract no. CX500041666, processed.

1979 Deevey, E. S., D. S. Rice, P. M. Rice, H. H. Vaughan, M. Brenner & M. S. Flannery. Mayan urbanism: impact on a tropical karst environment. Science, 206: 298–306.

1980 Deevey, E. S., G. B. Deevey & M. Brenner. Structure of zooplankton communities in the Peten lake district, Guatemala. pp. 669–678, in: Evolution and ecology of zooplankton communities, W. C. Kerfoot, ed., Univ. Press New England, Hanover, N.H.

1980 Deevey, E. S. & D. S. Rice. Coluviación y retención de nutrientes en el distrito lacustre del Petén Central, Guatemala. Biotica 5: 129–144.

1980 Deevey, E. S., Mark Brenner, M. S. Flannery & G. H. Yezdani. Lakes Yaxha and Sacnab, Peten, Guatemala: limnology and hydrology. Arch. Hydrobiol. Suppl., 57: 419–460.

1982 Deevey, E. S. Concluding remarks: Embryology continued by other means. pp. 485–489, in: Selected works of Gordon A. Riley, J. S. Wroblewski, ed., Dalhousie Univ., Halifax.

1983 Deevey, E. S., Mark Brenner & M. W. Binford. Paleolimnology of the Peten lake district, Guatemala, III. Late Pleistocene and Gamblian environments of the Maya area. Int. Symp. Paleolimnol., 3rd, Joensuu, Finland, 1981; Hydrobiologia, 103: 211–216.

1983 Hitchcock, D. R. & E. S. Deevey. Coastal and inland natural H_2S sources. pp. 162–171, in: Acid deposition, causes and effects: a state assessment model, A. E. S. Green & W. H. Smith, eds., Univ. Florida, Gainesville.

1983   Binford, M. W., E. S. Deevey & T. L. Crisman. Paleolimnology: An historical perspective on lacustrine ecosystems. Ann. Rev. Ecol. Syst. 14: 225–286.

1983   Rice, D. S., P. M. Rice & E. S. Deevey. El impacto de los Mayas en el ambiente tropical de la cuenca de los Lagos Yaxha y Sacnab, El Petén, Guatemala. Am. Indegena 43: 261–297.

1984   Deevey, E. S. Zero B. P. plus 34: 25 years of Radiocarbon. Radiocarbon 26: 1–6.

1984   Deevey, E. S. Stress, strain, and stability of lacustrine ecosystems. pp. 203–229, in: Lake sediments and environmental history, E. Haworth & J. W. G. Lund eds., Univ. Minnesota Press, Minneapolis.

1985   Vaughan, H. H., E. S. Deevey & S. E. Garrett-Jones. Pollen stratigraphy of two cores from the Peten lake district, with an appendix on two deep-water cores. pp. 73–89, in: Prehistoric lowland Maya environment and subsistence economy, M. Pohl, ed., Harvard Univ. Press, Cambridge.

1985   Rice, D. S., P. M. Rice & E. S. Deevey. Paradise lost: Classic Maya impact on a lacustrine environment. pp. 91–105, in: Prehistoric lowland Maya environment and subsistence economy, M. Pohl ed., Harvard Univ. Press, Cambridge.

1986   Deevey, E. S., M. W., Binford, M. Brenner & T. J. Whitmore. Sedimentary records of accelerated nutrient loading in Florida lakes. Hydrobiologia 143: 49–53.

1987   Binford, M. W., M. Brenner, T. J. Whitmore, A. Higuera-Gundy, E. S. Deevey & B. Leyden. Ecosystems, paleoecology and human disturbance in subtropical and tropical America. Quaternary Science Reviews, 6: 115–128.

1987   Coleman, J. M. & E. S. Deevey. Lacustrine sediment/groundwater nutrient dynamics. Biogeochemistry 4: 3–14.

1988   Deevey, E. S. Lake and bog sediments: hydrologic indications of short-term climatic oscillations. pp. 19–23, in: Paleoecology Workshop, convened 15–17 February 1988, Boston, Ma., G. Sharp & T. J. DeVries, Co-convenors. NOAA/NSF co-sponsors.

1988   Deevey, E. S. Estimation of downward leakage from Florida lakes. Limnol. Oceanogr. 33: 1308–1320.

in press   Brenner M, M. W. Binford & E. S. Deevey. Lakes. in: Ecosystems of Florida, J. J. Ewel & R. L. Myers eds., U. Central Fla. Press, Univ. Presses of Florida.

submitted Deevey, E. S., M. Brenner, A. M. Moore, M. W. Binford, K. T. Dorsey, X. Song & Z. Hu. Paleolimnology and limnology of Qilu Hu, Tonghai County, Yunnan. Journal of Yunnan University.

submitted Moore, A. M., M. Brenner, M. W. Binford, E. S. Deevey & X. Ou. Field notes on a paleolimnological expedition to five lakes on the Yunnan plateau. Journal of Yunnan University.

Hydrobiologia **214**: 9–24, 1991.
J. P. Smith, P. G. Appleby, R. W. Battarbee, J. A. Dearing, R. Flower, E. Y. Haworth, F. Oldfield & P. E. O'Sullivan (eds),
Environmental History and Palaeolimnology.
© 1991 *Kluwer Academic Publishers.*

Palaeolimnology in the English Lakes – some questions and answers over fifty years

Winifred Pennington (Mrs Tutin)
Department of Botany, University of Leicester, University Road, Leicester LE1 7RH, UK

Abstract

In a review of work on 10 of the very numerous lakes of the English Lake District – work which involved co-operation between palaeolimnologists from different disciplines and between palaeo- and neolimnologists – the questions asked by the researchers are listed and it is shown which kind of lake has provided constructive answers. Complete and conformable sediment profiles covering the last *ca* 14 000 years are found in these lakes, but no single profile is optimal for study of the whole period. Knowledge of the pattern of sediment accumulation is necessary before the record in microfossils and changing sediment composition can be interpreted with confidence. The demonstration that the sediment-source material of these open lakes must have been largely allochthonous established the importance of the sediment column as a record of changes through time in the composition and stability of the soils of each catchment. These changes illustrate the influence of:

i.   *Climatic changes*, from glacial to temperate interstadial followed by renewed glaciation (Younger Dryas) and then by a long temperate (postglacial) state, during which changes in stream run-off and in the pattern of deposition in lakes indicate changes in precipitation and wind-induced turbulence.
ii. *Processes of pedogenesis*, determining the composition of sediment-source material.
iii. *Changes in aquatic and terrestrial biota* in response to climatic changes interacting with biological processes; the major changes which coincide with changing land use by man over 5000 years, and the more recent changes associated with the dates for human discharges into lakes and the deposition of airborne pollutants.

Introduction

The work on the sediments of Lake District lakes which began on the initiative of C.H. Mortimer in 1937 has now been going on for more than 50 years. It seems reasonable to take a look at some of these much-studied lakes as individual sites and consider for which questions in palaeo-limnology they have provided apparently sensible and acceptable answers, and for which questions they have proved less rewarding. This approach, a critical overview, may provide an appropriate introduction to a conference taking place within the region. It will not attempt any comprehensive or descriptive review of the volume of work which has been published, for which references are available (Haworth & Long, 1989), but will give a highly selective comment on some questions and answers and types of lake which have proved most rewarding in each case. The thread throughout will be, 'to what extent do we appear to have done right?' 'How far does it appear that we did not ask the most appropriate question of each lake?' 'Which lakes have provided problems rather than answers?' This may provide some useful information for those now asking the initial question among palaeolimnologists – on which lake shall we begin work?

We are all greatly saddened by the death of Professor E.S. Deevey. Professor Livingstone's memorial lecture pays tribute to him as a person and a scientist. Any account of palaeolimnology over the last 50 years must gratefully acknowledge the work of Ed Deevey and its influence throughout the world. My own work on the sediments of Windermere began just before he published his first paper on Connecticut lake sediments in 1939, and I shall try to acknowledge the stimulation and inspiration which resulted from continuing dialogue with him, and contact with his flow of ideas, one after the other, from New World lakes.

Windermere

This large lake (14.8 km^2, max depth 64 m) has been studied continuously by limnologists since 1931 (see edit. Talling, 1986) and by palaeolimnologists since, in 1937, Clifford Mortimer drove in drainpipes to sample the near-shore sediments. He posed the question of whether the first reflection – indicative of sedimentary discontinuity – in the record produced by the Admiralty's Echo-sounding survey of that year, represented the boundary post-glacial mud/glacial clay? For practical reasons his simple corers were first put down in Low Wray Bay, near the Freshwater Biological Association's boathouse, in 3 m of water and *ca* 30 m from shore. Thus the answer to one question posed at the present meeting – 'Where shall we put in our corers?' was then based on severely practical grounds, but it proved a most fortunate choice which initiated 50 years of work on this site. The cores extruded from the drainpipes included, below the organic lake mud, two deposits of laminated (apparently varved) clay, separated by an organic sediment bursting with plant remains. The question 'Was this an interglacial deposit?' was answered within a year by the publication (Jessen & Farrington, 1938) of the first record from the British Isles of a lateglacial interstadial, a Late-Weichselian climatic

oscillation subsequently widely recognised in northern Europe. The extension of core-sampling to deeper water (made possible by the inventions of Mr B.M. Jenkin) identified the uppermost reflection in the Echo-sounding record as indeed the boundary between post-glacial mud and glacial clay, and provided the data on which a transverse section of the lake sediments was constructed (Fig. 1). The only chronology available in those pre-radiocarbon years was provided by a pollen chronology, comparable with what Deevey (1939) was producing from Linsley Pond and other New England lakes.

The pattern of sediment accumulation in Windermere gradually emerged. That for lateglacial time came to be understood in terms of subsequent investigations on lakes in arctic (Ostrem, 1975) and alpine (Sturm & Matter, 1978) environments. In early post-glacial time mud had clearly accumulated more rapidly at the margins than in the middle of the lake. Much spatial irregularity between cores from neighbouring positions was observed with respect to stratigraphic markers such as clay bands in the early post-glacial muds. The origin of irregularities and anomalies in some cores ('bad cores') could be understood from the Echo-sounding records of 1937 and later seismic surveys (Howell, 1971) which showed both 'draping' of sediment around projections of the basal bedrock and 'slumping' and 'overfolds' in the laminated clays on steep slopes. With respect to the organic muds, the process which came to be known as 'sediment focusing' (Likens & Davis, 1975) had clearly followed a different pattern in Windermere from any postulated by Lehman (1975), since preferential deposition in deep water only began at some intermediate date in post-glacial time.

Over the years of data accumulation from Windermere cores, it was the irregularity of the pattern which was impressive, and from the beginning this led to a sceptical approach to any idea of quantitative work based on estimation of the total annual deposition of sediment. Analyses for diatoms and pollen were carried out on a proportional basis, and this approach was con-

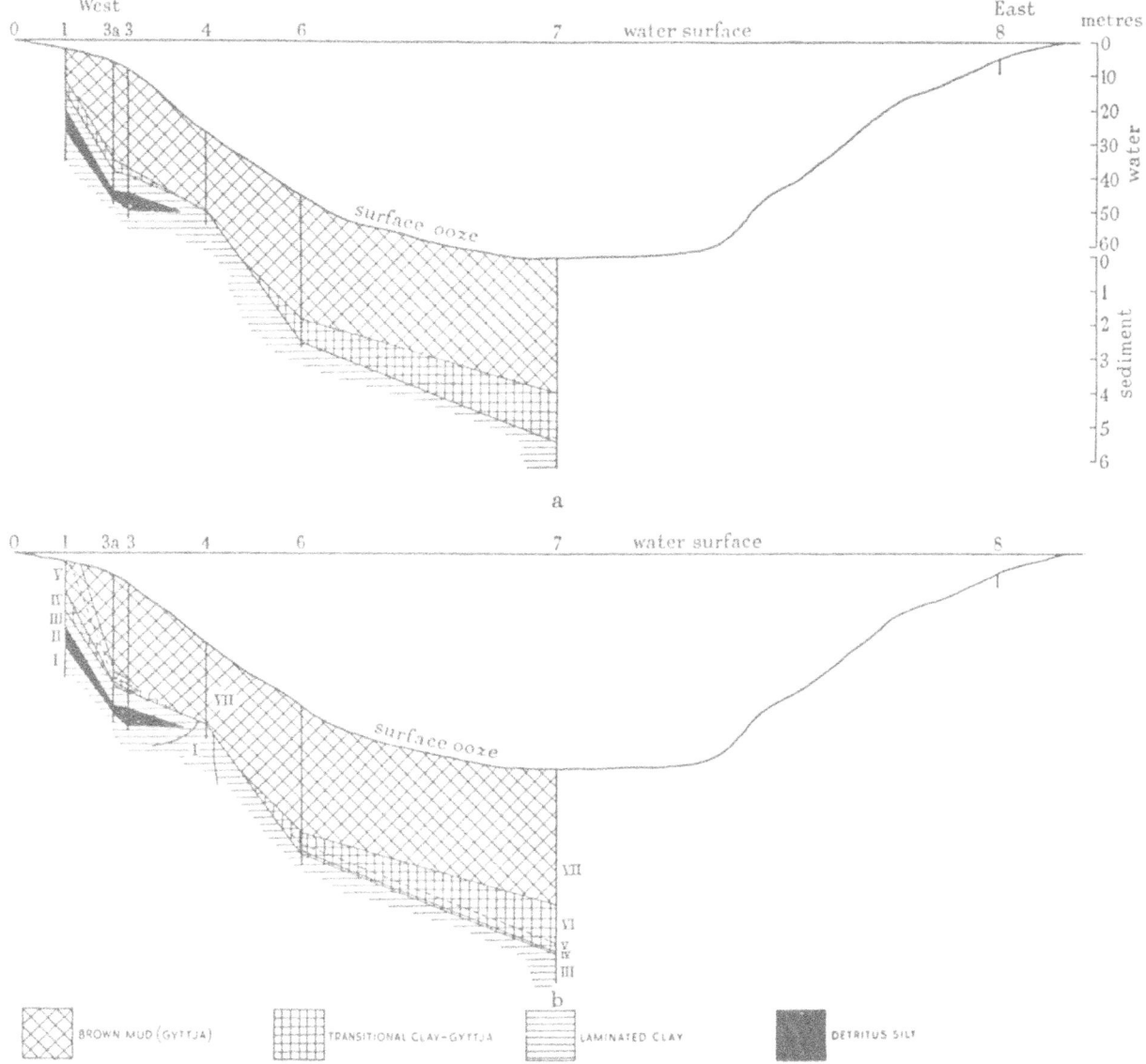

Fig. 1. Reproduced from *Phil. Trans.*, 1947, Fig. 2. Transverse section of Windermere, N. Basin. Lower figure shows boundaries of pollen zones.

tinued in the chemical investigations of Mackereth (1965; 1966), in which the changing composition (through time) of Windermere sediments was compared with, and found to be consistent with, that in other large lakes of the district.

This irregularity of the sediment pattern in a lake as large as Windermere constitutes good reason for preference for smaller lakes in the study of biological history (e.g. Berglund, 1986). Questions for which Windermere has provided

the most satisfactory answers have related to Quaternary geology, to the late-glacial palaeo-environment, and to geochemistry (Mackereth, op cit.) and geophysics, notably to questions of palaeomagnetism beginning with Mackereth's (1971) publication of the dated 'master curve' for declination changes in South Basin sediments.

The late-glacial section from Low Wray Bay, used as the reference site for the defined

12

Windermere Interstadial (Coope & Pennington, 1977) has been fully described (Pennington, 1943; 1977). A question asked in this volume 'How long was Younger Dryas?' is relevant. The upper (post-interstadial) varved clay in Windermere contains 400–450 varve couplets (described in Pennington, 1947) which have the features of annual glacial rhythmites and have been attributed to the outwash from Younger Dryas glaciers re-established in the mountains of the upper catchment. This suggests a period of this length for the time of active glaciation, i.e. the Loch Lomond Advance (Sissons, 1974). Sissons (1980) mapped the Lake District glaciers from geomorphological evidence. From biological evidence from sediments underlying and overlying the varved clay, the total length of the cold period (Younger Dryas or Loch Lomond Stadial) was considerably longer, approximating to the 1000 years (10000–11000 BP) now commonly accepted.

Figure 2 illustrates some detail from the Low Wray Bay section of the transition from glacial to temperate (interstadial) conditions. The fossil record in Lake District lakes is free from the complications of secondary deposition, since ice from the independent source in the central

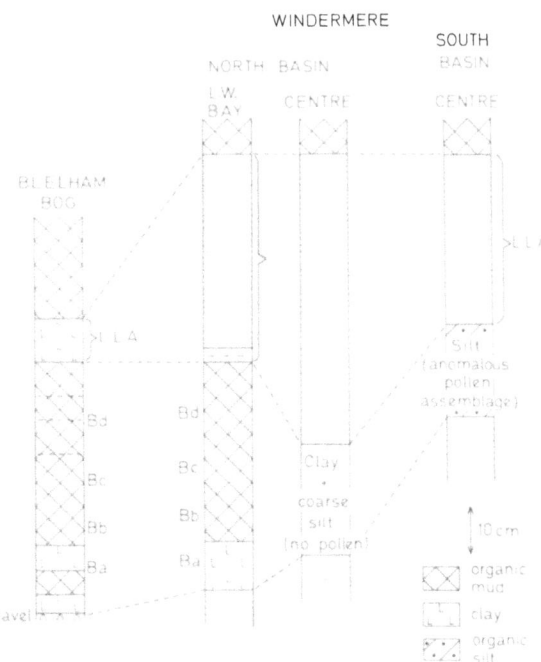

Fig. 3. Late-glacial stratigraphy and pollen zones, 3 positions in Windermere compared with Blelham Bog.

mountains moved radially outwards over a land surface already scoured of older soils. The response of other biota (e.g. algae, aquatic bryophytes, Cladocera) to the rising temperatures deduced from insect fossils (Coope, 1977) remains to be worked out in greater detail. Especially in view of current interest in the responses of organisms to rising temperatures, it seems that this is, for Windermere, a question as yet only partly answered which necessitates further work.

Late-glacial patterns of sediment accumulation were deduced from cores from increasing water depth, together with comparison with the waterlain deposits of a neighbouring small enclosed kettlehole, Blelham Bog (Fig. 3). The almost uniform Younger Dryas varved clay in Windermere is correlative, in the kettlehole (on pollen chronology and radiocarbon dating) with a narrow layer of clay attributed to solifluction (Fig. 3). Comparison with Sissons' map of reconstructed Loch Lomond Advance glaciers shows that the source material of the varve couplets must have entered Windermere via its two main inflow rivers

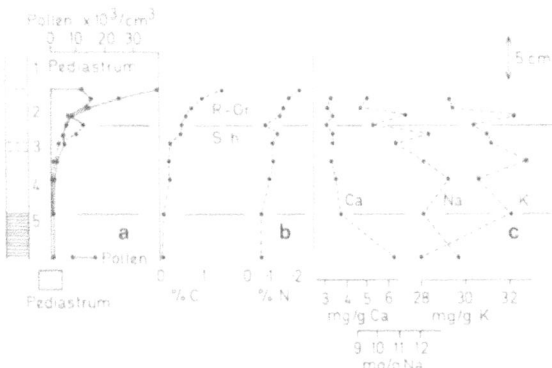

Fig. 2. Modified from *Striae*, 1981, Fig. 4a. The glacial/interstadial transition in Low Wray Bay. Deposits – 1 = interstadial woodland biozone, 2 = increasingly organic silt, 3 = *Fontinalis* stems, 4 = unlaminated clay, 5 = Lower Laminated Clay. (a) shows biota, total pollen and *Pediastrum*; (b) shows increasing organic content, S.h. = *Salix herbacea*, R-Gr = *Rumex*-Gramineae pollen zones; (c) Ca, Na & K as mg/g dry weight.

at the lake's head. Both of these pass through smaller lakes in their upper courses, between the limits of the glaciers and the entry of the rivers into Windermere. These smaller lakes cannot therefore be supposed to have functioned as complete sediment traps. Acceptance of this fact supported our reservations about the desirability of putting any great effort into attempts to estimate volumes of total annual increments to sediments of these lakes.

The biostratigraphy of organic interstadial deposits in Low Wray Bay was established as similar to that in the enclosed kettlehole, and (later) ^{14}C time-scales from these sites were found to correspond (Pennington, 1977). In Windermere there is however a complete discontinuity between marginal and central parts of the lake with respect to interstadial sediments (Figs. 1 and 3). There are no organic interstadial deposits in the central parts of North and South Basins – the correlative layer is mainly silt of very low organic content and where pollen is present, the anomalous assemblages of taxa suggest derivation of the pollen grains from the streamside vegetation of braided inflows. Questions with respect to the origin and transport of pollen to lakes in late-glacial time are only now being formulated and addressed. These questions carry implications with respect to palaeoclimate, since the spatial differentiation in organic late-glacial sediments and their pollen content implies little or no within-lake transport of organic sediment-source material, and therefore greatly reduced wind-driven turbulence compared with the present.

Sediment chronology. The post-glacial sediments needed a time-scale, to set their biostratigraphy within a chronological framework. Pollen analysis provided an approximate chronology via the boundaries between Godwin pollen zones, which at the time were accepted as probably synchronous. Until the development of radiocarbon dating from 1951, this was to a large extent relative dating, anchored in time only by correlation with the Swedish varve chronology. One feature of the diatom stratigraphy in Windermere stimu-

lated a special interest in the possibility of fitting a scale of years to recent sediments – the steep rise in the diatom *Asterionella*, at *ca* 20 cm below the mud surface in the centre of the North Basin (Pennington, 1943; 1974). Did this coincide in time with the onset of lake enrichment from sewage disposal?

At the same time, Deevey (using pollen) and Hutchinson were similarly seeking methods to fit a chronology to the sediments of Linsley Pond and other lakes in New England. In a paper published in 1940, Hutchinson & Wollack suggested a method for calculation of the relative length of periods of time represented by each successive unit of thickness in a sediment core – by assuming tentatively that unit mass of mineral material had been deposited in unit time. In Windermere this prompted the question 'What was the contemporary rate of deposition of mineral matter?' – to which a reasonably acceptable answer was provided by a programme of sediment trapping. Diatom counts as numbers per mg dry sediment made it possible to refine the data on mineral matter by subtraction of diatom silica. When data from the sediment traps for contemporary annual

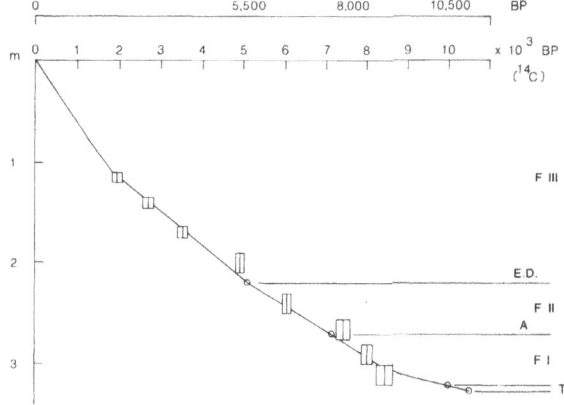

Fig. 4. Depth-time-scales for the post-glacial sediments of Windermere: 8 ^{14}C dates from a core taken in 1976 by R. Thompson from the South Basin. Circles indicate pollen horizons – E.D. = Elm Decline, A = *Alnus* rise, T = Transition zone. FI–III = Flandrian chronozones. The uppermost scale of ages BP shows the date assigned to the corresponding pollen horizon in a North Basin core, using the time-scale devised in 1943 on Hutchinson's model (see text).

14

deposition were included in Hutchinson and Wollack's model, a scale of years could be fitted to the Windermere data (Pennington, 1943; 1974, p. 366). It assigned an acceptable data of 100–150 years to the *Asterionella* horizon. The resulting time-scale is compared in Fig. 4 with that subsequently obtained from ^{14}C dating. The rather surprising degree of correspondence suggests that Hutchinson's assumption lay near the truth for Windermere. This can now be explained as due to the effects of a large lake in smoothing variations through time in deposition rates of mineral sediment – because the causes of variation were operating on only parts of its catchment.

Changes in sediment composition were originally investigated in relation to environmental changes suggested by the earliest pollen analyses (Pearsall & Pennington, 1947), and became the subject of an intensified research programme with the appointment of F.J.H. Mackereth to the FBA staff. His work (1965; 1966) used chemical changes – the distribution pattern of the major elements – to distinguish signals of a) leaching of catchment soils and b) the bodily erosion of such soils into lake basins. The main question arising – 'Was the organic component of the sediments mainly autochthonous or allochthonous?' – received the answer that in these lakes it was very largely allochthonous. Different conclusions with respect to Linsley Pond sparked off vigorous argument, prompting Deevey's well-remembered remark about 'the highly nervous landscape of the English Lake District'.

Lowland lakes of the Windermere catchment; Blelham Tarn, Esthwaite Water

When coring was extended to other lakes of the district, the first question to be asked was about the biological history of the most productive lakes of the Lake District series (a series of increasing productivity proposed by Pearsall (1921) and more recently discussed by Jones (1972)). In what respects had that differed from that of Windermere? Deevey (1942) had asked similar

questions about lake 'ontogeny'. In 2 productive lakes of the Windermere catchment, Blelham Tarn (10 h, max. depth 15 m) and Esthwaite Water (1 km^2, max. depth 15.5 m), the sedimentary record includes changes in biota and chemistry indicative of changing trophic status. Investigation of the sediments has been carried out at the same time and from the same institute as long-term studies of water chemistry, flora and fauna; these lakes are sites where 'palaeolimnology' has for a long time been integrated with what Smol (this volume) terms 'neolimnology'. Bridges between them have been described and discussed in many publications of the FBA.

The stratigraphy of the sediments of these 2 lakes could be related readily to that found in Windermere with respect to deep-water sediments (e.g. the survey of Esthwaite sediments (Franks & Pennington, 1961)). Unexpected stratigraphic complications were found in the late-glacial sediments over much of Blelham Tarn, with unconformities and reduplication of layers indicative of instability in late-glacial sandy deposits. This emphasised the degree of good fortune involved in the chance selection of Low Wray Bay for the initial experimental coring, for without prior knowledge of the undisturbed late-glacial sequence it would not have been possible to interpret the cores from Blelham Tarn.

A major question arose in these 2 lakes from apparent discrepancy between the biological and chemical indicators of increasing eutrophy in post-glacial deposits. Mackereth interpreted the distribution curves for iron and manganese, which follow a different course from those in Windermere and other large, unproductive lakes, in terms of seasonal deoxygenation throughout post-glacial time. In contrast, the faunal record for both lakes (Goulden, 1964, Harmsworth, 1968) showed that remains of the present Chironomid fauna – tolerant of seasonal anaerobiosis – are present only in the uppermost metre of sediment, which represents the deposits of the last *ca* 1000 years. Davison's recent (1981) work on the supply of iron and manganese to Esthwaite has made it possible to reconcile the chemical record with the history indicated by the

Time Years B.P.	Deposit	Vegetation of Catchment	Climate	Man	The lake	Sediment Chemistry
0 / 30	Black ooze			Tourists	Stephanodiscus Fragilaria	
	- - - - -					
50		Re-establishment of woods	(More	Detergents Sewage	Asterionella Enrichment	Rising organic & P
150		Landscaping	wind-induced	Estate change		Organic minimum
4-500		Ploughing - arable	turbulence	Subsistence farming		PEAK in Na & K
			deduced)			
1000	Organic mud	Extensive deforestation		Viking settlement	Chironomus seasonal ↕ anoxia Tanytarsus	
				? little local settlement		} Rising mineral
2000	Clay band	Temporary deforestation "Landnam"	Change to present - cool, wet "Sub-Atlantic"	Nomadic farming	Cyclotella comensis ↑ Cyclotella glomerata	
3000	Organic mud	Secondary forest, with Ilex, without Elm	? (peat formation on uplands)	No local settlement		} Stable
4000						
5000	Clay band	Elm Decline (no local change)	Increase in run-off	Neolithic settlements on coast; Axe-factories	Acidophilous diatoms increase Planktonic diatoms dominate	
6000	Organic mud	Continuous Climax Forest		? probably none locally		} Organic maximum
7000	Sand & clay band	Spread of Alder	Becomes wetter, with floods		Planktonic diatoms increase Decline in alkaliphilous diatoms	
8000	Organic mud		Dry, Rather Continental, warm summers		Melosira teres, open-(deep-)water areas	↑ Rising organic
9000					Temperate Cladocera Alkaliphilous diatoms	
10,000	- - - - -	Expansion of forest	Rapid rise in temperature			— — —
	Late-glacial clay & silt		Cold		High-alpine Cladocera	Inorganic

Fig. 5. Blelham Tarn. Environmental history of lake and catchment (modified from *Arch. Hydrobiol.* 1984 (Fig. 2); other data from Harmsworth, 1968; Evans, 1970 and Haworth, 1980).

faunal changes. From integration of all the palaeolimnological work on Blelham Tarn (Pennington & Lishman, 1984) it is now generally accepted that, at about 1000 years ago, the lake became more productive and seasonal deoxygenation began, coinciding with a sudden reduction in lake volume brought about by rapid input of sediment from a catchment in which deforestation was soon followed by ploughing. In the development of these ideas Deevey's (1955) paper on 'The obliteration of the hypolimnion' was a significant influence.

Both of these lakes show the effects of 'cultural eutrophication'. The rise in numbers of the diatom *Asterionella* which in Windermere was dated to *ca* 1850 is dated in Blelham to 1926–1930, and in Esthwaite a more gradual rise begins at some intermediate date (Haworth, 1985; Round, 1981; Evans, 1970). Other diatoms and algae indicative of more nutrient-rich waters appear subsequently in the sedimentary record. Detailed palaeolimnological studies of the uppermost sediments of Blelham Tarn, reported in the references cited, have related both biological and chemical changes to the historical record of local changes in sewage disposal, the changing usage of agricultural fertilisers, and the impact of air-borne pollution. Data from 'neolimnological' research have been used to interpret the fossil and geochemical record. In particular the phytoplankton records of Dr J.W.G. Lund which cover the last 40 years have made it possible to match the fossil diatom record with monitored changes in the plankton (Haworth, 1980). This gives considerable confidence in the accuracy of the stratigraphic and fossil record (cf. Fig. 5).

Talling & Heaney (1988) have discussed long-term changes in lakes monitored by the FBA over the period 1945–1986. There is good general correspondence between their record of changes in water chemistry and the changes in sediment composition (dated by ^{137}Cs) within the uppermost sediments of Windermere and Blelham Tarn. In Windermere the calculated present total annual addition of phosphorus to the sediments, based on values in the surface sediment (Pennington, 1978) shows that 'the sediments are a net sink for P', with good consistency between the calculated total and current values for input to and output from the lake (Talling & Heaney, op cit.).

Upland lakes in Southern Lakeland; Blea Tarn (Langdale), Angle Tarn

Some small lakes in the higher parts of the Windermere and adjoining catchments were investigated to provide answers to the following questions:

1. Could lakes whose basins had been occupied by ice of the Loch Lomond Advance (Younger Dryas) – i.e. those behind the fresh moraines mapped by Manley (1959) and Sissons (1980) – be identified by any common feature of their sediment sequence or biological development?
2. Could analysis of post-glacial sediments for a range of variables provide information on changes in the upland catchments resulting from prehistoric and later land use? – early questions on what came to be called 'lake-catchment relationships'.

Blea Tarn, (Langdale) (3.1 h, max. depth 8 m) proved one of the most rewarding sites with respect to both questions. Its sediments include the full sequence of late-glacial deposits, correlative with the lowland sites, showing that low moraines in the catchment must date from the main glaciation and that there was no local Younger Dryas ice. The post-glacial sequence of diatoms (Haworth, 1969), pollen, and chemical variables (Pennington & Lishman, 1971) for the years 10 000–4000 BP has proved informative as a record of soil history and forest development influenced by the natural process of an interglacial cycle, followed by the effects of human exploitation in the uplands (Pennington, 1973).

Two episodes of change appear in this record. At a horizon subsequently ^{14}C-dated to 7264 ± 70 BP (corresponding with the Boreal/Atlantic transition (Godwin, 1975)) there is a steep rise in influx (annual deposition cm^{-2}) of all arboreal pollen types to high peaks (Fig. 6), and a corresponding change in diatom assemblages with replacement of indicators of eutrophy by oligotrophic and rheophilous species (Haworth, 1969). These happenings are most readily explained by postulating an increase in run-off, an increased power of the inflows to transport pollen grains, and dilution of the tarn waters with reduction in nutrient status. This explanation in terms of climatic change is consistent with postulated rise in some lake levels at this time (Evans, 1970).

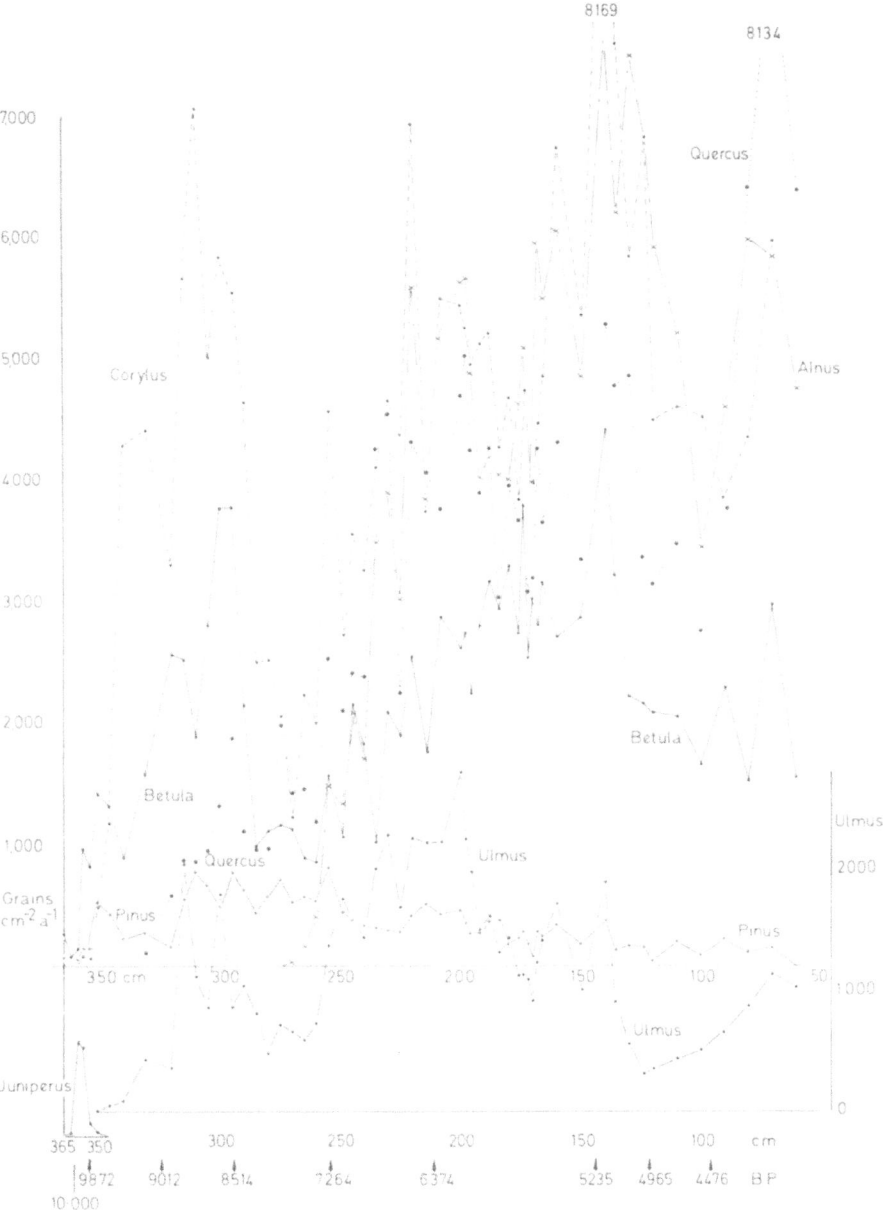

Fig. 6. Blea Tarn. Annual input of arboreal pollen taxa from 10 000 to 4000 BP. ¹⁴C dates correlated sediment depth (below mud surface) on horizontal axis. Pollen grains cm⁻² a⁻¹ on vertical axes.

The increased annual deposition of pollen cannot be explained in terms of annual production by the local forest. In general, annual pollen production can be shown to have been reduced at this time (Bennett, 1983) because the composition of the forest was changing from the early post-glacial domination by high pollen producers. Therefore the pattern shown in Fig. 6 must indicate a large increase at this date in the efficiency with which waterborne pollen grains were transported to the lake.

The second episode, at *ca* 5100 BP is recorded

18

by changes in pollen and chemical variables but by only minor changes in diatoms. It corresponds with the widely recognised but still puzzling Elm Decline of Northern Europe, which in parts of the Lake District is contemporaneous with, or immediately precedes, the first evidence of Early Neolithic settlement. Figure 7 shows firstly, the precision with which the pollen changes are recorded in Blea Tarn and the good replication of ^{14}C age in two cores, which is evidence against any significant 'bioturbation' in these deposits – and secondly how the major change in sediment composition at the Elm Decline appears as a shift in the component scores of a Principal Components analysis of the data matrix for 8 chemical elements. This is evidence for a sustained change from this date (*ca* 5100 BP) in sediment-source material – i.e. soils – from the catchment.

The episode of increase in pollen of grasses and

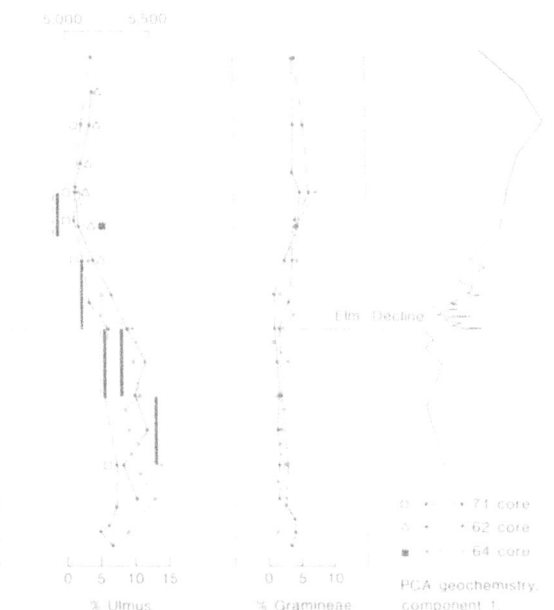

Fig. 7. Blea Tarn: Elm Decline detail. Three cores (taken in 1962, 1964 and 1971) correlated by position of steep fall in *Ulmus* pollen. 5 ^{14}C samples positioned on scale 5000–5500 BP, 3 (K-dates) from core 64 (single vertical stroke) and 2 (SRR-dates) from core 71 (double vertical stroke). Small squares and triangles = single grains of *Plantago lanceolata* in the core indicated. PCA geochemistry = scores of 1st component of PCA of data for % C, % mineral, Fe, Mn, Na, K, Ca, Mg as mg/g, and I as μg/g.

Plantago lanceolata which follows, within 100–200 years, the classical Elm Decline at *ca* 5000 BP (Figs. 6 and 7) coincides in ^{14}C date with local Early Neolithic activity, round the axe-factory sites on the other side of the Great Langdale. This activity was clearly associated with firstly, the formation of temporary forest clearings, and secondly with the onset of chemical changes in lake sediments indicative of an increased rate of soil erosion – cf. the arguments of Mackereth (1965; 1966) with respect to the large valley lakes.

Angle Tarn, Bowfell (3.4 h, max. depth 16 m) lies at an altitude of 579 m in a cirque associated with the course of one of the larger Loch Lomond Advance glaciers (Sissons, 1980, Fig. 6, no. 29). As expected, no late-glacial sequence was found in its sediments – the corer stopped in a stiff clay-silt interpreted as the product of englacial debris, at the base of post-glacial mud. In Angle Tarn this post-glacial mud yielded a complete and conformable sequence which on analysis provided useful data on upland soil history (Tutin, 1969; Pennington, 1984) but in other high corrie tarns there were gaps and irregularities in sediment stratigraphy suggesting that the sediment sequence in such small basins, in catchments of very high relief, should always be subject to critical appraisal. One question asked of the highest lakes referred to the pollen assemblages in the basal organic sediments of these basins behind the Loch Lomond Advance moraines. A tabulation of the basal pollen zones found in 14 such basins (Pennington, 1978b, Table 6) has been used to suggest a relative chronology for the time at which each lake began to accumulate organic sediment – i.e. became free of Loch Lomond Advance ice. It is consistent with the geomorphological evidence for glacier size that Angle Tarn should have become free of ice late in the sequence.

A detailed analysis of sediments covering the Elm Decline horizon showed a very precise record of changing sediment-source material in response to clearance of upland forest, including pine, at *ca* 5000 BP (Tutin, 1969, Fig. 2). The application of an e.s.r. technique for characterisation of humic compounds (Atherton *et al.*, 1967) showed that on these uplands soil acidification had already begun by this date (Pennington, 1984).

Lakes of the Western Lake District

The direct relationship between the presence of man, vegetation changes and variation in sediment composition, once established in these upland tarns in areas of intense Neolithic activity, was further investigated by work on lakes of the south-west Lake District where prehistoric remains of Neolithic and later date are more numerous than in the central valleys. The question asked was to what extent the accelerated input of soils from lake catchments, which appeared to have resulted from human settlement, had influenced lake biology. A related question referred to the nature of these inwashed soils – mineral detritus or predominantly organic (humus)? and if the latter, of what type?

Barfield Tarn near the coast lies among steep slopes formed by thick glacial drift. The sediment stratigraphy showed the striking feature that the organic mud of the early post-glacial period is overlain by pink clay-mud, the clay fraction of which is clearly derived from redeposited mineral soil from the adjacent slopes. Pollen analysis showed that the mud/clay contact coincides with the classical Elm decline of *ca* 5000 BP, and that here there is an equally steep decline in pollen of the oak, accompanied by a steep rise in pollen of grasses, herbs and cultivated cereals. This early example of the effects of cultivation in the immediate catchment remains unique among the Cumbrian lakes. The erosion resulting from cultivation practices (rendering unstable the particularly steep slopes around the tarn) would account for the complete change in composition of deposited sediment. The pollen diagram (Pennington, 1970, Fig. 11) shows a major change from closed forest to at least 50% cleared land in the area within up to *ca* 400 m of the lake, and no appreciable forest regeneration. This constitutes an effect differing by at least an order of magnitude from

the small temporary episodes recorded in the upland tarns for the same early Neolithic time.

It would be supposed that the inwash of mineral soils on this scale would lead to a marked increase in turbidity of the water of this small lake. The lithological change coincides with a reduction in numbers and species-diversity of diatoms (Haworth, 1985) which is explicable in these terms. It must be concluded that this episode was an example of what has been described by Binford, Deevey and Crisman (1984) as 'siltation', – 'a non-nutrient output of societal ecosystems'. There is no clear evidence from Cumbria of any nutrient contribution to lakes from the changes brought about by prehistoric human societies.

Devoke Water and Burnmoor Tarn, lakes of area 23–24 h and max. depth 14 m, both lie at *ca* 250 m O.D. in what are now moorland catchments, in a part of the south-western uplands where there are no traces of early Neolithic settlement. Their pollen diagrams show no evidence of vegetation change at this date, but from 5000 BP the rate of sediment accumulation increased (Pennington, 1981). In both lakes the pollen diagrams show that the transformation of the catchments from forest to moorland took place in two stages, one between 3000 and 4000 years ago and one in late Romano-British time, 200–400 A.D. In both lakes and in the nearby Seathwaite Tarn, chemical evidence from the sediments supported Mackereth's hypothesis that the organic fraction is largely allochthonous, derived from organic soils and shown by Atherton *et al.* (1967) to have become increasingly acid with the passage of post-glacial time (Fig. 8, from Pennington, 1984). The course of this natural acidification of upland soils is correlated with the vegetation change from deciduous forest to *Calluna* moorland and *Sphagnum* bog. It is inferred that the earlier (Bronze Age) episode of partial forest clearance led to soil changes – podzolization of vulnerable soils with formation of a superficial layer of acid mor, rich in *Calluna* pollen. At whatever subsequent date (Iron Age or later) these catchments were settled and farmed, the areas of podzols within the soil mosaic were liable to

BURNMOOR TARN   254 m

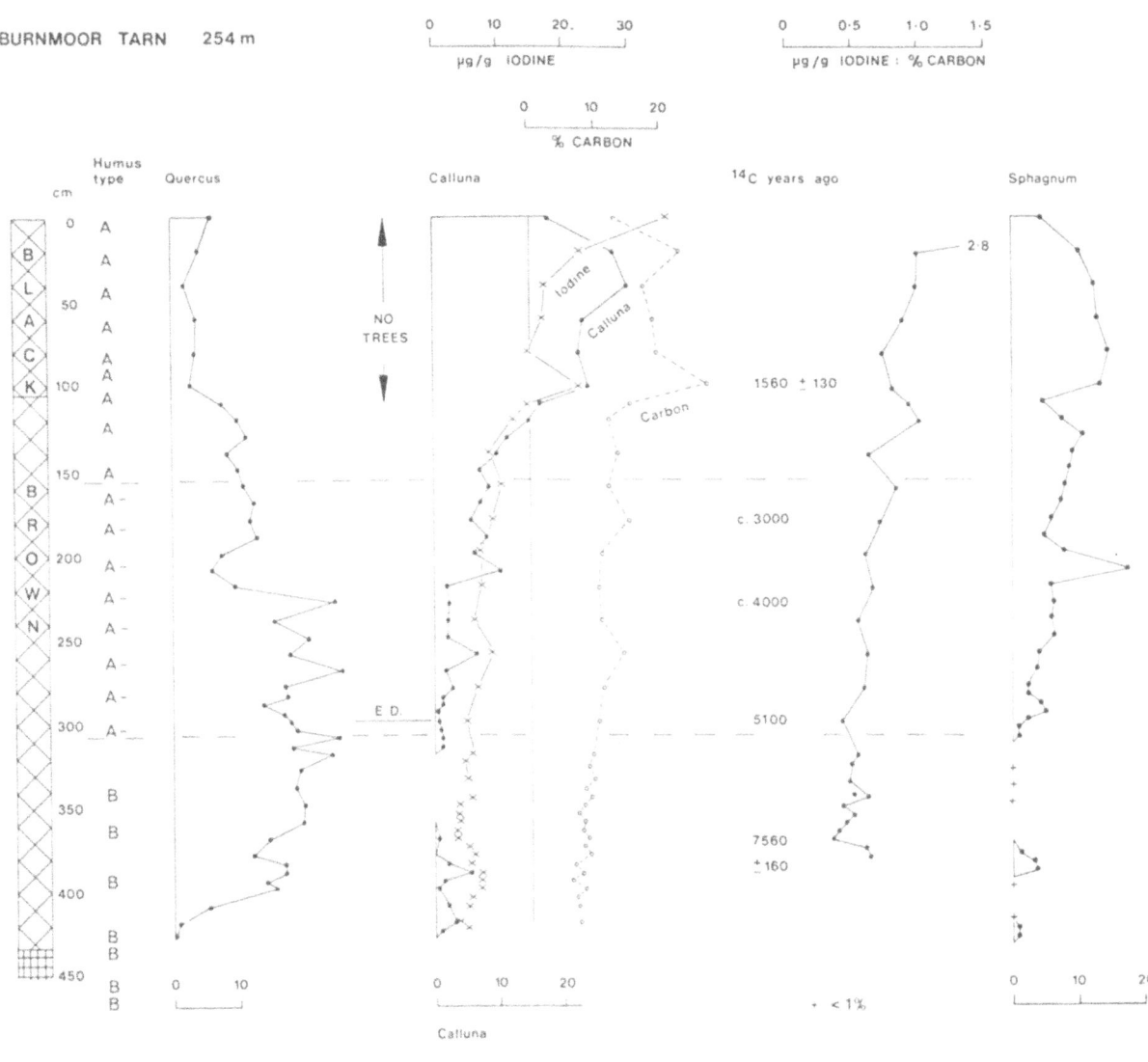

Percentage of total pollen

Fig. 8. Reproduced from *FBA Ann. Rept.* 1984, Fig. 6. See this for legend.

accelerated erosion with rapid input to the lakes of this mor humus and its contained *Calluna* pollen (Pennington, 1964; 1981). The overall effects on lake biota can be seen in the diatom sequence from Devoke Water (Evans, 1961). Figure 8 shows this history in the sediment profile from Burnmoor Tarn; the curve for *Quercus* as a percentage of total pollen shows the two main episodes of deforestation at *ca* 4000 B.P. and at the Romano-British horizon dated to 1560 ± 130,

and those for *Calluna* and *Sphagnum* show peaks at the latter date which correspond with peak input of acid organic matter (carbon curve and 'Humus type') at a time of upland cereal cultivation.

Deforestation of these upland catchments is therefore seen to have been important in accelerating the natural acidification of soils and waters. In the sediments of Devoke Water a clear reversal in ^{14}C age in a series of four samples dating from *ca* 1700–1400 B.P. (Tutin, 1969, Fig. 3) supports the interpretation of events at this time as a perturbation of orderly transfer of sediment-source material from catchment to lake, caused by soil erosion due to farming practices. These data show that 'siltation' (cf. p. 11), in lake/catchment ecosystems of this western upland type, featured the accelerated input to lakes of the organic material (mor humus) which had formed on the catchments as the result of earlier partial deforestation, with consequent retrogressive soil changes.

Ennerdale Water, a large (3 km²) valley lake in a catchment of high relief has a deep trough (42 m max. depth) and a large shallow area at the outflow end. Its pattern of sediment accumulation resembles that of Windermere in many ways, but in the 'trough' area both late- and early postglacial sediments show much disturbance by slumping on the steep underwater slopes. This stratigraphic complexity restricts the number of questions to be asked of this lake, but cores from the 'basin plain' of the trough have provided a complete and conformable record of the last 5000 years, from the horizon of the Elm Decline onward. This lake has clearly been oligotrophic throughout its history. The value of the sedimentary record relates to the interaction of soil and vegetation history as documented for a catchment of such high relief, where the human history has been similar to that of Burnmoor Tarn.

Figure 9 shows a closely dated depth-timescale for a central core from Ennerdale Water, illustrating the onset of reversals and anomalies in ¹⁴C age from *ca* 1000 B.P. This onset coincides with pollen evidence for extensive forest clearance with cereal cultivation, and with a peak in carbon content (in annual deposition over unit area as well as in percentage composition) attributed to greatly increased output of organic matter from the catchment following widespread deforestation (Pennington, 1978b, Fig. 2). The sediment of peak carbon content is rich in Calluna pollen and is interpreted as derived from the surface organic horizon of catchment soils which had been transformed from forest soils to heath during earlier periods of intense land use (Bronze Age and Romano-British). The sequence of changes

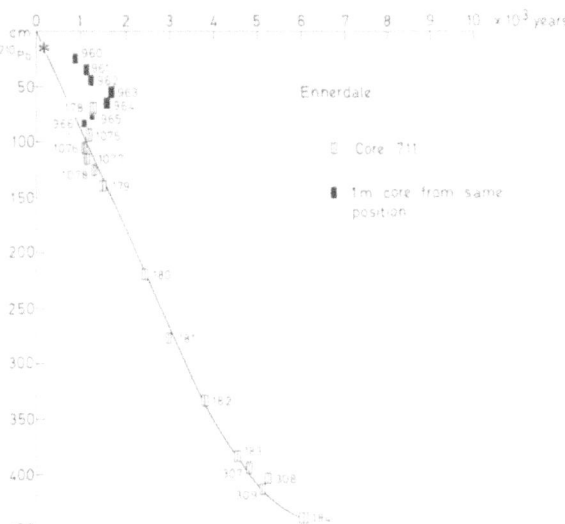

Fig. 9. Ennerdale Water. Depth-time-scale for the last 5000 years based on ¹⁴C from a 6 cm and a 1 m core, showing the onset of anomalies in ¹⁴C age at *ca* 1000 BP.

shown by closely spaced pollen analyses is conformable and credible, so the inversion of ¹⁴C age cannot be attributed to reworking of older sediments. It seems that the critical factor in this perturbation of orderly input of sediment-source material must have been the effect of cultivation practices involving a certain kind of soil disturbance, and the time of onset of ¹⁴C anomalies shows when these were first used in the immediate catchment. At Devoke Water this took place within the first few centuries A.D., while in the Ennerdale catchment it was not until *ca* 800 years later that the process became operative. Such reversals of ¹⁴C age within sediments of the historic period are of course very widely reported from the British Isles and Northern Europe. The precise mechanism involved must remain one of the as yet unanswered questions in the palaeolimnology of the English Lakes.

The feature of the western lakes is that this onset of anomalous ¹⁴C age first appears in sediment of distinct chemical composition (Pennington, 1977), indicative of a major perturbation of input from the catchment, but there is little evidence for biological change in the lake. The contrasted situation in Blelham Tarn, where a similar onset of ^{14}C anomaly occurs at approximately the same time in the record as in Ennerdale (Pennington *et al.*, 1976) is that here there is less change in sediment composition but evidence for major change in the lake environment (the onset of seasonal anaerobiosis: cf. Pennington & Lishman, 1984).

Conclusions

This study of a diversity of lakes within a region *ca* 20×20 km has produced answers to some questions of palaeolimnology and raised some problems. It has underlined the difficulty in predicting which lakes are most likely to provide answers to formulated questions.

In these lakes, of area from 3 to 1480 ha, the total depth of organic sediment is broadly similar except in those lakes of particularly high ratio of surface area to mean depth. There is evidence that, lying as they do as part of vigorous river systems, these lakes do not function as complete traps for waterborne sediment. There is no correlation between depth of sediment and size, of lake or catchment. The variables affecting sediment accumulation appear to have been:

(i) Available sources of the suspended-sediment load of streams: these were greater in late-glacial time (non-forest vegetation) and after deforestation of catchments by man. Therefore the pollen content of sediments is an index to the availability of such sources.

(ii) The transporting efficiency of inflow streams; the rate of total sediment accumulation was lower at times for which there is independent evidence of dry climate, and increased at times for which the evidence shows – (a) presence of flood debris and change in level of some lakes, and (b) increased run-off following deforestation and cultivation practices on the catchment.

(iii) Within-lake processes controlled by the pattern of water circulation and the degree of turbulence (driven by wind strength), and to some extent by the stability and shape of the shorelines.

The degree to which the profile in a lake sediment core can be accepted as a complete and conformable sequence, together with its contained microfossils and chemical indicators, depends on assessment of (i) the pattern of sediment accumulation and (ii) the extent to which it can be shown to have changed through time.

References

Atherton, N. M., P. A. Cranwell, A. J. Floyd & R. D. Haworth, 1967. Humic acid. 1. ESR of humic acids. Tetrahedron 23: 1653–67.

Bennett, K. D., 1983. Devensian late-glacial and Flandrian vegetational history at Hockham Mere, Norfolk, England. I. Pollen percentages and concentrations. New Phytol. 95: 457–487. II. Pollen accumulation rates. New Phytol. 95: 489–504.

Berglund, B. E., 1986. Palaeoecological reference areas and reference sites. In: Handbook of Holocene Palaeoecology and Palaeohydrology (edit. B.E. Berglund): 111–126. John Wiley & Sons.

Binford, M. W., E. S. Deevey & T. L. Gisman, 1983. Palaeolimnology: A historical perspective on lacustrine ecosystems. Ann. Rev. Ecol. Syst. 14: 255–286.

Coope, G. R., 1977. Fossil coleopteran assemblages as sensitive indicators of climatic changes during the Devensian (last) cold stage. Phil. Trans. r. Soc. Lond. B 280: 313–340.

Coope, G. R. & W. Pennington, 1977. The Windermere Interstadial of the Late Devensian. Phil. Trans. r. Soc. Lond. B 280: 337–339.

Davison, W., 1981. The supply of iron and manganese to an anoxic lake basin. Nature 290: 241–243.

Deevey, E. S., 1939. Studies on Connecticut lake sediments. I. A post-glacial climatic chronology for Southern New England. Am. J. Sci. 237: 691–724.

Deevey, E. S., 1942. Studies on Connecticut lake sediments. III. The biostratonomy of Linsley Pond. Am. J. Sci. 240: 233–64, 313–38.

Deevey, E. S., 1955. The obliteration of the hypolimnion. Mem. Ist. Ital. Idrobiol. Supl. 8: 9–38.

Evans, G. H., 1961. A study of the diatoms in a core from the sediments of Devoke Water. Univ. of Wales (Aberystwyth): MSc thesis.

Evans, G. H., 1970. Pollen and diatom analyses of Late-Quaternary deposits in the Blelham basin, north Lancashire. New Phytol. 69: 821–874.

Franks, J. W. & W. Pennington. The Late-glacial and Postglacial deposits of the Esthwaite basin, north Lancashire. New Phytol. 60: 27–42.

Godwin, H., 1975. History of the British Flora. Cambridge University Press, Cambridge.

Goulden, C. E., 1964. The history of the Cladoceran fauna of

Esthwaite Water (England) and its limnological significance. Arch. Hydrobiol. 60: 1–52.

Harmsworth, R. V., 1968. The developmental history of
Blelham Tarn (England) as shown by animal microfossils,
with special reference to the Cladocera. Ecol. Monogr. 38:
232–41.

Haworth, E. Y., 1969. The diatoms of a sediment core from
Blea Tarn, Langdale. J. Ecol. 57: 429–439.

Haworth, E. Y., 1980. Comparison of continuous phytoplankton records with the diatom stratigraphy in the recent
sediments of Blelham Tarn. Limnol. Oceanogr. 25:
1093–1103.

Haworth, E. Y., 1985. 'The highly nervous system of the
English Lakes': aquatic ecosystem sensitivity to external
changes, as demonstrated by diatoms. Ann. Rep.
Freshwat. Biol. Ass. 53: 60–79.

Haworth, E. Y. & J. Long, 1989. Lake District Palaeolimnology: a bibliography. Inst. Freshwat. Ecology.

Howell, F. T., 1971. A continuous seismic profile survey of
Windermere. Geol. J. 7: Pt. 2: 329–334.

Hutchinson, G. E. & A. Wollack, 1940. Studies on Connecticut lake sediments II. Chemical analyses of a core
from Linsley Pond. Am. J. Sci. 238: 493–517.

Jessen, K. & A. Farrington, 1938. The bogs at Ballybetagh,
near Dublin, with remarks on late glacial conditions in
Ireland. Proc. R. Ir. Acad. B 44: 205–260.

Jones, J. G., 1972. Studies on freshwater micro-organisms:
phosphatase activity in lakes of differing degrees of
eutrophication. J. Ecol. 60: 777–91.

Lehmann, J. T., 1975. Reconstructing the rate of accumulation of lake sediment: the effect of sediment focusing. Quat.
Res. NY 5: 541–50.

Likens, G. E. & M. B. Davis, 1975. Post-glacial history of
Mirror Lake and its watershed in New Hampshire, U.S.A.:
an initial report. Verh. int. Verein. Limnol. 19: 982–993.

Mackereth, F. J. H., 1965. Chemical investigations of lake
sediments and their interpretation. Proc. R. Soc. London
B 161: 295–309.

Mackereth, F. J. H., 1966. Some chemical observations on
post-glacial lake sediments. Phil. Trans. r. Soc. London B
250: 165–213.

Mackereth, F. J. H., 1971. On the variation in direction of the
horizontal component of remanent magnetisation in lake
sediments. Earth Planet. Sci. Lett. 37: 131–8.

Manley, G., 1959. The late-glacial climate of north-west
England. Liverpool and Manchester Geological Journal, 2:
Pt 2: 188–215.

Östrem, G., 1975. Sediment transport in glacial meltwater
streams. Glaciofluvial and Glaciolacustrine Sedimentation. S.P. Pub. No. 23: 101–122.

Pearsall, W. H., 1921. The development of vegetation in the
English Lakes, considered in relation to the general evolution of glacial lakes and rock basins. Proc. r. Soc. Lond. B
92: 259–84.

Pearsall, W. H. & W. Pennington, 1947. The ecological history of the Lake District. J. Ecol. 34: 137–148.

Pennington, W., 1943. Lake sediments: the bottom deposits
of the North Basin of Windermere, with special reference
to the diatom succession. New Phytol. 42: 1–27.

Pennington, W., 1947. Lake sediments: pollen diagrams from
the bottom deposits of the North Basin of Windermere.
Phil. Trans. r. Soc. London B 233: 137–175.

Pennington, W., 1970. Vegetation history in the north-west of
England: a regional synthesis. In: Studies in the vege-
tational history of the British Isles, edit. D. Walker & R. G.
West; pp. 41–79, Cambridge University Press.

Pennington, W. (Mrs T. G. Tutin), 1973a. Absolute pollen
frequencies in the sediments of lakes of different morphometry. Quaternary Plant Ecology (ed. by H. J. B. Birks
& R. G. West), pp. 79–104. Blackwell Scientific Publications, Oxford).

Pennington, W. (Mrs T. G. Tutin), 1973b. The recent sediments of Windermere. Freshwat. Biol. 3: 363–82.

Pennington, W. (Mrs T. G. Tutin), 1977. The Late Devensian
flora and vegetation of Britain. Phil. Trans. r. Soc. Lond.
B 280: 247–271.

Pennington, W., 1978a. Responses of some British lakes to
past changes in land use on their catchments. Verh. int.
Verein. Limnol. 20: 636–641.

Pennington, W., 1978b. Quaternary Geology. In: Geology of
the Lake District (edit. F. Moseley) 207–255. Yorkshire
Geological Society Occ. Publ. 3 Leeds.

Pennington, W., 1981a. Sediment composition in relation to
the interpretation of pollen data. IV International Palynological Conference. Lucknow (1976–77) 3: 188–213.

Pennington, W., 1981b. Records of a lake's life in time: the
sediments. Hydrobiologia 79: 197–219.

Pennington, W., 1984. Long-term natural acidification of
upland sites in Cumbria: evidence from post-glacial lake
sediments. Ann. Rep. Freshwat. Biol. Ass. 52: 28–46.

Pennington, W., R. S. Cambray, J. D. Eakins & D. D.
Harkness, 1976. Radionuclide dating of the recent sediments of Blelham Tarn. Freshwat. Biol. 6: 317–331.

Pennington, W. & J. P. Lishman, 1971. Iodine in lake sediments in northern England and Scotland. Biol. Reviews
Cambridge Phil. Soc. 46: 279–313.

Pennington, W. & J. P. Lishman, 1984. The post-glacial sediments of Blelham Tarn: geochemistry and palaeoecology.
Arch. Hydrobiol./Suppl. 69: 1, 1–54.

Round, F. E., 1961. The diatoms of a core from Esthwaite
Water. New Phytol. 60: 43–59.

Sissons, J. B., 1974. The Quaternary in Scotland: a review.
Scot. J. Geol. 10: 311–37.

Sissons, J. B., 1980. The Loch Lomond Advance in the Lake
District, northern England. Trans. r. Soc. Edin.v: Earth
Sciences 71: 13–27.

Sturm, M. & A. Matter, 1978. Turbidites and varves in Lake
Brienz (Switzerland): deposition of clastic detritus by density currents. pp. 145–168 in: Modern and ancient lake
sediments (International Association of Sedimentologists.
Special Publications: No. 2).

Talling, J. F. (Edit.), 1986. A general assessment of environmental and biological features of Windermere and their susceptibility to change. Rep. Freshwat. Biol. Ass.

Talling, J. F. & S. I. Heaney, 1988. Long-term changes in some English (Cumbrian) lakes subjected to increased nutrient inputs. In: Algae and the Aquatic Environment (edit. F. E. Round) 1 19, Biopress Ltd., Bristol.

Turner, G. H. & R. Thompson, 1981. Lake sediment record of the geomagnetic secular variation in Britain during Holocene times. Geophys. J. R. astr. Soc. 65: 703–725.

Tutin, W. (née Pennington), 1969. The usefulness of pollen analysis in interpretation of stratigraphic horizons, both late-glacial and post-glacial. Mitt. int. Ver. Limnol. 17: 154–64.

Hydrobiologia **214**: 25–34, 1991.
J. P. Smith, P. G. Appleby, R. W. Battarbee, J. A. Dearing, R. Flower, E. Y. Haworth, F. Oldfield & P. E. O'Sullivan (eds), 25
Environmental History and Palaeolimnology.

Accuracy and precision in sediment chronology

Ingrid U. Olsson
Uppsala University, Department of Physics, Box 530, S-751 21 Uppsala, Sweden

Abstract

Several complications are encountered in radiocarbon dating of limnological samples. The contamination problem is obvious. Since lake sediments are usually composed of both allochthonous and autochthonous material the origin must be considered. The organic content may give some guidance. Graphite in the catchment area and a low organic content indicate a potential risk of major errors in the dates. Organic material from the surroundings may be displaced into a lake by creeks, snowslides and wind. Such material may be older than, or contemporaneous with the sediment deposited at the same time. Water-level changes and wave action may cause erosion and thus admixture of old material. Such contamination can sometimes be traced via different fractions extracted mechanically or chemically. The accelerator technique has made it possible to date small samples and thus macrofossils can be selected as the material to be used for dating far more frequently than is possible with the conventional technique for activity measurement.

Another problem is the reservoir effect – thus the lower ^{14}C activity of the dissolved bicarbonate and carbon dioxide than of the atmosphere. There may be several reasons besides dissolved carbonates in hard-water lakes. Some plants even use carbon dioxide in the sediment. It is thus to be expected that submerged plants will be dated too old. Selection of plant remains for dating is thus no guarantee of an accurate date. The material to be preferred is terrestrial, such as tree leaves or remains of emergent plants.

Bioturbation must not be forgotten since this means that old material may be found too high in a sediment, and young material can be brought downwards in the sediment. A single date may be very misleading. Under rather stable conditions the sediment accumulation rate can be roughly estimated and acceptable dates obtained despite the complications.

It may be worth sieving gyttja samples to remove roots and rootlets. Roots in peat should also be removed. Diagrams are available of the errors due to certain degrees of contamination by younger or older material.

The ^{13}C content must be measured to allow a normalisation. The present knowledge of the secular variations of the ^{14}C activity allows calibration of dates made on wood etc. The same curves cannot be used for submerged material because of the lower ^{14}C activity. Sudden changes of the atmospheric ^{14}C activity are smoothed. Certain age ranges give better resolution than others. Recent curves indicate that strange results are to be expected for Holocene samples with a real age of 7000 years or older. Finally bacterial influence during storage may influence the isotopic composition.

Introduction

A ^{14}C age is normally given in ^{14}C years BP. The age calculation is then based on measurements yielding the ratio between the activity of a certain amount of carbon of the sample, and that of an international standard. The absolute activity is thus of little interest at age determinations. The half-life is, however, important but its value is eliminated as soon as a ^{14}C date is calibrated. Calibrations are needed for accurate work since the ^{14}C activity has varied. The standard is decaying but is defined as having had the correct activity in AD 1950. The activities of the standard and the sample have decreased with the same percentage from AD 1950 to the day of measurement. Thus the age of a sample is given before AD 1950 and called Before Present or BP. The standard was chosen to have an activity close to that of the atmosphere in 1850, before major contamination of the atmosphere by 'old' CO_2. The measured activities must be normalized because of the isotopic fractionation.

The quality of the final result is dependent not only on careful measurements but also on the collection, the possible contamination, the treatment in the laboratory and the reservoir effects. The activities may deviate slightly, regionally or locally, from the global average.

The global variations

Libby, who devised the ^{14}C method, collected contemporaneous samples from various places on earth and old samples with comparatively well known ages. Libby's technique with elemental carbon on the inner wall of the detector for the measurements (Libby, 1955) did not allow accurate results. The statistical uncertainties given by him were approximately ± 200 years. Upon the development of gas-filled counters with the sample converted to CO_2, or any other desired gas, it became apparent that the atmospheric ^{14}C/^{12}C ratio had varied during recent millennia (Münnich, 1957; de Vries, 1958). Since de Vries made the first attempt to explain these variations and correlate them with climatic changes this effect is called the de Vries effect and has been studied in many laboratories. Twenty years ago an interdisciplinary specialist symposium on this subject was held (Olsson, 1970), after which the results were still further improved until the first real high-precision measurements on tree rings from a few hundred year periods were released (Stuiver, 1978). Since then it has been possible for the ^{14}C community to accept data-sets allowing calibrations to be made (see Radiocarbon vol. 28 no. 2B, 1986). There are two very important com-plications to be recalled – the sample's own age and its reservoir age. Any contamination complicates the calibration still further.

The most suitable samples for determining the atmospheric secular ^{14}C/^{12}C variations are tree rings. The dendrochronology allows us to use samples with ages known to the year. *Sequoia* was frequently used earlier, but oak is usually to be preferred since the dendrochronology is very accurate and the rings may be comparatively broad. *Pinus aristata* has been very useful since a single tree can be very old.

The variations in the ^{14}C activity from the international standard are small for the last three thousand years, but a pronounced excess over the standard is seen earlier during the Holocene. The long-term variations are generally correlated with the variations in the earth's magnetic field (Bucha, 1970) and the short-term variations with those of the solar activity. The excess is increasingly larger back to c. 5000 BC and exhibits a complicated pattern still earlier. This means that the ^{14}C ages of samples older than three thousand years appear too young – with as much as almost one thousand years seven millennia ago. At five to seven thousand years BC a long-term dip in the excess has been detected. It may be very difficult to translate ages given in ^{14}C years into calendar years for these old samples. This is further treated by Lotter (this volume). Using laminated sediment he has succeeded in giving very accurate floating chronologies and lengths for the Younger Dryas and other periods. However, several other intervals cause problems at the calibration since a short ^{14}C-year range may correspond to a long calendar-year range. In such cases the expression 'radiocarbon plateau' has come into use.

General aspects on reservoir ages

Problems arise when samples incorporate carbon from other reservoirs than the atmosphere since, e.g., the exchange between the atmosphere and the surface waters is slow enough to cause the water to appear older, at ^{14}C dating, than contemporaneous terrestrial plants assimilating atmospheric carbon dioxide. Moreover old carbonates may be dissolved and dilute the ^{14}C present in waters. This was known very early (Deevey *et al.*, 1954). Aged groundwater may also contribute to the decrease of the ratio between ^{14}C and the sum of the stable isotopes ^{12}C and ^{13}C. It is seen that these processes must have a smoothing effect so that the atmospheric variations yield smaller variations in the CO_2 dissolved in waters. The calibrations will certainly be less accurate for lacustrine sediments than for charred grains. Indeed, samples with their own age, such as charcoal from a century or a slice of peat, need a smoothed calibration curve. If the sample is contaminated the problem will be still greater. The calibration volume of Radiocarbon (1986) contains curves for calibration of samples of marine origin. These should be used together with a map of values to be included at the calibration to consider the varying reservoir ages, dependent on the location. The curves and values on the map will certainly be revised when more measurements of recent material appear. In a revised computer program, based on that given in Radiocarbon (1986), there is an option to include an age range or span to consider the sample's own age. The reservoir age for a sample is difficult to determine since it may vary with time due to leaching and the degree of admixture of terrestrial material in the autochthonous material which may derive from a varying composition of submerged and emergent plants.

Isotopic fractionation

Although the carbon isotopes behave chemically in the same way, the slightly different weights will cause some isotopic fractionation to occur. The heavier isotope ^{13}C is enriched in one compound at certain processes and the lighter, ^{12}C, thus depleted in another compound in the same process. Since ^{12}C is about one hundred times more abundant than ^{13}C the depletion of ^{12}C is normally of little interest. ^{13}C is enriched when carbon dioxide is dissolved in water but depleted at the photosynthesis. The degree of depletion depends

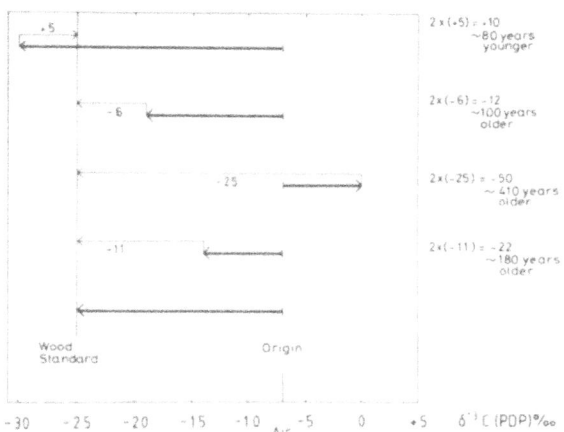

Fig. 1. Examples of the $\delta^{13}C$ normalization to -25 per mille and the corresponding ^{14}C and age corrections. The $\delta^{13}C$ normalization is intended to eliminate the effect of fractionation at the uptake of ^{14}C, and can be regarded as a bias, a positive or negative apparent age, when the carbon was incorporated. The bias will always have the same value in years.

on the type of plant. Three pathways are known for terrestrial plants. The plants growing in northern countries are mostly of the C_3 type with a ^{13}C content which is 25 per mille less than for marine carbonate. The standard used in ^{14}C work is a carbonate, a Cretaceous belemnite, *Belemnitella americana*, from the Peedee formation in South Carolina (Craig, 1957). The deviations are usually given as the per mille deviation from this standard, the PDB standard; i.e. the $\delta^{13}C$ values which should always be stated together with a ^{14}C result. The C_4 plants, e.g. *Zea mays*, are less depleted than the C_3 plants. Succulents with the CAM assimilation have an intermediate depletion. The thermodynamic laws postulate that the fractionation for ^{14}C is almost exactly double that for ^{13}C (Craig, 1954). There is a marked spread in the $\delta^{13}C$ values between plant species but also between different parts of a plant and chemical compounds extracted from a specific plant. At ^{14}C dating the main interest is the decay. It is thus necessary to compensate for the natural deviations arising from this fractionation via an adjustment to a normal $\delta^{13}C$ value. The natural value would have been that for air but since wood had been used as standard in many laboratories long before the need for normalization was generally accepted it was agreed to normalize to −25 per mille for all samples to be dated. It is thus absolutely essential that normalizations be performed (Fig. 1). The corrections due to this normalization are 410 years, to be added, for a sample with $\delta^{13}C = 0‰$, and 80 years, to be subtracted, for a sample with $\delta^{13}C = -30‰$. It was only a little more than two decades ago that the three photosynthetic pathways were recognized. Before that it was difficult to explain the observed spread in the $\delta^{13}C$ values. It is easily understood that the ^{13}C content of a sample is dependent on the value of the source and also on temperature, pH of water, excess or not of CO_2 etc.

Contamination

In order to obtain accurate results no carbon should have been added or removed from the sample so as to change the ratio between the radioactive ^{14}C and the stable isotopes, except for the radioactive decay – the system should have been closed. If the sample has been contaminated it is sometimes possible to remove the contaminants and thereby fulfill the requirements for a closed system. Contamination may occur not only when the sample is stored naturally and later but also when deposited. The allochthonous material may be very old. Varved clay is a typical example of material from which erroneous results may be expected because of the transport. When such material was submitted it was natural to test two fractions – the fraction soluble in NaOH and the insoluble remains, called SOL and INS respectively. It seemed plausible that the soluble fraction might contain remains from material grown when the varve was deposited or slightly earlier. The results from Lugnvik are given in Table 1. The insoluble fraction yielded very old, or infinitely old ages whereas the soluble fraction produced results which agreed within the limits of errors with the expected ages. Since the samples contained graphite it appears that the carbon in the insoluble fraction derived almost exclusively from graphite. The carbon content of the samples was very low and much smaller in the soluble fraction than in the insoluble. The expected age should be amended since the varve chronology was revised (Cato, 1985) but because of the large uncertainties of the finite ages it is of no importance here and the Table stands as compiled by Hörnsten & Olsson (1964).

Another example is Lake Rudetjärn (Olsson,

Table 1. Results of ^{14}C measurements of varved-clay samples from Lugnvik.

Lugnvik			Apparent ^{14}C age Radiocarbon years BP
Sample 1a			
U-213 Varves + 56 to + 82	INS	30,000	+ 2,500 − 2,000
U-215	SOL	9,000	+ 1,400 − 1,200
Sample 1b			
U-214 Varves + 56 to + 82	INS	34,000	+ 2,200 − 1,800
Sample 2			
U-260 Varves + 29 to + 55	INS	> 37,000	

Expected age *ca.* 8,800 years as determined by the old varve chronology (Hörnsten & Olsson 1964) before later revisions (Cato 1985).

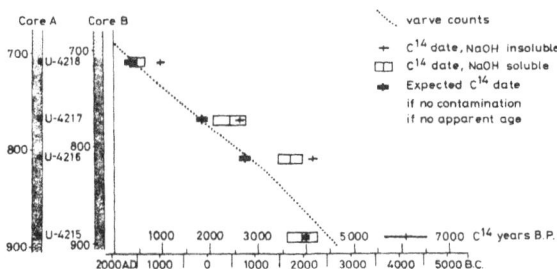

Fig. 2. Renberg's age determination by counting of layers in the laminated gyttja, the ^{14}C ages of the insoluble and soluble fractions of the four dated levels, and the radiocarbon ages expected from Renberg's estimate. The length of the bars and the horizontal length of the rectangles indicate $\pm \sigma$. The U-numbers given are those of the soluble fraction. The insoluble fractions were dated as U-4076 to U-4079 starting with the uppermost sample (from Olsson 1979a)

1979a) absolutely dated by laminated annual sediment layers by Renberg, and ^{14}C measured in Uppsala (Fig. 2). It is seen that the ^{14}C dates of the insoluble fraction of the sediment appear older than the layers. All ages received when the ages of the layers, determined by counting, are translated to expected ^{14}C ages, using a calibration curve, are younger than the measured ^{14}C ages for the insoluble fractions. The soluble fractions, however, are all younger than the corresponding insoluble fractions although significantly younger in only two of the four pairs. The differences between the ages of the soluble fraction and the corresponding expected ages have a mean value of 450 ± 130 years. The total carbon content was about 3%, the greater part was soluble. Graphite was detected in the samples.

Since a sediment usually contains an inorganic fraction deriving from terrestrial areas there is a great risk that old carbonaceous material was supplied to the sediment at the same time. This may be anything from small branches to graphite or elemental carbon and dissolved material. Mud-feeding animals may contribute to the breakdown and eventually to the reservoir age. Olsson (1986) filtered and chemically treated water from a lake, a fen, the lagg of a bog, and a hollow of a

raised bog. The various fractions had very different activities indicating either old constituents or old reservoir ages.

A diagram giving the error due to contamination with old or young material is presented (Fig. 3) to allow estimates thereof. Graphite is 'infinitely old', eroded material usually somewhat older than the real sample, whereas roots are younger. The error is given as a function of the age differences for various degrees of contamination. Thus an infinitely old sample will be dated at an age of two half-lives (11 140 years) if it is contaminated to 25% by modern material with the same activity as the standard. This figure will be smaller if the contaminant is affected by the atomic-bomb effect and thus has an excess activity over the international standard, but larger than 25% if the contaminant has less activity than the standard.

Carbonate often causes trouble unless carefully stored and treated. When wet, it may recrystallize; this can be detected with X-rays for material originally deposited as aragonite since the recrystallization is a transformation to calcite. There are numerous examples in the literature of infinitely old shell samples contaminated by atmospheric carbon dioxide during storage for years in contact with air. Only 1.5% of modern carbon (the activity of the standard) is needed to make the sample to appear about 34 000 ^{14}C years old. Since this age can be determined in most laboratories many shells have been ascribed a finite age although 'infinitely old'. Porous and thin shells are normally more susceptible of such a contamination than thick shells and lumps of carbonate. Their surfaces should be more contaminated than the insides. The contamination is enhanced by moisture according to Olsson et al. (1968). Consequently the normal procedure in the laboratories is to remove the outside and leach the rest to yield at least an inner and an intermediate part to allow comparison between the apparent ages for indications of a possible contamination (Olsson & Blake, 1962–62). The general advice (Olsson et al., 1968) is also to store forams, old mollusc shells and similar samples in an inert gas or a vacuum sealed from the atmosphere.

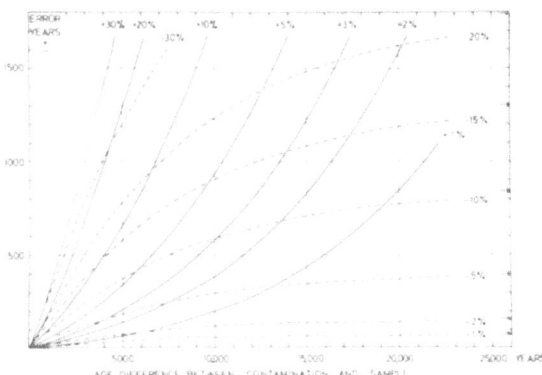

Fig. 3. The error (ordinate) obtained in a ^{14}C dating if a certain fraction of a sample, indicated by each curve, consists of contaminating material having a higher (continuous curves) or lower activity (dashed curves) than the original sample; expressed as an age difference (abscissa) between this sample and the contaminant.

Studies on the reservoir effect

If very recent material is used for studies of the reservoir effect it must be remembered that the atmospheric ^{14}C activity has declined during this century in consequence of the combustion of fossil fuels but that it rose about 35 years ago because of nuclear-bomb tests to reach double the activity at northern mid-latitudes in 1963 and 1964. The industrial effect, also called the Suess effect, first documented by Suess (1953 & 1955), caused the activity to diminish by about 2% until AD 1950, but a natural decrease was superimposed. The industrial effect varies regionally and locally, being greater in Central Europe than in Scandinavia. The excess, a balance between the artificially produced ^{14}C, the fossil-fuel contribution of stable carbon isotopes and natural variations, is now slightly less than 200 per mille in Sweden but still lower in Central Europe. In some sediments, however, the increase was limited to 20% or 30% as seen from some lakes close to Uppsala. Indeed a similar figure is valid for the surface water of the oceans. Pre-bomb lake sediments exhibit a deficiency in the activity, probably mainly due to a reservoir effect but also in varying degrees to contamination by old material. For Gillfjärden (Fig. 4) this is seen as older ages for a level corresponding to erosion following an artificial adjustment of the water-level.

Some plants collected in calcareous lakes on the Island of Öland in 1966 and 1968 were shown to have a significantly lower activity than the atmosphere. This applies not only to submerged but also to floating plants; moreover submerged plants from one lake behaved differently. Håkansson collected plants from several lakes in southern Sweden and all had appreciably lower ^{14}C activities than the atmosphere in 1974 and 1975. It must be stressed that this is the case regardless of whether or not the lake is calcareous. Very low figures for submerged plants – almost no excess at all in 1978 when the atmospheric excess was about 30% – collected in Lake Säynäjälampi are probably due to its very hard water (Olsson *et al.*, 1983). In 1970 submerged specimens of some *Potamogeton* plants in the same Lake exhibited an excess of $15.9 \pm 1.3\%$, or about 39% less than the atmosphere. Iversen warned already in 1949 against erroneous results when lake sediments were used for datings (Oana & Deevey, 1960).

Studies of plants from a few lakes close to Uppsala also reveal significant deficiencies in activity and a slower decrease in comparison with the atmosphere. *Lobelia dortmanna* uses the CO_2 in the sediment. This species seems to have different activities at two localities in Lake Siggeforasjön, and *Myriophyllum* still another at a third place therein. Lake Tarmlången is connected to Lake Siggeforasjön via a ditch which allows the water to flow from the former to the latter. *Myriophyllum* from Lake Tarmlången had much lower activity than *Myriophyllum* from Lake Siggeforasjön in 1981 and 1983 but insignificantly lower in 1985. The project will continue for a few years more.

Bartlett (1951) discussed roots as a source of contamination since they are younger than the level where they are found. Roots should preferably be avoided but some tests on roots have revealed much lower yield at the pretreatment than for the matrix of peat or gyttja (Olsson, 1985). Roots can often be removed by sieving. *Nuphar* and *Nymphea* have leaves with higher ^{14}C activity when still submerged than floating because of the decreasing atmospheric ^{14}C excess and the nutrients stored in the root and used for the early growth. Similarly, some resin from birch collected in spring-time had higher activity than the atmosphere from the same collection year.

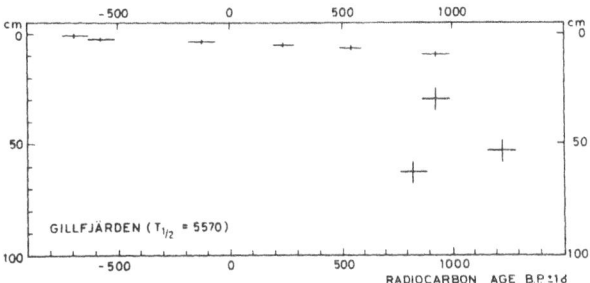

Fig. 4. ^{14}C results from Lake Gillfjärden. The sample collected at 50 cm depth was affected by erosion because of a water-level change.

The foregoing discussion indicates that it is difficult to predict an error in age, due to reservoir age and contamination, for a sediment from a lake. Not only autochthonous plants from the lake but also debris from local and emergent plants may contribute to the sediment as well as reworked sediment, dissolved humic acid and alluvial material. Although submerged plants normally reflect the activity of the water it is known that the CO_2 in the sediment, dissolved CO_2 and HCO_3^-, may be used by them and that the emergent plants rather reflect the atmospheric ^{14}C activity. Accelerators can measure the ^{14}C activity of small samples with nicety. In consequence it is possible to select terrestrial plant remains for the dating. It must also be emphasized that plant remains such as water mosses will always be affected by the reservoir effect.

Notwithstanding, it seems that a realistic figure for the reservoir age of lake sediments at least in Sweden is about three for four centuries rather than the zero years assumed by many authors. It might be much older. Old comparisons between different lakes with certain distinct pollen-analytical levels may suggest higher figures, but in many cases a lack of proper pretreatment must be recalled – the insoluble fraction has been used so that contamination has contributed to the age differences. The reservoir age for sea water and organisms living in the sea or feeding directly or indirectly on marine organisms is the same for rather large areas in the surface water because of the circulation of the water masses. Higher figures must be used for the deep water except for areas with surges. A simplified box model can serve to explain the reservoir age in sea water. The surface water down to a depth of about hundred metres is well mixed (one box) and the deep water is also well mixed (another box). The deeper box is two magnitudes larger so the decay of ^{14}C in this box will be significant before the dissolved carbon dioxide will mix with that in the upper box. When a certain fraction of the carbon dioxide in the upper box reaches the lower only a fraction, one hundred times smaller, will reach the upper box from the lower one at steady state. This gives a long residence time, allowing for the said decay.

This makes the deeper water appear old, and thus 'old' water is supplied to the surface water, causing this to appear too old, but younger than the deep water. The exchange between the atmosphere and the surface water is slow enough to maintain the reservoir age of the surface water.

A typical figure for Great Britain and Southern Scandinavia is about 330 ± 30 years (Olsson, 1980a). The reservoir age increases with the latitude along the Norwegian coast. It is about 515 ± 25 to the east of Greenland (Håkansson, 1983). He measured new samples from the Icelandic coast and revised the earlier value to 365 ± 20 years.

The reservoir ages of sea water are usually measured on pre-bomb shells from museums and estimates of expected ^{14}C ages from the atmospheric activity and the expected smoothing for the surface layer (Olsson, 1980a). Stuiver *et al.* (1986) discussed this in greater detail when constructing the calibration diagrams for marine samples. New results, such as Håkansson's (1983), are not included and thus a revision will certainly be welcome in future. Local deficiencies have been observed as a consequence of volcanic activity. Olsson presented a survey at the symposium Archaeology and ^{14}C held in Groningen in 1987 (in manuscript, to be published in PACT).

Pretreatment

Experience at the Uppsala Laboratory shows that a thorough pretreatment of the samples is essential. HCl is used to remove carbonate, washings with de-ionized water to remove fulvic acids and ions – especially Ca^{2+} which may otherwise affect the possibilities of extracting humic acids – NaOH to remove humic acids, new washings to remove the dissolved material, and finally acidification to pH = 3 to remove absorbed CO_2 from the atmosphere (Fig. 5). The humic acids are precipitated by adding conc. HCl. Elevated temperatures and long treatments are advisable, to allow the solvents to penetrate the samples. It was shown by Olsson (1979b & 1980b) that half an hour treatment with NaOH of heart-wood was far

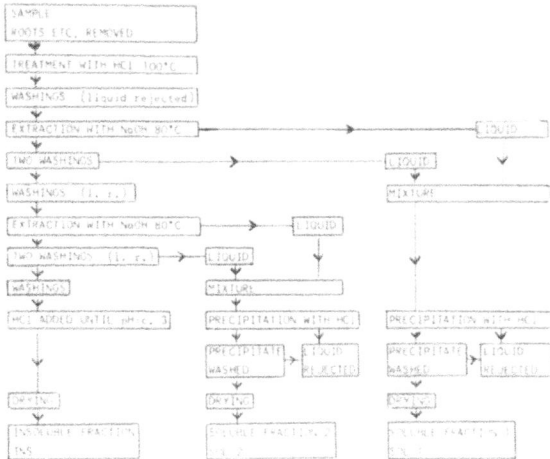

Fig. 5. Suggested treatment for charcoal etc. The number of extractions with NaOH may be increased if necessary. As a rule, however, one extraction is sufficient (from Olsson 1979b).

from sufficient. Two or more extractions over-night must be performed. The more elaborate extraction of cellulose is always a good alternative for wood samples. Many different procedures are used, yielding different types of cellulose. Since ^{14}C measurements with accelerators require samples weighing only about $1/1000$ of those needed for gas counting the development of the pretreatment technique could be expected. A better technique requires more starting material. The figures given are for carbon equivalents after pretreatment. Almost two decades ago lipid extraction was tried by Grant-Taylor (1973) for conventional measurements, a little later the technique was applied in Uppsala with good results (unpublished). Since organic solvents are used stricter rules for the treatment must be imposed. It is essential that any remains of the used solvents be removed to avoid erroneous results because of contamination at the pretreatment.

At Uppsala sediment samples always separated into the two fractions SOL and INS; in our experience the SOL fraction is more reliable for dy and gyttja than the INS fraction. For charcoal, wood and peat the INS fraction is preferred. Sometime both fractions are measured separately to yield a basis for the discussions. The yield is always estimated in a semi-quantitative manner to

facilitate the discussions of the validity of the results. Wet combustion was tried, but then graphite was oxidized more easily than at normal combustions of the INS fraction. Since graphite requires more oxygen an a hotter flame the combustion in Uppsala are always performed with a weak oxygen stream and a moderate flame.

Carbonate samples are always leached with acid to remove the outer parts which are more liable to contamination than the inner parts.

Other types of samples are not discussed here.

Bioturbation

Various bottom-living animals may contribute to the bioturbation of a sediment. In lakes with laminated sediment there is a deficiency of oxygen which essentially prevents biological life. Sometimes faeces are excreted at another level than where the mud was ingested. Often burrowing earthworms contribute by sorting the sediment particles when they select these as food. Burrows may be filled with sediment. Presumably there is no complete mixing down to a certain depth (the mixing depth). The situation might differ for, e.g., ^{14}C and ^{210}Pb, partly because of different half-lives since the steady state might not be reached for short-lived isotopes.

A simplified situation can be described for [14]C (Olsson, 1983) as an upper sediment layer, a 'box' some centimetres thick, which is completely mixed. The apparent ages for all depths within this box will then be the same. This means too old ages at the top and too young at the bottom. At a certain depth of this box the age determination yields a proper ^{14}C age. When more sediment is accumulated on top of the box a corresponding layer at the bottom will be sheltered from further bioturbation. Gradually the normal relationship between age and sediment accumulation rate is reached. Thus a younger age than the real ^{14}C is obtained at depths greater than the mixing depth. The result is that an extrapolation to zero depth will decrease the ^{14}C age whereas the reservoir age will increase it. Another consequence is that an injection at the surface (pollen, volcanic ash, oxygen-isotope changes etc.) will be detected too deep (too early), and also seen too high up in the sediment. The main part of a peak may be detected below the proper level if the injection time is short (i.e. if the proper depth corresponding to the time the supply lasted is small in comparison with the mixing depth).

The accuracy of ^{14}C measurements

With the development of the measuring technique when gas-filled counters were introduced early in the fifties, and the shields were improved so that low stable backgrounds could be recorded, the normal statistical uncertainties were reduced to about 100 years for Holocene samples dated in a typical laboratory. Scintillator counting was later introduced. Now a few high-precision laboratories claim statistical uncertainties of even less than 20 years for a modern samples. Measurements with accelerators yield results similar to those from good conventional laboratories. It has, however, been obvious that the results from the various laboratories differ more from each other than expected from the statistical uncertainties. The internal variations are also slightly too large. A few projects have been started to solve this problem. It has been found that most laboratories, although not all, claim smaller statistical uncertainties than the reality. The two laboratories responsible for the generally accepted calibration curves in *Radiocarbon* 28 (1986) have e.g. published multiplication factors of 1.26 and 1.60 respectively. The reports on the international comparisons suggest that there are a few laboratories with realistic estimates of the uncertainties although they do not belong to the group of high-precision laboratories. For many palaeolimnological investigations the reliability of the results may be of more interest than their high precision because of the problems met in nature. The laboratories working with scintillators usually have a larger multiplication factor than those working with gas-filled counters or accelerators. Attempts are now in progress to solve the problems.

A general figure for multiplying the given statistical uncertainties cannot be given. The figure must be found for each laboratory and will certainly vary with time.

Acknowledgements

I am very grateful to the Swedish Natural Science Research Council for supporting my research on the ^{14}C since AD 1955.

References

Bartlett, H. H., 1951. Radiocarbon datability of peat, marl, caliche, and archaeological materials. Science 114: 55–56.

Bucha, V., 1970. Influence of the Earth's magnetic field on radiocarbon dating. In: Radiocarbon Variations and Absolute Chronology. Twelfth Nobel Symposium I. U. Olsson (ed.): 501–510.

Cato, I., 1985. The definitive connection of the Swedish geochronological time scale with the present, and the new date of the zero year in Döviken, northern Sweden. Boreas 14: 117–122.

Craig, H., 1954. Carbon 13 in plants and the relationships between carbon 13 and carbon 14 variations in nature. J. Geol. 62: 115–149.

Craig, H., 1957. Isotopical standards for carbon and oxygen and correction factors for mass-spectrometric analysis of carbon dioxide. Geochim. Cosmochim. Acta 3: 53–92.

Deevey, E. S., M. S. Gross, G. E. Huthinson & H. L. Kraybill, 1954. The natural ^{14}C content of materials from hardwater lakes. Proc. Nat. Acad. Sci. U.S. 40: 285–288.

Grant-Taylor, T. L., 1973. The extraction and use of plant lipids as a material for radiocarbon dating. In: Proc. 8th Int. Conf. Radiocarbon Dating. Lower Hutt. 18–25 Oct. 1972, Royal Soc. of New Zealand, Wellington: 439–448.

Håkansson, S., 1983. A reservoir age for the coastal waters of Iceland. Geol. Fören. Stockh. Förh. 105: 64–67.

Hörnsten, Å. & I. U. Olsson, 1964. En ^{14}C-datering av glaciallera från Lugnvik, Ångermanland. Geol. Fören. Stockh. Förh. 86: 206–210.

Libby, W. F., 1955. Radiocarbon Dating. Univ. Chicago Press.

Münnich, K. O., 1957. Heidelberg Natural radiocarbon measurements. Science 126: 194–199.

Oana, S. & E. S. Deevey, 1960. Carbon 13 in lake waters and its possible bearing on paleolimnology. Am. J. Sci. 258 A: 253–272.

Olsson, I. U. (ed.), 1970. Radiocarbon Variations and Absolute Chronology. Twelfth Nobel Symposium, Uppsala, 11–15 Aug. 1969.

Olsson, I. U., 1979a. A warning against radiocarbon dating of samples containing little carbon. Boreas 8: 203–207.

Olsson, I. U., 1979b. The importance of the pretreatment of wood and charcoal samples. In: Proc. 9th Int. Conf. Los Angeles and La Jolla, 1976: Radiocarbon Dating Berger, R. & H. E. Suess (eds.): 613–618.

Olsson, I. U., 1980a. Content of ^{14}C in marine mammals from northern Europe. Radiocarbon 22: 662–675.

Olsson, I. U., 1980b. ^{14}C in extractives from wood. Radiocarbon 22: 515–524.

Olsson, I. U., 1983. Dating non-terrestrial materials. In ^{14}C and archaeology. Symp. Groningen, The Netherlands Hackens, T., Mook, W. G. & Waterbolk H. T. (eds.) PACT 8: 277–293.

Olsson, I. U., 1985. Radiometric dating. In: Handbook of Holocene Palaeoecology and Palaeohydrology B. Berglund (ed.): 273–312. John Wiley & Sons.

Olsson, I. U., 1986. A study of errors in ^{14}C dates of peat and sediment. Radiocarbon 28: 429–435.

Olsson, I. U. & W. Blake, Jr., 1961–62. Problems of radiocarbon dating of raised beaches, based on experience in Spitsbergen. Norsk Geog. Tidsskr. 18: 47–64.

Olsson, I. U., F. El-Daoushy & Y. Vasari, 1983. Säynäjälampi and the difficulties inherent in the dating of sediments in a hard-water lake. Hydrobiologia 103: 5–14.

Olsson, I. U., Y. Göksu & A. Stenberg, 1968. Further investigations of storing and treatment of Foraminifera and molluscs for ^{14}C dating. Geol. Fören. Stockh. Förh. 90: 417–426.

Stuiver, M., 1978. Radiocarbon timescale tested against magnetic and other dating methods. Nature 274: 271–273.

Stuiver, M., G. W. Pearson & Y. Braziunas, 1986. Radiocarbon age calibration of marine samples back to 9000 cal yr BP. Radiocarbon 28, No. 2B: 980–1021.

Suess, H. E., 1953. Natural radiocarbon and the rate of exchange of carbon dioxide between the atmosphere and the sea. Proc. Conf. on Nuclear Processes in Geologic Settings, Williams Bay 1953: 52–56.

Suess, H. E., 1955. Radiocarbon concentration in modern wood. Science 122: 415–417.

de Vries, Hl., 1958. Variation in concentration of radiocarbon with time and location on earth. Koninkl. Nederl. Akademie van Wetenschappen, Amsterdam, Proc. Ser B. 61: 94–102.

Hydrobiologia **214**: 35–42, 1991.
J. P. Smith, P. G. Appleby, R. W. Battarbee, J. A. Dearing, R. Flower, E. Y. Haworth, F. Oldfield & P. E. O'Sullivan (eds), 35
Environmental History and Palaeolimnology.
© *1991 Kluwer Academic Publishers.*

[241]Am dating of lake sediments

P. G. Appleby[1], N. Richardson[2] & P. J. Nolan[3]
*Departments of [1]Applied Maths & Theoretical Physics; [2]Geography; [3]Physics; University of Liverpool,
P.O. Box 147, Liverpool L69 3BX, UK*

Abstract

[241]Am derived from decay of fallout [241]Pu is now frequently detected in analyses of lake sediments by low-background gamma assay, and offers an alternative to weapons test ^{137}Cs in dating recent sediments at those sites where the ^{137}Cs record has been degraded by post-depositional mobility or obliterated by Chernobyl fallout. Calculations of the in-growth of [241]Am from [241]Pu indicate a nominal distribution broadly similar to that of ^{137}Cs, with the maximum [241]Am activity occuring in fallout dating from 1963. Results from a number of sites suggest that [241]Am is significantly less mobile in lake sediments than ^{137}Cs, and that its distribution in cores reflects more closely the fallout record. Since further decay of existing weapons debris will increase [241]Am concentrations by about 24% over the next 40 years, [241]Am is likely to play an increasingly important role in assessing the validity of ^{210}Pb dates at sites with varying sediment accumulation rates.

Introduction

Although ^{210}Pb is now routinely used for dating recent lake sediments problems frequently arise over the interpretation of data from sites with disturbed sediment records. These problems are inevitable in view of the manifold pathways by which ^{210}Pb is delivered to the sediments and the opportunities for post-depositional redistribution. Indeed, in view of these complexities it is surprising that the naive CRS model, which forms the basis of most ^{210}Pb age-depth calculations, works as well as it does.

There are various internal radiometric parameters, principally the ^{210}Pb inventory, which can be used as indicators of the validity or otherwise of the CRS model (Appleby & Oldfield, 1983). Although a critical evaluation of ^{210}Pb dates using such data is important, these can not always be relied on and positive independent well-dated markers are crucial to reliable dating. Such markers typically include visible stratigraphic features, pollen changes, records of atmospheric pollutants such as Pb, Zn, and soot particles, and artificial radioisotopes such as ^{137}Cs from nuclear weapons testing.

Since the early 1970's, ^{137}Cs measurements have provided valuable evidence for evaluating very recent accumulation rates. Until the Chernobyl accident the only widespread source of this isotope was fallout from atmospheric testing of nuclear weapons, and comparisons of the ^{137}Cs record in the sediments with the atmospheric fallout allowed in some cases confident identification of the depths representing 1954, the date in which ^{137}Cs fallout was first widely distributed, and 1963, the year of peak deposition in the northern hemisphere (cf. Pennington *et al.*, 1973). In many cases, however, the value of ^{137}Cs has been significantly reduced by the evident mobility of this isotope. Many recent studies have shown that peak ^{137}Cs activities occur too close to the surface to represent 1963 (Davis *et al.*, 1984), and continue to be measurable at depths well below those representing 1950. With the passage of time these problems will increase, and the problem has recently been exacerbated in some areas by fallout of ^{137}Cs from the Chernobyl accident. In regions of high Chernobyl fallout downwards diffusion of Chernobyl ^{137}Cs has in many cases largely obliterated the weapons-testing ^{137}Cs profile (Appleby et al., 1990).

Fallout from nuclear weapons testing contained a number of isotopes of plutonium, including ^{241}Pu (half-life 14.4 years). This decays by beta-emission to ^{241}Am (half-life 432 yr), which is readily measured through its gamma emissions at 59.5 keV using low background counting systems (Appleby et al., 1986). The in-growth of ^{241}Am has now reached a point where it is detected in a significant proportion of sediment cores. A growing data set from lakes with a wide range of pH values suggests that ^{241}Am is considerably less mobile than ^{137}Cs, and can provide a useful means of identifying sediments dating from the early 1960's in cases where the ^{137}Cs record has become degraded.

^{241}Am record in weapons test fallout

Widespread global dispersal and fallout of artificial radioisotopes from atmospheric testing of nuclear weapons began with the high-yield thermonuclear tests in November 1952. Radioactive debris from these tests was pushed into the stratosphere and around the world. The mean residence time in the atmosphere was estimated (Feely, 1960) to lie between 6 and 13 months, atmospheric circulation returning the debris to the troposphere at mid-latitudes, from which it was removed by dry deposition and rainfall.

Fallout of ^{137}Cs and ^{90}Sr has been closely monitored since 1954. Both isotopes have very similar fallout histories, and the record up to 1987 is summarised in Cambray et al. (1989). The record of plutonium fallout is much less complete, but can be reconstructed from the ^{137}Cs fallout record using the mean plutonium to ^{137}Cs ratio. Using data from a variety of sources (Krey et al., 1976; Cawse 1983; Eakins et al., 1984) the

activity ratios in fresh nuclear debris are estimated to be:

$^{239+240}$Pu/^{137}Cs	^{241}Pu/^{137}Cs
0.0125	0.177

The ^{241}Am of fresh nuclear weapons test debris is essentially zero (Krey et al., 1976), and its presence in older deposits is through in-growth from ^{241}Pu. If $\mathscr{P}(t)$ denotes the ^{241}Pu fallout history (uncorrected for decay), the ^{241}Am record at a later time τ is

$$\mathscr{A}(t, \tau) = \mathscr{P}(t) \frac{\lambda_{Am}}{\lambda_{Pu} - \lambda_{Am}} \times$$
$$\times (\exp\{-\lambda_{Am}(\tau - t)\} - \exp\{-\lambda_{Pu}(\tau - t)\}),$$

where

$$\lambda_{Am} = \ln2/432.7 = 0.00160 \text{ yr}^{-1},$$
$$\lambda_{Pu} = \ln2/14.4 = 0.0481 \text{ yr}^{-1}$$

are the ^{241}Am and ^{241}Pu radioactive decay constants. Table 1 shows a reconstruction of the cumulative decay corrected ^{241}Pu and ^{241}Am inventories (for the northern hemisphere) since 1954 as a percentage of nominal total ^{241}Pu fallout up to 1985, using the decay equations, and the fallout history given in Cambray et al., 1989. The calculations show that the ^{241}Am inventory doubled between 1972 and 1990, and (in the absence of further deposits) will increase by a further 24% up to the year 2037. Thereafter it will decay with a half-life of 432.7 years. Table 2 shows that calculations of ^{241}Am/$^{239+240}$Pu inventory ratios from these results are in good agreement with measured inventory ratios in soils.

Table 1 also shows the decay corrected ^{241}Am distribution functions for the years 1985 and 1990, given as a percentage of the ^{241}Am inventories for those two years. There is almost no change in the distribution during this period, and differences from the decay corrected ^{137}Cs distribution are negligible. In all cases the distribution is dominated by the 1963 peak, with ca. 20% of the inventory being attributable to that year.

Table 1. ^{241}Pu and ^{241}Am decay corrected fallout records in the Northern Hemisphere.

| | (a) Cumulative fallout (% of total nominal ^{241}Pu fallout to 1985[†]) | | (b) ^{241}Am distribution functions (% of total decay corrected ^{241}Am) for | | | |
| | | | (i) 1985 | | (ii) 1990 | |
	^{241}Pu	^{241}Am	Annual	Cumulative	Annual	Cumulative
1954	1.08	0.00	1.18	1.18	1.13	1.13
1955	5.12	0.00	4.75	5.93	4.54	5.67
1956	9.05	0.01	4.78	10.71	4.59	10.26
1957	12.62	0.03	4.54	15.25	4.37	14.63
1958	17.37	0.05	5.94	21.19	5.76	20.39
1959	27.23	0.08	11.66	32.85	11.36	31.75
1960	28.37	0.12	2.59	35.44	2.54	34.29
1961	29.88	0.17	2.98	38.42	2.94	37.23
1962	41.32	0.23	13.20	51.62	13.09	50.32
1963	59.72	0.31	20.45	72.07	20.40	70.32
1964	70.85	0.41	13.64	85.71	13.74	84.46
1965	72.27	0.52	4.54	90.25	4.60	89.06
1966	71.46	0.63	2.39	92.64	2.45	91.51
1967	69.39	0.74	1.11	93.75	1.16	92.67
1968	67.80	0.85	1.44	95.19	1.52	94.19
1969	65.54	0.95	0.75	95.94	0.81	95.00
1970	63.38	1.05	0.72	96.66	0.79	95.79
1971	61.57	1.15	0.88	97.54	0.98	96.77
1972	59.25	1.24	0.41	97.95	0.46	97.23
1973	56.89	1.33	0.29	98.24	0.33	97.56
1974	54.89	1.42	0.43	98.67	0.50	98.06
1975	52.89	1.50	0.34	99.01	0.42	98.48
1976	50.65	1.58	0.13	99.14	0.17	98.65
1977	48.77	1.66	0.25	99.39	0.33	98.98
1978	47.06	1.73	0.26	99.65	0.36	99.34
1979	45.14	1.80	0.11	99.76	0.16	99.50
1980	43.19	1.87	0.05	99.81	0.09	99.59
1981	41.64	1.93	0.13	99.94	0.23	99.82
1982	39.83	1.99	0.03	99.97	0.07	99.89
1983	38.11	2.05	0.02	99.99	0.06	99.95
1984	36.41	2.10	0.01	100.00	0.03	99.98
1985	34.78	2.15	0.00	100.00	0.02	100.00
1986	33.15	2.20				
1987	31.59	2.25				
1988	30.11	2.29				
1989	28.69	2.33				
1990	27.34	2.37				

[†] Fallout from nuclear weapons testing fell below limits of detection in 1985.

^{241}Am in lake sediments

During the past five years we have analysed over 70 lake sediment cores by gamma assay in the Liverpool University Environmental Radiometric Laboratory in the course of an extensive program of ^{210}Pb dating. The germanium detectors used in this program have been described in Appleby *et al.* (1986). In over half of the cores analysed we have observed traces of ^{241}Am through its gamma emissions at 59.5 keV. The peak is very weak compared to the ^{137}Cs peak at 661.6 keV, but

38

Table 2. Comparison of measured and calculated ^{241}Am/$^{239+240}$Pu ratios in weapons test fallout.

^{241}Am/$^{239+240}$Pu inventory ratios			Reference to measured ratio
Date	Measured	Calculated	
1973	0.25	0.20	Krey *et al.*, 1976
1974	0.22	0.21	Krey *et al.*, 1976
1979	0.23	0.26	Cambray & Eakins, 1982
1981	0.24	0.28	Hallstadius *et al.*, 1986

unambiguous. Figure 1 compares the ^{210}Pb, ^{241}Am and ^{137}Cs peaks in the gamma spectrum of a sediment sample from Devoke Water in Cumbria, England. Figure 2 shows ^{241}Am peaks in gamma spectra of sediment samples from three other sites. With the Liverpool system, ^{241}Am can be detected at levels of ca. 0.02 pCi g^{-1}.

In the majority of cases the principal source of ^{241}Am in lake sediments is undoubtedly fallout from atmospheric testing of nuclear weapons.

Graphs of ^{241}Am activity vs depth (Appleby *et al.*, 1988 & 1990) typically reveal a maximum concentration at the same depth as the maximum ^{137}Cs activity. These peaks are almost invariably dated by ^{210}Pb measurements to the mid 1960's, confirming that they are indeed a record of the 1963 fallout maximum from nuclear weapons testing. Results from Lake Consultation in California, U.S.A. are shown in Fig. 3. ^{137}Cs and ^{241}Am activities have a clearly defined maximum at a depth of ca. 5 cm. The ^{210}Pb dates for this core, shown in Fig. 4, are in good agreement with the inferred date of 1963 for the ^{137}Cs, ^{241}Am peak.

A comparison of weapons test ^{137}Cs and ^{241}Am inventories and their ratios for sediments cores from 28 U.K. lakes measured at the Liverpool University Environmental Radiometric Laboratory is shown in Table 3. In post-1986 cores, the weapons test ^{137}Cs inventory has been determined by subtracting Chernobyl ^{137}Cs from the total ^{137}Cs inventory, making use of the ^{134}Cs/^{137}Cs ratio of 0.6 in Chernobyl fallout (Cambray *et al.*, 1987). Calculations based on the

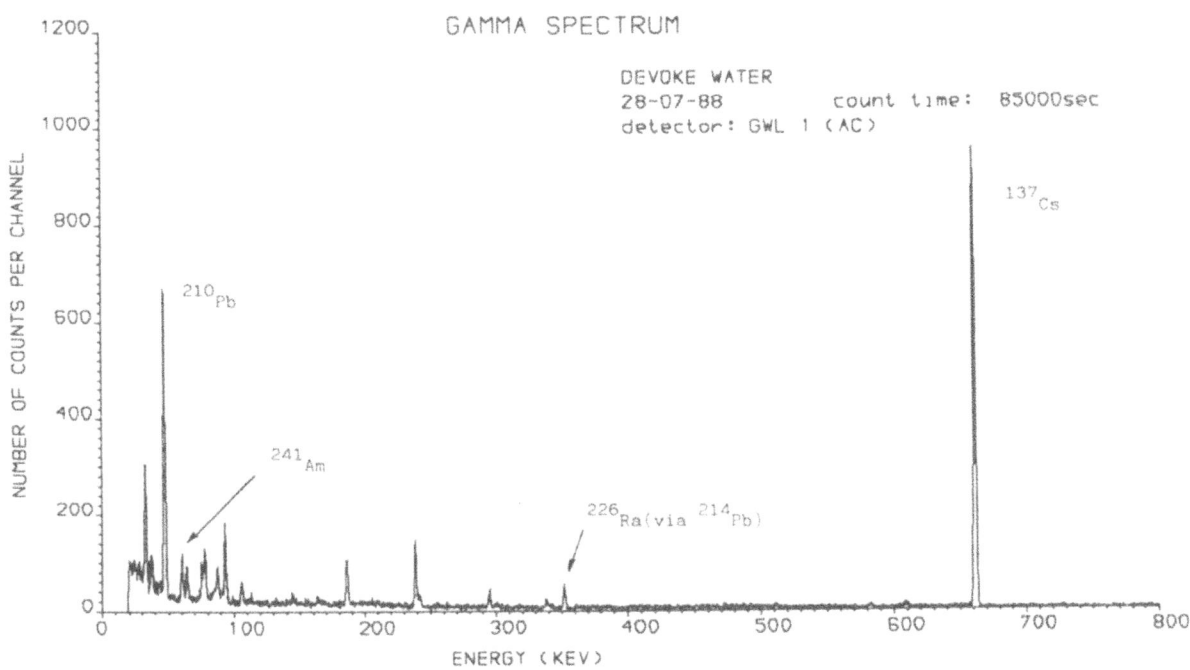

Fig. 1. Gamma spectrum of a sediment sample from Devoke Water (Cumbria) showing ^{210}Pb, ^{241}Am and ^{137}Cs gamma peaks.

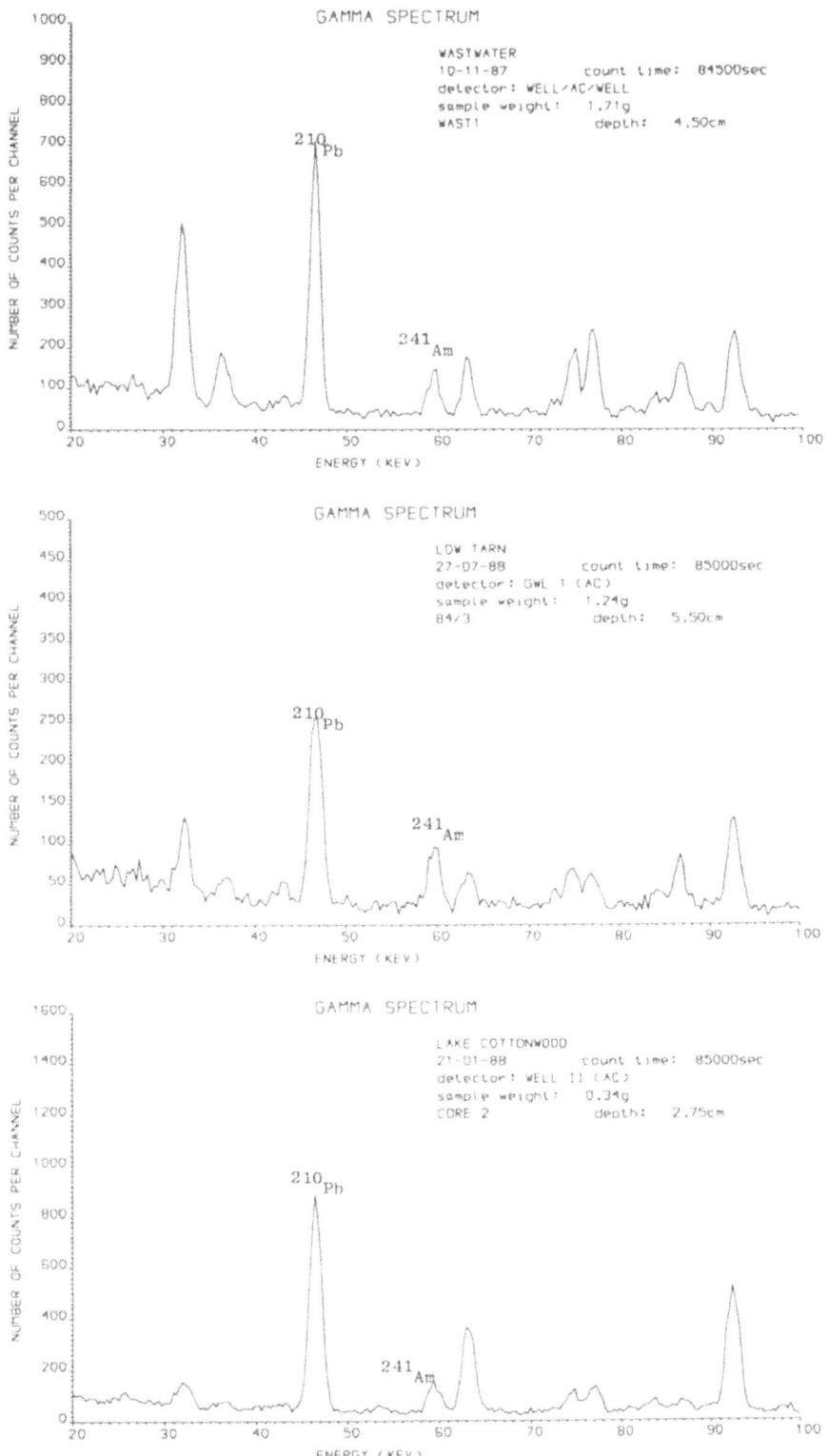

Fig. 2. ²⁴¹Am peaks in gamma spectra of sediment samples from Wastwater and Low Tarn (Cumbria, U.K.), and Lake Cottonwood, (Sierra Nevada mountains, California, U.S.A.)

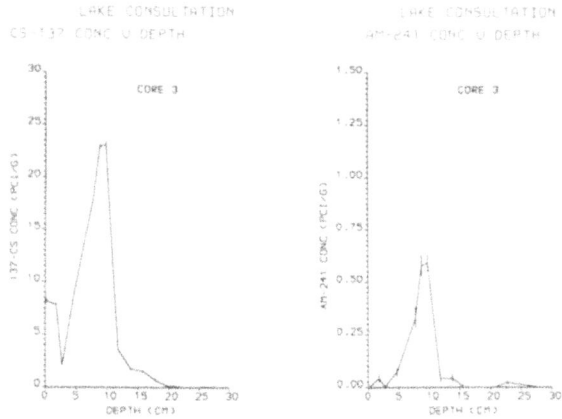

Fig. 3. ^{137}Cs and ^{241}Am activities vs depth in a core from Lake Consultation, (Sierra Nevada mountains, California, U.S.A.).

^{137}Cs fallout record (Cambray *et al.*, 1989) and the ^{241}Am record given in Table 1 indicate that by 1986, the date of the majority of these cores, the ^{241}Am/^{137}Cs inventory ratio from weapons test fallout should be ca. 0.66%. The mean value for the ten Scottish sites and four East Cumbrian sites is comparable to this figure. The higher ratios for the three West Cumbrian sites closest to the Sellafield nuclear reprocessing plant, Wastwater, Low Tarn and Devoke Water, are presumably due to discharges from the Sellafield plant. The values given here are in good agreement with the

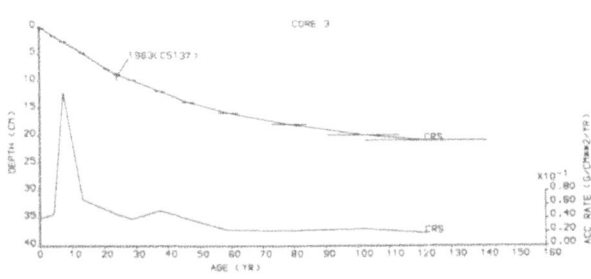

Fig. 4. A comparison of CRS model ^{210}Pb dates with the 1963 ^{137}Cs, ^{241}Am date in a core from Lake Consultation, U.S.A.

Table 3. Inventories of weapons test fallout ^{137}Cs and ^{241}Am in U.K. lake sediment cores, and their ratios.

Site	Weapons test ^{137}Cs pCicm^{-2}	^{241}Am Inventory pCicm^{-2}	Ratio %
Scotland			
Loch Sionascaig	10.0	0.06	0.60
Lochan Dubh	7.9	0.07	0.89
Loch Doilet	14.3	0.07	0.49
Loch Laidon	9.5	0.05	0.53
Lochan na h'Achlaise	5.4	0.02	0.37
Loch Chon	14.2	0.04	0.28
Loch Tinker	23.4	0.10	0.43
Round Loch of Glenhead	6.1	0.06	0.98
Loch Dee	7.7	0.04	0.52
Loch Grannoch	8.5	0.04	0.47
Mean	10.7	0.06	0.54
Cumbria, England			
Wastwater	12.8	0.17	1.33*
Low Tarn	5.5	0.20	3.64*
Devoke Water	29.3	0.51	1.74*
Esthwaite Water	20.4	0.08	0.39
Blelham Tarn	14.9	0.13	0.87
Windermere	21.2	0.15	0.71
Brotherswater	14.5	0.12	0.83
Mean	16.9	0.19	0.69
Wales			
Llyn Geirionydd	10.1	0.10	0.99
Llyn Clyd	4.3	0.05	1.15
Llyn Llagi	12.1	0.19	1.57
Llyn Conwy	42.8	0.40	1.37
Llyn Cwm Mynach	1.9	0.02	1.06
Llyn Irddyn	13.9	0.07	0.51
Llyn Hir	7.3	0.05	0.69
Llyn Llygad Rheidol	8.2	0.08	0.98
Mean	12.6	0.12	0.97
Northern Ireland			
Lough Blue	5.0	0.25	5.04*
Lough Neagh	17.1	0.09	0.55
Lough Catherine	5.7	0.03	0.53
Mean	9.3	0.12	0.54

* These values have been excluded in calculations of the mean ratios (see text).

41

results given in Cambray & Eakins (1980 & 1982) for West Cumbrian lake sediments and soil samples. The latter study, based on soil samples collected in 1977, found enhanced ^{241}Am/^{137}Cs inventory ratios ranging from 11.8% on the coast to 1.04% at a site 10.9 km inland. The enrichment was attributed to the presence of actinides in seaspray. The enhanced ^{241}Am/^{137}Cs inventory ratio for the core from Lough Blue in Northern Ireland, also within 10 km of the Irish Sea coast, may have a similar origin. The sediment cores from the Welsh lakes have on average a higher ^{241}Am/^{137}Cs inventory ratio than the Scottish and East Cumbrian sites. The highest values occur in small high altitude tarns, and may reflect different residence times of ^{137}Cs and ^{241}Am in these lakes prior to incorporation in the sediments or their proximity to the Irish Sea coast.

Post-depositional mobility of ^{241}Am

The mobility of ^{137}Cs in lake sediments and its consequent limitation as a dating tool is well documented (Davis et al., 1984). There is growing evidence that ^{241}Am is significantly less mobile in lake sediments than ^{137}Cs. Because ^{241}Am activities are so much lower than ^{137}Cs activities exact comparisons are difficult to make. The ab-

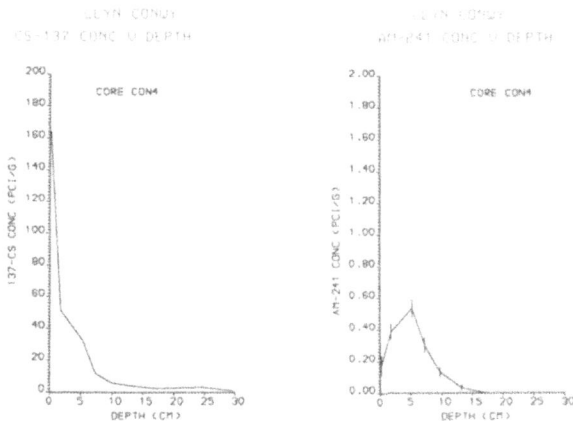

Fig. 6. ^{137}Cs and ^{241}Am activities vs depth in a core from Llyn Conwy, Wales. This core post-dates the 1986 Chernobyl accident.

sence of a downward tail in most ^{241}Am profiles may simply reflect the fact that it is below the limit of detection. ^{241}Am is however rarely seen in near surface sediments, which frequently have quite high (pre-Chernobyl) ^{137}Cs activities. Figure 5 compares ^{137}Cs and ^{241}Am profiles in a pre-Chernobyl core from Loch Tinker in Scotland. The maximum ^{137}Cs activity occurs in the surficial sediments. In contrast, the ^{241}Am activity has a clearly defined maximum at a depth of 2.5 cm below the surface of the core. Figure 6 shows results from a post-Chernobyl core, in which the near-surface ^{137}Cs record is dominated by Chernobyl-derived ^{137}Cs. The well-defined ^{241}Am peak identifies sediments from the mid 1960's.

The relative mobility of ^{137}Cs and ^{241}Am can be compared using the parameter.

$$\frac{\text{peak activity}}{\text{inventory}}.$$

Enhanced post-depositional mobility will reduce the height of the maximum activity in the sediment record without affecting the inventory, so reducing the value of this parameter. Values of this parameter for ^{241}Am in cores are invariably greater than the corresponding values for weapons test fallout ^{137}Cs, supporting the hypothesis that ^{241}Am is relatively less mobile.

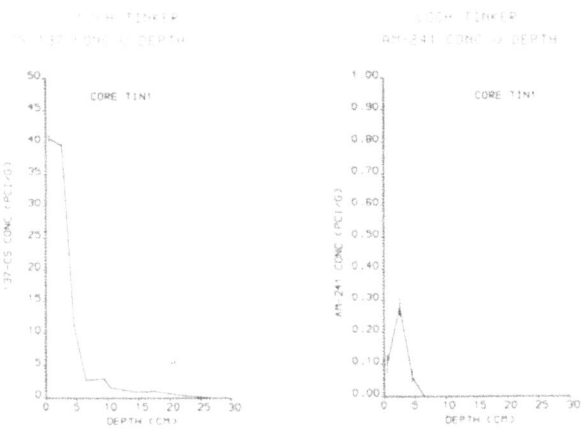

Fig. 5. ^{137}Cs and ^{241}Am activities vs depth in a core from Loch Tinker, Scotland. This core pre-dates the 1986 Chernobyl accident.

Conclusions

Although ^{210}Pb is likely to remain the principal means of dating recent sediment accumulations, with the passage of time a reliable 1963 date will become increasingly valuable in discriminating between different ^{210}Pb models. The importance of such independently dated points is illustrated by results from Devoke Water in Cumbria (Atkinson & Haworth, 1990) and Llyn Llygad Rheidol (Bonnett & Appleby, 1991). In the Devoke Water core, an incompatibility between the CRS model and CIC model ^{210}Pb dates was resolved by the use of a well defined 1963 ^{137}Cs, [241]Am date. This pointed to the existence of a hiatus in the sediment record which invalidated the CRS model. In the Llyn Llygad Rheidol example, the CRS model dates, which suggested a significant post-1963 acceleration in sediment accumulation rates, were validated by the independent 1963 date.

At sites with a well preserved ^{137}Cs record the value of [241]Am will be only marginal. Where however the ^{137}Cs record is degraded, either through post-depositional mobility or obliteration by Chernobyl fallout, [241]Am measurements may prove to be of significant value. Its half-life of 432 years ensures that it will remain detectable in lake sediments for several centuries.

Acknowledgements

We gratefully acknowledge financial support from the U.S. National Science Foundation, Grant BSR 86-17622.

References

Appleby P. G., P. J. Nolan, D. W. Gifford, M. J. Godfrey, F. Oldfield, N. J. Anderson & R. W. Battarbee, 1986. [210]Pb dating by low background gamma counting. Hydrobiologia 143: 21–27.

Appleby, P. G., P. J. Nolan, F. Oldfield, N. Richardson & S. R. Higgitt, 1988. [210]Pb dating of lake sediments and ombrotrophic peats by gamma assay. Sci. Tot. Envir. 69: 157–177.

Appleby, P. G. & F. Oldfield, 1983. The assessment of ^{210}Pb dates from sites with varying sediment accumulation rates. Hydrobiologia 103: 29–35.

Appleby, P. G., N. Richardson, P. J. Nolan & F. Oldfield, 1990. Radiometric dating of the United Kingdom SWAP sites. Phil. Trans. R. Soc. Lond. B 327: 233–238.

Atkinson, K. M. & E. Y. Haworth, 1990. Devoke Water and Loch Sionascaig: recent changes and the post-glacial overview. Phil. Trans. R. Soc. Lond. B 327: 349–355.

Bonnett, P. J. P. & P. G. Appleby, 1991. Deposition and transport of radionuclides within an upland drainage basin in mid-Wales. In J. P. Smith et al. (eds), Environmental History and Palaeolimnology. Kluwer Academic Publishers, Dordrecht: 71–76. Reprinted from Hydrobiologia 214.

Cambray, R. S., P. A. Cawse, J. A. Garland, J. A. B. Gibson, P. Johnson, G. N. J. Lewis, D. Newton, L. Salmon & B. O. Wade, 1987. Observations on radioactivity from the Chernobyl accident. Nuclear Energy 26: 77–101.

Cambray, R. S. & J. D. Eakins, 1980. Studies of environmental radioactivity in Cumbria, Part 1. Concentrations of plutonium and ^{137}Cs in environmental samples from West Cumbria and a possible maritime effect. AERE Report R-9807.

Cambray, R. S. & J. D. Eakins, 1982. Pu, [241]Am and ^{137}Cs in soil in West Cumbria and a maritime effect. Nature 300: 46–48.

Cambray, R. S., K. Playford, G. N. J. Lewis & R. C. Carpenter, 1989. Radioactive fallout in air and rain: results to the end of 1987. AERE Report R-13226, Harwell.

Cawse, P., 1983. The accumulation of ^{137}Cs and [239 + 240]Pu in soils of Great Britain, and transfer to vegetation. In P. J. Coughtrey, (ed.), Ecological aspects of radionuclide release. Blackwell, Oxford, 47–61.

Davis, R. B., C. T. Hess, S. A. Norton, D. W. Hanson, K. D. Hoagland & D. S. Anderson, 1984. [137]Cs and ^{210}Pb dating of sediments from soft-water lakes in New England (U.S.A.) and Scandinavia, a failure of ^{137}Cs dating. Chem. Geol. 44: 151–185.

Eakins, J. D., R. S. Cambray, K. C. Chambers & A. E. Lally, 1984. The transfer of natural and artificial radionuclides to Brotherswater from its Catchment. In E. Y. Haworth & J. W. G. Lund (eds) Lake Sediments and Environmental History: 125–144. Leicester Univ. Press.

Feely, H. W., 1960. [90]Sr content of the stratosphere. Science, 131: 645.

Hallstadius, L., A. Aarkrog, H. Dahlgaard, E. Holm, S. Boelskifte, S. Duniec & B. Persson, 1986. Plutonium and americium in Arctic waters, the North Sea and Scottish and Irish coastal zones. J. Envir. Radioact., 4: 11–30.

Krey, P. W., E. P. Hardy, C. Pachuki, F. Rourke, J. Coluzza & W. K. Benson, 1976. Mass isotopic composition of global fall-out plutonium in soil. In Proc. Conf. Transuranium Nuclides in the Environment, IAEA, 671–677.

Pennington, W., R. S. Cambray & E. M. R. Fisher, 1973. Observations on lake sediments using fallout ^{137}Cs as a tracer. Nature 242: 324–326.

Hydrobiologia **214**: 43–52, 1991.
J. P. Smith, P. G. Appleby, R. W. Battarbee, J. A. Dearing, R. Flower, E. Y. Haworth, F. Oldfield & P. E. O'Sullivan (eds),   43
Environmental History and Palaeolimnology.

Accuracies in Po-210 determination for lead-210 dating

F. El-Daoushy[1], K. Olsson[1, 2] & R. Garcia-Tenorio[1, 3]
[1]Department of Physics, Box 530, S-751 21 Uppsala; present address: [2]Department of Quaternary
Geology, Box 555, S-751 22 Uppsala, Sweden; [3]Facultad de Fisica, Apartado 1065, E-410 80 Sevilla,
Spain

Key words: aquatic deposits, $^{210}Pb/^{210}Po$, speciation, dating, alpha spectrometry, accuracies in alpha-counting

Abstract

Various laboratory techniques have been utilized worldwide for measuring lead-210 in sub-recent deposits through its grand-daughter product polonium-210. Isotope dilution alpha spectrometry proved a suitable tool for absolute determination of lead-210 for the dating of aquatic deposits. Moreover, isotope dilution alpha spectrometry along with speciation experiments can be used to resolve depositional anomalies arising from supported lead-210/Ra-226 disequilibrium levels and unsupported lead-210 mobile fractions. Isotope dilution alpha spectrometry of sub-recent sediment and peat deposits has been critically evaluated for more than ten years. Our results show that type, size and composition of deposits analyzed as well as radiochemical procedures used, together with alpha counting techniques, are important factors influencing lead-210 determinations and tailing corrections using its granddaughter product polonium-210. Optimization of these parameters is of prime importance to achieve economic and accurate analyses, especially at low lead-210 concentrations and small sample sizes.

Introduction

Several analytical problems are usually associated with the determination of low levels of alpha-emitting nuclides in environmental materials especially in the presence of large quantities of interfering stable and radioactive elements (e.g. Eakins, 1984). Chemical separation techniques have to be adapted to the counting facilities, the type, number and size of samples available as well as the accuracy level required to accomplish the final goal. In lead-210 dating of sub-recent aquatic deposits enormous efforts were devoted to improve the chemical yield of polonium-210 extraction for the determination of lead-210. However, only a few critical studies have been carried out to

evaluate accuracies obtained for different materials, especially aquatic deposits, in relation to the chemical and physical procedures utilized to measure polonium-210 (e.g. Fleer & Bacon, 1984; Eakins & Morrison, 1978). In absolute determinations of alpha activities the utilization of internal yield tracers in isotope dilution is considered to give more accurate results compared with gross-alpha counting. Isotope dilution alpha spectrometry requires the application of chemical procedures that remove interfering species and provide thin alpha sources with well-resolved alpha peaks. In addition to the determination of lead-210 in bulk materials, isotope dilution could be utilized to study lead-210 speciation in nature. Such studies yield additional information on the

44

geochemistry of supported and unsupported lead-210 (El-Daoushy & Garcia-Tenorio, 1988, 1990) thus improving our understanding of the global biogeochemical cycle of lead-210 (e.g. El-Daoushy, 1988).

Extraction and measurement of polonium-210 – general background

Polonium-210 represents a suitable alternative for accurate determination of lead-210 providing that polonium-210 exists in secular equilibrium with lead-210. It is generally accepted that Po-210 and Pb-210 are in secular equilibrium for sub-recent deposits older than 2 yr (e.g. Eakins & Morrison, 1978; Smith & Walton, 1980). Chemical extractions of polonium-210 currently used by various laboratories could be grouped into two main procedures which were originally developed by Flynn (1968) and Eakins & Morrison (1978).

Although the Flynn method was developed to monitor polonium-210 in effluent, natural waters, rocks and minerals which indeed contain minor amounts of organic matter, it was later applied to aquatic deposits. The Flynn method is based on acid-leaching with no provision for the oxidation of organic matter. Following Flynn's method, aquatic deposits such as sediment, soil and peat in dried and ground forms are either leached or totally dissolved by strong acids alone or in different combinations (HNO_3, HCl, $HClO_4$, HF) to extract polonium-210. In this procedure, which is widely utilized (e.g. Häsänen, 1977; Benninger et al., 1979; Smith & Walton, 1980; Dominik et al., 1981; Van der Wijk & Mook, 1987), some laboratories do not always use yield tracers (e.g. Persson, 1970; Robbins & Edgington, 1975; Erlenkeuser & Pederstad, 1984; Chanton et al., 1983) as they claim that their overall recoveries were more or less the same. Along with the polonium almost, if not, all other stable and radioactive elements in the sample analyzed would be found in the extracted solution. Some organic remnants may also exist in the solution. Under favourable conditions, self-deposition on silver discs (Flynn, 1968) seems to be a quantita-

tive and selective approach to separate the polonium from weak hydrochloric acid solutions. However, in the presence of oxidants, organic materials and elements that also deposit on silver it is absolutely necessary to obtain thin alpha sources suitable for isotope dilution alpha spectrometry, especially in cases where polonium-208 (5.114 MeV; $T(1/2) = 2.93$ yr) is used as an internal yield tracer. Several cocktails containing sodium citrate, hydroxylamine hydrochloride and ascorbic acid have been utilized in the self-deposition step to minimize interferences from oxides and other elements. Despite the mentioned precautions, the self-deposition remained a critical step (e.g. Dominik et al., 1981), particulary for marine sediments with high inorganic contents (El-Daoushy, unpublished data). Furthermore, the polonium extraction for marine, lacustrine and peat deposits showed a wide range of chemical yields, 30–98%, especially in the case of marine sediments (e.g. Smith & Walton, 1980; Erlenkeuser & Pederstad, 1984; Dominik et al., 1981; Chanton et al., 1983; El-Daoushy & Tolonen, 1984; Van der Wijk & Mook, 1987).

Eakins & Morrison (1978) developed a purification procedure to eliminate the organic and inorganic impurities in marine and lacustrine sediments prior to the self-deposition step. Modified versions of the Eakins and Morrison's procedure (as described below) are now in current use by some laboratories (e.g. Bloesch & Evans, 1982; El-Daoushy, 1981a). Deposits are first spiked and treated with concentrated HCl to transform polonium to chloride and dried overnight. The polonium chloride along with the organic matter is distilled at 550 °C. Strong HNO_3 is then used to destroy the organic matter by mild oxidation under reflux. HNO_3 is carefully removed and a 1.5–2 M HCl solution is prepared for self-deposition. The extracted polonium is not always pure as the oxidation step is not efficient enough and thick polonium deposits are occasionally obtained. This applies particularly to lacustrine sediments with high organic contents (El-Daoushy, unpublished data). The overall chemical yield seems always to be higher than 60% (El-Daoushy, 1981a; Bloesch & Evans, 1982).

The presence of HNO_3 is considered critical for the self-deposition step (e.g. Eakins & Morrison, 1978). However, despite Häsänen (1977) showing that mixtures of HCl and HNO_3 give the same yields as HCl, his results do not appear to give adequate resolutions, especially if polonium-208 is to be used as yield tracer. In any case, HNO_3 in combination with other impurities may deteriorate yields and resolutions.

Despite the excellent resolution of polonium-210 (5.305 MeV; T(1/2) = 138.4 d) and polonium-209 (4.882 MeV; T(1/2) = 103 yr), and the long half-life of the latter isotope, polonium-208 is still much more popular (in isotope dilution alpha spectrometry) because it is easily available. Nevertheless, calibrated tracers should be stored and used under controlled conditions as systematic changes (other than radioactive decay) may occur in their specific activities. Spikes are usually calibrated using specially delivered Po-210 standards (Amersham, United Kingdom; Chemapol, Czechoslovakia). It is obvious that routine gross-counting of polonium-210 (e.g. Flynn, 1968; Robbins & Edgington, 1975; Persson, 1970; Erlenkeuser & Pederstad, 1984) can give erroneous determinations of Pb-210 as the described chemical extractions are not likely to give the same overall chemical yield (Schell, 1977). For pure organic fractions, ombrotrophic peat and lichen materials, both extraction procedures may give very high overall yields, $\geq 90\%$. However, the polonium resolution becomes poor due to interferences in the self-deposition when using the Eakins and Morrison version.

Materials and methods

During the period 1977–1984 the Uppsala lead-210 laboratory utilized a modified version of the Eakins and Morrison method (El-Daoushy, 1981a, 1986). It was observed that deposits with high organic content, especially lake sediments (material analyzed was about 1 g), may produce thick alpha sources. This method gave satisfactory results for most deposits but it required large amounts of strong acids and tedious efforts to remove these acids. This procedure may create technical problems unless suitable fumehoods and ventilation systems are properly used. Oxidation did not remove all organic remnants which deposited together with polonium on silver discs and caused poor alpha resolution (polonium-210 and polonium-208). Tail corrections did not work properly for thick alpha sources, especially in cases where the polonium-210/polonium-208 ratios deviated much from one. Some improvements were obtained by using smaller amounts of materials and/or prolonged periods of oxidation.

We thought that the Flynn method might give us an easier and better alternative if some additional purification steps were introduced prior to self-deposition. However, we wanted to explore the conditions under which both methods in modified forms give the best possible results, to explain the variability of the overall chemical yields and to optimize the required chemicals, efforts and difficulties in these methods. Among available combinations of strong acids, HNO_3 and H_2SO_4 seem to be reasonable for normal laboratory routines. In addition, limited amounts of HNO_3 and H_2SO_4 are able to perform quick and effective oxidation of aquatic deposits. These acids have not been frequently used because of polonium evaporation risks or problems in subsequent separations (e.g. Joshi & Durham, 1976). Polonium sulphates and nitrates are comparatively involatile up to 400 °C and no significant losses were observed in the H_2SO_4/HNO_3 oxidation (e.g. Eakins & Morrison, 1975).

In 1985 we developed a new method, 'wet-oxidation/distillation' labelled as (A) in Tables 1 and 2, to make Flynn's procedure applicable to aquatic deposits with high organic contents. Moreover it must be mentioned that the total metal contents in sediments could be much higher than those used by Flynn in his recovery experiments. These methods (A and B) have been applied to various marine, lacustrine and peat deposits (El-Daoushy, 1988) with a wide range of organic and inorganic composition. A summary of these versions is given in Table 1 and a comparison between chemicals required and time needed is shown in Table 2.

Table 1. A brief description of the chemical extraction used for the Uppsala alpha-spectrometric analyses of ^{210}Pb.

Step	(A) Oxidation/distillation (new)	(B) Distillation/oxidation (old)
(1) Physical pretreatment	Freeze-drying provides homogeneous and fine-powdered samples with easy and reliable handling properties. In addition, matrix compounds are kept undecomposed and volatile materials are not lost, thus allowing quantitative/speciation studies to be carried out.	
(2) Chemical pretreatment	(2A) Direct addition of ^{208}Po tracer followed by gentle drying at about 90 °C in a sand-bath for 2–8 hr.	(2B) Transformation of the inorganic polonium of the sample to a chloride form using mild/concentrated HCl followed by addition of the ^{208}Po tracer. Gentle drying at 90 °C in a distillation tube for about 24 hr in a block-heater.
(3)	(3A) Wet-oxidation in a 100 ml flask using conc H_2SO_4 and conc HNO_3 in a sand-bath at about 270 °C. HNO_3 is added stepwise, to allow evolution of brown fumes to cease, until complete removal of organic matter is observed (10–30 min). The final product is clean inorganic residue in a clear yellow solution. Allow to cool (5–10 min).	(3B) Distillation of organic and inorganic polonium at 550 °C for about 30 min. Such polonium is trapped in a quartz collection-tube, with wetted glass wool, which is allowed to cool and then carefully transferred to a 3 l beaker (~30 min).
(4)	(4A) Separation of polonium in chloride form by successive dissolutions of polonium in 2–4 M HCl at about 100 °C. Liquids from 2–3 extractions are separated by centrifugation, neutralized by NH_4OH until complete precipitation of $Fe(OH)_3$ is achieved. $Fe(OH)_3$ is centrifuged and dissolved in a minimum amount of conc HCl (~100 min). The solution is allowed to dry in the distillation tube (6–12 hr) by gentle heating (block-heater).	(4B) Oxidation of the organic remains under reflux using HNO_3 (≥1–2 hr). This oxidation does not remove all organic matter even after long time (up to 16 hr). Allow to cool and filtrate to a 400 ml beaker. Reduce the filtrate and collect washings by evaporation to ~5 ml (~3 hr).
(5)	(5A) Distillation at 400 °C (30 min), allow the distillation tube (with wetted glass wool) to cool, transfer condensed matter by 2 M HCl and filtrate to a deposition cell (15 min).	(5B) Remove all HNO_3 using conc HCl and evaporate but never to dryness. Complete elimination of HNO_3 is necessary for proper self-deposition and gentle heating is required to avoid loss of polonium. Dissolve in 2 M HCl and filtrate to the deposition cell (~30 min).
(6) Self-deposition	Self-deposition (Flynn, 1968) is carried out on silver discs from about 60 ml of 1.5–2 M HCl in a water-bath. Plating takes place at 85–90 °C with continuous stirring using air bubbling or revolving propeller for about 3 hr. Masking cocktails have been utilized in some experiments.	

Table 2. Time and chemicals required for accurate and absolute determination of $^{210}Pb/^{210}Po$ in aquatic deposits. Methods given are those utilized by the Uppsala ^{210}Pb laboratory.

Chemical method	Chemical used [ml]		Treatment time [hr]		Technical remarks
			Total	Net	
(A) Wet-oxidation/distillation (new)	Conc H_2SO_4	2.5	14–24	5–6*	[1] Normally 0.5 g sediment is enough. However, amounts of 0.05 g are quite possible to analyze.
	Conc HNO_3	10			[2] Two samples could be easily treated in parallel.
	Conc HCl	8			[3] Smooth, easy and less risky steps.
	Conc NH_4OH	10			[4] Efficient and normal vapour-hood is sufficient.
	Total	~30			[5] Low distillation temperature (~400 °C) is enough and normal glass is possible.
(B) Eakin's distillation (old) "distillation/oxidation"	Conc HNO_3	150	34–48	5–6*	[1] About 1 g sediment is needed.
	Conc HCl	40			[2] Major difficulties to carry out parallel treatments.
	Total	190			[3] Extreme caution is required.
					[4] Relatively high distillation temperature (~550 °C) and therefore quartz or vicor-glass is needed.
					[5] Very efficient and large vapour-hood is recommended.
					[6] Frequent cleaning and maintenance of vapour-hood system.

* Including cleaning glass-ware, writing protocols and calculating sample activity.

Results and discussion

It is clear from Tables 1 and 2 that chemical procedure (A) is much more economic and convenient than (B). This applies to aquatic deposits with high organic content, but for low organic content sediments, especially marine ones, method (B) remains the best choice. Recently we were able to apply method (B) on marine sediments using almost the same quantities of strong acids as in method (A), however the total treatment time was somewhat longer compared to method (A).

Although high yields were given by method (B), about 80% on the average (El-Daoushy, 1981a; Bloesch & Evans, 1982), the tail of polonium could be higher than 10% in about 25% of the samples analyzed. Generally speaking the polonium tail is about ≤6–7% for about 45% of the samples and marine sediments tend to have smaller tails while freshwater and peat deposits always give higher tails depending on sediment matrix and composition. According to our calculations, tail corrections are quite simple if tails are below 9–10% and the polonium-210/polonium-208 ratio is between 0.5 and 2. However, beyond these limits tail correction may become invalid or complicated. This means that slightly less than 25% of the samples have to be reanalyzed as calculations may yield questionable results.

Method (A) was first applied to humic and humate fractions (El-Daoushy & Garcia-Tenorio, 1988), but for fulvic fractions this method was simplified by using HNO_3 instead of H_2SO_4. Method (A) was later tested on several lacustrine and marine sediments. Table 3 gives the composition of the lacustrine sediments studied here. Sediments from lakes Krageholmssjön and Bjäresjö were analyzed with methods (A) and (B). Lakes Koltjärn, Gulspettvann & Holmevatn allowed us to apply method (A) on sediments with very low and very high organic contents. The hardwater lakes Bjäresjö & Krageholmsjö have a factor of 20 higher accumulation rates compared to the softwater lakes Gulspettvann & Holmevatn. Sediments of these lakes have different concentrations and forms of elements which have been introduced through atmospheric and landuse activities. Sediments from a laminated lake with a wide range of organic content (3–26%) and high concentrations of Al (7%), Fe (5%) and Zn (4%) were also utilized. However, higher iron concentrations (up to 30%) could be found in some Swedish inland lakes (Johansson, 1988). As ombrotrophic peats are exclusively fed on atmospheric precipitation the corresponding concentrations of elements are very much lower, especially for Fe.

When we started to apply method (A) we were very surprised to have variable and low overall recoveries. Through a series of systematic studies some important factors were selected for critical examination. Figure 1 shows that the time of wet

Table 3. Lakes used in this study and chemical parameters of their sediments. Materials represent the central parts of lakes given and cover the past 200 yrs except for Lake Koltjärn where recent sediments from the whole lake were used. Values given for Lake Koltjärn are average of all sediments apart from a marginal area, with organic content as low as 3%, and the central sediments with occasionally higher organic content (up to 26%). Materials are mainly clayey-sediments (Koltjärn, Bjäresjö and Krageholmssjön), however fine sand/minerals are also available in Gulspettvann, Holmevatn and some cores from Koltjärn.

Lake	Organic content [%]	Carbonate content [%]	Special features
Koltjärn	10–20	Not measured	High Fe, Al and Zn contents
Bjäresjö	18–35	3–15	Hardwater lake, pH ≥ 9
Krageholmssjön	25–35	5–15	Hardwater lake, pH ≥ 8
Gulspettvann	45–54	2–5	Acidified lake, pH ~ 4.5
Holmevatn	46–61	2–5	Acidified lake, pH ~ 4.5

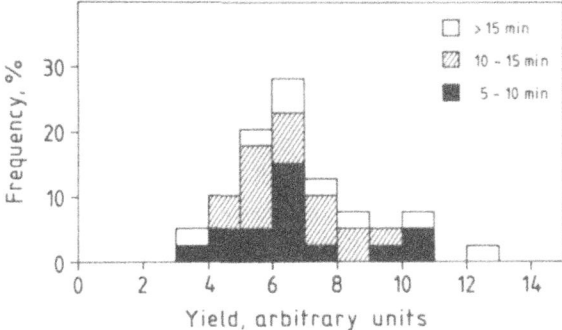

Fig. 1. The influence of the oxidation time (step 3A, Table 1) on the overall recovery of polonium. Although yields are usually given in percentage it was rather convenient to use 'arbitrary units' as no calibrations were done to get absolute values. Results given here are collected from routine dating experiments and the counting geometries (sample/detector) were more or less the same during the course of this study (typical variations in counting efficiencies are $\pm 5\%$).

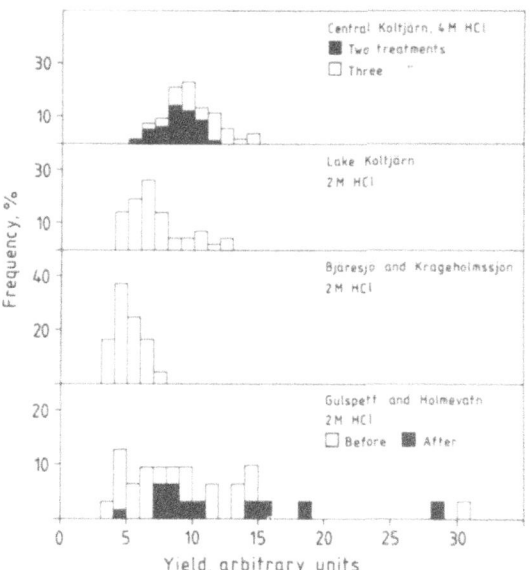

Fig. 2. Yield distribution of the polonium extraction from different samples (Table 3). Lakes Gulspett and Holmevatn are acid lakes with high concentrations of heavy metals while lakes Bjäresjö and Krageholmssjön are influenced by landuse activities. 'Before' and 'After' in the lower part of the figure refers to the addition of the polonium-208 spike (step 3A, Table 1). However, in routine extractions the spike is always added at step 2A (Table 1). The upper part of the figure shows the influence of the HCl concentration and number of HCl treatments (step 4A, Table 1) on the yield distribution. Lake Koltjärn refers to the whole lake (shallow and deep regions) excluding the central part which is shown in the upper part.

oxidation (method A) does not influence the yield of polonium extraction although the temperature could be as high as 300 °C and oxidation time up to 30 min in some cases. This demonstrates that polonium sulphates are relatively involatile. The overall yield for polonium extraction (method A) from the sediments (Table 1) is given in Fig. 2. It is quite clear that the sediment composition, organic/inorganic, is a major factor influencing the overall yields. Moreover, for every set of sediments there is a frequency distribution with a characteristic 'mean value' and spread. Generally speaking, the distributions are shifted towards higher overall recoveries with increasing organic contents. However, the spread seems to depend on how the organic matter is incorporated in the sediments. The minimum overall yield is about 15–25% (corrected for variations in the distance between source and detector) depending on sediment composition and the concentration of HCl (step 4A in Table 1). For organic-rich sediments (Gulspettvann & Holmevatn) the overall yield could be about $\geq 95\%$ in some cases (equivalent to the 30% of arbitrary unites). The very high overall yield obtained for these acid lakes may indicate that the organic matter in these sediments is loosely bound to minerals. Figure 2 shows that steps before and during oxidation are not respon-

sible for the variations in the yield as the addition of the spike, polonium-208, before (step 2A) and after the oxidation (3A/4A) does not seem to influence the distribution. There are some improvements to the overall yield by increasing the concentration of HCl (from 2 M to 4 M) and the number of HCl treatments. This indicates that the major part of polonium is in the insoluble form and the mean overall polonium recovery from the sediment could reach 35% (maximum value c. 50%) by improved HCl extraction. It is possible that the overall yields could be increased by extraction with distilled water after step (4A).

The great improvement to the resolution of the polonium peaks is also of importance. Figure 3 shows the influence of distillation, step (5A), on polonium tailing. With the distillation step, typical

50

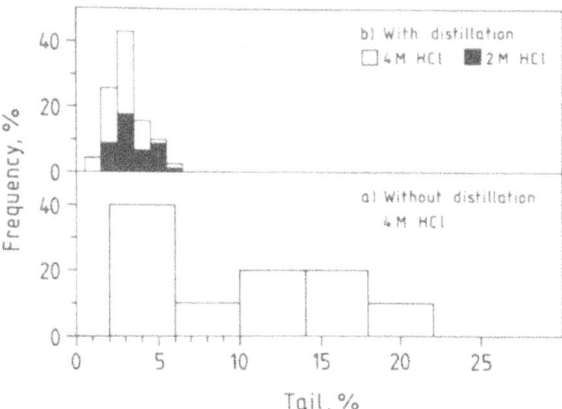

Fig. 3. The influence of distillation (step 5A, Table 1) on polonium tailing.

tails are only 3% and tail corrections are seldom required. We were not able to improve the results of Fig. 3(a) by using the cocktails mentioned.

For marine sediments the tails are larger than those obtained for lake sediments and tail corrections have to be made more often. However, the application of method (B) allowed us to obtain excellent resolution although chemicals and treatment time were very similar to method (A). This is easy to understand since marine deposits contain relatively low contents of organic matter which, under natural conditions, exist in less complicated structures. The distillation products (including the wetted glass wool) of step (3B) is transferred by minimum amounts of conc. HNO_3 to a small beaker (100–250 ml) and the organic remnants are oxidized (under reflux) on a sand-bath. The oxidation process needs about 30 min but this could be accelerated by adding suitable amounts of conc. HCl. The maximum amount of HCl and HNO_3 required is about 30–40 ml in total, this is about 15–20% of the amounts used by Eakins & Morrison (1978). Such modifications made method (B) superior to method (A) and the overall yields and resolutions were generally excellent for marine sediments.

For routine isotope dilution alpha spectrometry it is advisable to use sediments less than 1 g and preferably about 0.5 g, which yields very good resolutions and the self-deposition step can be kept long.

In routine laboratory work, memory effects on glassware are a serious problem and great caution is necessary. Polonium in weak acid solutions is very easily adsorbed on glassware. It is difficult to give exact limits, but 6 M HCl proved to be enough to keep polonium in solutions for very long periods of time (at least 5 yrs). However, we have observed that polonium was adsorbed during self-deposition (2 M HCl) if gentle washing was applied. Iron in natural deposits, especially sediments, acts as a carrier and thus minimizes adsorption of polonium on glassware, particulary in the absence of strong acids. We have also observed that in speciation experiments the lack of Fe causes Po to be totally lost from solution. The most suitable cleaning agent is dichromate-sulfuric acid, other agents are either inefficient (Extran, Merck) or require additional treatment under reflux (HNO_3). Moreover, it is quite easy to know when dichromate-sulfuric acid is exhausted. Memory effects and purity of chemical reagents have to be checked now and then through control tests and blank runs, and glassware and chemicals have to be replaced if necessary.

Isotope dilution alpha spectrometry of polonium is usually carried out in simple alpha counting chambers such as the Canberra Quad Alpha Spectrometer 7404. The Canberra unit is very easy to handle and requires minimum efforts of the operator particulary if ion-implanted silicon detectors are used. The increase of the detector background due to recoil and volatilization of the decaying polonium is kept to a minimum through a small negative voltage to sample holders. Unlike silicon surface barrier detectors the passivated ion-implanted silicon detectors provide very stable counting characteristics and excellent resolution at room temperature because of their extremely low leakage currents. Serious variations in the polonium-210/polonium-208 ratios could occur if silicon surface barrier detectors are operated at temperatures above 24°–26 °C, especially when daily temperature variation exceeds ± 1 °C (El-Daoushy, 1981b). Furthermore, the stability of ion-implanted silicon detectors allows us to keep good records on their background which is

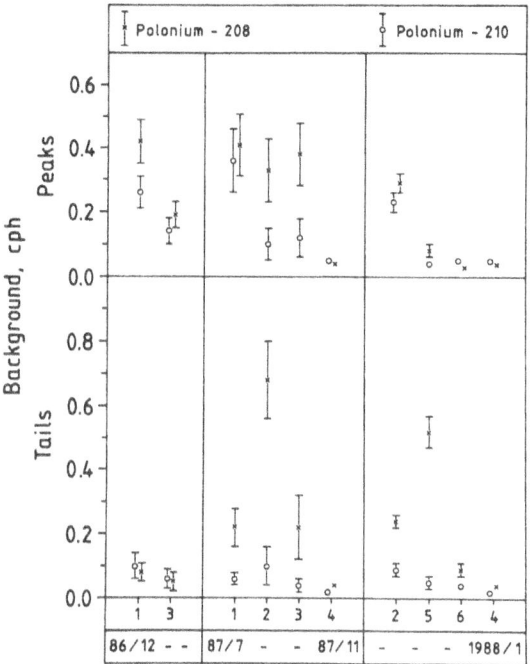

Fig. 4. Background characteristics of detectors used. Detectors 1 and 3 were used in the major part of this study. The detectors are placed (from left to right) in counting chambers as shown in the figure. During 86/12–87/7 detectors 1 and 3 were placed in the homemade chamber and were later moved to the Canberra chamber. Detector 4 was always in the fourth section, detector 2 was moved from the second section to the first during the later part of this study. The second and third sections are presently occupied by detectors 5 and 6.

of importance at very low activities such as those measured in speciation studies (El-Daoushy & Garcia-Tenorio, 1988). Figure 4 shows background records of six detectors; three of them (4, 5 and 6) are ion-implanted detectors (Enertic) and have been used since 1987 (detector 4) and 1988. Before July 1987 the silicon detectors were housed in a homemade chamber instead of the present canberra Quad Alpha Spectrometer. Since 1986 the chamber assemblies are kept at constant temperature (20°–22 °C) in an air-conditioned room. Detectors 4, 5 and 6 have lower background compared with 1, 2 and 3 (Ortec) which were utilized during 1976/78 and 1985/87 (1); 1982/85 and since 1987 (2); and 1986/87 (3). The background values are due to the silicon detectors and not the counting chamber

except for the second section of the Canberra unit which gave a high value in the energy region of the polonium-208 tail. With these specifications, activities as low as 1 mBq could be determined which enabled us to carry out speciation studies on sensitive fractions (El-Daoushy & Garcia-Tenorio, in preparation). Method (A) proved to be the best choice for polonium extraction from fulvic, humic, humin and humate fractions (El-Daoushy & Garcia-Tenorio, 1988, 1990). Polonium extraction from particulate matter and water samples (with secular equilibrium between Pb-210 and Po-210) could be done in a similar way as the fulvic fraction if the organic component exists in simple form. In any case, water samples have to be handled in acid solutions containing suitable amounts of Fe-carrier (Benninger, 1978; Schell, 1977). Fresh aquatic deposits should not be kept in wet conditions for long periods of time as sensitive fractions could be lost by decomposition and adsorption. Drying (freeze-drying has to be used if deposits are to be studied by speciation), is recommended to be carried out at 80 °C and upon arrival to the laboratory. We have utilized speciation experiments to evaluate levels of supported lead-210 in cases where radium-226 seems to be mobile/variable (El-Daoushy & Garcia-Tenorio, 1990). Furthermore, speciation could be used to study the geochemistry of unsupported and supported lead-210 in aquatic reservoirs, especially early diagenetic processes and interactions between organic and inorganic matters.

It has been observed that polonium-208 contains trace amounts of polonium-209 (e.g. Fleer & Bacon, 1984) and should be considered in tail corrections. Because of the long half-life time of polonium-209 its relative concentration increases if polonium-208 is used during a long period of time. The net polonium tails are then utilized to calculate the correct activity of the polonium peaks. The correction for polonium tailing using computer iteration (Fleer & Bacon, 1984) is quite satisfactory providing that the polonium-210/polonium-208 ratio is in the range 0.5–2 and polonium tails are $\leq 10\%$. For polonium ratios differing by one, correction for polonium tailing seems to fail if tails are higher than 10%. In cases where self-deposition yields inhomogeneous samples (variable thickness) due to improper stirring conditions, corrections for polonium tailing may not be valid. At low counting rates correction for polonium tailing is most reliable when the polonium-210/polonium-208 ratio is kept close to one.

Acknowledgements

This work was carried out at the Department of Physics, Uppsala University and supported by the Swedish Natural Science Research Council as well as Spanish research funds, contracts CAICYT No. 2849/83 and PB86/0207.

References

Benninger, L. K., 1978. ^{210}Pb balance in Long Island Sound. Geochim. Cosmochim. Acta 42: 1165–1174.

Benninger, L. K., R. C. Aller, J. K. Cochran & K. K. Turekian, 1979. Effects of biological sediment mixing on the ^{210}Pb chronology and trace metal distribution in a Long Island Sound sediment core. Earth Planet. Sci. Lett. 43: 241–259.

Bloesch, J. & R. D. Evans, 1982. Lead-210 dating of sediments compared with accumulation rates estimated by natural markers and measured with sediment traps. Hydrobiol. 92: 579–586.

Chanton, J. P., C. S. Martens & G. W. Kipphut, 1983. Lead-210 sediment geochronology in a changing coastal environment. Geochim. Cosmochim. Acta 47: 1791–1804.

Dominik, J., A. Mangini & G. Muller, 1981. Determination of recent deposition in Lake Constance with radioisotopic methods. Sedimentology 28: 653–677.

Eakins, J. D., 1984. The application of radiochemical separation procedures to environmental and biological materials. Nucl. Instr. & Meth. 223: 194–199.

Eakins, J. D. & R. T. Morrison, 1975. The determination of polonium-210 in urine by coprecipitation with manganese dioxide. United Kingdom Atomic Energy Authority Reports, AERE-R 7923.

Eakins, J. D. & R. T. Morrison, 1978. A new procedure for the determination of lead-210 in lake and marine sediments. Int. J. Appl. Radiat. Isot. 29: 531–536.

El-Daoushy, F., 1981a. An ionization chamber and a Si-detector for lead-210 chronology. Nucl. Instr. & Meth. 188: 647–655.

El-Daoushy, F., 1981b. Stability of two silicon detectors in low-level alpha counting. Methods of Low-Level Counting and Spectrometry. International Atomic Energy Agency, Vienna: 151–160.

El-Daoushy, F., 1986. Scandinavian limnochronology of sediments and heavy metals. Hydrobiol. 143: 267–276.

El-Daoushy, F., 1988. A summary on the lead-210 cycle in Nature and related applications in Scandinavia. Environ. Inter. 14: 305–319.

El-Daoushy, F. & R. Garcia-Tenorio, 1988. Speciation of Pb-210/Po-210 in aquatic systems and their deposits. Sci. Tot. Envir. 69: 191–209.

El-Daoushy, F. & R. Garcia-Tenorio, 1990. ^{210}Pb (^{210}Po) speciation of aquatic deposits: Refinement and utility. J. Radioanal. Nucl. Chem. 138: 5–15.

El-Daoushy, F. & K. Tolonen, 1984. Lead-210 and heavy metal contents in dated ombrotrophic peat hummocks from Finland. Nucl. Instr. & Meth. 223: 392–399.

Erlenkeuser, H. & K. Pederstad, 1984. Recent sediment accumulation in Skagerrak as depicted by ^{210}Pb-dating. Norsk Geologisk Tidsskrift 64: 135–152.

Fleer, A. P. & M. P. Bacon, 1984. Determination of ^{210}Pb and ^{210}Po in seawater and marine particulate matter. Nucl. Instrum. Methods 223: 243–249.

Flynn, W. W., 1968. The determination of low levels by polonium-210 in environmental samples. Analytica Chimica Acta 43: 221–227.

Häsänen, E., 1977. Dating of sediments, based on ^{210}Po measurements. Radiochem. Radioanal. lett. 31: 207–214.

Johansson, K., 1988. Heavy metals in Swedish forest lakes – factors influencing the distribution in sediments. Doctoral thesis, Institute of Limnology, Uppsala University.

Joshi, S. R. & R. W. Durham, 1976. Determination of ^{210}Pb, ^{226}Ra and ^{137}Cs in sediments. Chem. Geol. 18: 155–160.

Persson, B. R., 1970. ^{210}Pb-Atmospheric deposition in lichen-carpets in northern Sweden during 1961–1969. Tellus 22: 564–571.

Robbins, J. A. & Edgington, 1975. Determination of recent sedimentation rates in Lake Michigan using Pb-210 and Cs-137. Geochim. Cosmochim. Acta 39: 285–304.

Schell, W. R., 1977. Concentrations, physico-chemical states and mean residence times of ^{210}Pb and ^{210}Po in marine and estuarine waters. Geochim. Cosmochim. Acta 41: 1019–1031.

Smith, J. N. & A. Walton, 1980. Sediment accumulation rates and geochronologies measured in the Saguenay Fjord using the Pb-210 dating method. Geochim. Cosmochim. Acta 44: 225–240.

Van der Wijk, A. & W. G. Mook, 1987. ^{210}Pb dating in shallow moorland pools. Geol. Mijnb. 66: 43–55.

Hydrobiologia **214**: 53–57, 1991.
J. P. Smith, P. G. Appleby, R. W. Battarbee, J. A. Dearing, R. Flower, E. Y. Haworth, F. Oldfield & P. E. O'Sullivan (eds),
Environmental History and Palaeolimnology.
© 1991 *Kluwer Academic Publishers.*

How long was the Younger Dryas? Preliminary evidence from annually laminated sediments of Soppensee (Switzerland)

André F. Lotter
Geobotanik, Universität Bern, Altenbergrain 21, CH-3013 Bern, Switzerland

Key words: varve chronology, Younger Dryas, Late-Glacial, AMS-^{14}C-datings

Abstracts

Despite extensive AMS-^{14}C dating series on Late-glacial terrestrial plant remains, a precise estimate of the duration of the Younger Dryas biozone (*sensu* Ammann & Lotter, 1989) is hampered by the occurrence of a period of constant ^{14}C-age at 10 000 yr B.P. However, varve counts at Soppensee suggest that the Younger Dryas biozone comprises approx. 680–720 varves, and that the phase of constant radiocarbon age includes between 270–310 varves.

Introduction

The Late-Glacial and especially the Younger Dryas with its severe climatic oscillations, is a period of high environmental dynamics. In order to interpret changes in the sedimentary record of this period, a precise time-scale with which to determine the speed and duration of such events is required. The establishment of an accurate and high-resolution chronology is, in fact, one of the major issues for the Late-Glacial period. On one hand, the dendrochronological record does not yet extend to the Late-Glacial, and on the other Late-glacial sediments either lack datable organic matter, or the ^{14}C-dates include a hard-water error of unknown magnitude. Nevertheless, it was possible to circumvent these problems recently by AMS (Accelerator Mass Spectrometry) ^{14}C-dating of terrestrial plant remains: three series of more than 90 AMS ^{14}C-datings allowed us to establish a Late-Glacial high resolution radiocarbon chronology at Rotsee (Lotter, 1988; Lotter & Zbinden 1989) and at Lobsigensee (Andrée *et al.*, 1986; Ammann & Lotter, 1989). However, owing to two distinct phases of constant radiocarbon age recorded at both sites, new problems emerged. These so called '^{14}C-plateaux' are located at ca 12 700 B.P. (i.e. at the onset of reforestation) and at 10 000 B.P. (i.e. at the transition from Late-glacial to Holocene sediments) – two periods of high interest. The plateaux seem to be generated by changes in the global carbon cycle (Andrée *et al.*, 1986; Zbinden *et al.*, in press) and their implications for palaeoecology are discussed in detail by Ammann & Lotter (1989). Yet, the problems related to these plateaux can only be mastered by ^{14}C-independent dating techniques, e.g. by a varve chronology. Given the high degree of precision and time resolution, the information available from such a varve chronology is an exceptional tool for assessing rates at which limnic and terrestrial processes, as well as climatic and environmental variations, have occurred.

54

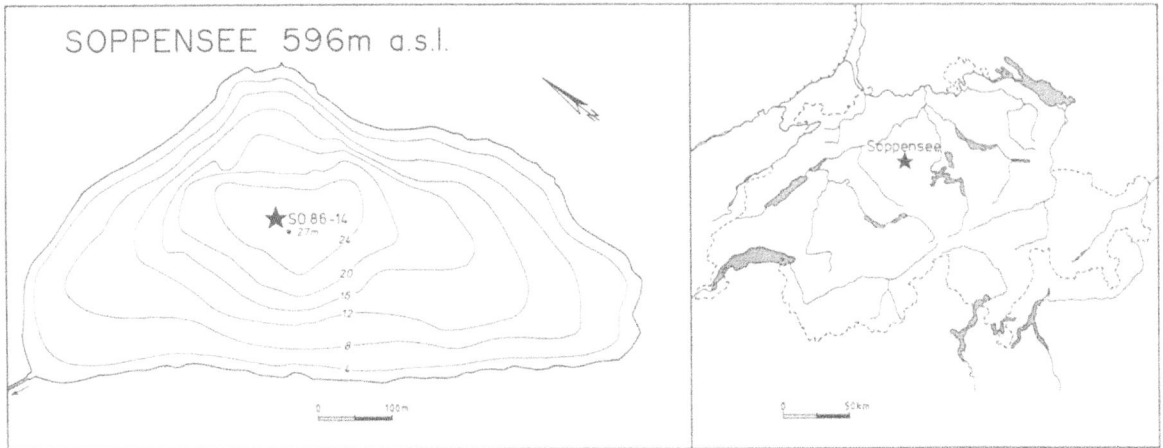

Fig. 1. Map of Switzerland indicating the geographical location of the investigation site and bathymetric map of Soppensee, indicating the position of the core SO86-14.

Investigation site

Soppensee (47° 5′ 30″ N, 8° 5′ E, 596 m a.s.l.) is located 20 km northwest of Lucerne on the Central Swiss Plateau (Fig. 1). It is a kettle lake 800 m long and up to 400 m wide, with a maximum water-depth of 27 m (Fig. 1). The lake drains a 1.6 km² catchment, and was formed towards the end of the last glaciation. Its sediment yields information on the environmental history of the last 15 000 years. Moreover, annual laminations, which occur between the Bölling and the Atlantic (Lotter, 1989), include approx. 5 600 varves.

Methods

Several cores were taken with a modified Kullenberg sampler (Kelts *et al.*, 1986). The core chosen for this study was taken in the deepest part of the lake at a water depth of 27 m, as indicated in Fig. 1. Pollen analysis was carried out as described in Lotter (1988). Varve counts were performed on 11 cm long and 2 cm wide sediment thin-sections (Merkt, 1972) using a polarisation microscope.

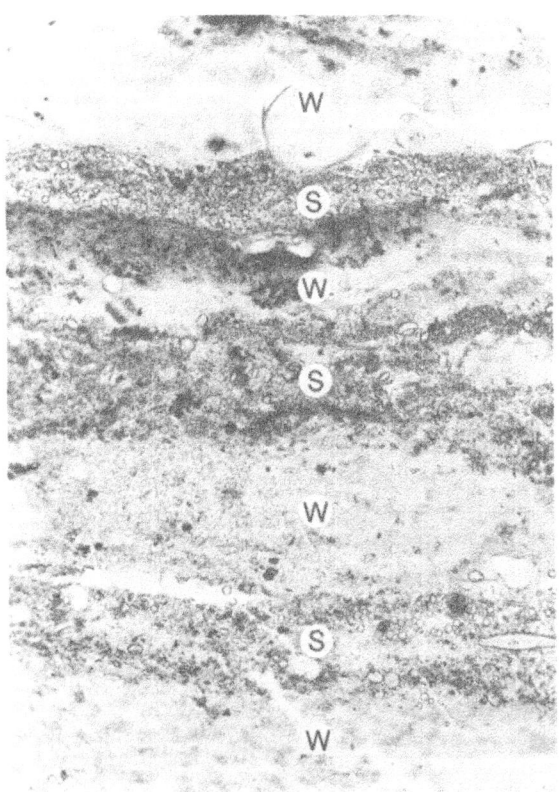

Fig. 2. Sediment thin section of a sample taken at 486 cm of core-depth (Alleröd), showing three consecutive varves. W = autumn/winter layer, consisting of organic fine-detritus. S = spring/summer layer consisting of diatoms and chrysophyte cysts.

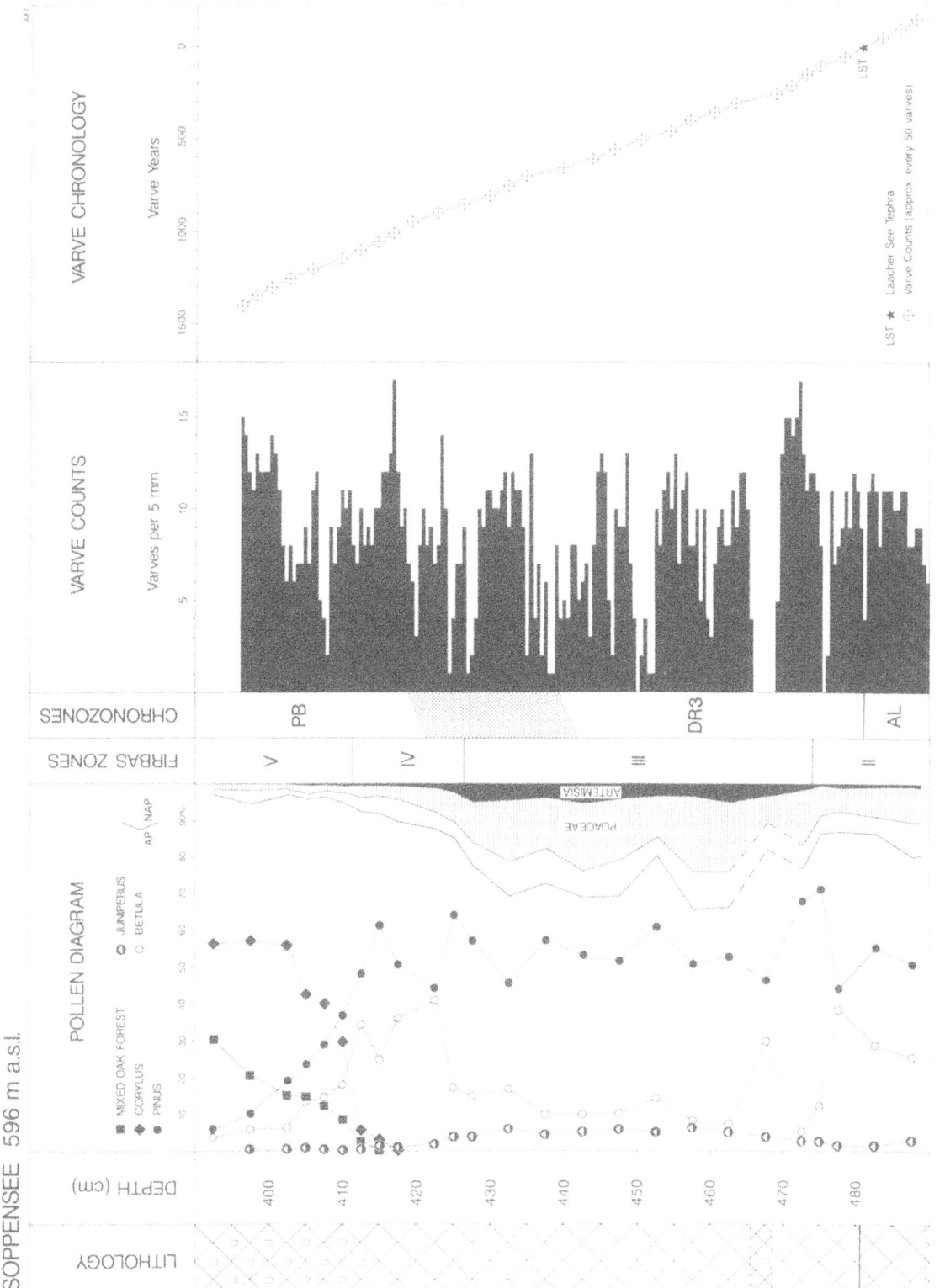

Fig. 3. Palynostratigraphy and varve chronology at **Soppensee**. The shaded area marks the phase of constant radiocarbon age at 10000 yr **B.P.**

Results and discussion

The Late-glacial (non-glacial) varves at Soppensee (Fig. 2) consist of a spring/summer diatom and chrysophyte cyst layer including few very fine-grained carbonate crystals, and a large autumn/winter layer of organic fine detritus. This is in contrast to the more conspicuous Holocene varves that are chiefly built up of calcium carbonate (Lotter, 1989). Varve counts in core SO86-14 allowed me to establish a high precision and high resolution absolute chronology (Fig. 3).

The sediment record of Soppensee includes the Laacher See Tephra (LST) usually given a radiocarbon age of 11 000 years B.P. (Van den Bogaard & Schmincke, 1985; Ammann & Lotter, 1989). Because the varve chronology is floating, this stratigraphic time marker was used as year 0 in the Soppensee varve chronology (marked by * in Fig. 3).

An answer to the question 'How long was the Younger Dryas' is not simple and straightforward. First of all, we have to agree on what we mean by the term 'Younger Dryas'. If it is the chronozone according to Mangerud *et al.* (1974) or Welten (1982), then the Younger Dryas (DR3 in Fig. 3) has by definition a length of 1 000 radiocarbon years, starting at 11 000 B.P. and ending at 10 000 B.P. However, in Central Europe the Firbas (1949, 1954) pollen zonation system, with its regional adaptions (e.g. Welten, 1982; Ammann & Lotter, 1989), is commonly used. In this system the Younger Dryas as a biozone (III in Fig. 3), comprises the response of the vegetation to the last large-scale climatic change of the Late-Glacial; it is therefore shorter in duration than the chronozone of the same name.

The local pollen assemblage zones (PAZ) at Soppensee were assigned to the Firbas pollen zonation system (Roman numerals in Fig. 3). The Younger Dryas (III) biozone corresponds to the *Pinus*-Poaceae-NAP regional PAZ (Lotter, 1988; Ammann, 1989). The pollen stratigraphies of Rotsee (Lotter, 1988) and Soppensee (Fig. 3, and Lotter unpubl.) are easily comparable. Therefore, it was possible to correlate the 10 000 B.P. plateau phase (indicated by the shaded area in Fig. 3)

which is located within the laminated part of the sedimentary record. The varve counts suggest that this period of constant ^{14}C-age encloses a time-span of 270–310 varves. The phase of climatic oscillation of the Younger Dryas biozone (III) yields 680–720 varves. This result is in good agreement with the classic investigation at Faulenseemoos (Welten, 1944), where approx. 690 have been counted for the same period.

The sediment between the deposition of the LST and the response of the vegetation to the climatic change of the Younger Dryas (III) includes 185–235 varves whereas the radiocarbon dates suggest an interval of more than 300 ^{14}C-years. Further studies will show whether this discrepancy between the sidereal and the radiocarbon time-scale is real.

Acknowledgements

The help of the following persons is gratefully acknowledged: B. Ammann, J. Hofmann, G. Lang, J. Merkt, J. P. Smol, M. Sturm, H. Zbinden. This investigation was financially supported by Swiss National Science Foundation.

References

Ammann, B., 1989. Late-Quaternary palynology at Lobsigensee – regional vegetation history and local lake development. Diss. Bot. 137: 1–157.

Ammann, B. & A. F. Lotter, 1989. Late-Glacial radiocarbon- and palynostratigraphy on the Swiss Plateau. Boreas 18: 109–126.

Andrée, M., H. Oeschger, U. Siegenthaler, T. Riesen, M. Möll, B. Ammann & K. Tobolski, 1986. ^{14}C dating of plant macrofossils in lake sediment. Radiocarbon 28/2A: 411–416.

Bogaard, P. Van den & U. Schmincke, 1985. Laacher See tephra: a widespread isochronous late Quaternary tephra layer in central and northern Europe. Geol. Soc. Am. Bull. 96: 1554–1571.

Firbas, F., 1949. Spät- und nacheiszeitliche Waldgeschichte Mitteleuropas nördlich der Alpen. Vol. 1. Fischer Verlag, Jena. 480 pp.

Firbas, F., 1954. Die Synchronisierung der mitteleuropäischen Pollendiagramme. Danm. Geol. Unders II, 80: 12–21.

Kelts, K., U. Briegel, K. Ghilardi & K. Hsu, 1986. The

limnogeology-ETH coring system. Schweiz. Z. Hydrol. 48: 104–115.

Lotter, A., 1988. Paläoökologische und paläolimnologische Studie des Rotsees bei Luzern. Pollen-, grossrest-, diatomeen- und sedimentanalytische Untersuchungen. Diss. Bot. 124: 1–187.

Lotter, A. F., 1989. Evidence of annual layering in Holocene sediments of Soppensee, Switzerland. Aquat. Sci. 51: 19–30.

Lotter, A. F. & H. Zbinden, 1989. Late-Glacial palynology, oxygen isotope record, and radiocarbon stratigraphy from Rotsee (Lucerne), central Swiss Plateau. Eclogae Geol. Helv. 82: 191–202.

Mangerud, J., S. T. Andersen, B. E. Berglund & J. J. Donner, 1974. Quaternary stratigraphy of Norden, a proposal for terminology and classification. Boreas 3: 109–128.

Merkt, J., 1972. Zuverlässige Auszählungen von Jahresschichten in Seesedimenten mit Hilfe von Gross-Dünnschliffen. Arch. Hydrobiol. 69: 145–154.

Welten, M., 1944. Pollenanalytische, stratigraphische und geochronologische Untersuchungen aus dem Faulenseemoos bei Spiez. Veröff. Geobot. Intst. Rübel Zürich 21: 1–201.

Welten, M., 1982. Vegetationsgeschichtliche Untersuchungen in den westlichen Schweizer Alpen: Bern – Wallis. Denkschr. Schweiz. Naturf. Ges. 95: 1–104.

Zbinden, H., M. Andrée, H. Oeschger, B. Ammann, A. Lotter, G. Bonani & W. Wölfli. Atmospheric radiocarbon at the end of the Last Glacial: an estimate based on AMS radiocarbon dates on terrestrial macrofossils from lake sediments. Radiocarbon: in press.

.

Hydrobiologia **214**: 59–61, 1991.
J. P. Smith, P. G. Appleby, R. W. Battarbee, J. A. Dearing, R. Flower, E. Y. Haworth, F. Oldfield & P. E. O'Sullivan (eds), 59
Environmental History and Palaeolimnology.

Absolute dating of late Quaternary Lacustrine sediments by high resolution varve chronology

Bernd Zolitschka
Abt. Geologie, Universität Trier, D-5500 Trier, Germany

Key words: varves, absolute dating, Late Quaternary, West Germany

Abstract

A high precision absolute timescale has been developed from annually laminated lake sediments from lakes in the Eifel area, West Germany. Calibration of relative dating methods (palynology, paleomagnetism) was carried out successfully. In addition palecological and astronomical information was obtained from varve thickness measurements, and the composition of annual layers.

Introduction

Lacustrine sediments record palecological information from lakes and their catchment areas. Every kind of investigation on these sediments will be substantially improved, if fine laminations occur and have been shown to be annual layers (varves). This paper will describe varved sediments from the lakes Holzmaar and Meerfelder Maar situated in the West Eifel Volcanic Field (West Germany). Both lakes are eutrophic, dimictic and holomictic soft water lakes. They possess small surface areas, steep sides and a flat bottom. Together with high organic productivity this favours the formation and preservation of annually laminated sediment layers.

Materials and methods

Sediment cores were obtained with a modified Livingstone piston corer. Overlapping sequences from the centre of each lake were used for thin section preparation. Microscopical examination of these thin sections was carried out in 1 cm intervals, recording the absolute number of laminations as well as their composition.

The sediments of both lakes are divided into two parts: an upper section of finely laminated diatomaceous gyttja with a 4 to 8 cm thick bipartite ash layer at its base, assigned to the eruption of Laacher See Volcano (Negendank, 1984). The lower parts of the sediments consist of more coarsely laminated silt and clay layers.

The annual rhythm

Microstratigraphical examination of the organic sediments from Holzmaar and Meerfelder Maar reveal the seasonal character of all these laminations. Organic varves begin with a layer of chrysophycean cysts in early spring, followed by a bloom of planktonic diatoms in spring and early summer. During late Glacial and early Holocene summer layers, primary calcites occur in a gradation from large to small crystals, typical for authigenic precipitation (Eugster & Kelts, 1983; Kelts & Hsü, 1978). Under certain conditions calcite may be substituted by siderite. Fall and winter laminae are composed of littoral diatoms, organic and minerogenic detritus.

Microstratigraphic demonstration of the annual character is feasible in three ways:

1. Diatomological evidence: At present diatom blooming occurs from March until June in the maar lakes of the Eifel area (Hustedt, 1954; Post, 1980).

2. Palynological evidence: Shortly after the onset of the diatom maximum, *Corylus* pollen occurs in the Eifel, *Corylus avellana* blooms from February until April. At the top of the diatoms layer, *Pinus* pollen appears. *Pinus sylvestris* blooms from May until June.

3. Sedimentological evidence: Authigenic calcite crystals are precipitated at the end or just after the diatom maximum in summer, when photosynthetic activity of phytoplankton reaches its yearly optimum.

Varve chronology

By counting the varves from top to base of the upper organic sediment it is possible to build up an absolute time scale for the last 12 800 years. The lower clastic laminites have also been counted, but since the annual rhythm of deposition has not yet been demonstrated, they will be left out of consideration in this study.

Age determinations have been carried out in 1 cm intervals for varves in Holzmaar, so that more than 1000 single dates exist. Using the so-called sediment increase rate (= thickness increase of the sediment sequence for a certain time interval expressed in mm a^{-1}), every desirable depth may be absolutely dated. This is important for every application, because sediment increase rate is not linear, but ranges between 0.3 and 10 mm a^{-1}.

Microscopic analysis of thin sections supplies the most precise varve counting, therefore dating accuracy varies about ± 1.1%. Indistinct and ambiguous annual laminae are the main reason for this error. Subdivision into three periods of different dating accuracy, e.g. varying distinctness of varve formation is possible: Late Glacial varves (10 800 BC to 8 000 BC) lead to high errors of ± 5.5%, whereas early and mid-Holocene laminites (8 000 BC to 1 600 BC) are well developed with errors of only ± 0.8. The youngest laminae (last 3 600 years) are more indistinct and counting error is rising to ± 3.5%.

Applications

Dating of the Laacher See Tephra

The most direct application of absolute varve dating is to date prominent layers of the sediment sequence. In Eifel maar lakes the isochrone of Laacher See Tephra is present, which is radiocarbon-dated all over Central Europe to 11 000 years BP. Using the varve chronology developed here, the Laacher See tuff is dated to 9200 years BC. This means that varve dating yields an age of about 200 years older than the radiocarbon dates.

Calibration of relative dating methods

One of the main tasks of varve chronology is of course the calibration of relative dating methods (Zolitschka, 1989). The palynological record (Usinger & Wolf, in prep.) has been dated with the varve chronology. A comparison of absolute dated pollen zones with the chronozones of Mangerud *et al.* (1974) shows slight deviations, first of all due to differences between bio- and chronozones. Other reasons may be local factors and inaccuracies or calibration problems of radiocarbon dating. This weak correspondence was expected.

Calibration of other dating techniques

Calibration has also been carried out for paleomagnetic data (Haverkamp, 1988). Varve chronologically calibrated curves of declination and inclination measurements have been compared with the radiocarbon dated United Kingdom master curve of Thompson & Turner (in Creer, 1985).

Evidence for the correctness of varve chronology is delivered by comparing the varve dated paleosecular variation record from Holzmaar for the last 1 000 years with the geomagnetic paleosecular variation measured at Paris (Thellier, 1981) for the same time interval. After correcting for the 'lock-in' effect both curves are in good agreement. Calibration of relative dating methods using varve chronology seems most promising. The next step will be to generate a calibration curve for radiocarbon dating of the last 12 800 years.

Astronomical data

A geophysical signal stored in the sediment of Holzmaar has been recovered by spectral analysis of about 500 single varve thickness measurements at the Boreal/Atlantic transition (time span from 5959 BC to 6471 BC). These data have been smoothed and submitted to a Fast Fourier Transform (FFT) algorithm. The spectra obtained are similar to cycles of sun spot activities and to astronomical periodicities modifying climate on earth. These findings corroborate Anderson & Koopmans (1963), who regard varved sediment sequences as imperfect meteorological time series.

Conclusions

Detailed microstratigraphical examination of annually laminated lake deposits demonstrate the potential of high resolution sedimentary records. The annual nature of laminations has been demonstrated by palynological, diatomological, sedimentological and paleomagnetic methods. It is confirmed by correct dating of the well-known isochrone of Laacher See Tephra and by dis-

covering astronomical periodicities from varve thickness measurements.

Composition of these varves is a valuable tool for reconstruction of past climates and their environmental responses. A careful reading of all components will provide environmental and climatic informations not available by any other method.

References

Anderson, R. Y. & L. H. Koopmans, 1963. Harmonic analysis of varve time series. J. Geophys. Res. 68: 877–893.

Creer, K. M., 1985. Review of lake sediment paleomagnetic data (part I). Geophys. Surveys 7: 125–160.

Eugster, H. P. & K. Kelts, 1983. Lacustrine chemical sediments. In Goudie, A. & K. Pye (eds), Chem. sed. geomorphol.: 321–361.

Haverkamp, B., 1988. Magnetostratigraphische Untersuchungen an frischen Seesedimenten aus Maaren der Westeifel. Nachr. Dt. Geol. Ges. H. 39: 24–25.

Hustedt, F., 1954. Die Diatomeenflora der Eifelmaare. Arch. f. Hydrobiol. 48: 451–496.

Kelts, K. & K. J. Hsü, 1978. Freshwater carbonate sedimentation. In Lerman, A. (ed.), Lakes – Chemistry, Geology, Physics: 295–323. New York.

Mangerud, J., S. T. Andersen, B. E. Berglund & J. J. Donner, 1974. Quaternary stratigraphy of Norden, a proposal for terminology and classification. Boreas 3: 109–128.

Negendank, J. F. W., 1984. Die Untersuchung der Schwerminerale der Seesedimente des Meerfelder Maares und des 'Laacher Bims-Tuffes' in den Sedimenten des Meerfelder Maares, des Hinkelsmaares und der Hitsche. Cour. Forsch. Inst. Senckenberg 65: 41–47.

Post, M., 1980. Beitrag zur Primärproduktion des Weinfelder und des Schalkenmehrener Maares. Mitt. Pollichia 68: 143–155.

Thellier, 1981. Geomagnetic palaeosecular variation at Paris for the last 1 000 years. Phys. Earth Planet. Inter. 24: 89–132.

Zolitschka, B., 1989. Jahreszeitlich geschichtete Seesedimente aus dem Holzmaar und dem Meerfelder Maar (Westeifel). Z. Dt. Geol. Ges. 140: 25–33.

Hydrobiologia **214**: 63–70, 1991.
J. P. Smith, P. G. Appleby, R. W. Battarbee, J. A. Dearing, R. Flower, E. Y. Haworth, F. Oldfield & P. E. O'Sullivan (eds), 63
Environmental History and Palaeolimnology.
© *1991 Kluwer Academic Publishers.*

The record of deposition of radionuclides in the sediments of Ponsonby Tarn, Cumbria

P. J. P. Bonnett[1,2] & R. S. Cambray[2]
[1] *Dept. of Geography, University of Liverpool, P.O. Box 147, Liverpool L69 3BX, UK;* [2] *Modelling and Assessments Group, Environmental and Medical Sciences Division, B364, Harwell Laboratory, United Kingdom Atomic Energy Authority, Oxon OX11 0RA, UK*

Key words: radionuclide invertories, Sellafield, Chernobyl, weapons fallout, ^{134}Cs, ^{137}Cs, activities, diatom analysis

Abstract

A study has been made of the radionuclide content of sediments from Ponsonby Tarn in Cumbria to examine the pattern of deposition of radiocaesium and actinides upon the catchment. Sediment cores obtained from the tarn in 1986 were dated by core correlation and compared with results obtained from a previous study in 1980 (Eakins & Cambray, 1985).

Sediments from the tarn contained actinides derived mainly from discharges to atmosphere. The 1986 cores contain greater quantities of weapons fallout derived caesium than may be accounted for by direct atmospheric input but less attributable to the Chernobyl accident than expected. Diatom analysis together with increased sediment accumulation rates post 1980 suggest that complex patterns of sedimentation have contributed to the changes evident in the sediment.

Introduction

The surface of a lake and its surrounding catchment area is a system subject to a flux of both natural radionuclides resulting from cosmic ray interactions with stable elements in the upper atmosphere or by decay of radioelements in the earth's crust, and artificial (man-made) radioactivity. Such artificial radionuclides may originate from 3 main sources:

a) weapons fallout resulting from the atmospheric testing of distant nuclear weapons,
b) discharges from nuclear installations
c) the return to the atmosphere and subsequent disintegration of satellites containing nuclear power sources.

A fraction of the radioactivity deposited on the catchment is subject to weathering and, via trans-port processes is transferred to the lake (Eakins *et al.*, 1984) and, together with material entering the lake as direct atmospheric input by rain is deposited at the sediment/water interface by sedimentation and exchange processes. The lake watershed ecosystem and in particular the lake sediment record thus provides a framework within which to evaluate the impact of human activity and its present and future implications (Oldfield, 1977; O'Sullivan, 1983).

Sediment cores were taken from Ponsonby Tarn in Cumbria in 1980 and 1986 to examine the changing depositional record of radionuclides contained in the tarn's sediments. Earlier work (Pattenden *et al.*, 1980; Peirson *et al.*, 1982) suggests that the deposit of actinides and ^{137}Cs near Sellafield is derived mainly from two sources, maritime and atmospheric sources. Ponsonby Tarn relates to the latter category. Eakins &

64

Cambray (1985) compared the results from such water bodies with those unaffected or only marginally affected by discharges from BNFL, Sellafield.

To examine the history of radionuclide deposition contained in the sediments of a water body the establishment of an accurate chronology is essential. The two principal radiometric techniques for dating recent lake sediments are ^{210}Pb and ^{137}Cs. The ^{210}Pb method can be used to date sediments spanning the past 100–150 years. There are two main models for calculating ^{210}Pb dates, the CRS (or Constant Flux) model and the CIC (or Constant Activity) model (Appleby & Oldfield, 1978; Robbins, 1978; Eakins, 1983). The CRS model assumes a constant rate of supply of ^{210}Pb from the atmosphere to the sediments irrespective of changes in the net dry mass sedimentation rate. The CIC model assumes that sediments have a constant initial ^{210}Pb activity. Factors governing model choice are discussed in Appleby & Oldfield (1983) and Oldfield & Appleby (1984). The ^{137}Cs method (Pennington *et al.*, 1973, 1976) is applicable to sites where the sedimentary record reflects the fallout history from atmospheric weapons testing. In many sediment cores the ^{137}Cs activity versus depth profile has a well defined peak which can be attributed to the 1963 fallout maximum. In some cases (Miller & Heit, 1987) a secondary peak corresponding to 1959 may be observed. The application of this technique to Ponsonby Tarn is however vitiated by discharges of ^{137}Cs to the atmosphere from the nearby Sellafield Works.

Study area

Ponsonby Tarn (National Grid Reference NY046045) lies on the border of the Lake District National Park ~1.5 km east of Sellafield and ~3 km inland from the Irish Sea (Fig. 1). The tarn is an artificial lake constructed towards the end of the nineteenth century by the widening of Newmill Beck which rises on Ponsonby Fell ~15 km to the east.

The area is underlain by Permian and Triassic

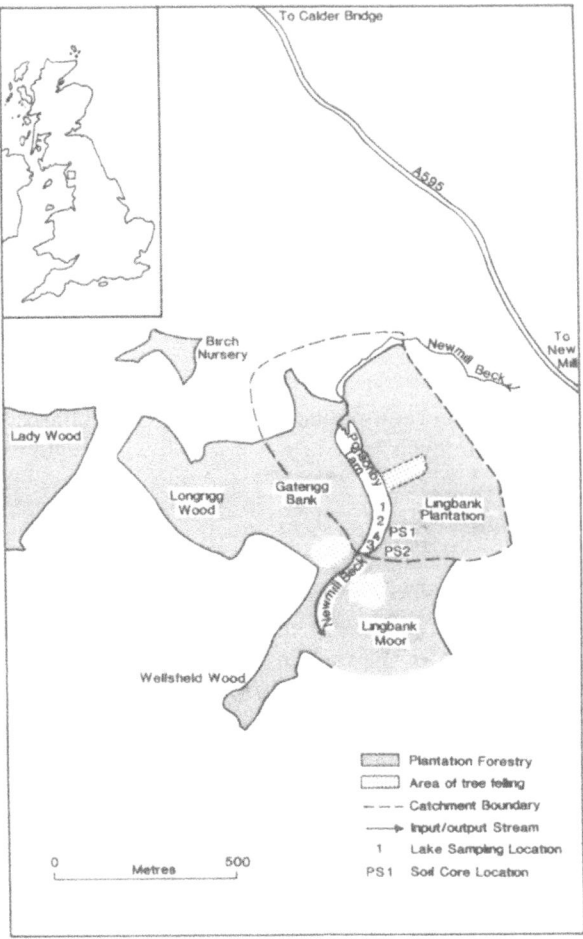

Fig. 1. Location of study area and sampling sites.

(New Red) sandstone. The tarn is underlain by Quaternary alluvium. On the higher part of the catchment there are peat deposits, and some Pleistocene fluvio-glacial sands and gravels are exposed in the northernmost area of the catchment. (Institute of Geological Sciences, 1978). The tarn is surrounded by managed woodland, conifers and rhododendrons with slopes adjoining the tarn being clear felled between 1981 and 1986 (E. Y. Haworth, 1988 pers. comm.).

Catchment characteristics are given below:

Catchment and Tarn Characteristics

Tarn altitude	~50 m
Max relief	~85 m

Lake area	~ 0.02 km^2
Catchment area	~ 0.30 km^2
Length	~ 0.5 km
Max width	~ 0.05 km
Max depth	~ 2 m
Catchment area to lake ratio	~ 15

Methods

Sampling

The first sediment core (PT1) was obtained in 1980 at location 1 in Fig. 1 using a Gilson corer (Macan, 1970). Three further cores (PT862, PT863, PT861) were obtained in October 1986 from locations 2, 3 and 4 using a short pneumatically driven Mackereth corer (Mackereth, 1969) of diameter 6.3 cm. Sediments from PT862 and PT863 were extruded using a hydraulically driven piston (Mackereth, op. cit.) and sectioned at intervals of one to two centimeters. Core PT861 was analysed as a bulk sample. Soil cores to a depth of 15 cm were taken from two separate locations (PS1 and PS2), near the tarn.

Analysis

^{137}Cs (half-life 30.0 years) and ^{134}Cs (half-life 2.06 years) were determined in samples of dried and ground sediment by gamma ray spectrometry. The techniques of analysis used are described by Salmon et al. (1983) in greater detail. Plutonium analysis was carried out upon dried, ground and ashed samples. The determination was performed on aliquots of the ash by acid leaching and ion exchange (Lally & Eakins, 1978) using ^{242}Pu as an internal yield tracer. Due to the normally indistinguishable energies of ^{239}Pu and ^{240}Pu (5.15 meV and 5.16 meV respectively) the activities of the two nuclides are expressed as the sum of their total. ^{241}Am was determined sequentially with $^{239+240}$Pu by further radiochemistry and alpha spectrometry using ^{243}Am as an internal yield tracer.

^{210}Pb was determined by dry distillation of its

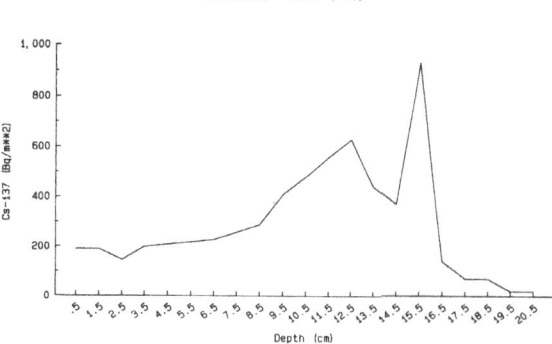

Fig. 2. ^{137}Cs vs depth (Core PT1) in 1980.

^{210}Po daughter using the methods laid down by Eakins & Morrison (1978). ^{226}Ra in sediment was determined by a method similar to that described by Pennington et al., (1976) using ^{133}Ba as a yield tracer.

Percentage weight loss on ignition was calculated by heating pre-weighed sub-samples in a muffle furnace for two hours at 550 °C.

Results

Actinide and radiocaesium activities in cores PT1 and PT863 are shown in Figs. 2 to 5. Radiocaesium activities in core PT862 are shown in Fig. 3. The short-lived isotope ^{134}Cs derives almost entirely from Chernobyl fallout. Using the ^{134}Cs/^{137}Cs ratio in Chernobyl fallout of 0.6

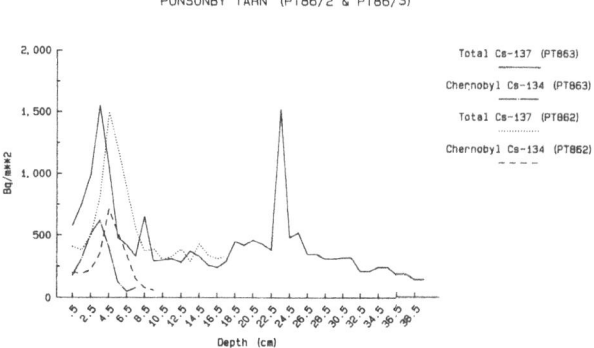

Fig. 3. ^{137}Cs and ^{134}Cs vs depth (Cores PT862 and PT863) in 1986.

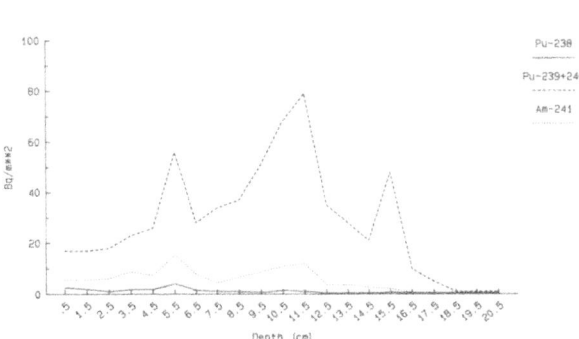

Fig. 4. ^{238}Pu, $^{239+240}$Pu and ^{241}Am vs depth (Core PT1) in 1980.

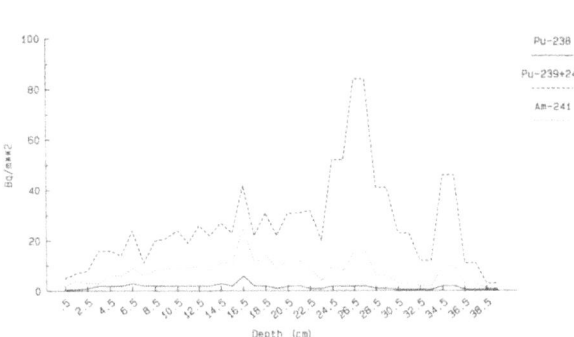

Fig. 5. ^{238}Pu, $^{239+240}$Pu and ^{241}Am vs depth (Core PT863) in 1986.

(Cambray *et al.*, 1987), ^{134}Cs activities have been used to partition the total ^{137}Cs activity into a Chernobyl derived component and a pre-Chernobyl component.

The ^{210}Pb results were inconclusive. In core PT1 ^{210}Pb equilibrium is apparently reached at a depth of ca. 20 cm. Since the artificial radionuclide profiles indicate that sediments above 16 cm cannot be dated earlier than ca. 1950, it would appear that there has been a rapid acceleration in accumulation rates over the past 30–35 years. In core PT863 the base of the core did not reach ^{210}Pb equilibrium, and the artificial radionuclides suggest that the entire core is post 1950. Dating the base of the actinide profiles to 1950, mean accumulation rates for the past 30–35 years are calculated to be:

Core PT1

0.60 cm y^{-1} or 0.082 g cm^{-2} y^{-1}

Core PT863

1.11 cm y^{-1} or 0.20 g cm^{-2} y^{-1}

Assuming that accumulation rates have been reasonably constant during this period, the elevated ^{137}C and $^{239+240}$Pu activity in Core PT1 at 15.5 cm is dated ca. 1955, and later peaks at 10.5–11.5 cm are dated 1960–65. The decline in activity above 8.5 cm is dated 1966–70. In core PT863, the general decrease in activity above 18.5–23.5 cm is dated at 1966–71. The existence of a peak in Chernobyl derived ^{137}Cs in PT863 at a depth of 3.5 cm below the mud-water interface confirms the very high accumulation rate in this core.

Table 1. Radionuclide inventories in the sediment cores.

Core	← ^{137}Cs (Bq m^{-2}) →			^{134}Cs (Bq m^{-2})	$^{239+240}$Pu (Bq m^{-2})
	Total	Pre-Chernobyl	Chernobyl	Chernobyl	
PT1	6100	6100			600
PT861	21000	14500	6500	4600	
PT862	>9700	>5000	4700	2800	
PT863	16800	13000	3800	2300	1060
Estimated deposition					
1980	7100	7100			890
1986	~15800	6130	9700	5800	920

Radiocaesium inventories in all four sediment cores are given in Table 1. Values for the bulked core are in good agreement with those for PT863.

Discussion

In order to evaluate temporal changes in the deposition of radionuclides recorded within lake sediments the assessment of the expected contribution from differing sources is essential. The establishment of such baseline values for various radionuclides allows some assessment of the environmental significance of events such as the Chernobyl accident.

Caesium-137 and Caesium-134 input values

Sediment from Ponsonby Tarn could contain ^{137}Cs from the following sources:-
(1) Fallout from the atmospheric testing of nuclear weapons,
(2) Pre-1957 discharges of particulate material from Sellafield (NRPB, 1984),
(3) The accidental release of radioactivity to the atmosphere from Windscale (Sellafield) in 1957 (Command 302, 1957),
(4) Atmospheric discharges from Sellafield since 1957,
(5) Maritime discharges via sea-spray,
(6) Fallout arising from the Chernobyl accident.

In the United Kingdom the accumulated deposit, from weapons testing, of ^{137}Cs to a depth of 30 cm in soil is estimated to be 3700 Bq m^{-2} (in 1980) and 3200 Bq m^{-2} (in 1986) per 1000 mm of annual rainfall (Cambray et al., 1987). Of this 75% is retained in the upper 15 cm of the soil (Cawse, 1980). Whilst no rain gauges are situated on the catchment, annual rainfall contour maps for Cumbria based upon the period 1941–70 (Cawse, 1980) are available. These, supplemented by average annual rainfall data for three nearby sites, (Sellafield, Braystones and Wastwater [1.5 km, 5 km and 9 km distant respectively]) between 1970 and 1984 (Meteorological Office, 1972–1986) suggest a mean annual rainfall for the catchment of ~1350 mm/year. This would give an expected deposit of weapons fallout, integrated to a depth of 30 cm of 5000 Bq m^{-2} and 4320 Bq m^{-2} in 1980 and 1986 respectively.

From the data quoted in Booker (1962) and Cambray et al. (1987), the total expected deposition of ^{137}Cs at Ponsonby Tarn up to 1986 (corrected for decay) is as follows:

i) Sellafield discharges prior to 1957	=	1530 Bq m^{-2}
ii) Windscale fire of 1957	=	285 Bq m^{-2}
iii) Fallout derived from atmospheric weapons testing	=	4320 Bq m^{-2}
iv) Chernobyl-derived ^{137}Cs	= ~	9665 Bq m^{-2}
Total		~15800 Bq m^{-2}

This figure compares reasonably with the mean value derived from the two soil cores taken upon the catchment (17400 Bq m^{-2}) and the values obtained under nearly deciduous woodland (NY036046) by Hursthouse (pers. comm., 1989). Significant variations between soil core inventories may be expected due to differences in interception by vegetation, microtopographic features etc.

A total of ~16000 Bq m^{-2} of ^{137}Cs may thus be expected to have been deposited directly onto the surface of Ponsonby Tarn by the end of 1986. At the time of the previous study in 1980, the total deposition of ^{137}Cs at Ponsonby Tarn was ~7100 Bq m^{-2}.

Plutonium input values

Estimates of the accumulated deposit of actinides at Ponsonby Tarn suggest that discharges from Sellafield are the principal source. Estimates of $^{239+240}$Pu deposition between 1952 and 1983 are given in Howorth and Eggleton (1988) and are available until 1988 (J.M. Howorth, 1989, pers. comm.), and suggest a total deposit from Sellafield discharges of 876 Bq m^{-2} until the start of 1979 and 918 Bq m^{-2} until the beginning of 1986. Approximately 50% of the material was deposited prior to 1960. The increase in the inventory between 1980 and 1986 is largely due to marine sources, stack discharges and discharges to sea from the Magnox ponds (Howorth & Eggleton, 1988).

Cawse (1980) estimated weapons fallout $^{239+240}$Pu in UK soils (to 30 cm depth) to be 67 Bq m^{-2} per 1000 mm mean annual rainfall, giving a cumulative deposit of 91.5 Bq m^{-2} at Ponsonby Tarn. This is in reasonable agreement with that quoted by Wade *et al.* (1985) at Seascale of 81 Bq m^{-2} of weapons fallout $^{239+240}$Pu and Howorth & Eggleton's (1988) estimate of 83 Bq m^{-2}.

The deposition of Chernobyl-derived plutonium in the UK was negligible, even in areas of greatest deposit from Chernobyl the addition of $^{239+240}$Pu did not exceed 0.5 Bq m^{-2} (Cambray *et al.*, 1987). Previous studies (Bonnett, 1990) suggest that any contribution to the lake sediment record of plutonium labelled material from erosion of the surrounding catchment will be low.

Radionuclide profiles

Despite the problems associated with calculating an accurate ^{210}Pb chronology for PT863, a number of features present in PT1 may be used to cross validate and date the sediment record of PT863. These features are summarised below:-

	PT1	PT863	~Date
Base of actinides	18 cm	40 cm	early 1950's
Fall of ^{210}Pb below			
75 Bq kg^{-1}	18.5 cm	35 cm	early 1950's
^{137}Cs peak	15.5 cm	23.5 cm	~1954
$^{239+240}$Pu peak	11.5 cm	16.5 cm	~1961
1980 surface	0 cm	~5 cm	1980

Eakins & Cambray (1985) noted that the elevated ^{137}Cs deposit present at 15.5 cm in PT1 was due to a single particle ascribed to particulate material containing ^{137}Cs discharged to the atmosphere from the Sellafield Works prior to 1957 (NRPB, 1984). Subdivision of the 23–24 cm sample in PT863 revealed a similar single particle which accounted for the elevated ^{137}Cs level. This section also contained low ^{241}Am levels which

Eakins & Cambray (1985) attributed to pre-1957 particulate material with a very low burn-up in core PT1. It is therefore likely that the peak in PT863 at 23.5 cm is due to similar material of a similar age.

The differences between the depths at which the base of the actinides occurs and ^{210}Pb falls below 75 Bq kg^{-1} suggests that rapid accumulation of sediment may have occurred at PT863 location in the early 1950's possibly due to sediment slumping or focussing at this site during this period.

Radionuclide inventories

A comparison of the radionuclide inventories (Table 1) shows that there are considerable differences between the 1980 core (PT1) and the 1986 cores.

In core PT1 the total ^{137}Cs inventory is in close agreement with that expected from weapons fallout and Sellafield discharges. In core PT863 the inventory from these sources is twice the expected value. Conversely, in core PT862 it is only ~80% of the expected value. In contrast, Chernobyl derived ^{137}Cs and ^{134}Cs inventories are markedly below expected levels.

These anomalies may be due to a number of factors:
1) Marked changes in the atmospheric flux of radionuclides to the catchment.
2) Catchment inputs to the tarn.
3) Hydraulic flushing.
4) Sediment focusing.

The variation cannot be attributed to changes in atmospheric deposition. During the period 1980–1986 fallout of ^{137}Cs was negligible, and the cumulative deposit of $^{239+240}$Pu increased by only ~2% (Cambray *et al.*, 1990; Howorth & Eggleton, 1988). Natural radioactive decay will have reduced the ^{137}Cs inventory by ~15%.

Catchment inputs of radionuclides in particle-associated form possibly resulting from tree felling carried out around the tarn could account for the elevated weapons test ^{137}Cs inventories in the

1986 sediment record. However, diatom analysis of cores PT863 and PT862 does not display the marked fluctuations in diatom concentrations commonly associated with the episodic inwash of catchment material resulting from catchment disturbance (E.Y. Haworth, pers. comm. 1988: A. Kreiser, pers. comm. 1988). Similarly the stable loss on ignition values indicate a homogeneity in the physical characteristics of the sediment in PT862 and this also suggests the lack of inwash of either organic or minerogenic material from the catchment.

Whilst the mean residence time of water within Ponsonby Tarn is not known it is conceivable, given the size and depth of the tarn, that it may be quite short. Hydraulic flushing during high flow conditions would reduce this further and may account for the lower than expected ^{137}Cs inventories if this process occurred between initial deposition of Chernobyl radionuclides (May, 1986) and sampling (October, 1986). Hilton *et al.* (1989) noted that this process resulted in significant losses of Chernobyl-derived ^{137}Cs and ^{134}Cs from Windermere and Esthwaite Water via their lake outflows. A similar initial 'pulse' of ^{134}Cs occurred in Haweswater and Llyn Cwmystradllyn in the months following the Chernobyl accident (Bonnett, 1990). Some of the Chernobyl derived radiocaesium deposited onto the surface of the tarn may therefore have been transported out of the system before scavenging from the water column by settling particles occurred.

Differences in the weapons fallout and Sellafield derived ^{137}Cs inventories apparent between PT1, PT861 and PT863 may reflect the complex sedimentation patterns noted in other small lakes (e.g. Davis, 1976; Dearing *et al.*, 1981; Dearing, 1983). Sediment focusing (Lehman, 1975) could account for the greater than expected inventories present in these cores with differences between the cores reflecting differences in the intensity of sediment focusing at each sampling site.

Conclusions

Examination of the sediment record points to considerable changes having occurred between the 1980 and 1986 studies at Ponsonby Tarn. Patterns of caesium and actinide deposition suggest a complex pattern of sediment accumulation over the tarn basin. Quantities of weapons fallout and Sellafield derived radionuclides are present in greater quantities in the 1986 cores, compared with the core taken in 1980. However, the 1986 cores show no evidence of any input of Chernobyl labelled material from the catchment. Hydraulic flushing together with variable sediment focusing of sediments within the tarn are thought to account for some of these features.

Acknowledgements

The authors thank the following: Mr P. Stanley for allowing access to the tarn; Dr E. Y. Haworth, Mr P. V. Allen (Institute of Freshwater Ecology) and Mr K. Playford for their assistance in sampling the tarn. Dr E. Y. Haworth for water chemistry data and diatom analysis (PT863); Dr A. Kreiser (Palaeoecology Research Unit, University College, London) for diatom analysis (PT862). Messrs G. N. J. Lewis, P. J. Burton & R. T. Morrison for gamma and actinide analysis. Dr P. G. Appleby (University of Liverpool) for his invaluable comments on an earlier draft of this paper. This work was financed by BNFL (1980 study) and the Department of Health (1986 study) and forms part of the UKAEA Radiological Protection Research Core Programme.

References

Appleby, P. G. & F. Oldfield, 1978. The calculation of lead 210 dates assuming a constant rate of supply of unsupported ^{210}Pb to the sediment. Catena 5: 1–8.

Appleby, P. G. & F. Oldfield, 1983. The assessment of ^{210}Pb dates from sites with varying sediment accumulation rates. Hydrobiologia 103: 29–35.

Bonnett, P. J. P., 1990. A review of the erosional behaviour of radionuclides in selected drainage basins. J. Envir. Radioactivity 11: 251–266.

Booker, D. V., 1962. Caesium 137 in soil in the Windscale area. AERE-R4020, HMSO.

Cambray, R. S., P. A. Cawse, J. A. Garland, J. A. B. Gibson, P. Johnson, G. N. J. Lewis, D. Newton, L. Salmon & B. O. Wade, 1987. Observations on radioactivity from the Chernobyl accident. Nuclear Energy, 26: 77–101.

Cambray, R. S., K. Playford, G. N. J. Lewis & R. C. Carpenter, 1990. Radioactive fallout in air and rain: results to the end of 1988. AERE R13575, HMSO.

Cawse, P. A., 1980. Studies of environmental radioactivity in Cumbria Part 4: Caesium 137 and plutonium in soils of Cumbria and the Isle of Man. AERE R9851, HMSO.

Cawse, P. A., 1983. The accumulation of Caesium 137 and Pu 239 + 240 in soils of Great Britain and transfer to vegetation, In Coughtrey P. J. (ed) Ecological aspects of radionuclide release, special publication No. 3 of British Ecological Society.

Davis, M. B., 1976. Erosion rates and land-use history in Southern Michigan. Envir. Conserv. 3: 139–148.

Dearing, J. A., J. K. Elner & C. M. Happey-Wood, 1981. Recent sediment flux and erosional processes in a Welsh upland lake catchment based on magnetic susceptibility measurements. Quat. Res. 16: 356–372.

Dearing, J. A., 1983. Changing patterns of sediment accumulation in a small lake in Scania, Southern Sweden. Hydrobiologia 103: 59–64.

Eakins, J. D. & R. T. Morrison, 1978. A new procedure for the determination of lead 210 in lake and marine sediments. Int. J. Appl. Radiation, Isotopes 29: 531–536.

Eakins, J. D., 1983. The ^{210}Pb techniques for dating sediments and some applications. AERE R10821, HMSO.

Eakins, J. D., R. S. Cambray, K. C. Chambers & A. E. Lally, 1984. The transfer of natural and artificial radionuclides to Brotherswater from its catchment. In Haworth, E. Y. & Lund, J. W. G. (eds) Lake sediments and environmental history. Leicester University Press: pp. 125–144.

Eakins, J. D. & R. S. Cambray, 1985. Studies of environmental radioactivity in Cumbria Part 6: The chronology of discharges of caesium 137, plutonium and americium 241 from BNFL Sellafield, as recorded in lake sediments. AERE R11182, HMSO.

HMSO (1957). Accident at Windscale No. 1 Pile on 10 October, 1957. HMSO, London.

Hilton, J., W. Davison, F. Livens, M. Kelly & J. Hamilton-Taylor, 1989. Transport mechanisms and rates for the long-lived Chernobyl deposits. DoE PECD 7/9/385.

Howorth, J. M. & A. E. J. Eggleton, 1988. Studies of environmental radioactivity in Cumbria Part 12: Modelling of the sea-to-land transfer of radionuclides and an assessment of the radiological consequences. AERE R11733, HMSO.

Institute of Geological Sciences, 1978. British Regional Geology, Northern England. HMSO, London.

Lally, A. E. & J. D. Eakins, 1978. Some recent advances in environmental analysis at AERE Harwell. Symposium on the determination of radionuclides in environmental and biological materials. CEGB, Sudbury House, London. 9–10 October 1978.

Lehman, J. T., 1975. Reconstructing the rate of accumulation of lake sediment: the effect of sediment focusing. Quat. Res. 5: 541–550.

Macan, T. T., 1970. Biological studies of the English lakes. Longman, London.

Mackereth, F. J. H., 1969. A short core sampler for sub-aqueous deposits. Limnol. Oceanogr. 14: 145–151.

Meteorological Office, 1986. Monthly and annual totals of rainfall in 1984 for the United Kingdom. Meteorological Office, Bracknell.

References for Met Office 1972–1985 are similar in nature to that for 1986 and contain results for the year two years prior to the publication date.

Miller, K. M. & M. Heit, 1986. A time resolution methodology for assessing the quality of lake sediment cores that are dated by ^{137}Cs. Limnol. Oceanogr. 31: 1292–1300.

National Radiological Protection Board, 1984. The risks of leukaemia and other cancers in Seascale from radiation exposure. R171, HMSO.

Oldfield, F., 1977. Lakes and their drainage basins as units of sediment based ecological study. Progr. Physic. Geogr. 1: 460–504.

Oldfield, F. & P. G. Appleby, 1984. Empirical testing of ^{210}Pb dating models for lake sediments. In Haworth, E. Y. & J. W. G. Lund (eds). Lake sediments and environmental history. Leicester University Press. pp. 93–124.

O'Sullivan, P. E., 1983. Annually laminated lake sediments and the study of Quaternary environmental changes – a review. Quat. Sci. Rev. 1: 245–313.

Pattenden, N. J., R. S. Cambray, K. Playford, J. D. Eakins & E. M. R. Fisher, 1980. Studies of environmental radioactivity in Cumbria Part 3 – Measurements of radionuclides in airborne and deposited material. AERE R9857, HMSO.

Peirson, D. H., R. S. Cambray, P. A. Cawse, J. D. Eakins & N. J. Pattenden, 1982. Environmental radioactivity in Cumbria. Nature 300: 27–31.

Pennington, W., R. S. Cambray & E. M. Fisher, 1973. Observations on lake sediments using fallout ^{137}Cs as a tracer. Nature 242: 324–326.

Pennington, W., R. S. Cambray, J. D. Eakins & D. D. Harkness, 1976. Radionuclide dating of the recent sediments of Blelham Tarn. Freshwat. Biol. 6: 317–331.

Robbins, J. A., 1978. Geochemical and geophysical applications of radioactive lead, In Nriagu, J. O. (ed.), Biogeochemistry of Lead in the Environment. Elsevier, Holland, pp. 285–393.

Salmon, L., M. M. Davies & G. N. J. Lewis, 1983. The automatic analysis of gamma-ray spectra from the natural and working environments: IVth Symposium on the Determination of Radionuclides in Environmental and Biological Materials. Laboratory of the Government Chemist, Teddington, Middlesex.

Wade, B. O., A. E. J. Eggleton & D. H. Peirson, 1985. Studies of environmental radioactivity in Cumbria Part 7: Summary of progress to December 1984. AERE R11743, HMSO.

Hydrobiologia **214**: 71–76, 1991.
J. P. Smith, P. G. Appleby, R. W. Battarbee, J. A. Dearing, R. Flower, E. Y. Haworth, F. Oldfield & P. E. O'Sullivan (eds),
Environmental History and Palaeolimnology.
© 1991 *Kluwer Academic Publishers.*

Deposition and transport of radionuclides within an upland drainage basin in mid-Wales

P. J. P. Bonnett [1,3] & P. G. Appleby [2]
Departments of [1] Geography; [2] Applied Maths and Theoretical Physics; University of Liverpool, P.O. Box 147, Liverpool, L69 3BX, UK; [3] present address: Modelling and Assessment Group, Environmental & Medical Sciences Division, Harwell Laboratory, United Kingdom Atomic Energy Authority, Oxon OX11 0RA, UK (address for correspondence)

Abstract

The deposition of radiocaesium from nuclear weapons testing and the Chernobyl accident upon the Llyn Llygad Rheidol catchment in mid-Wales is described. Inventories of soil cores from the catchment support estimates of total atmospheric fallout. The mean inventory of weapons testing ^{137}Cs in lake sediment cores is broadly similar to that in soil cores. The inventory of Chernobyl fallout in sediment cores is significantly lower and raises questions concerning the residence time of ^{137}Cs in the catchment soils and lake waters. ^{137}Cs and ^{241}Am activities in a sediment core record the 1963 peak of fallout from nuclear weapons testing. The association of the peak activities of ^{137}Cs and ^{241}Am in the sediments with the fallout maximum is confirmed by ^{210}Pb dating. The ^{210}Pb dates also reveal a significant increase in sediment accumulation rates over the past 20 years.

Introduction

The lake-watershed ecosystem is subject to a flux of both natural and artificial radionuclides from the atmosphere. These radionuclides constitute ready made tracers for studying transport processes within the catchment. The principal natural fallout isotope is ^{210}Pb, a decay product of radioelements in the earth's crust. Fallout of the artificial radionuclide ^{137}Cs has arisen from atmospheric testing of nuclear weapons, and from the nuclear reactor accident at Chernobyl.

Peirson & Salmon (1959) have shown the deposition of weapons test ^{137}Cs to be a linear function of rainfall. Areas experiencing greatest annual rainfall in the United Kingdom include the uplands regions (Chandler & Gregory, 1976). The same mountainous areas also experienced the highest deposition of Chernobyl derived radionuclides (Clark & Smith, 1988; Smith & Clark,

1989). The upper Rheidol was one of a number of upland drainage basins in mid-Wales selected to examine the spatial distribution and rates of removal of fallout from these systems.

The lake sediment record forms an important component of any study of fallout deposition. Lake sediments incorporate direct fallout onto the lake, together with a proportion of fallout onto the catchment transferred to the lake. The sediments also record a range of ecological and physiographic changes in the catchment, and these may be dated using ^{210}Pb or ^{137}Cs.

The study area

The River Rheidol rises on the north-west slopes of Plynlimon (at ~600 m) and flows 47 km before discharging to Cardigan Bay at Aberystwyth. The upper drainage basin containing Llyn Llygad

72

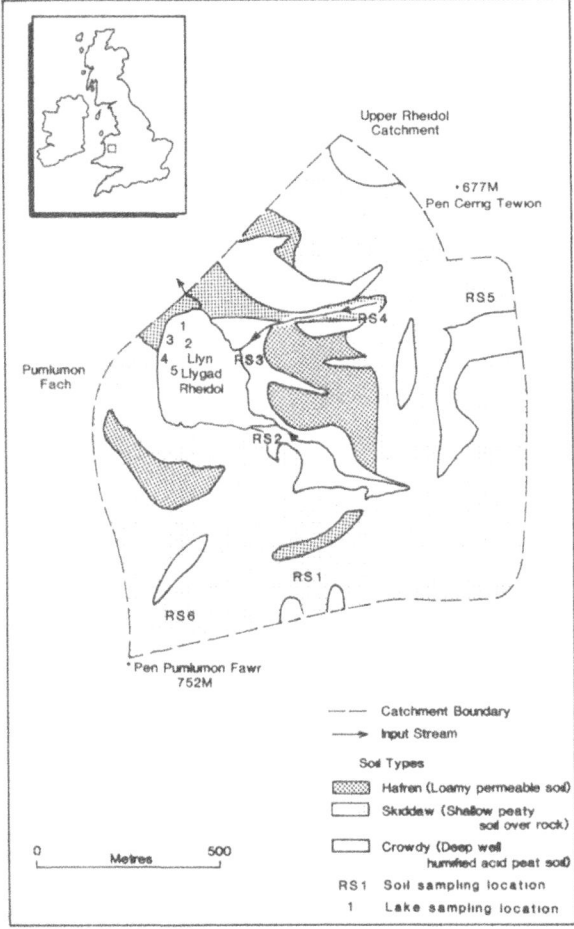

Fig. 1. Upper Rheidol catchment soil types and sampling locations.

Rheidol consists of Ordovician and Silurian mudstones and shales with superficial deposits of boulder clay. Soils (Fig. 1) range from loamy permeable soils of the Hafren series close to the lake to shallow and deep acid peats in the upper reaches of the basin. Average annual rainfall at the lake outflow is ~2000 mm (Welsh Water Authority, pers. comm. 1987) although some var-

Table 1. Catchment and Tarn Characteristics.

Altitudinal Range	510–750 m
Lake Area	~0.13 km²
Cathment Area	~1.02 km²
Catchment area to lake ratio	~8

iation in rainfall across the catchment is to be expected. Other catchment features are detailed in Table 1 (below).

Methodology

Surface soil samples measuring $25 \times 25 \times 5$ cm were taken in May 1988 from six locations on the catchment (Fig. 1). Sediment cores were obtained in March 1988 from five locations (Fig. 1) using a short pneumatically driven Mackereth corer (Mackereth, 1969).

Analysis

^{137}Cs and ^{134}Cs in dried soil samples were determined by gamma-ray spectrometry at Harwell. The techniques of analysis are described by Salmon *et al.* (1983) in greater detail.

Lake sediment samples were analysed at the University of Liverpool for ^{210}Pb, ^{226}Ra, ^{137}Cs, ^{134}Cs and ^{241}Am by gamma-ray spectrometry using a well-type coaxial low background intrinsic germanium detector fitted with a NaI(T1) escape suppression shield (Appleby *et al.*, 1986).

Baseline values

In the United Kingdom the accumulated deposit from weapons testing of ^{137}Cs to a depth of 30 cm in soil is 3300 Bq m^{-2} (in 1987) per 1000 mm of rain (Cawse & Horrill, 1986; Cambray *et al.*, 1989). Associated work in the region (Bonnett *et al.*, 1989) together with rainfall data for the catchment in May 1986 (Welsh Water Authority, pers. comm. 1987) suggests that ~620 Bq m^{-2} of ^{137}Cs and 370 Bq m^{-2} of ^{134}Cs could be expected to have been deposited on the catchment as a result of Chernobyl accident. This is in good agreement with other estimates of Chernobyl fallout in the area (Smith, pers. comm., 1988; Clark & Smith, 1988). With an annual rainfall of 2000 mm the total deposit of weapons test fallout ^{137}Cs to 30 cm depth on the catchment would be

6600 Bq m^{-2}. Studies in the adjoining Wye and Severn basins ~ 5 km to the east of Llygad Rheidol suggest that all the Chernobyl-derived caesium and $\sim 90\%$ of weapons fallout is contained in the top 5 cm of the soil (Bonnett *et al.*, 1989) giving a deposit of 5940 Bq m^{-2} derived from weapons fallout and 620 from Chernobyl (total = 6560 Bq m^{-2}) in the top 5 cm of the soil on the catchment.

Results

Inventories of ^{137}Cs and ^{134}Cs in the soil samples are shown in Table 2(a). ^{134}Cs is a short-lived isotope characteristic of Chernobyl fallout. Using the ^{134}Cs it is possible to partition the total ^{137}Cs

into a component derived from weapons testing fallout and a component derived from Chernobyl fallout. The corresponding inventories in the lake sediment cores are shown in Table 2(b). Fig. 2 shows the variation of ^{137}Cs and ^{134}Cs activity with depth in core 2 together with ^{241}Am, another fallout radionuclide deriving from atmospheric nuclear weapons testing. The ^{137}Cs and ^{241}Am activities both have a well defined peak at 5–6 cm and this depth is assumed to represent 1963, the year of maximum fallout from nuclear weapons testing.

The ^{210}Pb age v depth curve for core 2 calculated using the CRS ^{210}Pb dating model is shown in Fig. 3. The model is based upon an assumed constant rate of supply (CRS) of ^{210}Pb from the lake waters to the sediment irrespective

Table 2(a). ^{137}Cs and ^{134}Cs in soils from the Upper Rheidol catchment (Bq m^{-2}).

Location	Total ^{137}Cs[2]	Fallout ^{137}Cs[2]	Chernobyl-derived	
			^{137}Cs [1,2]	^{134}Cs [2]
RS1	4780 ± 2%	4350 ± 2%	430 ± 20%	260 ± 20%
RS2	2980 ± 2%	2495 ± 2%	485 ± 18%	290 ± 18%
RS3	4730 ± 1%	4430 ± 1%	300 ± 30%	180 ± 30%
RS4	2980 ± 2%	2460 ± 2%	520 ± 9%	310 ± 9%
RS5	2650 ± 2%	2330 ± 2%	320 ± 16%	190 ± 16%
RS6	4200 ± 2%	3660 ± 2%	540 ± 12%	325 ± 12%
Mean values	3720	3290	430	260

[1] Derived assuming a ratio of 0.6 for ^{134}Cs/^{137}Cs as measured in Chernobyl deposit immediately after the accident.
[2] Corrected for decay to 3 May 1986.
Sampling date: 24 May 1988.

Table 2(b). ^{137}Cs and ^{134}Cs inventories in cores from Llyn Llygad Rheidol (Bq m^{-2})

Location	Total ^{137}Cs[2]	Fallout ^{137}Cs[2]	Chernobyl-derived	
			^{137}Cs [1,2]	^{134}Cs [2]
2	3260 + 2%	3158	105 ± 12%	65 ± 12%
3	1335 ± 4%	>1185	<150	<90
5	4915 + 2%	>4715	<200	<120
Mean values	3170	>3020	<150	<90

[1] Derived assuming a ratio of 0.6 for ^{134}Cs/^{137}Cs as measured in Chernobyl deposit immediately after the accident.
[2] Corrected for decay to 3 May 1986.
Coring date: 8–9 March 1988.

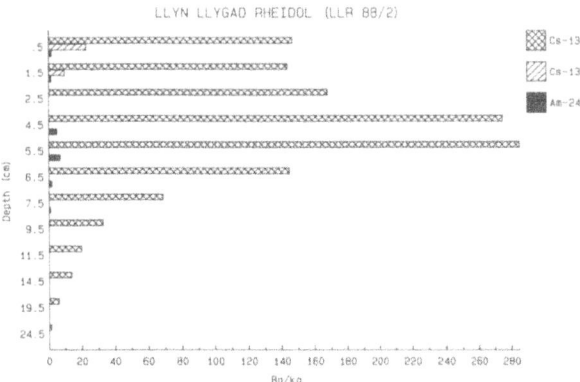

Fig. 2. Cs-137, Cs-134 and Am-241 vs depth (LLR 88/2).

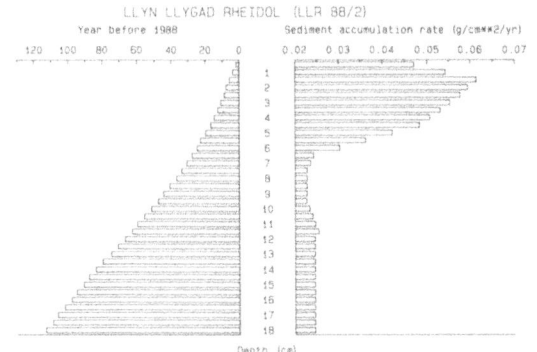

Fig. 3. Pb-210 chronology and sediment accumulation rate vs time (LLR 88/2).

of changes in the net dry mass sedimentation rate (Oldfield and Appleby, 1984). Rates of sediment accumulation are also shown in Fig. 3.

Discussion

Comparison of the radionuclide inventories in the soils of the catchment reveals marked variations. ^{137}Cs inventories range from 2650 Bq m^{-2} to 4780 Bq m^{-2}, whilst ^{134}Cs inventories range from 180 Bq m^{-2} to 330 Bq m^{-2}. The mean ^{137}Cs inventory is 3720 Bq m^{-2}, of which 3290 Bq m^{-2} is attributable to nuclear weapons testing. This is only about 63% of notional value based on the mean rainfall on the catchment and is discussed below. The 430 Bq m^{-2} of ^{137}Cs attributed to

Chernobyl fallout represents about 70% of the estimated atmospheric flux.

^{137}Cs inventories in the lake sediment cores reveal similar marked variations. The mean value from weapons fallout of ca. 3080 Bq m^{-2} is, however, remarkably similar to that in soils. In contrast, the mean inventory from Chernobyl fallout of < 150 Bq m^{-2} is only about a third of that in soils.

The ^{137}Cs inventories in the soil sampled from locations RS1 and RS6 appear to confirm that the total fallout of weapons testing ^{137}Cs on the catchment is ca. 5000 Bq m^{-2}. Both of these cores are from relatively flat sites near the boundary of the catchment and are unlikely to receive ^{137}Cs other than by direct fallout. The reduced inventory at RS5, also near the boundary, may be explained by the nature of the soil in this part of the catchment. The enhanced mobility of ^{137}Cs in acid peats is well documented (e.g. Oldfield *et al.*, 1979; Schell & Massey, 1987: Sheppard *et al.*, 1989). The reduced inventories at the mid slope sites RS2 and RS4, and enhanced inventory at the base site RS3, may well be due to transport within the catchment. The similarity of the mean weapons ^{137}Cs inventories in the lake and catchment would, however, appear to preclude significant transfers of that material from the catchment to the lake. The Chernobyl ^{137}Cs inventories of the soil sampling sites confirm a total atmospheric flux of ca. 500 Bq m^{-2}. The reduced inventories of Chernobyl ^{137}Cs in the lake sediments, as least in the short term, confirm to a pattern observed at other sites.

The unsupported ^{210}Pb inventory of core 2 was calculated to be 6070 Bq m^{-2}, representing a constant ^{210}Pb flux to the sediments of 189 Bq m^{-2} yr^{-1}. This is well within the range of values expected if the ^{210}Pb supply is dominated by atmospheric fallout, and is one of the conditions for using the CRS ^{210}Pb dating model (Appleby & Oldfield, 1978). The ^{210}Pb chronology dates the peak ^{137}Cs and ^{241}Am activities at 5–6 cm (Fig. 2) to the period 1963–68 and confirms the association of these features with the 1963 maximum in atmospheric fallout from nuclear weapons testing. The ^{210}Pb also reveals an abrupt acceleration in sediment accumulation rates since the late 1960's, but that prior to that there was a more or less constant rate of accumulation of sediment of ca. 0.025 g cm^{-2} yr^{-2}, or ca. 0.14 cm yr^{-1}.

A comparison of the ^{137}Cs and ^{241}Am profiles for core 2 confirms the relatively greater mobility of the former. Significant ^{137}Cs activities are recorded in sediments which clearly predate the onset of large-scale nuclear weapons testing in 1954. The mobility of caesium is further indicated by the penetrations of Chernobyl derived ^{137}Cs down to 2 cm (dated 1980) in core 2 and 3 cm in core 4. This has also been observed in Cumbrian lakes, e.g. Ponsonby Tarn (Bonnett & Cambray, 1989) and Esthwaite Water and Windermere (Hilton et al., 1989).

Conclusions

Soil radiocaesium inventories broadly confirm estimates of the total fallout from both nuclear weapons testing and the Chernobyl accident. Variations in the inventories suggest post-depositional transport, either through erosion processes or through groundwater flow. Inventories of weapons test ^{137}Cs in the lake sediment cores provide no evidence of significant transfers from the catchment to the lake, though the lower Chernobyl inventories raise questions as to the immediate fate of direct fallout into the lake. Hilton et al. (1989) noted that hydraulic flushing and consequent loss via the lake outflow is responsible for significant losses of ^{137}Cs and ^{134}Cs from Windermere and Esthwaite Water in Cumbria. The weapons fallout inventories raise the possibility that ^{137}Cs remains in residence in the lake waters. More detailed investigations over a period of time are required to determine the pathways by which ^{137}Cs is transported through the system.

The ^{210}Pb inventory of core 2 is broadly in agreement with estimates of the atmospheric flux. Very good agreement exists between dates determined by ^{210}Pb (CRS model) and artificial fallout radionuclides. The study provides further evidence of the applicability of ^{241}Am as a dating tool in certain lake environments. The presence of a clearly defined 1963 ^{137}Cs peak, in spite of the obvious mobility of this isotope, suggests that Chernobyl ^{137}Cs will eventually leave a well-defined 1986 marker at many sites.

Acknowledgements

Thanks are due to the Welsh Water Authority for granting access to the site and to Mr S. M. Hutchinson (University of Liverpool) and Mr R. J. O. Miller (Institute of Arctic and Alpine Research, University of Colorado) for their help with lake coring. Gamma-ray spectrometry at Harwell was carried out by Mr G. N. J. Lewis. Mr J. A. Garland, Mr R. S. Cambray (Harwell) and Professor F. Oldfield (University of Liverpool) provided valuable comments on an earlier draft of this paper. This work was funded by CEGB as part of the UKAEA Radiological Protection Research Core Programme.

References

Appleby, P. G. & F. Oldfield, 1978. The calculation of ^{210}Pb dates assuming a constant rate of supply of unsupported ^{210}Pb to the sediment. Catena 5: 1–8.

Appleby, P. G., P. Nolan, D. W. Gifford, M. J. Godfrey, F. Oldfield, N. J. Anderson & R. W. Battarbee, 1986. ^{210}Pb dating by low background gamma counting. Hydrobiologia 143: 21–27.

Bonnett, P. J. P. & R. S. Cambray, 1989. The chronology of deposition of radionuclides as recorded in the sediments of Ponsonby Tarn, Cumbria. Paper presented at the Vth International Symposium on Palaeolimnology, Ambleside, Cumbria, September 1989.

Bonnett, P. J. P., G. J. L. Leeks & R. S. Cambray, 1989. Transport processes for Chernobyl-labelled sediments: preliminary evidence from upland mid-Wales. Land Degradation and Rehabilitation 1: 39–50.

Cambray, R. S., P. A. Cawse, J. A. Garland, J. A. B. Gibson, P. Johnson, G. N. J. Lewis, D. Newton, L. Salmon & B. O. Wade, 1987. Observations on radioactivity from the Chernobyl accident. Nucl. Energy 26: 77–101.

Cambray, R. S., K. Playford, G. N. J. Lewis, R. C. Carpenter & J. A. B. Gibson, 1989. Radioactive fallout in air and rain: results to the end of 1987. AERE-R13226, HMSO, London.

Cawse, P. A. & A. D. Horrill, 1986. A survey of caesium-137 and plutonium in British soils in 1977. AERE-R10155, HMSO, London.

Chandler, T. J. & S. Gregory, 1976. The climate of the British Isles. Longmans, London.

Clark, M. J. & F. B. Smith, 1988. Wet and dry deposition of Chernobyl releases. Nature 332: 245–249.

Hilton, J., W. Davison, F. Livens, M. Kelly & J. Hamilton-Taylor, 1989. Transport mechanisms and rates for the long-lived Chernobyl deposits. DoE PECD 7/9/385.

Mackereth, F. J. H., 1969. A short core sampler for sub-aqueous deposits. Limnol. Oceanogr. 14: 145–151.

Oldfield, F. & P. G. Appleby, 1984. Empirical testing of ^{210}Pb dating models for lake sediments. In E. Y. Haworth & J. W. G. Lund (eds) Lake sediments and environmental history. Leicester University Press: 93–124.

Oldfield, F., P. G. Appleby, R. S. Cambray, J. D. Eakins, K. E. Barber, R. W. Battarbee, G. R. Pearson & J. M. Williams, 1979. ^{210}Pb, ^{137}Cs and ^{239}Pu profiles in ombrotrophic peat. Oikos 33: 40–45.

Peirson, D. H. & L. Salmon, 1959. Gamma radiation from deposited fallout. Nature 184: 1678–1679.

Salmon, L., M. M. Davies & G. N. J. Lewis, 1983. The automatic analysis of gamma-ray spectra from the natural and working environments: IVth Symposium on the Determination of Radionuclides in Environmental and Biological Materials, Laboratory of the Government Chemist. Teddington, Middlesex.

Schell, W. R. & C. D. Massey, 1987. Radioactive waste disposal in simulated peat bog repositories. Trans. Am. Nucl. Soc. 54: 76–79.

Sheppard, M. I., D. H. Thibault & P. A. Smith, 1989. Iodine dispersion and effects on groundwater chemistry following a release to a peat bog, Manitoba, Canada. Appl. Geochem. 4: 423–432.

Smith, F. B. & M. J. Clark, 1989. The transport and deposition of airborne debris from the Chernobyl nuclear power plant accident with special emphasis on the consequences to the United Kingdom. Meteorological Office Scientific Paper 42, HMSO, London.

Hydrobiologia **214**: 77–84, 1991.
J. P. Smith, P. G. Appleby, R. W. Battarbee, J. A. Dearing, R. Flower, E. Y. Haworth, F. Oldfield & P. E. O'Sullivan (eds), 77
Environmental History and Palaeolimnology.
© *1991 Kluwer Academic Publishers.*

Progress in understanding the chemical stratigraphy of metals in lake sediments in relation to acidic precipitation

Stephen A. Norton & Jeffrey S. Kahl
Department of Geological Sciences, University of Maine, Orono, Maine 04469, USA

Abstract

Sediment cores, dated by ^{210}Pb and/or varves, from lakes that do not receive point or non-point source discharge of pollutant metals and metalloids from within their catchments have been used to:
1. Develop a chronology of atmospheric deposition of trace elements related to air pollution.
2. Identify sources of these elements.
3. Estimate net fluxes of trace metals from both natural and anthropogenic sources.
4. Determine the extent of sediment focussing of metals (e.g. Pb) relative to the atmospheric flux of that metal.
5. Assess long term variations in the input of dry deposition of selected elements to lakes.
6. Establish if the water column has acidified.
7. Determine the maximum possible net increase in alkalinity generation attributable to cation release from the sediments of lakes which have undergone acidification.
8. Establish that fluxes of some metals (e.g. Al and Fe) from the catchment to the sediments have increased in many systems undergoing acidification.
9. Determine the net maximum alkalinity generation represented by the net sulfate reduction and storage in the sediment.
10. Estimate temporal variations in the speciation of metals retained in the sediment, caused by altered chemical conditions in the catchment soils, streams, and lakes.

Introduction

The chemistry of sediment cores from lakes has been used for the documentation of change associated with physical disturbance of catchments (e.g., Davis & Norton, 1978), the eutrophication of lake waters from fertilization, and point and non-point discharge of pollutants in the catchment (including the streams) (e.g., Bartleson, 1971), and alteration of the hydrology of the lakes by a variety of activities, including de- or afforestation, re-routing of water, and alteration of the infiltration characteristics of catchments (Engstrom & Wright, 1984). Since about 1975, there has been considerable study of the short and

long term effects of the atmospheric deposition of acidic and acidifying substances and associated pollutants from industrial activities on lake catchment vegetation and soil, and on surface and groundwaters (National Academy of Sciences, 1986). Recently, attention has started to shift to the paleolimnological chemical record as influenced by climate change (Engstrom & Wright, 1984), including precipitation amounts and temperature. To isolate the influences of atmospheric pollution and climate change from the influence of disturbances within catchments, researchers have studied sediment cores from relatively pristine upland or headwater, generally small, lake-catchment systems (e.g., the PIRLA project in the USA, Whitehead *et al.*, 1990; and the DOE and SWAP projects in the UK, e.g., Battarbee *et al.*, 1988). This chapter focusses on lakes that have been studied to understand the issue of atmospheric pollution. The underlying assumptions behind these studies are that: (1) the net chemistry of the sediments is determined by a combination of factors in the catchment and lake, coupled with atmospheric inputs; (2) a change in any of the factors is expected to induce a signal in the chemical record, and (3) a single undisturbed core from the profundal part of the lake will accurately record these changes. Unless the signal to noise ratio is considerable, recognition of catchment disturbances and other altered processes requires steady state sediment chemistry prior to the event of interest. This situation is commonly the case.

A typical study lake for such studies has sediment accumulation rates, as measured from cores taken in the profundal area (and thus a maximum rate), of less than $10\,000\ \mu g\ cm^{-2}\ yr^{-1}$ (Norton *et al.*, in review), or 100 metric tons $km^{-2}\ yr^{-1}$. In the context of a small lake/catchment complex, this amount of material is small. For example, all of the annually deposited sediment for a 10 hectare lake with a 100 hectare (1 km^2) catchment could be derived from as little as 5 m^3 of soil (with an assumed specific gravity of 2). Viewed on an areal basis this amount would be equivalent to removing 0.0005 cm of soil yr^{-1} from the entire watershed, a highly improbable process. Except for alpine situations, this rate of erosion may be less than that of chemical weathering. Physical erosion is typically heterogeneous, spatially and temporally. Soil chemistry is also typically heterogeneous, even in regions where the bedrock lithologies are homogeneous. Thus one might expect that sediment chemistry would typically vary considerably and with high frequency as the products of isolated erosional events reach the lake and are deposited. However, wide chemical variation in downcore sediment chemistry is the exception suggesting that the sedimentation process smooths out variation because of a lag in deposition in profundal areas. The result is an integration of many different erosional events and a

smoothing of chemical variation. Similar arguments apply to biological material derived from the catchment. An alternative or additional cause of the smoothing of the chemical record is that a considerable fraction of the profundal sediment may be derived from shoreline or littoral zone erosion. This source of sediment must be important because seepage lakes, with no input of catchment-originated inorganic detritus commonly have accumulation rates in the 1000 to $5\,000\ \mu g\ cm^{-2}\ yr^{-1}$ range (Norton *et al.*, in review). Trace element concentrations are typically less than 1000 ppm (0.1%). For seepage lakes, this value is equivalent to 1 to $5\ \mu g\ cm^{-2}\ yr^{-1}$. Atmospheric deposition rates for trace elements range from 0 to as much as $50\ \mu g\ cm^{-2}\ yr^{-1}$, depending on the element and location of the lake. All factors being equal, seepage lakes should provide a better record of atmospheric deposition of trace metals.

Differences in techniques of coring, dating (radiometric or varve), sediment preparation (bulk analysis, acid digestion, or speciation), chemical analysis, and data manipulation exist among laboratories and researchers. Nonetheless, the chronology and magnitude of inferred air pollution and atmospheric deposition of trace metals determined by different researchers is surprisingly consistent (Charles & Norton, 1986; Norton *et al.*, 1990a). Furthermore, the paleolimnologic evidence for ecosystem acidification and related processes is also consistent among researchers. Below, we discuss and illustrate some of the major types of chemical changes in lake sediment inferred to be caused by atmospheric pollution, and the uses of the sediment chemical data. We focus on Pb because of its relative immobility, relatively high abundance in the pollution record, and the frequency with which it has been studied.

Discussion

The chronology of air pollution

Increased concentrations of trace metals in sediments have typically been used as an indication of

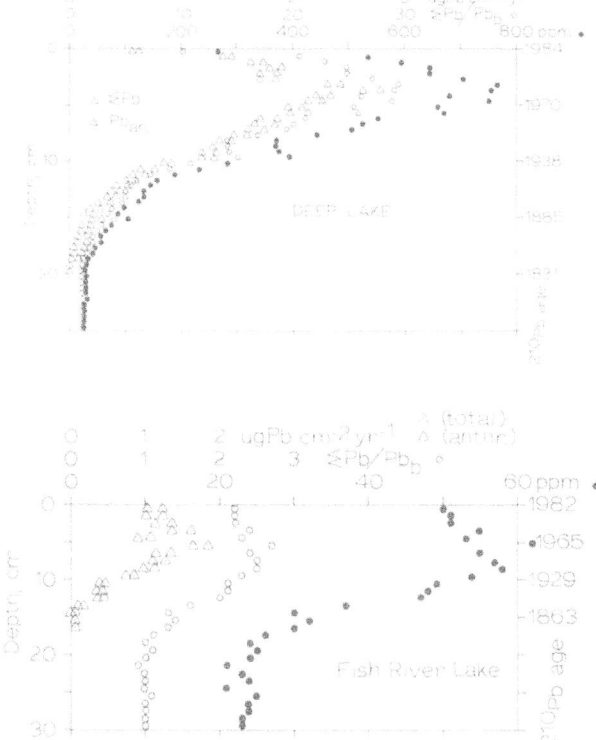

Fig. 1. Concentration, enrichment factor, and total and anthropogenic net accumulation rates for Pb for ^{210}Pb-dated sediment cores from Deep Lake (New York, USA) and Fish River Lake (Maine, USA). Individual data points from sediment sub-samples are shown. Methods are detailed in Norton *et al.*, in review.

air pollution (Charles & Norton, 1986; Norton *et al.*, 1990a). Figure 1 shows the concentration of Pb in sediment in two remote lakes as a function of sediment depth (and ^{210}Pb age). Deep Lake in the Adirondack Mountains, New York has been acidified by acidic precipitation (Charles *et al.*, 1990); Fish River Lake, northern Maine, is circumneutral. Background concentrations for Pb are similar (20 to 25 ppm) in sediment from the two lakes. Concentrations increase about 1850 (Deep) and 1860 (Fish River), dates which appear to consistently represent the first recognizable air pollution in eastern North America, caused by industrial development and the burning of coal (Norton *et al.*, 1990a). Maxima are reached in 1965–1975 and decline thereafter, probably as a

result of the diminishing use of leaded gasoline. As for Deep and Fish River Lakes, the values for the concentration maxima are very different among lakes in lake districts such as the Adirondack Mountains.

In other parts of North America, pollution by Pb starts later and is from different sources of pollution. For example, Pb concentrations increase in the 1940–1950 period in sediment from lakes in the Sierra Nevada Mountain region of California, presumably relating to the onset of the heavy use of leaded gasoline west of that region. In contrast, the northwestern USA has an early Pb pollution history related to development of the mining and smelting industry.

The flux of all elements probably has increased through the atmosphere as a consequence of agricultural and industrial activities. However, recognition of these increases for each element is only possible when and where the signal to noise ratio in the sediment chemistry is favorable. Thus increases in the atmospheric flux of Fe can not be recognized in lake sediments where the autochthonous sediment flux may average 100 μg cm^{-2} yr^{-1}, and the atmospheric flux is 1 to 10. However, in sediments from ombrotrophic peatland, the anthropogenic component of the total atmospheric Fe flux may be a substantial increase and easily recognizable (Battarbee *et al.*, 1988).

The list of elements for which anthropogenic atmospheric inputs are significant on a regional basis, based on undisturbed well dated cores, includes As, Nriagu, 1983; Be, Miller *et al.*, in review; Cd and Zn, Evans *et al.*, 1983; Cu and Ni, Wong *et al.*, 1984; Pb and Hg, Norton *et al.*, 1990a; V, Norton, 1985). Other elements have been shown to be enriched in some lakes.

Concentration data for a particular time (i) may also be expressed as a dimensionless enrichment ratio, generally relative to background (b = pre-anthropogenic) concentrations (e.g. Pb_i/Pb_b, see Figure 1), or relative to another element believed to be unaffected such as Al (e.g. $[Pb_i/Al_i]/[Pb_b/Al_b]$). This approach suffers if increases in the gross sedimentation rate dilute the effect of absolute enrichment of a pollutant. In Fig. 1, the enrichment in Pb in Deep reaches 30 ×

whereas it has a maximum of only 2.7 × in Fish River Lake. As is shown below, this approach may lead to misinterpretation.

Fluxes of pollutants

Since about 1975, paleolimnology has seen the development of techniques of 'continuous' dating of cores of recent sediment using ^{210}Pb (e.g. Appleby and Oldfield, 1983; Appleby *et al.*, 1986), and varves (e.g. Renberg, 1979). We have thus moved away from the use of chronostratigraphic markers such as pollen (e.g. Davis & Norton, 1978), charcoal or shoot (e.g., Renberg & Wik, 1985), and radionuclide pulse inputs (Pennington, 1976) as primary dating techniques. Continuous dating with these techniques can only be based on interpolation and extrapolation. Chronostratigraphic markers are now typically only used to corroborate ^{210}Pb and varve dating (e.g. Appleby *et al.*, this volume).

Net accumulation rates for metals (or other substances) are calculated from knowledge of water content, organic content, and concentrations in intervals of dated sediment. For trace metal pollutants such as Pb, it is desirable to distinguish the component of the total flux which originated within the watershed (geologic material) from the component that has apparently been derived from atmospheric inputs directly to the lake or indirectly via the catchment (the anthropogenic atmospheric component).

The calculated total fluxes are shown on Figure 1 for the two lakes. Although the values for the concentration profiles for the two lakes differ by an order of magnitude at the maximum concentration reached, the maximum total net accumulation rates for Pb for the lakes are similar. Without further analysis, this difference could be attributed to a much higher recent gross accumulation rate of Pb derived from the watershed of Fish River Lake; background concentrations of Pb (ca. 25 ppm) are similar to those of Deep but the maximum recent concentration is only 0.1 as high. However, if one calculates the Pb attributed to geologic sources, a different interpretation

emerges. For calculations shown in Figure 1 we have assumed that there are two components that comprise the total flux of Pb to the sediment, a geologic source and an anthropogenic source. The formulation of the calculation is:

$$Pb_{a(x)} = Total\ Pb_{(x)} - [Ti_{(x)}/Ti_{(b)}]\ [Pb_b]$$

or

Anthropogenic flux =
Total flux – Geologic flux,

where a = anthropogenic, (x) = any depth interval x, b = background, and all values are fluxes. The geologic component of the total Pb is assumed to be contributed in constant proportion to Ti. Examination of background ratios indicates there is only minor variation. This inherent variation in Pb_b/Ti_b limits our ability to recognize the onset of pollution for Pb, as it does for all elements. We also assume that all the Ti is geologic. Thus, as the flux of Ti to the sediment varies, the geologic contribution of Pb is assumed to vary. However, it has been established that the flux of Ti from the atmosphere has increased in recent decades (Norton *et al.*, 1990b). If the atmospheric flux of Ti is significant with respect to the geologic flux, the above calculation over-corrects for geologic Pb (or any other constituent), yielding an anthropogenic atmospheric component which is too low or even negative. For typical drainage lakes with normal sediment accumulation rates, the method works well. It fails to varying degrees if the signal to noise ratio (for the pollutant) is near one and if the gross accumulation rate of Ti is not much larger than the atmospheric flux of Ti.

Results of these calculations are shown on Fig. 1. Two conclusions emerge: (1) For sediments deposited over the last 100 + years, the total accumulation rate of Pb is dominated by atmospheric inputs, and (2) The atmospheric accumulation rates for Pb are similar for these two widely separated lakes. The latter observation represents a fortuitous situation. Maximum atmospheric accumulation rates ($\mu g\ cm^{-2}\ yr^{-1}$) and maximum atmospheric accumulations

(μg cm^{-2}) range over an order of magnitude among and within lake districts (such as the Adirondack Mountains of New York, or in Maine). There is strong co-variance among lake burdens of atmospheric Pb, atmospheric V, unsupported (largely atmospheric) ^{210}Pb, and gross sediment accumulation rates (Norton *et al.*, in review). These relationships suggest that, for Pb, sediment focussing is important in determining the concentrations in sediment cores, the enrichment factors, and particularly all variables which have rate units. It also enables preliminary attempts at relating anthropogenic Pb (and V) sediment accumulation rates to atmospheric

deposition rates, through the linkage with unsupported ^{210}Pb deposition. The latter is believed to be relatively consistent averaged over yearly periods.

Chemical evidence related to acidification

Declines in concentrations of metals (especially Ca) in younger sediments have been interpreted as indicating leaching of the sediment column by an acidifying water column, desorption of metals prior to sedimentation, and less effective scavenging of metals from the water column due to the effect of lowered pH on partitioning of metals

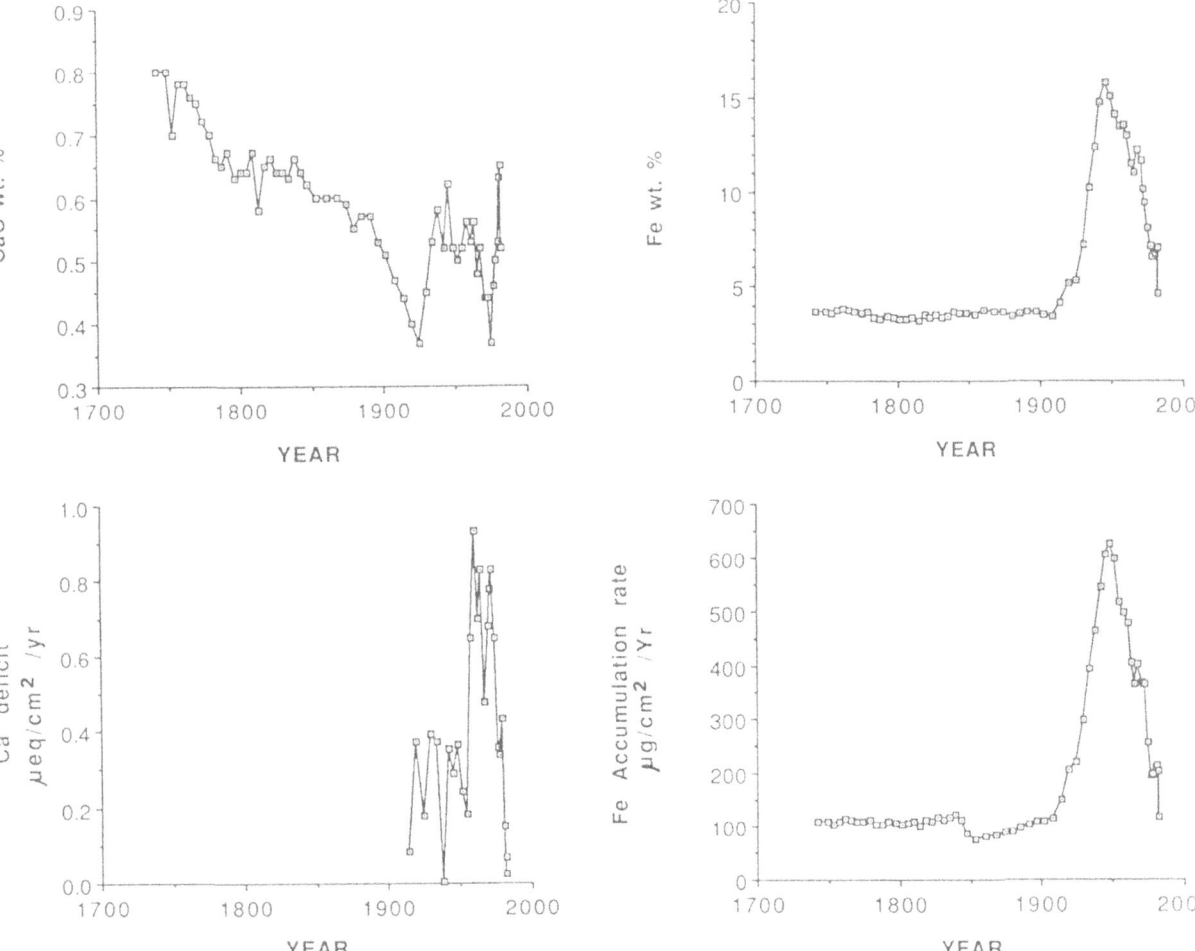

Fig. 2. Selected chemical parameters for a sediment core from Deep Lake – New York: (a) Concentration of Ca as a function of depth. (b) Ca deficit in the sediment as defined in the text. (c) Fe concentration as a function of sediment age. (d) Net Fe accumulation rate as a function of sediment age.

between solid phases and the water (Kahl *et al.*, 1984; Schafran & Driscoll, 1987; Norton *et al.*, 1990c). Because of the discussion presented above, it is clearly important that data be analyzed in units other than concentration. For example, Fig. 2a shows a general decline in the concentration of Ca for sediment from Deep Lake over the period 1750 to 1945, with a partial recovery in younger sediment. However, all base metal concentrations decline over this time frame (see Charles *et al.*, 1990). But from about 1945 to the present the base metals, except Ca, increase substantially, i.e. Ca is depleted relative to the other base metals. This relative loss of Ca occurs in a time during which fossil diatom communities indicate a declining pH for the water column (Charles *et al.*, 1990). Losses of Ca (or deficits) from the sediments can be established by normalizing accumulation rates to relatively immobile constituents such as Ti, again with caution (see e.g. Norton *et al.*, 1990c, in review). The anticipated metal accumulation rate is calculated as for the geologic component (see above) and then subtracted from the measured accumulation rate. A negative value suggests possible metal leaching ($\mu g\ cm^{-2}\ yr^{-1}$) and that value can be converted into μ-equivalents of alkalinity released per unit area of sediment per unit time. The results of such calculations for Deep Lake (Fig. 2b) indicate that alkalinity generation by the increased release of metals is relatively modest compared with cation release that normally occurs in the absence of acidification. Only when the catchment : lake area ratio is small does accelerated leaching become important.

The reduction of water column sulfate to sulfide, fixation of the sulfide by precipitation of FeS or FeS_2, and net sedimentation also results in the production of alkalinity. Although many workers have measured *in situ* sulfate reduction at the sediment-water interface of lakes receiving considerable excess sulfate from acidic precipitation, there is little evidence that increased net storage of reduced S, and thus significantly increased alkalinity production, has occurred (Norton *et al.*, 1988).

Some elements may be deposited by diagenetic processes in sediment not equivalent in age to the time of deposition (e.g. Zn and S; Carignan and Tessier, 1985). Correct reconstruction of atmospheric fluxes based on profiles of such elements has been attempted but the trends typically do not match known emission trends.

Drainage lake systems acidify from the 'top' down. The process first involves the lowering of base saturation in soils with release of Ca to downstream waters. This Ca may be re-sorbed by downstream systems which have higher pH. Subsequently, the catchment starts to release Al and Fe as acidification becomes chronic; Ca may then be desorbed in sedimenting material while Al and Fe mobilized from the catchment may be precipitated and sedimented in lakes. Enrichment of Fe in the sediment record of recently acidified lakes has been interpreted to originate in this manner and may play an important role in the fixation of reduced sulfur (Carignan and Tessier, 1988; Norton *et al.*, 1989). Such enrichment is shown for Deep Lake in Fig. 2c and 2d; the increases in the concentration and accumulation rate of Fe are matched by increases in sediment S. Enrichment of Al in sediments as a result of mobilization in the catchment and subsequent precipitation in the higher pH downstream lakes is probable but not easily demonstrated from increased concentration of Al in the sediment. However, soluble Al (probably a form of $Al(OH)_3$) does increase in lakes whose catchments have acidified. Such speciation studies hold considerable promise for distinguishing different sources of metals, sources which are indistinguishable using bulk sediment analysis or analysis of strong acid leachate.

Conclusions

Paleolimnological techniques of sampling, sample processing, dating, chemical analysis, and interpretation have been developed to distinguish the impacts of acidic precipitation, atmospheric deposition of materials, changes in land use or vegetation, and climate change. Critical to the success of such interpretations are undisturbed cores, high quality and high resolution dating, and complete chemical analyses. Establishment of the chronology of pollution, and its direct and indirect effects on lake sediment chemistry make possible quantitative statements about atmospheric deposition and in-lake processing of sediment. Diagenesis of sediments obfuscates the interpretations but in some cases diagenetic effects may be useful in understanding differences between the sedimentary records of proximal lakes. Interpretations of sediment chemistry are strengthened if analyses of fossil biota are completed on the same core, thereby constraining the interpretation of the chemistry.

Acknowledgements

We are very grateful to numerous people who have discussed with us the material and ideas presented above and from whom we have learned much. Many of these people have pioneered the modern techniques employed in paleolimnological studies. They include Peter Appleby, Michael Binford, Donald Charles, Ronald Davis, Daniel Engstrom, Frank Oldfied, John Smol, and Jeffrey White. The data presented here were gathered with funds provided by the U.S. EPA, the Electric Power Research Institute, and the Superior Mining Company.

References

Appleby, P. G. & F. Oldfied, 1983. The assessment of ^{210}Pb dates from sites with varying sediment accumulation rates. Hydrobiologia 103: 29–35.

Appleby, P. G., P. J. Nolan, D. W. Gifford, M. J. Godfrey, F. Oldfield, N. J. Anderson & R. W. Battarbee, 1986. ^{210}Pb dating by low background gamma counting. Hydrobiologia 143: 21–27.

Appleby, P. G., N. Richardson & P. J. Nolan, 1990. ^{241}Am dating of lake sediments. Hydrobiologia. (this volume.)

Bartleson, G. C., 1971. The chemical investigation of recent lake sediments from Wisconsin lakes and their interpretation. Water Pollution Control Research Series 16010EHR03/71. U.S. EPA, 278 p.

Battarbee, R. W., N. J. Anderson, P. G. Appleby, R. J. Flower, S. C. Fritz, E. Y. Haworth, S. Higgitt, V. J. Jones, A. Kreiser, M. A. R. Munro, J. Natkanski, F. Oldfield, S. T. Patrick, N. G. Richardson, B. Rippey & A. C. Stevenson, 1988. Lake acidification in the United Kingdom (1988–1986). ENSIS, London, 68 p.

Carignan, R. & A. Tessier, 1985. Zinc deposition in acid lakes – the role of diffusion. Science 228: 1524–1526.

Carignan, R. & A. Tessier, 1988. The co-diagenesis of sulfur and iron in acid lake sediments of southwestern Quebec. Geochim. Cosmochim. Acta 52: 1179–1188.

Charles, D. F. & S. A. Norton, 1986. Paleolimnological evidence for trends in atmospheric deposition of acids and metals. In National Academy of Sciences, Acid Deposition – Long Term Trends. National Academy Press, Washington: 335–431.

Charles, D. F., M. W. Binford, E. T. Furlong, R. A. Hites, M. J. Mitchell, S. A. Norton, F. Oldfield, M. J. Paterson, J. P. Smol, A. J. Untala, J. R. White, D. R. Whitehead & R. J. Wise, 1990. Paleoecological investigation of recent lake acidification in the Adirondack Mountains, N.Y. Paleolimnology 3: 195–242.

Davis, R. B. & S. A. Norton, 1978. Paleolimnologic studies of human impact on lakes in the United States, with emphasis on recent research in New England. Proc. II International Symposium on Paleolimnology, Poland. Pol. Arch. Hydrobiol. 25: 99–115.

Engstrom, D. R. & H. E. Wright, Jr., 1984. Chemical stratigraphy of lake sediments as a record of environmental change. In E. Y. Haworth & J. W. G. Lund, (eds), Lake Sediments and Environmental History. Studies in Paleolimnology and Paleoecology. Leicester University Press: 11–67.

Evans, H. E., J. P. Smith & P. J. Dillon, 1983. Anthropogenic zinc and cadmium burdens in sediments of selected southern Ontario lakes. Can. J. Fish. aquat. Sci. 40: 570–579.

Kahl, J. S., S. A. Norton & J. S. Williams, 1984. Chronology, magnitude and paleolimnologic record of changing metal fluxes related to atmospheric deposition of acids and metals in New England. In O. P. Bricker (ed.), Geological Aspects of Acid Deposition. Butterworth Pub, Boston, 23–35.

National Academy of Sciences, 1986. Acid Deposition – Long Term Trends. National Academy Press, Washington, 506 p.

Norton, S. A., 1985. The sedimentary record of atmospheric pollution in Jerseyfield Lake, Adirondack Mountains, New York. In D. D. Adams & W. Page, (eds), Acid Deposition – Environmental Economic, and Policy Issues. Plenum, New York, pp. 95–107.

Norton, S. A., M. J. Mitchell, J. S. Kahl & G. F. Brewer, 1988. In-lake alkalinity generation by SO_4 generation: A paleolimnological assessment. Water, Air, and Soil Pollut. 39: 33–45.

Norton, S. A., 1989. Watershed acidification – A chromatographic process. In J. Kamari, D. F. Brakke, A. Jenkins, S. A. Norton & R. F. Wright (eds), Regional Acidification

Models: Geographic Extent and Time Development. Springer-Verlag, Heidelberg: 89–102.

Norton, S. A., P. J. Dillon, R. D. Evans, G. Mierle & J. S. Kahl, 1990a. The history of atmospheric pollution and deposition of Cd, Hg, and Pb in North America. In S. E. Lindberg, A. L. Page & S. A. Norton (eds), Sources, Deposition, and Canopy Interactions. 3. Springer-Verlag, New York: 73–102.

Norton, S. A., M. Verta & J. S. Kahl, 1990b. Relative contribution to lake sediment chemistry by atmospheric deposition. Soc. Int. Limnol. Verhandlungen. (in press)

Norton, S. A., J. S. Kahl, A. Henriksen & R. F. Wright, 1990c. Buffering of pH by sediments in streams and lakes. In S. A. Norton, S. E. Lindberg & A. L. Page (eds), Soils, Aquatic Processes, and Lake Acidification. Springer-Verlag, New York. 4: 133–157.

Nriagu, J. O., 1983. Arsenic enrichment in lakes near the smelters at Sudbury, Ontario. Geochim. Cosmochim. Acta. 47: 1523–1526.

Pennington, W., R. S. Cambray, J. D. Eakins & D. D. Harkness, 1976. Radionuclide dating of the recent sediments of Blelham tarn. Freshwat. Biol. 6: 317–331.

Renberg, I., 1979. Environmental monitoring by chemical, physical, and biological analyses of annually laminated lake sediments – the possibilities of the method in the use of ecological variables in environmental monitoring. The National Swedish Environment Protection Board, Report PM 1151, 318 p.

Renberg, I. & M. Vik, 1984. Dating recent lake sediments by soot particle counting. Verh. Int. Ver. Limnol. 22: 712–718.

Schafran, G. C. & C. T. Driscoll, 1987. Comparison of terrestrial and hypolimnetic sediment generation of acid neutralizing capacity. Envir. Sci. Tech. 21: 988–993.

White, J. R. & C. P. Gubala, 1990. Sequentially extracted metals in Adirondack lake sediment cores. J. Paleolimnology 3: 187–194.

Whitehead, D. R., D. F. Charles & R. A. Goldstein, 1990. The PIRLA Project (Paleoecological Investigation of Recent lake Acidification). J. Paleolimnology 3: 187–194.

Wong, H. K. T., J. O. Nriagu & R. D. Coker, 1984. Atmospheric input of heavy metals chronicled in lake sediments of the Algonquin Provincial Park, Ontario, Canada. Chemic. Geol. 44: 187–201.

Hydrobiologia **214**: 85–90, 1991.
J. P. Smith, P. G. Appleby, R. W. Battarbee, J. A. Dearing, R. Flower, E. Y. Haworth, F. Oldfield & P. E. O'Sullivan (eds),
Environmental History and Palaeolimnology.

Spheroidal carbonaceous particles as a marker for recent sediment distribution

Maria Wik & Ingemar Renberg
Department of Ecological Botany, University of Umeå, S-901 87 Umeå, Sweden

Key words: sediment distribution, carbonaceous particles, combustion products, sediment traps, lake acidification

Abstract

We have tested the hypothesis that spheroidal carbonaceous particles (SCP), emitted from oil and coal combustion, can be used as a marker for recent sediment distribution in lake basins. Sediment traps have been used to study when and how the particles are introduced into a lake, a laboratory experiment has been made to study the depositional behaviour of the particles in relation to other sedimentary constituents, and spatial SCP distributions have been surveyed in three lakes. The results support the hypothesis that SCP can be used as a marker, and also indicate that acidification can change sediment distribution in lakes.

Introduction

Sediments are distributed unevenly in lakes, and can be resuspended several times before final deposition. Knowledge of sediment accumulation and the resultant sediment distribution has important implications for the interpretation of palaeolimnological data and for lake geochemistry and biology. Sediment accumulation distribution has previously been studied using several approaches such as: ^{210}Pb (Evans & Rigler, 1983), pollen (Davis, 1973; Davis & Ford, 1982) magnetics (Dearing, 1983), volcanic ash (Kimmel, 1978) and chemical stratigraphy (Davison *et al.*, 1985). There are also models predicting sediment distribution (Håkansson & Jansson, 1983), and the occurrence and mechanisms of sediment focusing and redistribution (Hilton, 1985) in lakes. Unfortunately these models are not fully applicable to very small lakes,

such as the lakes studied in this investigation (all ≤ 0.3 km^2).

We suggest that spheroidal carbonaceous particles (SCP) can be used as a marker for the recent (post 1950's) sediment distribution in lake basins. SCP are emitted to the atmosphere during oil and coal combustion. They are mainly composed of a porous structure of elemental carbon and are, therefore, chemically inert. These kind of particles have been found in considerable numbers in sediments from various parts of the world (Griffin & Goldberg, 1983; Renberg & Wik, 1985a; Battarbee *et al.*, 1988).

We have studied the spatial SCP distribution in three lakes, using multiple coring along profiles or transects. In another lake, sediment traps were used to study the temporal accumulation of these particles and a laboratory experiment was undertaken to study the redepositional behaviour of SCP in relation to some other sedimentary constituents.

Methods and study areas

The spatial particle distribution was studied in three small dimictic lakes, Koltjärn, Gårdsjön and Stora Galten. Cores were collected with a Kajak-corer (tube $\emptyset$ = 6.4 cm). Recovery represents at least the 20th century but as the majority of the SCP have been deposited after World War II (Renberg & Wik, 1985a) the results mainly reflect the particle accumulation at each sampling site since the 1950's. Each whole-core was put into a plastic bag, weighed and homogenized (mixed for 5 min) after which a subsample was taken and weighed. Subsamples were prepared and the number of particles ($>5\ \mu$m) counted following Renberg & Wik (1985b). The number of SCP accumulated per square metre were then calculated.

Gårdsjön is situated in S.W. Sweden. The total lake area is 0.3 km^2 and the maximum depth 18.6 m. Gårdsjön acidified during the 1950's (Renberg & Wallin, 1985), and was limed initially in 1982. In 1982, before liming, a total of 55 sediment cores were collected of which 33 were taken along three profiles. In 1985, after liming, 30 new cores were sampled along the same profiles (Fig. 5).

Stora Galten is situated only 20 km north of Gårdsjön. It has an area of 0.3 km^2 and, unlike Gårdsjön, is not acidified. 12 cores were sampled along a profile from a well defined small basin with a maximum depth of 13.5 m (Fig. 4).

Koltjärn is situated in N. Sweden. The lake area is 0.16 km^2, maximum depth 18.5 m and the deeper parts of the basin have varved sediments. A total of 39 cores were sampled in 1985 along transects in the northern half of the lake (Fig. 3).

Temporal variability in particle deposition was studied in the dimictic Lake Lövösundet which is situated 135 km north of Koltjärn. Lövösundet has an area of 0.4 km^2, a maximum depth of 14.5 m and has varved sediment. Sediment traps,

attached to stands, were placed 0.5 m above the bottom (Renberg unpublished). SCP deposition was investigated from May 1979 until May 1982, but the traps were sampled with the closest time interval during the last year, and it is the results from this period that are reported here. Results are expressed as no. cm^{-2} day^{-1}.

The redeposition experiment was simple and based on the principles of the pipette method, but in contrast to this method, organic material, iron and salts were not removed. Two tests where done, using 100 g of wet, fresh totally untreated sediment collected from deep-water in Gårdsjön. The sediment was suspended in water and homogenized carefully in a 500 ml graduated glass. The suspension was left to settle and small subsamples were taken at a certain depth at close-time intervals. The subsamples were dried and weighed. SCP samples were prepared following the method described by Rose (1990) but counted according to Renberg & Wik (1985b) and the amount of total inorganic and organic material determined by loss-on-ignition. The second test followed the same procedures, although 50 ml of 0.05 M Na$_4$P$_2$O$_7$ was added in order to minimize coagulation. Results are expressed as percentages compared to concentration at time 0.

Results and discussion

In order to evaluate the use of SCP as a marker of general sediment distribution it is important to understand when and how particles are introduced into the lake, and how they behave in relation to other sedimentary components.

The intensive sediment trap sampling in Lövösundet during May 1981-May 1982 clearly demonstrated the considerable temporal variation in SCP input into the lake (Fig. 1). During the warm summer period (June-August) oil combustion was low and consequently the SCP deposition in the traps was low (1.3–1.6 particles cm^{-2} day^{-1}). In the autumn, heating requirements increased and this is also recorded in the sample covering the major part of September and beginning of October (1.8 particles cm^{-2}

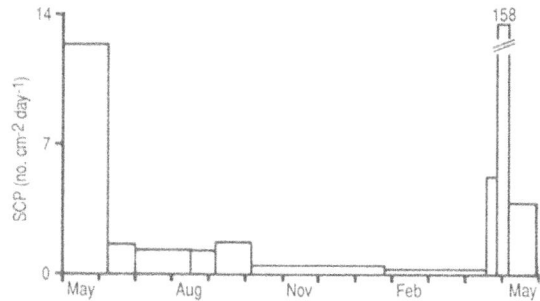

Fig. 1. SCP accumulation, May 1981 to May 1982, in sediment traps in Lake Lövösundet.

day^{-1}). During the latter period, the probability of autumn circulation resuspending material from shallow areas, must be considered as well. During winter, with about six months of ice-cover, accumulation was low and probably consists of suspended particles settling under the ice.

In springtime the ice becomes porous and full of pipes. We have brought 'spring-ice' from the lake into the laboratory and observed that the SCP that have accumulated on the ice during winter, have formed aggregates with mineral and organic matter. These aggregates melt into the ice, and trap results suggest that a lot of the aggregates melt through the ice and settle rapidly, just before ice clearance. In 1982 the ice cleared from Lövösundet on May 4 and during the period April 26 to May 5 an extremely high value of 158 particles cm^{-2} day^{-1} was recorded. Also the sample from 1981 (May 1 to June 8) reflects the high deposition rate associated with ice clearance although the sampling period is too long to disclose the real peak. It is obvious that a substantial part of the annual SCP accumulation takes place during a few weeks in spring.

The results from Lövösundet are applicable to all the lakes in this investigation. Koltjärn normally has six months of ice cover and Gårdsjön and Stora Galten four months. In other climatic regions with no or just a short period with ice cover, the temporal variations in deposition will be different and probably not so dramatic. In these climatic regions direct deposition onto the water surface will be dominant, and the mechanism for deposition of SCP to the sediments will probably differ from those described here.

A preliminary study for the settlement of SCP (from an oil fired power plant) in water, demonstrates that there are great variations between individual particles. Some settle rapidly, some very slowly, while others, apparently with many internal airpockets, float on the surface. In the natural environment SCP might be incorporated into aggregates and if the floating type of SCP is to reach the sediment they must aggregate with other material on the ice or in the water surface film.

The resuspension experiment shows that SCP settled together with the organic and inorganic material in the untreated sediment (Fig. 2). When dispersing agent was added the SCP settled fastest, followed by the inorganic and organic fractions (Fig. 2). As the inorganic fraction of this sediment is dominated by diatom frustules and very fine mineral grains, this fraction was behaving in a similar way as the organic fraction, even when dispersing agent was added. A substantial part of these two fractions stayed in solution for several weeks. None of these tests are perfect reflections of natural conditions and the most relevant situation probably occurs somewhere in between. However, they indicate that SCP can be used as a marker for general sediment distribution. The tests indicate also, that pine and spruce pollen settles in a way very similar to SCP. Davis (1973) found that pollen assemblages were resuspended and redeposited in association with the sediment in which it is contained, without sorting of individual particles.

A relationship between sediment accumulation and water depth has been found in several lakes

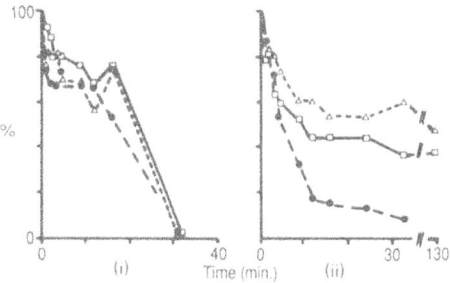

Fig. 2. Material left compared to concentration at time 0 for SCP (●), inorganic (□), and organic material (△). (i) Pure sediment. (ii) Dispersing agent added to sediment.

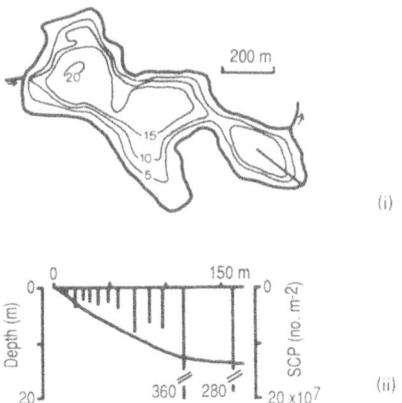

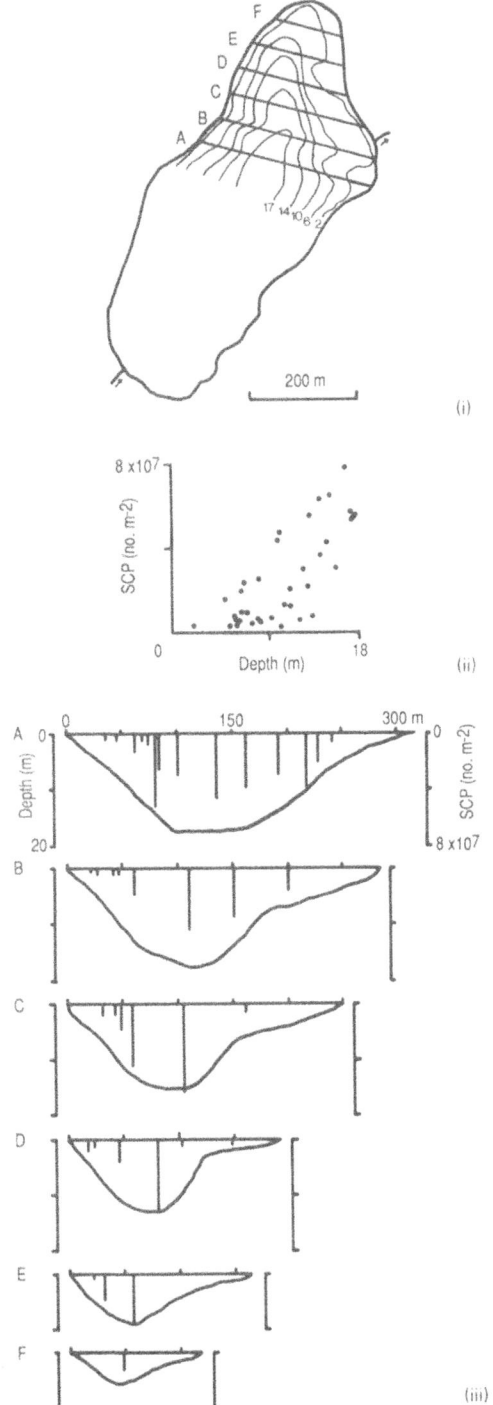

Fig. 4. Lake Stora Galten. (i) Bathymetric map with sampling profile indicated. (ii) SCP accumulation at each sampling point.

(e.g. Evans & Rigler, 1983; Hilton *et al.*, 1986; Kimmel, 1978) and follows the concept of sediment focusing (Likens & Davis, 1975). In both Koltjärn and Stora Galten there is a correlation between particle accumulation and water depth ($r = 0.78$ and $r = 0.76$ respectively $p < 0.001$), showing that SCP are focused to the deep parts of the basins, as other sedimentary components (Fig. 3 and 4.).

Important factors governing sediment redistribution are wind/waves, water circulation, and slopes (Hilton, 1985). Koltjärn and Stora Galten are small, relatively deep lakes, protected from wind action by hilly terrain and the direct influence of waves is small. The lakes are dimictic and not affected by large inflows. Both lakes have slopes exceeding 4%, an inclination above which the sediment accumulation is less efficient (Håkansson, 1977). However, the deep areas the Koltjärn have varved sediment and there are no signs of slumping or sliding in the varves.

At Gårdsjön, in contrast to the other two lakes studied, there was no correlation between depth and SCP accumulation (Fig. 5), before liming (1982). It seems as if SCP had been deposited more or less evenly all over the lake bottom. A major difference between Gårdsjön at the sampling in 1982 and the other two investigated lakes were the extensive bottom areas covered by a mat of filamentous algae and *Sphagnum* spp, typical for acidified lakes (Lazarec, 1982; Grahn,

Fig. 3. Lake Koltjärn. (i) Bathymetric map with sampling transects indicated. (ii) SCP accumulation plotted vs. water depth. (iii) The individual transects with SCP accumulation at each sampling point.

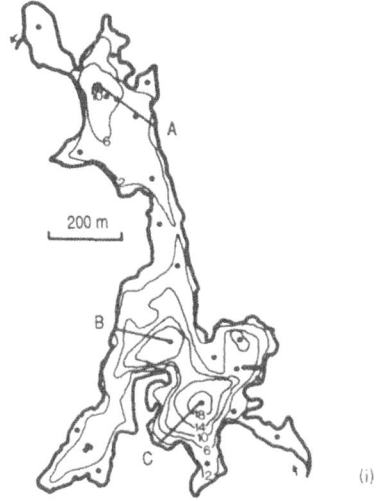

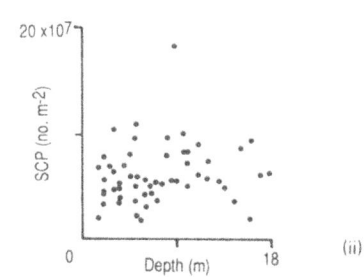

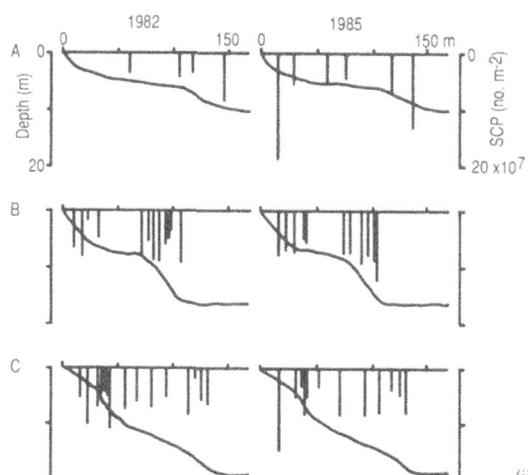

Fig. 5. Lake Gårdsjön. (i) Bathymetric map with sampling profiles and sampling points indicated. (ii) SCP accumulation plotted vs. water depth for all samples taken in 1982. (iii) Profiles with SCP accumulation at each sampling point for 1982 and 1985.

1985). De March (1978) found that bottom areas with moss growth accumulated more sediment than other areas with comparable depths but no moss. The distribution of soft surficial sediment (0–2 cm) in Gårdsjön was investigated by Andersson (1985) before the lake was limed initially. He found that soft sediment was distributed all over the lake and that fine-grained organic particles where predominant even on shallow bottoms. Only very restricted highly exposed areas were classified as erosion or transport bottoms. The even SCP distribution that we found in 1982 agrees with Andersson's observations. He suggested that the normal sediment deposition pattern was altered as a result of successive incorporation of fine particulates by the rapidly growing filamentous algae and *Sphagnum* beds, and that they acted as a rough surface where small particles were trapped, and finally, also stabilized the surface preventing resuspension of material. Field observations of a clear gradation in the thickness of total post-glacial sediment from shallow to deeper water (a few dm to m, respectively), demonstrate that sediment focusing had occurred earlier in Gårdsjön. This indicates that the distribution of soft sediment reported by Andersson (1985) and the SCP results from 1982 reflect a recent change in the sediment accumulation pattern.

The liming in 1982 severely reduced the abundance of *Sphagnum* spp and filamentous algae, and the dead material started to decompose (Grahn & Sangfors, 1988). In 1985, profile B showed a correlation between depth and SCP accumulation (Fig. 5, $r = 0.81$, $p < 0.001$). This change might be triggered by the disappearance of the bottom felt-like structure releasing material previously attached into the structure, and additionally letting normal sediment processes start again, resulting in a focusing of the released material into deeper regions. For the other two profiles, however, no such trend could be found although some change is indicated in profile A (Fig. 5). If the shallowest extremely high sample is removed there is a significant relationship ($r = 0.87$, $p < 0.002$). The number of samples from this profile is, however, low ($n = 6$). To be able to confirm or reject the hypothesis of changed deposition patterns in acidified lakes with surficial growth of *Sphagnum* spp and filamentous algae, it is necessary to investigate more acidified lakes before and after liming. If sediment distribution is changed this must have consequences for the flora and fauna as well as lake chemistry.

Compared to Stora Galten and Koltjärn, Gårdsjön also has a more complex morphometry which generally is considered to generate more complex sediment distributions. However, the morphology of the individual basins in Gårdsjön are not complex.

The strong correlation between SCP accumulation and depth that we found in both Koltjärn and Stora Galten shows that SCP are focused into the deep parts of the basin, as have been observed for other sedimentary components (eg. Evans & Rigler, 1983). The results, including those from Gårdsjön and the resuspension test, support the hypothesis that SCP can be used as a marker to study recent sediment distribution. We believe that the different sediment distribution in the acidified Gårdsjön, as recorded by SCP analyses, is caused by trapping, incorporation and stabilization of sediments by the mat of filamentous algae and *Sphagnum* spp that covered large bottom areas before the lake was limed. Because of the important implications this has for nutrient cycling, faunal habitats and surface-sediment processes etc., further work is required to clarify the specific processes involved.

Acknowledgement

Special thanks to Ingvar Andersson who sampled all the cores in 1982. This study was financially supported by the National Swedish Environmental Protection Board.

References

Andersson, B. I., 1985. Properties and chemical composition of surficial sediments in the acidified Lake Gårdsjön, SW Sweden. Ecol. Bull. 37: 251–262.

Battarbee, R. W., N. J. Anderson, P. G. Appleby, R. J. Flower, S. C. Fritz, E. Y. Haworth, S. Higgit, V. J. Jones, A. Kreiser, M. A. R. Munro, J. Natkanski, F. Oldfield, S. T. Patrick, N. G. Richardson, B. Rippey & A. C. Stevenson, 1988. Lake acidification in the United Kingdom 1800–1986. ENSIS Publishing, London, 68 pp.

Davis, M. B., 1973. Redeposition of pollen grains in lake sediment. Limnol. Oceanogr. 18: 44–52.

Davis, M. B. & J. Ford, 1982. Sediment focusing in Mirror Lake, New Hampshire. Limnol. Oceanogr. 27: 137–150.

Davison, W., J. Hilton, J. P. Lishman & W. Pennington, 1985. Contemporary lake transport processes determined from sedimentary records of copper mining activity. Envir. Sci. Technol. 19: 356–360.

De March, L., 1978. Permanent sedimentation of nitrogen, phosphorus, and organic carbon in a high arctic lake. J. Fish. Res. Bd Can. 35: 1089–1094.

Dearing, J. A., 1983. Changing patterns of sediment accumulation in a small lake in Scania, southern Sweden. Hydrobiologia 103: 59–64.

Evans, R. D. & F. H. Rigler, 1983. A test of Lead-210 dating for the measurement of whole lake soft sediment accumulation. Can. J. Fish. aquat. Sci. 40: 506–516.

Grahn, O., 1985. Macrophyte biomass and production in Lake Gårdsjön – an acidified clearwater lake in SW Sweden. Ecol. Bull. 37: 203–212.

Grahn, O. & O. Sangfors, 1988. A comparative study for macrophytes in Lake Gårdsjön, during acid and limed conditions. In W. Dickson (ed.), Liming of Lake Gårdsjön, and acidified lake in SW Sweden. National Swedish Environmental Protection Board. Report 3426: 281–308.

Griffin, J. J. & E. D. Goldberg, 1983. Impact of fossil fuel combustion on sediments of Lake Michigan: A reprise. Envir. Sci. Technol. 17: 244–245.

Hilton, J., 1985. A conceptual framework for predicting the occurrence of sediment focusing and sediment redistribution in small lakes. Limnol. Oceanogr. 30: 1131–1143.

Hilton, J., J. P. Lishman & P. V. Allen, 1986. The dominant process of sediment distribution and focusing in a small, eutrophic, monomictic lake. Limnol. Oceanogr. 31: 125–133.

Håkansson, L., 1977. The influence of wind, fetch, and water depth on the distribution of sediments in Lake Vänern, Sweden. Can. J. Earth Sci. 14: 397–412.

Håkansson, L. & M. Jansson, 1983. Lake sedimentology. Springer-Verlag, Berlin Heidelberg, 316 pp.

Kimmel, B. L., 1978. An evaluation of recent sediment focusing in Castle Lake (California) using a vulcanic ash layer as a stratigraphic marker. Verh. int. Ver. Limnol. 20: 393–400.

Lazarek, S., 1982. Structure and function of a cyanophytan mat community in an acidified lake. Can. J. Bot. 60: 2235–2240.

Likens, G. E. & M. B. Davis, 1975. Post-glacial history of Mirror Lake and its watershed in New Hampshire, USA: An initial report. Int. Ver. Theor. Angew. Limnol. Verh. 19: 982–993.

Renberg, I. & J.-E. Wallin, 1985. The history of the acidification of Lake Gårdsjön as deduced from diatoms and Sphagnum leaves in the sediment. Ecol. Bull. 37: 47–52.

Renberg, I. & M. Wik, 1985a. Carbonaceous particles in lake sediments – pollutants from fossil fuel combustion. Ambio 14: 161–163.

Renberg, I. & M. Wik, 1985b. Soot particle counting in recent lake sediments: An indirect dating method. Ecol. Bull. 37: 53–57.

Rose, N. L., 1990. A method for the extraction of carbonaceous particles from lake sediment. J. Palaeolim. 3: 45–53.

Hydrobiologia **214**: 91–97, 1991.
J. P. Smith, P. G. Appleby, R. W. Battarbee, J. A. Dearing, R. Flower, E. Y. Haworth, F. Oldfield & P. E. O'Sullivan (eds), 91
Environmental History and Palaeolimnology.

Magnetic spherules in recent lake sediments

D. McLean

*Department of Environmental Science, Crewe and Alsager College of Higher Education, Crewe CW1
1DU, Cheshire, England*

Key words: magnetic spherules, lake sediments, dating method

Abstract

The record of magnetic spherules in sediments from Crummock Water (the English Lake District) is presented and a method for obtaining magnetic extracts described. It is shown that magnetic spherule concentrations in the Crummock Water sediments increases by nearly two orders of magnitude following the Industrial Revolution, and this can be attributed to increases in fossil fuel combustion particularly after about 1900 A.D. Associated with the increase in spherule numbers are changes in their size distributions, which may reflect the changing sources for the spherules.

It is demonstrated that magnetic spherules are distributed regionally and hence form widespread cryptic marker horizons in lacustrine sediments. Location of this horizon in a lake sediment profile provides an approximate dating method and this horizon may be located using susceptibility measurements corroborated by microscopic spherule counts.

Introduction

Magnetic spherules are small spheres or spheroids, with diameters ranging from less than one to several hundred microns. They are composed mostly of magnetite or other iron oxides either alone or in combination with silicates, and are produced during exposure of fossil fuels to prolonged and high combustion temperatures (Fisher *et al.*, 1978) or by rapid grinding of metals, (Parkin *et al.*, 1980). These conditions are produced in only two natural environments (volcanic plumes and cosmic dusts) as well as during high temperature fossil fuel combustion processes, (Puffer *et al.*, 1980). This group of particles include a significant fraction below 10 μm in diameter which may travel in excess of 3000 miles from source (Parkin *et al.*, 1980), having previously been recorded in remote Finnish bogs (Oldfield *et al.*, 1981), and mid-Atlantic sediments (Parkin *et al.*, 1970). Spherules should not therefore be geographically restricted to sediments close to industrial sources, (contrary to Puffer *et al.*, 1980), but should be at least regionally distributed. Ease of identification in contemporary sediments potentially allows their use as stratigraphic markers, providing an approximate sediment dating method and a record of industrial activity.

This use of magnetic spherules is analogous to the use of carbonaceous spheres, a more abundant by-product of oil and coal combustion, as an indirect dating method for lake sediments, (Renburg & Wik, 1984). Concentrations of such 'soot spherules' in lake sediments have been shown to closely match regional increases in coal and oil combustion from power stations in the U.K., USA, and Sweden, although local sources may add to the spherule record, (Wik *et al.*, 1986).

Site description

Crummock Water is situated in the north-west of the English Lake District, occupying part of the Buttermere glacial trough, and has a catchment area of approximately 44 km². The solid geology of the catchment is predominantly Skiddaw Slates of the Ordovician Skiddaw Group with smaller areas of green slate of the Borrowdale Volcanic Group and granophyre, (Arthurton *et al.*, 1980). Glacial diamicts overlie bedrock throughout much of the catchment and the soils developed on this are generally thin and peaty.

Crummock Water is oligotrophic, with a mean depth of 36.7 m, maximum length of 4 km, and a surface area of 2.5 km². The lake basin is steep sided with a relatively flat bottom.

Methods

Cores were taken from Crummock Water in 1987 using six metre Mackereth and one metre mini-Mackereth corers. The long cores were extruded in the field and wrapped in plastic sheeting, the short cores being extruded and sliced in the laboratory. All cores were sliced into 2 cm thick sections and over dried at ≤40 °C. One 6 m and one 1 m core from the central, deepest part of Crummock Water were used in this study, with the 1 m core used to provide surface sediment which is lost during the six metre coring process, (Mackereth, 1969).

Laboratory analyses included the following:

i) Natural Remanent Magnetism measurements for core dating: two 6 m cores from Crummock Water were measured in sections in a whole core spinner magnetometer for declination and intensity variation which allows the construction of a relative declination change curve, (Molyneux *et al.*, 1973). Comparison of the declination curve with an independently dated regional declination curves allows transfer of dates to matched declination change features. Data for Crummock Water, Windermere, and Loch Lomond (from Turner & Thompson, 1979) are illustrated on Fig. 1A. Approximate ages for samples selected from between dated horizons were calculated using sediment accumulation rates based on adjacent pairs of dated horizons assuming a constant sediment accumulation rate between them. A date of *circa* 10 000 B.P. for the top of the microlaminated clay (Fig. 1) was obtained by comparison with the similar independently-dated Lake Windermere stratigraphy (for example, Pennington, 1981), where the change from laminated glacial to more organic post-glacial sediments represents the regional waning of Late Glacial corrie glacier activity.

ii) Magnetic measurements: samples of dried sediment of known mass were tightly packed into 10 cm³ plastic containers for measurement of magnetic susceptibility using a Bartington MS1 dual frequency susceptibility meter. The results, (Fig. 1B) allow identification of sediment of relatively higher ferrimagnetic content and thus provided a guide for sub-sampling for magnetic extracts and spherule counting.

iii) Magnetic extract: dried sub-samples, from which any sediment which had been in contact with the core tubes was excluded, were lightly crushed and a weighed amount (≤1 g) mixed with a solvent (acetone was used in this study) and dispersed by ultrasound for fifteen minutes. Aliquots were pipetted into a glass petri dish attached to the pole of a large (55 mT) permanent magnet. The magnetic gradient was increased by placing small iron bolts on the pole surface underneath the petri dish. The magnet was gently agitated allowing the magnetic particles in the aliquot to gather above the pole and the remaining liquid gently pipetted off. This process was repeated until the magnetic fraction appeared free of clay and other non-magnetic sediment. The magnetic extract solution was left to allow the acetone to evaporate and then transferred to a well slide and covered with a cover slip.

The slides were examined at ×100 magnification under a light microscope using strong side-

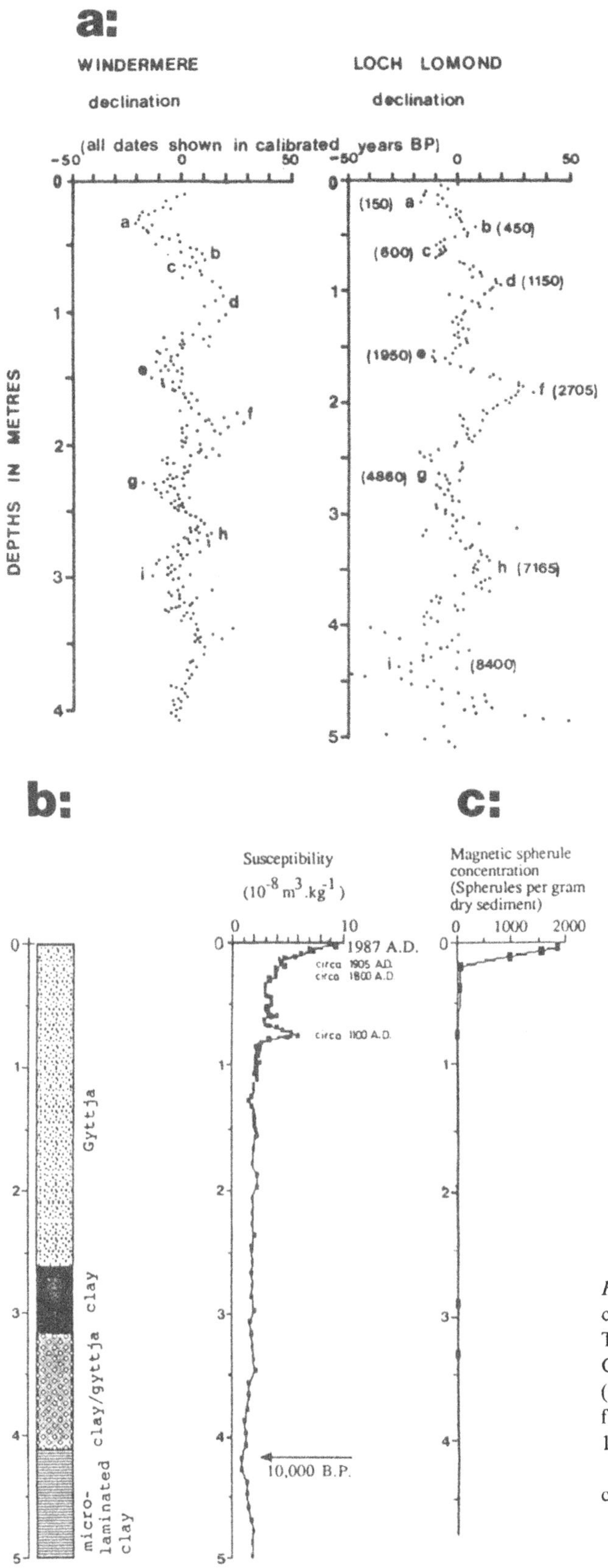

Fig. 1. A – Dated natural remanent magnetism declination curves from Windermere and Loch Lomond (Turner and Thompson, 1979), and relative declination curves for two Crummock Water cores, (C2L and C3L). Dated features (years B.P.) are: a – 150; b – 450; c – 600; d – 1150; e – 1950; f – 2705; g – 4860; h – 7165; i – 8400, (Turner & Thompson, 1979).

B – sedimentary log and magnetic susceptibility profile for central Crummock Water core, C2L;

C – magnetic spherule concentration for central Crummock Water core, C2L.

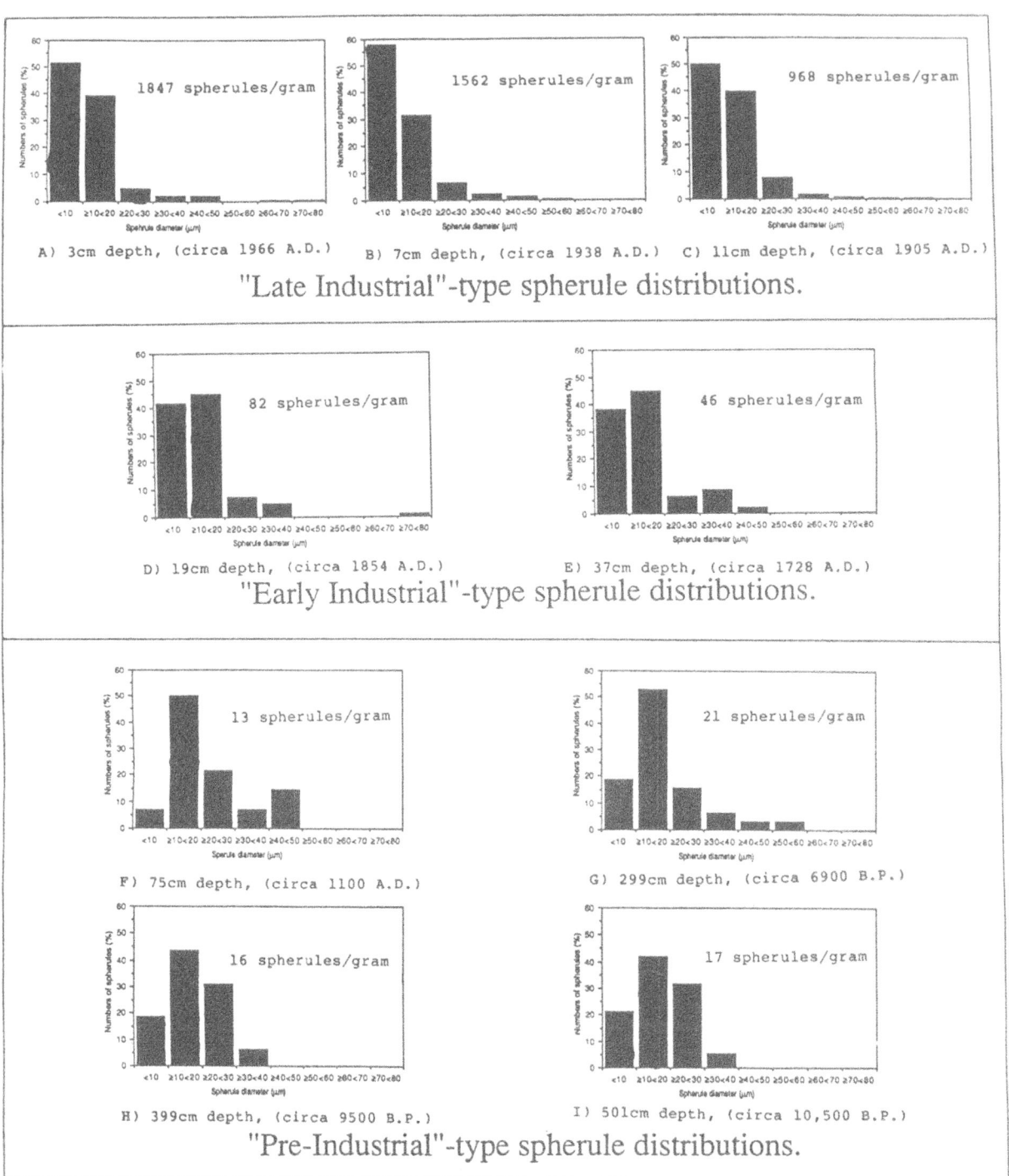

Fig. 2. Size distributions of magnetic spherules in magnetic extracts from a central Crummock Water core.
Graphs A, B, and C show distributions characteristic of high levels of industrially-derived spherules.
Graphs D and E show transitional distributions, a mixture of low levels of industrially-derived spherules with cosmic spherules.
Graphs F, G, H, and I are from samples predating industrial activity and therefore represent distributions characteristic of cosmic spherules alone.

lighting for illumination. This allows the most satisfactory view of the spherules, which are easily distinguished from any remaining detritus by their shape and metallic lustre. Total numbers of spherules were counted into size groups of 10 μm intervals, (estimated using an eyepiece graticule).

Results

Results of magnetic susceptibility measurements and spherule concentration (expressed as numbers of spherules per gram of dry sediment) are shown on Fig. 1B and 1C, with their size distributions on Fig. 2. Magnetic susceptibility is dominated by the concentration of ferrimagnetic minerals in a given sample (for example Locke & Bertine, 1986), thus the increases in magnetic susceptibility towards the core top indicate large increases in the proportion of ferrimagnetic minerals present in the sediment.

Magnetic spherules are present in all the nine samples examined from along the core, which included four samples ranging in age from *circa* 750–10000 B.P. Total spherules numbers are generally low (<30 spherules g^{-1}) but numbers increase rapidly by two orders of magnitude in the uppermost sediment, (approaching 2000 spherules g^{-1}). Recent sediment samples (post-1900 A.D.) show size distributions dominated by the <10 μm size fraction; from *circa* 1700 to 1900 AD the 10–20 μm size fraction is dominant, but the <10 μm fraction is still large; samples prior to *circa* 1700 show size distributions dominated by the 10–20 μm fraction with a relatively small <10 μm fraction.

Discussion

Oldfield *et al.* (1978) demonstrated that, for ombrotrophic peat bogs, magnetic measurements alone could be used to monitor the record of magnetic atmospheric inputs, large increases in concentration-dependent magnetic parameters mainly reflecting deposition of particulate iron in the form of urban and industrially-generated magnetic spherules. Limnic systems are more complex however, as lacustrine sediments are substantially derived from terrestrial catchment sources which usually contain additional sources of magnetic oxides. A lake sediment susceptibility profile, which records changes in concentration of all ferrimagnetic particles, may therefore record changes in terrestrial sources in addition to changing atmospheric inputs. It is therefore necessary to identify what ferrimagnetic particles are contributing to particular susceptibility features.

Susceptibility measurements from the Crummock Water samples in this study show two marked increases, both in the top metre of sediment. As discussed above, these increases may potentially reflect increased inputs of any one or all sources – primary or secondary source-derived iron oxides or magnetic spherules.

The lower susceptibility peak, centred at around 0.75 m, is dated to approximately 1100 A.D., (Fig. 1A). This predates industrialisation and is not associated with increases in total numbers of spherules, (Figs. 1C, and 2F). Pollen analysis of sediments from other lakes in the Lake District has indicated this period as one of the regional forest clearance and hence of soil disturbance, (Pennington, 1981). Licences granted in the early 1300's for fortification of buildings situated on the south-western edge of Crummock Water (Collingwood, 1928, p. 105) also demonstrate increasing local catchment disturbance from human settlement around Crummock Water during this period, which is also likely to have caused accelerated rates of soil erosion, and hence greater input of magnetically-enhanced soil-derived material. This circumstantial evidence – the coincidence with regional deforestation and additional more local soil-disturbing events – suggests that an increased input of magnetically-enhanced topsoil to Crummock Water is the most likely source for this lower increase in susceptibility.

The upper increase in susceptibility starts at approximately 11 cm, and interpolation between the NRM-dated 1800 A.D. horizon and the core top (1987 A.D.) suggests an age of *circa*

96

1905 A.D. for the start of this increase. Comparison of Figs. 1B and 1C shows this increase in susceptibility as synchronous with dramatically increasing numbers of magnetic spherules. Rising numbers of spherules therefore contribute at least in part to the observed upper increase in susceptibility. As susceptibility responds to all ferrimagnetic material, and the amount of non-spherical magnetic material was not quantified in this study due to extreme variation in size and shape of this material, the exact contribution of the spherules alone to the susceptibility values of the total magnetic assemblage is not known. However, the close similarity of both the susceptibility and spherule concentration profiles suggests that there, magnetic spherules provide a major contribution to total susceptibility values.

Known sources of magnetic spherules were described in the introduction: *natural*, from cosmic 'micrometeorites' and volcanic plumes, and generally occurring in relatively low numbers; *industrial*, mainly from high temperature combustion of coal in coal-burning power stations (Locke & Bertine, 1986), where spherules are produced and, particularly if unfiltered, released into the atmosphere in large numbers. Sediments post-dating industrialisation may therefore contain a mixture from both natural and industrial sources, with spherules of industrial origin by far outnumbering natural spherules, and the increasing release of spherules into the atmosphere from the early 1900's (when coal-burning power stations became more common) likely to result in greater total numbers of spherules deposited in more recent sediments. Results from other studies of a variety of different sediments have shown large increases in spherule numbers coincident with industrialisation (for example, Oldfield et al., 1978) and particularly the widespread commissioning of coal-fired power generating stations after about 1900 A.D., (Locked & Bertine, 1986).

The increase in spherule concentration towards the present observed in the extracts from this Crummock Water core conforms to the pattern predicted from the progression of industrialisation, as discussed above. Low numbers (<30 spherules g^{-1}) are present in all 'Preindustrial'

samples (Fig. 2, F–I); increasing numbers ($>30 < 100$ spherules g^{-1}) are found in post-1700 – pre-1900 A.D., 'Early Industrial', samples; high numbers (>900 spherules g^{-1}) are found in post-1900 A.D., 'Late Industrial', samples when, particularly, the number and size of coal-burning power stations increased. It therefore seems most likely that the source for the additional spherules over the low 'background' values is industrial, mainly from coal-fired power stations.

There are many suitable industrial sources close to the Lake District which could have contributed to the industrial atmospheric pollution apparently recorded in these sediments, (Liverpool, Manchester, Leeds, Newcastle, and Glasgow are all within a 100 mile radius of the Lake District). Since the finest of the spherules ($<10 \mu m$), which are actually present in greatest numbers in the post-1900 A.D. sediments may travel by wind over 3 000 miles from source (Parkin et al., 1980), sources are not only restricted to such 'local' areas however. Large increased in total numbers of magnetic spherules from industrial outputs have been observed in studies of sediment profiles from a variety of environments and locations, (for example Locke & Bertine, 1986; Oldfield et al., 1978; Oldfield et al., 1981). Together, these studies demonstrate the widespread nature of increasing spherule numbers and hence of its potential as a regional dating horizon.

Figure 2 shows that the spherule size distributions change in the more recent sediments, becoming increasingly dominated by the finer ($<10 \mu m$) fraction. This may reflect the additional industrial sources for spherules following the Industrial Revolution, where the largest producers of magnetic spherules – coal-fired power generating stations – typically show a peak modal particle size in the $10 \mu m$ region, (McElroy et al., 1982). The sorting effect of long-distance wind transport, if more distant sources are contributing to the spherules observed in the recent Crummock Water samples, may additionally increase the proportion of finer-sizes of spherules deposited in the sediments.

Summary

The numbers of magnetic spherules in post-1900 A.D. sediments from Crummock Water increase by nearly two orders of magnitude over low, natural, background values. The most likely sources for these additional spherules are industrial, particularly coal-burning power stations.

Increasing magnetic spherule numbers in recent sediments, due to industrial atmospheric discharge, provide a regional horizon with the potential for use as an approximate dating horizon in other lake sediments. Susceptibility measurements can help identify this horizon, and such measurements are both rapid and non-destructive. As lake sediments also record inputs from other terrestrial sources, it will often be necessary to confirm increasing spherule numbers by microscopic counting. Such counts are readily performed using a permanent magnet for extraction and an optical microscope.

Acknowledgements

This work was undertaken during a research studentship at Luton College of Higher Education financed by Bedfordshire Education Authority and Luton College of Higher Education, with additional funds for sample collection from the Bill Bishop Memorial Trust and the Quaternary Research Association Grants to Young Research Workers. The studentship was supervised by Dr. A.A. Morrison (Crewe and Alsager College) Dr. C. Eccles (Luton College) and Prof. F. Oldfield (Liverpool University). All are gratefully acknowledged. Thanks also to Liverpool University Geography Department for the use of coring and magnetics equipment, the Freshwater Biological Association for efficiently taking the 6 m Mackereth cores, and Newcastle University Department of Geophysics and Planetary Physics for use of their whole core spinner magnetometer.

References

Arthurton, R. S., D. V. Frost, D. C. Greig & A. A. Wilson, 1980. The Lake District (Sheet 54 N 04 W) Solid Edition. Institute of Geological Sciences.

Collingwood, W. G., 1928. Lake district history. Wilson & Son, Kendal, 175 pp.

Fisher, G. L., B. A. Prentice, D. Silberman, J. H. Ondov, A. H. Bierman, R. C. Ragaini & A. R. McFarland, 1978. Physical and morphological studies of size-classified coal fly ash. Envir. Sci. Technol. 12: 447–451.

Locke, G. & K. K. Bertine, 1986. Magnetite in sediments as an indicator of coal combustion. Appl. Geochem. 1: 345–356.

Mackereth, F. J. H., 1969. A short core sampler for subaqueous deposits. Limnol. Oceanogr. 14: 145–151.

McElroy, M. W., R. C. Carr, D. S. Ensor & G. P. Markowski, 1982. Size distribution of fine particles from coal combustion. Science 215: 13–18.

Molyneux, L., R. Thompson, F. Oldfield & M. E. McCallan, 1972. Rapid measurement of the remanent magnetisation of long cores of sediment. Nature 237: 42–43.

Oldfield, F., R. Thompson & K. E. Barber, 1978. Changing Atmospheric Fallout of Magnetic Particles Recorded in Recent Ombrotrophic Peat Sections. Science 199: 679–680.

Oldfield, F., K. Tolonen & R. Thompson, 1981. History of particulate atmospheric pollution from magnetic measurements in dated Finnish peat profiles. AMBIO 10: 185–188.

Parkin, D. W., D. R. Phillips, R. A. L. Sullivan & L. Johnson, 1970. Airborne Dust Collections over the North Atlantic. J. Geophysic. Res. 75: 1782–1793.

Parkin, D. W., R. A. L. Sullivan & J. N. Andrews, 1980. Further studies on cosmic spherules from deep-sea sediments. Phil. Trans. R. Soc. Lond. A. 297: 495–518.

Pennington, W., 1981. Records of a lakes life in time: the sediments. Hydrobiologia 79: 197–219.

Puffer, J. H., E. W. B. Russel & M. R. Rampino, 1980. Distribution and origin of magnetic spherules in air, waters, and sediments of the Greater New York City area and the North Atlantic Ocean. J. Sediment. Petrol. 50: 247–256.

Renburg, I. & M. Wik, 1984. Dating recent sediments by soot particle counting. Verh. Int. Ver. Limnol. 22: 712–718.

Turner, G. M. & R. Thompson, 1979. Behaviour of the Earth's magnetic field as recorded in the sediment of Loch Lomond. Earth Planet. Sci. Lett. 42: 412–426.

Wik, M., I. Renburg & J. Darley, 1986. Sedimentary records of carbonaceous particles from fossil fuel combustion. Hydrobiologia 143: 387–394.

Hydrobiologia **214**: 99–106, 1991.
J. P. Smith, P. G. Appleby, R. W. Battarbee, J. A. Dearing, R. Flower, E. Y. Haworth, F. Oldfield & P. E. O'Sullivan (eds),
Environmental History and Palaeolimnology.
© *1991 Kluwer Academic Publishers.*

Lake sediment records of erosional processes

J. A. Dearing
Centre for Environmental Science Research and Consultancy, Department of Geography, Coventry Polytechnic, Coventry CV1 5FB, UK

Key words: lake-catchment systems, erosional processes, sediment sources, erosion models

Abstract

Accumulations of sediment at the beds of lakes, estuaries and reservoirs provide partial records of materials transported from the surrounding water catchment areas. Physical, chemical, biological and magnetic analyses, with data for accumulation rates, have been used in a range of environmental settings to infer the rate, form, cause and source of erosion. This paper is a brief review of these studies from a hydrological perspective, setting sediment studies within a lake-catchment system. The need for long term erosional records is discussed in terms of the type of erosional data which may be obtained. Alternative approaches to studying short and long term erosion are assessed with regard to their cost-effectiveness and their levels of precision and accuracy. Finally, some suggestions are made about how these erosion records may be used to model hydrological, pedological and geomorphological processes, thus linking together long term, short term and contemporary timescales of process operation.

Introduction

Numerous palaeolimnological studies have shown how sediment properties may be interpreted in terms of processes acting within the surrounding catchment (or paralimnion). Only recently, though, have there been specific and deductive studies on how the sediment column may allow an extension of contemporary studies of erosional processes. Figure 1 represents a generalized lake-catchment system and shows the major subsystems which exert control on the rate and mode by which material enters a water body before it becomes deposited as bottom sediment. Notable features of the system are its essentially cascading nature and the variety of potential sediment sources which exist in a catchment. These two features alone ensure that it will normally be difficult to identify the precise form of processes,

such as stream channel scour or hillslope rilling, from an examination of the bottom sediments. Similarly, an estimation of the quantity of material deposited in the lake may be difficult to interpret in terms of specific catchment-erosion relationships. Sediment reaching the lake bed will reflect not only the transporting energy in the catchment but also the availability of sediment for erosion and the timelags in the system as material moves towards the lake bed via temporary sediment storage zones such as floodplains and deltas. Despite these problems, however, several studies have recently produced apparently reliable data on erosion, sediment yields and sediment sources, and have shown that it is possible to quantify many of the sediment flows (Figure 1) through time.

The need for long records of erosion processes is dependent upon the questions being asked, and

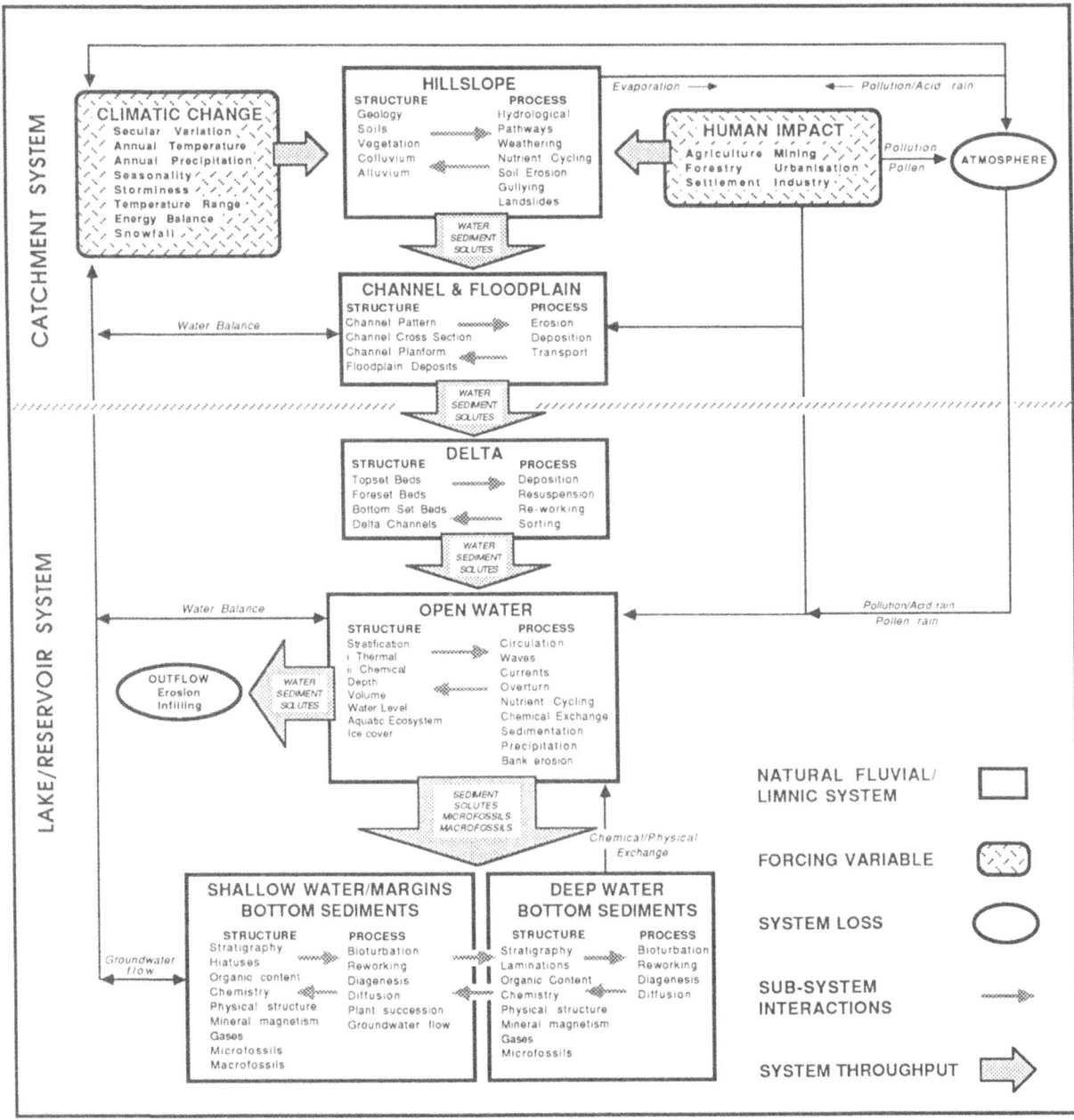

Fig. 1. The lake-catchment system (modified from Foster *et al.*, 1988), showing the range of environmental processes and factors which control the movements of materials between sub-systems.

it is useful here to list the range of applications:
1. Defining 'background' or 'natural' and 'accelerated' gross rates of erosion.
2. Identifying the erosional responses to rapid environmental change, usually human impact.
3. Reconstructing catchment sediment budgets.
4. Identifying erosion processes and long term pedological changes.
5. Creating or testing erosional models over different timescales.

Before reviewing some of the recent developments in each of these categories, it is necessary to assess the different approaches to obtaining erosional data from lake sediment records.

The type of data

Erosion data are drawn from either the quantity or quality of bottom sediment. For many lakes, accumulation rates in central, deep water cores will normally provide a good approximation of gross changes in the rate of allochthonous inputs, especially where rates are converted to influx rates of allochthonous inorganic material. Such data will have the same time resolution as the chronology, producing mean influx data over hundreds of years (^{14}C) to a few years (^{210}Pb) and, where annual laminations are present, over one year. Several studies, however, consider the variability of sedimentation patterns through time (e.g. Dearing, 1983; Dearing et al., 1981; Foster et al., 1985; Foster et al., 1986) and the mechanisms of sediment focusing and redistribution (Hilton, 1985; Hilton et al., 1986; Lehman, 1975; Likens & Davis, 1975), and illustrate the difficulties of predicting the location of a core with the mean sedimentation rate for the lake.

These difficulties have led to several studies where sediment yields have been calculated by correlating synchronous levels (Dearing, 1986) in a large number of sediment cores to a master chronology; using density measurements to convert basin-wide estimates of sediment volume to sediment mass deposited over a specific time period from the catchment area (eg. t km^{-2} yr^{-1}). This approach overcomes many of the problems of unrepresentative sedimentation records but increases the accuracy of sediment yield calculations at the expense of temporal resolution; core correlations are not normally possible on one or two centimetre levels, and the resulting sediment yield record is usually in terms of decadal or century mean values. High density sampling is also time-consuming, expensive and potentially damaging to aquatic ecosystems and future sediment studies. In many cases it may be more useful to use the accumulation rates in a central, deep water core, corrected for major deviations from the mean sedimentation value after correlation to a few other cores. The definition of past and present sedimentation areas in small lakes is likely to represent a large source of error in sediment yield calculations, especially where there is evidence that the sediment limit has changed because of water-level fluctuations (cf. Dearing & Foster, 1986a; Dearing et al., 1990). Assessment of the true catchment-derived sediment yield will require subtraction from the gross sediment yield of those components originating from lake margin erosion, the atmosphere and autochthonous production (see Dearing & Foster, 1986a).

The quality of sediment, as defined by chemical (eg. Thompson et al., 1986), radiometric (eg. Wasson et al., 1987), magnetic (eg. Dearing & Foster, 1986b) and microfossil (eg. Oldfield & Clark, 1990) properties, has been used to infer the source of deposited material, notably topsoil/subsoil and weathered/non-weathered sources. On their own, these data can only be used to infer, rather than to measure, the rate and processes of erosion, but when linked to a good master chronology may produce useful estimates of sediment influx from different sources.

Soil erosion and environmental change

Sediment data may be used to confirm evidence for past catchment disturbances, or to quantify the soil losses under 'natural' and 'accelerated' forms of erosion. Trends in soil erosion are difficult to determine on the basis of short term contemporary monitored data in instrumented catchments. Ergodic reconstructions of erosional and related processes inevitably lose the detail of the transition phases between different systems in apparent steady state, as may be seen in a comparison of the sediment yield records obtained by Davis (1976) and Wolman (1967) (Fig. 2). Recent summaries of sediment yield records (Dearing, in press; Duck & McManus, 1990; Foster et al., 1990) show clearly the 'accelerated' rates of

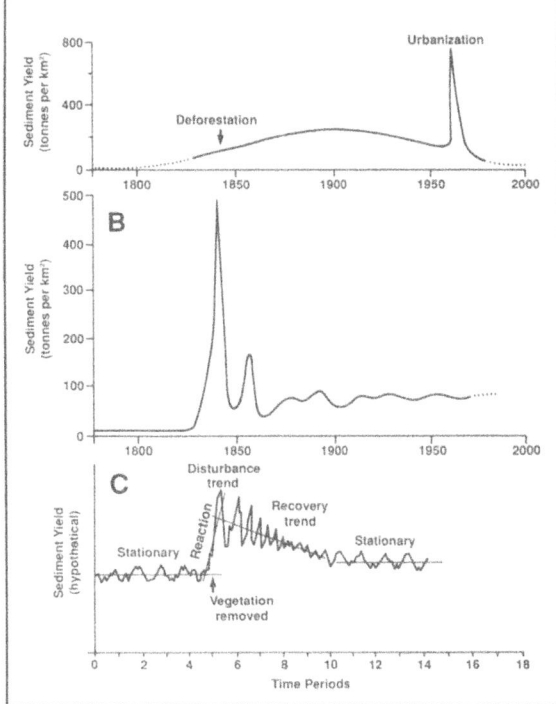

Fig. 2. Sediment yields derived from (A) ergodic (time-space) sequences of instrumented catchment records in the Piedmont region, U.S.A. (Wolman, 1967), (B) lake sediment records from Frain's Lake, Michigan, U.S.A. (Davis, 1976), and (C) a hypothetical response to devegetation (Thornes and Brunsden, 1977). Note the high resolution of the lake based record, especially through the early phase of deforestation, and the potential for quantifying the hypothetical 'background' and 'contemporary' stationary yields, reaction and recovery rates.

erosion in recent centuries which exist at many sites worldwide, but several studies also show the more dramatic erosional impact of earlier societies, especially in Mediterranean and sub-tropical environments (eg. Deevey, *et al.*, 1979; Hutchinson *et al.*, 1970). Many of the records show increases in sediment yield in excess of one order of magnitude over the yields from relatively undisturbed catchments (see also Binford *et al.*, 1983), but care should be taken in interpreting these records in terms of soil loss. Where sediments have been matched to sources, or where sediment budgets have been constructed (see below), the evidence frequently points to catchment disturbances increasing the efficiency of erosion processes acting within the stream channels, rather than increasing the rate of soil erosion from hillslopes directly.

Sediment budgets

Some recent studies have shown the ways in which lake sediment data can be linked to contemporary erosion data in the catchment. 'Fingerprinting' techniques have been used with estimates of sediment yields to combine suspended stream sediment, soils and lake sediment into a sediment budget (Foster *et al.*, 1988). Such case-studies help to link together different scales of operation of processes, both in space and time, and tend to highlight the most important aspects of the sediment system (eg. by showing the importance of sediment storage in floodplains). One study has reconstructed partial chemical budgets (phosphorus) for late-Medieval and contemporary periods in a Swedish agricultural catchment (Dearing *et al.*, 1987), a conceptual approach which has been extended by Deevey *et al.* (1979) who were able to study the sustainability of agriculture in Mayan societies (reviewed by Binford *et al.*, 1983).

Erosion processes and soil development

The majority of studies under this heading have focused on the nature of the sediment; linking sediment qualities to particular events in the catchment, such as deforestation or early farming. Many studies, following the work and ideas of Mackereth (1966), have explained changes in the bulk chemistry of sediments in terms of weathering or soil development, but there has been no attempt to evaluate some of the common conclusions in terms of what is known about Holocene weathering rates or the chemical differentiation of soil horizons. Recent papers by Engstrom and Wright (1984) and Engstrom & Hansen (1985) describe sequential digest techniques which are able distinguish between allogenic and authigenic chemical phases in lake sediment.

Their results for high latitude sites indicate that for some lake sediment the allogenic component originating in the catchment has changed little in its chemical composition over millenia, and that the effects of weathering and pedogenesis may be imperceptible in the sediment record. Their arguments represent an important shift in thinking about the chemical linkages between lake sediment and catchment sources.

Where variations do exist between the allogenic chemistry at different sediment depths there is the further possibility that the chemical properties are controlled by variations in particle-size. There is little direct chemical evidence for this proposition, although Thompson and Morton (1979) were able to show that the bulk chemistry was significantly different (by up to 17% for Al_2O_3) in clay/silt and sand fractions of Loch Lomond sediment. However, relationships between magnetic susceptibility and particle-size in a range of soils and sediments show steep gradients, especially between the silt and clay fractions which normally dominate lake sediments (Dearing et al., 1985). In a study of modern trapped sediment in Lough Neagh, Dearing & Flower (1982) were able to demonstrate that the magnetic properties of the sediment were correlated to rainfall, which implied that long sediment records of magnetic susceptibility (Thompson et al., 1975) could be interpreted in terms of processes acting on coarse particles in stream channels rather than on hillslopes. There are substantial problems of determining particle-size distributions in sediments, particularly with respect to the degree to which aggregation and disaggregation in limnic and fluvial environments takes place (Holmes, 1968; Walling & Moorehead, 1988) and the need to identify the effective or absolute particle-sizes (Walling, 1988); but such data may prove to be decisive in evaluating chemical and magnetic records, and in providing strong indications of past fluvial conditions.

Modelling and explanation

Recently, several studies have tackled the problem of how to model sediment sources, largely through the use of mineral magnetic techiques. Frequency dependent magnetic susceptibility has been used to model the contributions from topsoil and channel/subsoil sources (Dearing & Foster, 1986b), and other magnetic parameters have been used to calibrate lake sediment properties against the properties of artificial mixtures of sediment sources (Foster et al., 1988; Stott, 1987). This approach has been extended by Yu & Oldfield (1989) who used multivariate statistics to reconstruct the contributions from a number of magnetically distinct source materials in estuarine cores from Chesapeake Bay, U.S.A., and by Thompson et al. (1986) who were able to model the erosion of volcanic ash, chemically, in Icelandic sediments.

Some studies point to the possibilities for empirical modelling of erosion processes. The record of sediment yield for the past 250 years at Seeswood Pool in Midland England was shown to be significantly correlated to annual rainfall over the same period (Dearing & Foster, 1986b). Studies at several sites in Scania, Sweden, show that sediment yields are inversely related to lake-catchment size thus illustrating the importance of sediment storage within this landscape (Dearing, in press). Other studies (eg. Davis, 1976; Fig. 2B) indicate how it would be possible to quantify reaction and recovery rates to, and following, environmental change (Fig. 2C), and there are also several records of sediment yield data which could be used to make retrospective tests of empirical/statistical models of erosion and sediment yield. Unfortunately, much of the sediment data is not matched in terms of continuity, accuracy or precision by complementary records of other environmental data such as climate and land use, and future studies would do well to assess the quality of supporting records before deciding upon a suitable lake site. Inevitably, the success of long term empirical models of erosion and sediment yield will not only be limited by the unavailability of critical environmental data, but also by the fact that inference of cause and effect relationships will be difficult given that small variations in causative processes may give rise to a variety of outcomes (Thornes, 1987). This implies that more emphasis should be given to how lake sediment data may be used deductively to test models or hypotheses. As Thornes (1987) suggests, 'it is easier to refute inadequate models on the basis of very sparse historical data than to build them for such data' (p. 18).

Although the literature on lake acidification (eg. Battarbee, 1985) points the way ahead for palaeo-limnological data to provide deductive explanations of process-response mechanisms, there have been no attempts to use similar data to test deductive models of erosion based on physical reasoning. Specifically, models need to be developed which can be tested by sediment yield data; these data are likely, at present, to be more reliable than records of sediment source or palaeodischarge. Over long timescales it may be possible to parameterize the models with data from pollen diagrams. Not only do recent studies of the controls on sediment movements emphasize the role

of vegetation, but there are signs that pollen data may soon provide much more than descriptions of past vegetation; data for the vegetation structure of forests (eg. Bradshaw, 1981) and whole landscapes (J. B. Birks pers. comm.) could provide the means for establishing historical records of cover, biomass, productivity and soil organic matter, and thus the means for calibrating the equations of runoff and sediment mobilization in predictive models (Dearing *et al.*, 1987).

One area of model application would be to predict sediment yields under 'natural' conditions, for which a growing number of early and mid-Holocene records of sediment yield or sediment accumulation from temperate zones could be used as surrogate data sets. There is evidence to suggest that the 'natural' trend for sediment yield from the early Holocene to the present is decreasing or stationary (Fig. 3A-B). Decreasing trends imply either a reduction in sediment supply or a

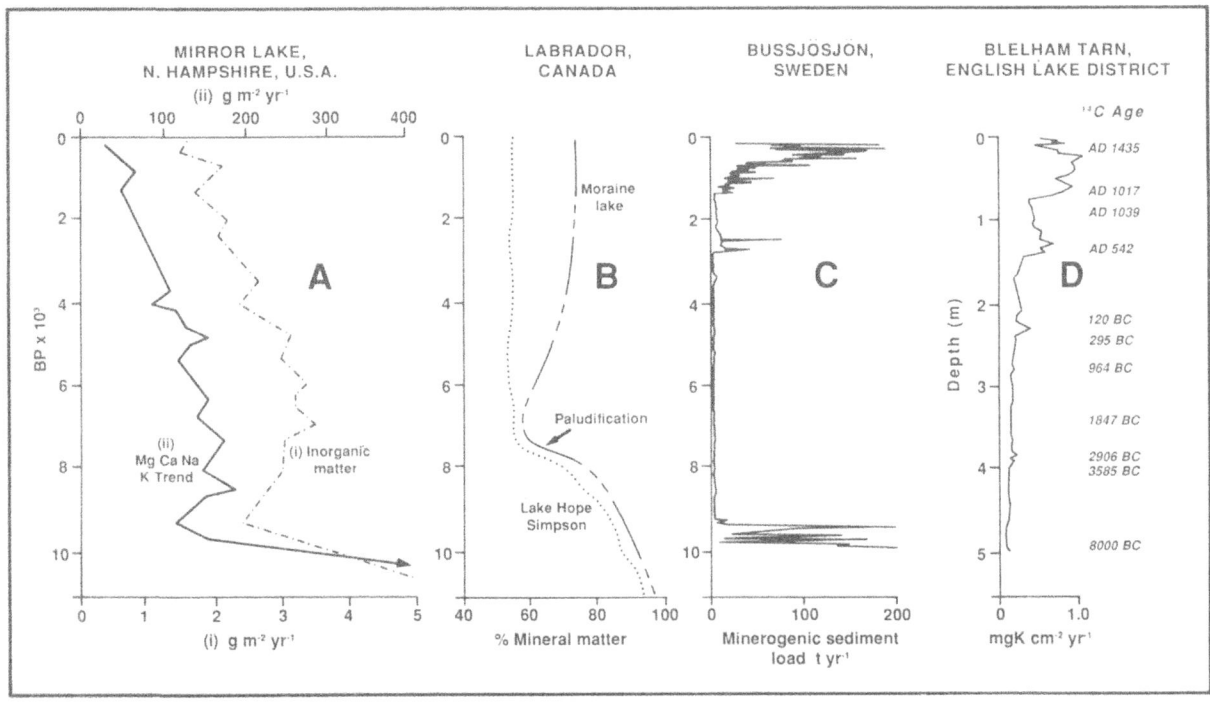

Fig. 3. Holocene records of erosion from (A) Mirror Lake, N. Hampshire, U.S.A. (Likens & Davis, 1975) showing accumulation of total alkali metals and inorganic sediment accumulation, (B) Lake Hope Simpson and Moraine Lake, Labrador (Engstrom & Hansen, 1985) showing the percentages of mineral matter and the onset of paludification, (C) Bussjösjön, S. Sweden, showing minerogenic sediment load from the catchment (Dearing *et al.*, 1990), and (D) Blelham Tarn, English Lake District, showing accumulation of total K, used here as an indicator of minerogenic input from the catchment (Pennington, 1981).

reduction in transporting capacity of the fluvial system. The data suggest long term and widespread shifts of the systems towards the equilibrium conditions set by ecological, pedological and hydrological controls which govern runoff and sediment supply. At all the sites there is some evidence that the evolution of organic-rich soil horizons plays a vital role in controlling sediment losses from the catchments. This is especially true at the Labrador sites (Fig.3B) where the process of paludification is correlated with both reduced and increased sediment losses in different catchments, implying a marked contrast in the hydrological response to a common pedological process (cf. Thornes, 1987). Although the authors (Engstrom and Hansen, 1985) propose a regional climatic shift to account for the different responses, there remains the possibility of complex feedback mechanisms controlling the nutrient availability for forest growth and hence the levels of forest cover, evapotranspiration and runoff. Elsewhere, at sites where human impact has been 'significant' (Fig. 3C-D), there are stepwise reversals in the trend from the mid to late Holocene onwards; trends which pose numerous questions about the magnitude of sediment responses to similar catchment disturbances, the processes of sediment delivery and the ability of the vegetation-soil systems to reach new quasi-equilibrium states. These examples illustrate that modelling the sediment yield record over long timescales for different kinds of environment, as defined by calibrated equations, might produce well-founded generalizations about the nature of controls on sediment movements in cool temperature ecosystems, and their sensitivity to human impact and climatic change.

References

Battarbee, R. W., R. J. Flower, A. C. Stevenson & B. Rippey, 1985. Lake acidification in Galloway; a palaeoecological test of competing hypotheses. Nature 314: 350–352.

Binford, M. W., E. S. Deevey & T. L. Crisman, 1983. Paleolimnology; an historical perspective on lacustrine ecosystems. Ann. Rev. Ecol. Syst. 14: 255–286.

Bradshaw, R. H. W., 1981. Quantitative reconstruction of local woodland vegetation using pollen analysis from a small basin in Norfolk, England. J. Ecol. 69: 941–955.

Davis, M. B., 1976. Erosion rates and land use history in Southern Michigan. Envir. Conserv. 3: 139–148.

Dearing, J. A., 1983. Changing patterns of sediment accumulation in a small lake in Scania, southern Sweden Hydrobiologia 103: 59–64.

Dearing, J. A., 1986. Core correlation and total sediment influx. In, B. E. Berglund (Ed), Handbook of Holocene Palaeoecology and Palaeohydrology, Wiley.

Dearing, J. A., 1991. Erosion and land use. Chap 5 In, B. E. Berglund et al. (Eds) The Cultural Landscape of the Ystad area 0-6000 BP, Ecobulletin Special Volume, in press.

Dearing, J. A., K. Alström, A. Bergman, J. Regnell & P. Sandgren, 1990. Past and present erosion in southern Sweden. In, J. Boardman, I. D. L. Foster, J. A. Dearing (Eds), Soil erosion on agricultural land, Wiley.

Dearing, J. A., J. K. Elner & C. M. Happey-Wood, 1981. Recent sediment flux and erosional processes in a Welsh upland lake catchment based on magnetic susceptibility measurements Quat. Res. 16: 356–372.

Dearing, J. A. & R. J. Flower, 1982. The magnetic susceptibility of sedimenting material trapped in Lough Neagh, Northern Ireland, and its erosional significance. Limnol. Oceanogr. 27: 969–975.

Dearing, J. A. & I. D. L. Foster 1986a. Lake sediments and palaeohydrological studies. In, B. E. Berglund (Ed), Handbook of Holocene Palaeoecology and Palaeohydrology, Wiley, 706 pp.

Dearing, J. A. & I. D. L. Foster, 1986b. Limnic sediments used to reconstruct sediment yields and sources in the English Midlands since 1765. In, V. Gardiner (Ed), Int. Geomorphol. I, Wiley.

Dearing, J. A., H. Håkansson, B. Liedberg-Jönsson, A. Persson, S. Skansjö, D. Widholm & F. El-Daoushy, 1987. Lake sediments used to quantify the erosional response to land use change in southern Sweden. Oikos 50: 60–78.

Dearing, J. A., B. A. Maher & F. Oldfield, 1985. Geomorphological linkages between soils and sediments: the role of magnetic measurements. In, K. S. Richards, R. R. Arnett & S. Ellis (eds), Geomorphology and Soils, Allen and Unwin.

Deevey, E. S., D. S. Rice, P. M. Rice, H. H. Vaughan, M. Brenner & M. S. Flannery, 1979. Mayan urbanism: impact on a tropical karst environment. Science 206: 298–306.

Duck, R. W. & J. McManus, 1990. Relationships between catchment characteristics, land use and sediment yield in the Midland valley of Scotland. In, J. Boardman, I. D. L. Foster & J. A. Dearing (eds), Soil erosion on agricultural land. Wiley, 706 pp.

Engstrom, D. R. & B. C. S. Hansen, 1985. Postglacial vegetational change and soil development in southeastern Labrador as inferred from pollen and chemical stratigraphy Can. J. Bot.: 63543–561.

Engstrom, D. R. & Jr., H. E. Wright, 1984. Chemical stratigraphy of lake sediments as arecord of environmental

106

change. In, E. Y. Haworth and J. G. W. Lund (eds), lake sediments and environmental history, Leicester University Press, U.K.: 11–67.

Foster, I. D. L., J. A. Dearing & P. G. Appleby, 1986. Historical trends in catchment sediment yields: a case study in reconstruction from lake-sediment records in Warwickshire, U.K. Hydrol. Sci. J. 31: 427–443.

Foster, I. D. L., J. A. Dearing & R. Grew, 1988. Lake-catchments: an evaluation of their contribution to studies of sediment yield and delivery processes Sediment Budgets (Proc. Porto Alegre Symp.) I.A.H.S. Publn. no. 174: 143–424.

Foster, I. D. L., J. A. Dearing, R. Grew & K. Orend 1990. The sedimentary data base; an appraisal of lake and reservoir of studies of sediment yield. Erosion Transport and Deposition Processes (Proc. Jerusalem Workshop) I.A.H.S. Publ. no. 189: 119–143.

Foster, I. D. L., J. A. Dearing, A. Simpson, A. D. Carter & P. G. Appleby, 1985. Lake catchment based studies of erosion and denudation in the Merevale catchment, Warwickshire, U.K. Earth Surf. Processes and Landfs. 10: 45–68.

Hilton J., 1985. A conceptual framework for predicting the occurrence of sediment focusing and sediment redistribution in small lakes. Limnol. Oceanogr. 30: 1131–1143.

Hilton, J., J. P. Lishman & P. V. Allen, 1986. The dominant processes of sediment distribution and focusing in a small, eutrophic, monomictic lake. Limnol. Oceanogr. 31: 125–133.

Holmes, P. W., 1968. Sedimentary studies of Late Quaternary material in Windermere Lake (Great Britain). Sediment. Geol. 2: 201–224.

Hutchinson, G. E., E. Bonatti, V. M. Cowgill & C. E. Goulden, 1970. Ianula: an account of the history and development of the Lago di Monterosi, Latium, Italy. Trans. Am. Philos. Soc. (NS) 60: 1–78.

Lehman, J. T., 1975. Reconstructing the rate of accumulation of lake sediments: the effect of sediment focusing. Quat. Res. 5: 541–550.

Likens, G. E. & M. B. Davis, 1975. Post-glacial history of Mirror Lake and its watershed in New Hampshire, U.S.A.: an initial report; Int. Ver. Theor. Angew. Limnbol. Verh. 19: 982–993.

Marckereth, F. J. H., 1966. Some chemical observations on post-glacial lake sediments. Phil. Trans. R. Soc. London, ser. B: 2500165–213.

Olfdield, F. & R. Clark, 1990. Sediment-based studies of soil erosion. In, J. Boardman, I. D. L. Foster and J. A. Dearing (Eds), Soil erosion on agricultural land, Wiley.

O'Sullivan, P. E., 1983. Annually-laminated lake sediments and the study of Quaternary environmental changes. Quat. Sci. Rev. 1: 245–313.

Pennington, W. (Mrs T. G. Tutin), 1981. Records of a lakes's life in time: the sediments. Hydrobiologia 79: 197–219.

Stott, A. P., 1987. Medium-term effects of afforestation on sediment dynamics in a water supply catchment: a mineral magnetic interpretation of reservoir deposits in the Macclesfield forest. Earth Surf. Proc. Landf. 12: 519–630.

Thompson, R., R. W. Battarbee, P. E. O'Sullivan & F. Oldfield, 1975. Magnetic susceptibility of lake sediments Limnol. Oceanogr. 20: 687–698.

Thompson, R., R. H. W. Bradshaw & J. E. Whitley, 1986. The distribution of ash in Icelandic lake sediments and the relative importance of mixing and erosional processes. J. Quat. Sci. 1: 3–11.

Thompson, R. & D. J. Morton, 1979. Magnetic susceptibility and particle-size distribution in recent sediments of the Loch Lomond drainage basin Scotland. J. Sedimentary Petrol. 49: 801–812.

Thornes, J. B., 1987. Models for palaehydrology in practice. Palaeohydrology in Practice. (Eds) K. J. Gregory, J. Lewin & J. B. Thornes, Wiley.

Thornes, J. B. & D. Brunsden, 1977. Geomorphology and time. Methuen.

Walling, D. E., 1988. Erosion and sediment yield research – some recent perspectives. J. Hydrol. 100: 113–141.

Walling, D. E. & P. W. Moorehead, 1988. The particle size characteristics of fluvial suspended sediment: an overview. Proc. Fourth Int. Symp. Interaction Sediments Water, Melbourne, 1987.

Wasson, R. J., R. L. Clark, P. M. Nanninga & J. Walters, 1987. [210]Pb as a chronometer and tracer, Burrinjuck Reservoir, Australia, Earth Surf. Proc. and Landf. 12: 399–414.

Wolman, M. G., 1967. A cycle of sedimentation and erosion in urban river channels, Geografiska Annaler 40A: 385–395.

Yu, L. & F. Oldfield, 1989. A multivariate mixing model for identifying sediment source from magnetic measurements Quat. Res.

Hydrobiologia **214**: 107–114, 1991.
J. P. Smith, P. G. Appleby, R. W. Battarbee, J. A. Dearing, R. Flower, E. Y. Haworth, F. Oldfield & P. E. O'Sullivan (eds), 107
Environmental History and Palaeolimnology.
© 1991 *Kluwer Academic Publishers.*

A multidisciplinary study of the lake Bjäresjösjön (S Sweden): land-use history, soil erosion, lake trophy and lake-level fluctuations during the last 3000 years

M.-J. Gaillard[1], J.A. Dearing[2], F. El-Daoushy[3], M. Enell[4] & H. Håkansson[1]
[1]*Dept. of Quaternary Geology, University of Lund, Lund Sweden;* [2]*Dept. of Geography, Coventry Polytechnic, Coventry, UK;* [3]*Institute of Physics, University of Uppsala, Sweden;* [4]*Swedish Environmental Research Institute, Stockholm, Sweden*

Key words: Holocene, South Sweden, cultural landscape, land-use history, soil erosion, lake trophy, lake-level fluctuations, pollen, diatoms, chemical analysis, palaeomagnetic measurements, sediment yields

Abstract

The lake Bjäresjösjön, Southern Scania, Southern Sweden, was studied in the context of the project 'The cultural landscape of the past 6000 years in Southern Sweden'. Pollen, plant macrofossils, diatoms, physical and chemical analysis, magnetic measurements and radiometric methods (^{210}Pb, ^{14}C) have been used to study palaeoecological changes, i.e. climate, land use, lake trophy and soil erosion during the past 3000 years. This multidisciplinary study shows striking responses of diatom communities, physical and chemical characteristics, sediment yields and magnetic parameters to land-use changes and lake-level fluctuations. Moreover, the latter are closely related to the settlement history at the site, inferred from archaeological records and historical sources.

Before 650 AD, the limnological development was affected mainly by lake-level fluctuations, but partly also by human impact (extensive forest clearings and dominant pastoral farming). With the expansion of arable farming (around 650 AD), human impact on the landscape was the major factor influencing soil erosion processes in the catchment and limnological changes in the lake.

Introduction

This paper aims to show the advantages of multidisciplinary studies of lake sediments in an attempt to reconstruct palaeoecological conditions and processes. We also illustrate the importance of studying sediments from small lake basins and catchments for a more precise description of local processes.

Similar approaches for palaeoenvironmental reconstructions over the Holocene period have been published by e.g. Dearing *et al.* (1987), Peglar *et al.* (1989) and Fritz (1989). Our investi-

gation of Bjäresjösjön was part of a large Swedish research project 'The cultural landscape of the past 6000 years in southern Sweden' (Berglund, 1988). The locality provides a particularly rich source of information which has contributed to our understanding of past ecological processes.

Pollen, plant macrofossils, diatoms, physical and chemical analyses, magnetic measurements and radiometric methods (^{210}Pb, ^{14}C) have been used to infer palaeoecological changes, such as climate, land use, lake trophy and soil erosion during the past 3000 years. This paper presents the lake sediment-based studies, but archaeological and historical records are mentioned, in order to provide a broader comprehension of the changes described, and for a better evaluation of the possible interactions between human activity and the environment. Only the most significant results are presented and discussed. A complete synthesis of the results will be published elsewhere (Gaillard *et al.*, in prep.).

Site description

Bjäresjösjön (55° 27′ N, 13° 45′ E, 48 m asl) is situated in the outer hummocky landscape of the project area and belongs to one of four key areas of investigation (Fig. 1). It is a small, more

or less circular lake, with a maximum diameter of 200 m, a surface area of ca 2 ha and a maximum depth of 1.7 m. It is of eutrophic to hypertrophic status, with a pH of 8,5. A natural outlet drained the lake to the river Svartån until sometime between 1812 and 1888. Then the outlet was canalized, and drainage has been through a subsoil pipe since 1970. The former catchment area (ca 25 ha) was reduced to ca 15 ha between 1809 and 1915. Bjäresjö village lies on the low hills north of the lake and cultivated fields prevail in the surroundings.

The Quaternary deposits of the area are dominated by clayey till. Peat-growth started in the central part of the Bjäresjö basin in the early Holocene (9700 BP), and peat accumulation (5 m) lasted until 2700 BP when a significant rise in the water table resulted in the creation of a lake. During the last 2700 years, *ca.* 4 metres of lake sediments have accumulated in the deepest part of the lake.

Methods

Detailed descriptions of the sampling methods employed will be given in Gaillard *et al.* (in prep.). For the purpose of calculating historical sediment yields (t ha^{-1} yr^{-1}) and measuring magnetic parameters (HIRM/χfd) of sediments to show their source, the lake sediments were sampled on a 25 m × 25 m grid (Dearing in press). Reconstruction of past lake-level fluctuations followed the research strategy proposed by Digerfeldt (1986) and is based on the study of a series of cores on a W–E transect (Gaillard & Digerfeldt, in press).

The 4 metres of lake sediments from the central core (C3) were investigated for pollen, plant macrofossil, diatom, magnetic parameters and physical and chemical stratigraphy. Dating of the uppermost metre of sediment was performed in two cores (C3 and C4) by ^{210}Pb dating (e.g. El-Daoushy, 1981; El Daoushy *et al.*, 1989). Layers with high organic content were ^{14}C-dated (Håkansson, 1986).

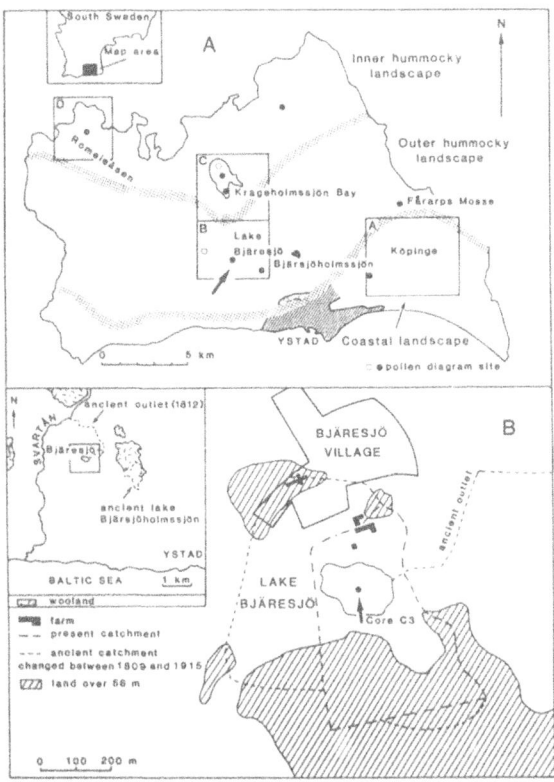

Fig. 1. Location of Lake Bjäresjö. A. Study area of the research project 'The cultural landscape of the past 6000 years in Southern Sweden' with the four major areas (A–D). Bjäresjösjön is situated in area B (from Berglund 1988). B. Bjäresjösjön site and location showing the land over 56 m, farms, Bjäresjö village and the catchment boundary.

109

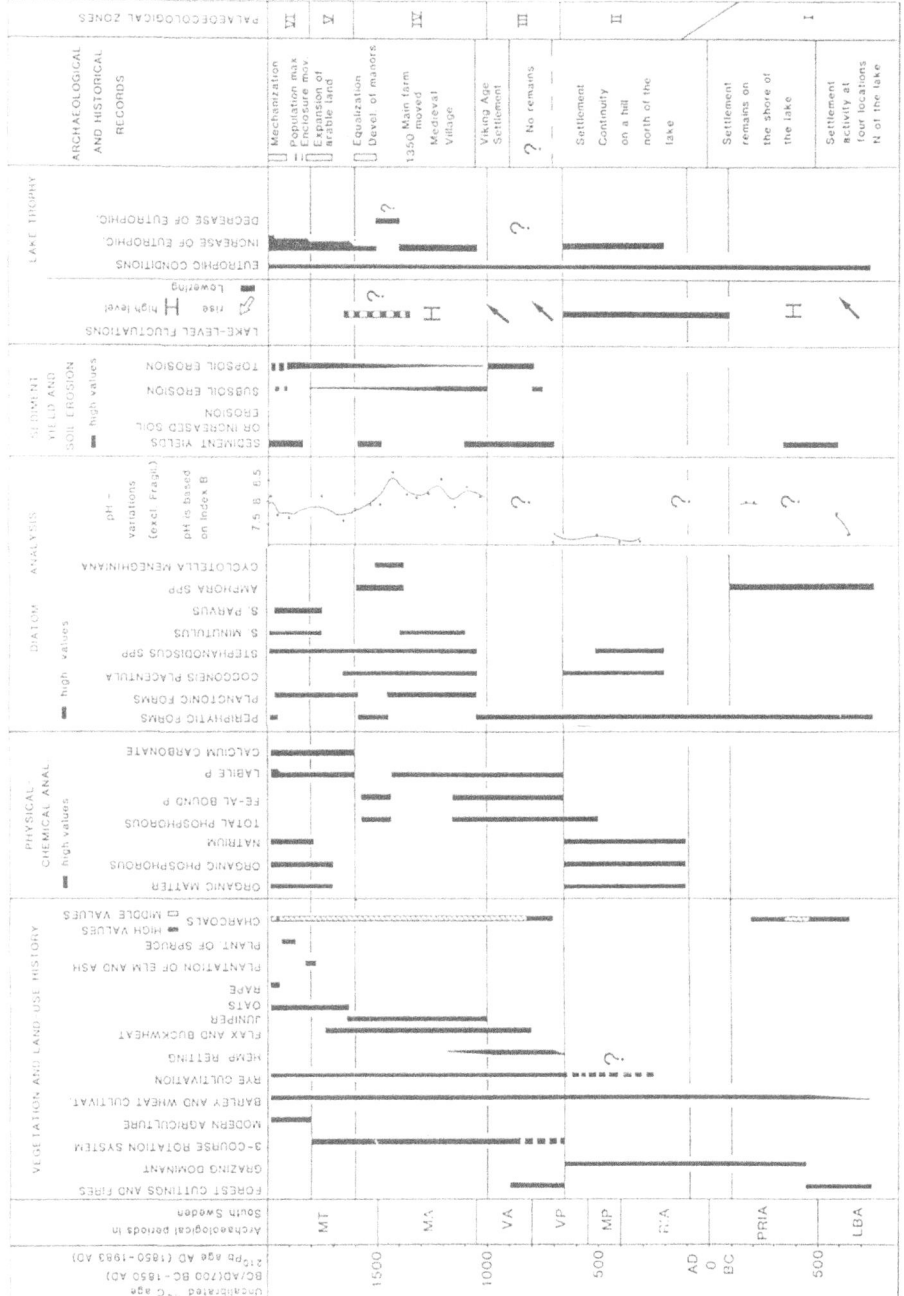

Fig. 2. Summary diagram of the most significant results from the lake sediment-based studies of Bjäresjösjön. Black segments indicate either high values of particular taxa and chemical elements, or the occurrence of an inferred process such as subsoil erosion, lake-level fluctuation, etc. The time scale is based on ^{210}Pb and ^{14}C dates. Archaeological periods follow Berglund *et al.* (in press). LBA: Late Bronze Age, PRIA: Pre-Roman Iron Age, RIA: Roman Iron Age, MP: Migration Period, VP: Vendel Period, VA: Viking Age, MA: Middle Ages, MT: Modern Times. Archaeological and historical records are based on the investigations of Callmer *et al.* (in press), Skansjö (1986), Persson (1986) and Germundsson & Olsson (in press). For more explanations, see text section 4.

Results

Detailed presentation of the results from the individual studies are or will be published elsewhere (Gaillard & Berglund, 1988; Dearing, in press; Gaillard & Digerfeldt, in press; Gaillard & Göransson, in press; Gaillard et al., in prep; Gaillard, in prep). Figure 2 attempts to summarise the major results of the multidisciplinary study. The occurrence as high values of pollen, chemical elements and diatom taxa, as well as inferred data, such as land-use type, pH variations, intensity and type of soil erosion, lake-level fluctuations and trophic conditions, are included. This figure represents a first step in the analysis of the possible relationships between the observed palaeoecological changes. As a further step, canonical correspondence analysis (Ter Braak, 1986), performed on the whole data set, will be attempted. It will probably provide a more rigorous synthesis of the relationships between the different palaeoecological variables, and estimates of the magnitude and rate of changes.

Dating and chronology

In order to compare the lake sediment-based studies to the archaeological record, the establishment of a reliable chronology of core C3 was essential. Because of the relatively high carbonate content of the lake sediments, ^{14}C dating was reliable only from the bottom peat (dated at 2700 BP) and a layer of coarse detritus *gyttja* deposited between 1810 and 1240 BP. The ^{210}Pb chronology covers the upper 70 cm of sediments and provides 24 dates between 1835 and 1980. This chronology could be partly confirmed by pollen-stratigraphical levels of known age such as the plantations of spruce around 1850 and the introduction of rape for cultivation after 1945 (Fig. 2).

The loss-on-ignition curves between 1200 BP and 300 BP record constant organic and calcium carbonate content, and support the assumption of more or less constant accumulation rates (Gaillard & Berglund, 1988). Therefore, the inter-polation of the ^{210}Pb chronology (1835–1980 AD) with the ^{14}C chronology (2700–1240 BP) is believed to be correct. Moreover, this interpolation provides plausible ages for land-use changes demonstrated by pollen analysis and known from historical documents, such as the reclamation of outfields with juniper for the cultivation of oats during the seventeenth century (Fig. 2).

In Fig. 2, palaeoecological changes are plotted against this time-scale. The limits of the archaeological periods for southern Sweden (Berglund et al., in press) were established on the basis of archaeological data, independently of any stratigraphical results. They are also plotted in Fig. 2 against the Bjäresjö time-scale.

Land-use history

The land-use history was reconstructed on the basis of pollen and plant-macrofossil analyses, with the addition of historical documents dating back to the sixteenth century (Gaillard & Berglund, 1988; Gaillard & Göransson, in press; Gaillard et al., in press). The landscape around Bjäresjösjön was very open from the beginning of Late Bronze Age (2700 BP). Extensive forest cuttings and fires occurred during the course of the Late Bronze Age. Pastures and meadows were dominant until the middle of the Vendel Period (650 AD). At that time, a very significant change occurred with the introduction of rye as a major crop. Rye and barley have been cultivated according to a rotation system (probably the 3-course rotation) since that time. The lake was used as a retting pond for hemp from the middle of the Vendel Period until the Early Middle Ages. During the Middle Ages and Modern times, the area of cultivated land progressively increased at the expense of meadows and during the course of the seventeenth century, all outfields with juniper were reclaimed for cultivation. The nineteenth century was characterized by the break-up of the old cultivation systems and by the development of modern agriculture.

Lake-level fluctuations

Palaeohydrological reconstruction (Gaillard & Berglund, 1988; Gaillard & Digerfeldt, in press) suggests that a rise in the local water table resulted in the creation of Bjäresjösjön around 2700 BP. Around 2200 BP (end of the Pre-Roman Iron Age), lake level fell, and very low levels prevailed until 1300 BP (middle of the Vendel Period), when a second rise in water level occurred. Maximum levels were reached around 750 BP during the Middle Ages. There is some indication (plant macrofossils, diatoms) of a lower water level between 700 and 400 BP. Comparison of the present reconstruction with those from the project area and elsewhere in South Sweden shows that the fluctuations of Bjäresjösjön correspond to regional palaeohydrological changes, most probably caused by regional climatic shifts towards wetter (high levels) or drier (low levels) conditions.

Diatom stratigraphy

The diatom stratigraphy may be divided into four main zones. The first (Late Bronze Age – Viking Age) contains high percentages of periphytic forms and several levels with high fragmentation of the diatoms. The second zone (Middle Ages – *ca.* 1750 AD) exhibits generally higher percentages of planktonic diatoms until *ca.* 1400 AD, and a marked decrease of those forms between 1400 and 1600 AD, with a temporary abundance of *Cyclotella meneghiniana* Kütz. (1450–1500 AD). The third (1750–1945) contains comparatively high percentages of planktonic forms, and increases of *Stephanodiscus parvus* Stoermer & Håkansson and *S. minutulus* (Kütz.) Cleve & Möller. Finally, the fourth zone (1945 to the present) includes a significant rise of *Achnanthes* spp. and *S. minutulus*, and high percentage of periphytic forms.

Variations in pH were inferred using Index B (Renberg & Hellberg, 1982). They suggest four main phases:
(1) After an increase at the end of the Late Bronze Age, pH values around 8 were reached during the Pre-Roman Iron Age.

(2) The pH values were comparatively lower (around 7.2) during the period of low lake-levels (Roman Iron Age – Vendel Period).
(3) An increase in pH occurred sometime at the Vendel/Viking Age transition period and the values fluctuated between 8 and 8,5 until 1400 AD.
(4) The pH values decreased after 1400 AD and reached *ca.* 7.5 around 1650 AD. A recent increase to pH 8 is recorded after 1945 AD.

Physical and chemical analyses

The major changes shown by the physical and chemical analyses may be summarized as follows:
(1) Organic matter, organic phosphorus and nitrogen all increase significantly during the period of lake-level lowering and maintain high values during the whole period of low levels (2200–1300 BP).
(2) Labile phosphorus and Fe- and Al-bound phosphorus reach very high values during the period of intensive hemp retting.
(3) A renewed increase of the Fe- and Al-bound phosphorus is noted during the fifteenth century.
(4) Modern times (1600 AD to the present) are characterized by increased content of organic matter, labile and organic phosphorus and a strong increase in calcium carbonate. A further increase in labile phosphorus is recorded during the present century.

Sediment yields and soil erosion

The curve of sediment yields (t ha^{-1} yr^{-1}) is used to infer rates of soil erosion through time (Dearing *et al.*, 1987; Dearing, in press). Three periods of increased soil erosion may be distinguished: a minor one during Late Bronze Age, a second (more significant) during Late Iron Age, and the last and most significant between 1840 and 1900 AD. The period between *ca.* 1000 AD and 1800 AD is characterized by the stabilization of sediment yield, with slight decreases between *ca.*

1000 and 1400 AD and between *ca.* 1600 and 1800 AD.

The ratio of the magnetic remanence at a high magnetic field (HIRM) to the frequency dependent magnetic susceptibility (χfd) is a measure of the changing proportions of topsoil erosion (low values) and subsoil erosion (high values) (Dearing, in press). This variable was determined for the period 1300 BP to the present. Dominance of topsoil erosion is indicated during the Viking Age and after 1700 AD. Subsoil erosion was prevalent at the end of the Vendel Period and during Early Middle Ages; but topsoil erosion became progressively greater during the Late Middle Ages until 1700 AD.

Discussion and conclusions

Seven periods of significant change are chosen for discussion:

(1) The rise in water table leading to the creation of the lake at Bjäresjö around 2700 BP was probably the result of a regional climatic shift towards wetter conditions. The very high breakage of diatoms can probably be related to this rise in water-level, and to the input of coarse mineral particles through erosion of the lake margins. The archaeological evidence of major settlement activity on the hills north of the lake during the Late Bronze Age and Pre-Roman Iron Age, together with the pollen-stratigraphic record of forest clearance and cereal cultivation, indicates that soil erosion may have increased significantly during that period, and contributed to the slight increase in sediment yields.

During this period, two major factors have influenced the local environment and may explain the observed palaeoecological changes: climate and human impact. These are difficult to separate, and may both influence the changes in sediment yield and diatom flora observed.

(2) The drier conditions prevailing during the second part of the Early Iron Age are not believed to have caused the observed land-use changes. The climatic shift resulted mainly in local landscape changes. The lake at Bjäresjö was transformed into a very shallow pond. The high values of sedimentary nitrogen and the increase in sedimentary phosphorus are almost certainly related to the lake-level lowering, whereas the higher content of organic matter is probably due to increased lake trophic status. A decrease in pH values to around 7 may tentatively be explained by decomposition of organic matter and, therefore by chemical processes due to a low water-level.

During this period, lake-level lowering was the major factor influencing chemical conditions and the diatom flora in the lake. Human impact was rather constant during the Early Iron Age and did not produce any significant palaeoecological changes.

(3) The climatic shift towards wetter conditions during the Late Iron Age led to a general rise of water table and to higher lake levels. This is accompanied at Bjäresjö by evidence of hemp retting in the lake, which probably led to a depletion of dissolved oxygen, and a reduction of pH levels in the lake water. Such changes in the water quality would explain the high concentrations of labile phosphorus and Fe- and Al-bound phosphorus in the sediments of that age. The higher sediment yields during the Late Iron Age are interpreted in terms of increased soil erosion. Separation of the results of a rise of lake level, and an increase of the area of cultivated land, is difficult. However, topsoil erosion dominated between 800 and 1000 AD and was probably due to surface erosion processes on open fields. Evidence of an important Viking Age settlement of village type situated on the hills north of the lake suggests a very intensive use of the landscape at that time.

Therefore, human activities at the site were certainly the decisive factor for the palaeoecological changes occurring at the Vendel/Viking Age transition period. The role of climate change on human settlements and on land use cannot be deciphered from this study.

(4) Hemp retting in the lake was abandoned sometime between 1000 and 1200 AD. The lake seems to have recovered quickly from anoxis, and values of Fe- and Al-bound phosphorus fell to comparatively low values. Sediment yields were

relatively low during the Middle Ages, which indicates rather stable conditions. However, topsoil erosion increased steadily, which may be related to the progressive increase in the area of cultivated land at the expense of meadows, as suggested by pollen analysis.

(5) The renewed increase of Fe- and Al-bound phosphorus and high percentages of periphytic forms of diatoms during the period 1400–1600 AD are difficult to interpret. There is some indication of a minor lake-level lowering. The decrease in organic and labile phosphorus indicates slightly less eutrophic conditions. However, other processes may have been involved, such as temporary inputs of unknown origin.

(6) The seventeenth and eighteenth centuries are characterized by a further expansion of arable land. Topsoil erosion increased. This period is also marked by a progressive increase in eutrophication, as shown by diatoms and chemical analyses.

(7) The nineteenth and present centuries represent a period of considerable ecological change. The agricultural development is registered by dominant topsoil erosion. The strong increase in soil erosion is also indicated by high sediment accumulation rates and ^{210}Pb influx. Eutrophication has also increased steadily and reached very high levels after 1960 AD, which is indicated by the diatom flora and the significant increase in sedimentary labile phosphorus.

Comparison of our results with those from other sites in the project area (Dearing, in press) and in Southern Scania (Dearing et al., 1987) are in generally good agreement. However, our study of Bjäresjösjön provides more precise dating of the most significant changes during the last 3000 years, such as those occurring during the Late Iron Age. The decrease in sediment yields during the period 1300–1550 AD found at Havgårdssjön and ascribed to the early part of the 'Little Ice Age' (Dearing et al., 1987) has no equivalent at Bjäresjö. Moreover, the pollen record does not indicate any reduction in the area of cultivated land at this time. There was, however, a stabilisation (between ca. 1300 and 1450 AD) of the

steady increase seen from ca. 700 AD. Neither historical nor pollen-analytical data from the project area (Fig. 1) show any significant change in settlement or land use during the period known as the 'agrarian crisis' in N.W. Europe (Gaillard et al., in press). It seems, therefore, that Havgårdssjön is an 'outlier', and that the decrease in sediment yields found there is a result of local changes (demise of the castle, see discussion in Dearing et al.). It also implies that sediment mobilisation and transport during this period was controlled by land use rather than by climatic change.

Until now, no other sites in South Sweden have been studied with the same combination of methods, but a similar investigation is on progress at another site in the project area (Bussjö situated 5 km east of Bjäresjö), with a complete Holocene sequence (Dearing, Håkansson, Regnell in progress). The multidisciplinary study of Bjäresjösjön shows clearly how lake development and soil erosion are closely related to land-use and settlement history during the last 3000 years. It also shows the sensitivity of small lakes to changes within the catchment, a fact stressed earlier by Dearing (in press) and Fritz (1989). As it has been noted at Diss Mere (Norfolk, U.K.) for diatom assemblages (Fritz, 1989), there is also at Bjäresjösjön striking responsiveness of diatom communities, chemical characteristics, sediment yields and magnetic properties to land-use changes documented by pollen analysis, and to lake-level changes. This study shows, therefore, the usefulness of such parameters as indicators of past limnological and land-use changes.

Acknowledgements

We are very grateful to Björn Berglund (coordinator of the project) for his continuous support and interest in our results, and to our colleagues Anders Persson, Sten Skansjö, Gunilla Olsson and Johan Callmer for useful information concerning their investigations at Bjäresjö. We are much indebted to Thomas Persson and other colleagues for their invaluable help during field work, and to H. John B. Birks and Hilary H. Birks for reading the manuscript and improving it. Figures were drawn by Bitten Arvidson. The study was supported by the Bank of Sweden Tercentenary Foundation.

References

Berglund, B. E., 1988. The cultural landscape during 6000 years in south Sweden – an interdisciplinary study. In Birks, H. H., H. J. B. Birks, P. E. Kaland & D. Moe (eds), The Cultural Landscape – Past, Present and Future. Cambridge University Press: 241–254.

Berglund, B. E., L. Larsson, N. Lewan, G. Olsson, M. Riddersporre & S. Skansjö, Interdisciplinary terms and concepts. In Berglund, B. E. (ed.), The Cultural landscape of the past 6000 years in southern Sweden. Ecol. Bull., in press.

Callmer, J., D. Olausson & G. Olsson, The Bjäresjö area – Bronze Age and Early Iron Age. Late Iron Age and Early Middle Ages. In Berglund, B. E. (ed.), The cultural landscape of the past 6000 years in southern Sweden. Ecol. Bull., in press.

Dearing, J. A., Erosion and land use. In Berglund, B. E. (ed.), The cultural landscape of the past 6000 years in southern Sweden. Ecol. Bull., in press.

Dearing, J. A., H. Håkansson, B. Liedberg-Jönsson, A. Persson, S. Skansjö, D. Widholm & F. El-Daoushy, 1987. Lake sediments used to quantify the erosional response to land use in southern Sweden. Oikos 50: 60–78.

Digerfeldt, G., 1986. Studies on past lake-level fluctuations. In Berglund, B. E. (ed.), Handbook of Holocene Palaeoecology and Palaeohydrology. John Wiley & Sons, Chichester: 127–143.

El-Daoushy, F., 1981. An ionization chamber and a Si-detector for lead-210 chronology. Nucl. Instrum. Meth. 188: 647–655.

El-Daoushy, F., K. Olsson & R. Garcia-Tenorio, 1991. Accuracies in Po-210 determination for lead-210 dating. In J. P. Smith et al. (eds), Environmental History and Palaeolimnology. Kluwer Academic Publishers, Dordrecht: 43–52. Reprinted from Hydrobiologia 214.

Fritz, S. C., 1989. Lake development and limnological response to prehistoric and historic land-use in Diss, Norfolk, U.K. J. Ecol. 77: 182–202.

Gaillard, M.-J. & B. E. Berglund, 1988. Land-use History during the Last 2700 Years in the area of Bjäresjö, Southern Sweden. In Birks, H. H., H. J. B. Birks, P. E. Kaland & D. Moe (eds), The cultural landscape – past, present and future. Cambridge University Press: 409–428.

Gaillard, M.-J. & G. Digerfeldt, Palaeohydrological studies and their contribution to palaeoecological and palaeoclimatic reconstruction. In Berglund, B. E. (ed.). The cultural landscape of the past 6000 years in southern Sweden. Ecol. Bull, in press.

Gaillard, M.-J. & H. Göransson, The Bjäresjö area – vegetation and landscape through time. In Berglund, B. E. (ed.). The cultural landscape of the past 6000 years in southern Sweden. Ecol. Bull., in press.

Gaillard, M.-J., D. Olausson & S. Skansjö, The Bjäresjö area – Conclusions. In Berglund, B. E. (ed.), The cultural landscape of the past 6000 years in southern Sweden. Ecol. Bull., in press.

Germundsson, T. & G. Olsson, The Bjäresjö area – The nineteenth and twentieth centuries. In Berglund, B. E. (ed.) The cultural landscape of the past 6000 years in southern Sweden. Ecol. Bull., in press.

Håkansson, S., 1986. University of Lund radiocarbon dates XIX. Radiocarbon 28: 1111–1132.

Peglar, S.-M., S. C. Fritz & H. J. B. Birks, 1989. Vegetation and land-use history at Diss, Norfolk, U.K. J. Ecol. 77: 203–222.

Persson, A., 1986. Agrar struktur i Bjäresjö by 1570 och 1699. En trendanalys utförd med ADB-stöd. Arbetsrapport från kulturlandskapet under 6000 år. Dept. of Plant Ecology, Lund University: 64 pp.

Renberg, I. & T. Hellberg, 1982. The pH history of Lakes in Southwestern Sweden, as Calculated from Subfossil Diatom Flora of the Sediments. Ambio 11/1: 30–33.

Skansjö, S., 1986. Från Vikingatida stormansgård till Renässansslott. Arbetsrapport från kulturlandskapet under 6000 år. Dept. of History, Lund University: 82 pp.

Ter Braak, C. J. F., 1986. Canonical correspondence analysis: a new eigenvector technique for multivariate direct gradient analysis. Ecology 67: 1167–1179.

Hydrobiologia **214**: 115–124, 1991.
J. P. Smith, P. G. Appleby, R. W. Battarbee, J. A. Dearing, R. Flower, E. Y. Haworth, F. Oldfield & P. E. O'Sullivan (eds). 115
Environmental History and Palaeolimnology.
© 1991 *Kluwer Academic Publishers.*

Paleolimnology of Slapton Ley, Devon, UK

P.E.O'Sullivan[1]*, A.L. Heathwaite[2], P.G. Appleby[3], D. Brookfield[1], M.W. Crick[4], C. Moscrop[1],
T.B. Mulder[1], N.J. Vernon[1] & J.M. Wilmshurst[5]
[1]*Department of Environmental Sciences, Polytechnic South West, Plymouth PL4 8AA, UK (*author for
correspondence);* [2]*Department of Geography, University of Sheffield, Sheffield S10 2TE, UK;*
[3]*Department of Applied Mathematics and Theoretical Physics, University of Liverpool, Liverpool L69
3BX, UK;* [4]*Nature Conservancy Council, Foxhold House, Cookham Common, Newbury RG15 8EL,
UK;* [5]*Department of Zoology, University of Canterbury, Christchurch* [1], *New Zealand*

Key words: paleolimnology, ^{210}Pb dating, magnetic susceptibility, core correlation, erosion, diatom
analysis, chlorophyll a, eutrophication

Abstract

Slapton Ley, a coastal lake in SW England, has been shown by a variety of paleolimnological studies,
to have become increasingly eutrophic in the period since 1950 AD. Since that time, intensification of
agriculture has resulted in increased erosion of topsoil from fields in the catchment of the Ley. Sediment
accumulation rates, as estimated by ^{210}Pb-dating and multiple core correlation of peaks in whole core
volume magnetic susceptibility, are equivalent to a catchment erosion rate of $13.4 \, \mathrm{t \, km^{-2} \, a^{-1}}$, which
figure agrees well with directly monitored data. Diatom and chlorophyll a analysis of the uppermost
sediments shows that the Ley has recently experienced a major shift in its trophic status, changing from
a clear water, macrophyte lake to one dominated by plankton in a hypertrophic system. This last point
is further amplified in the paper by Heathwaite & O'Sullivan (1991).

Introduction

Slapton Ley (Fig. 1) is a coastal wetland, 10 km
south of Dartmouth, Devon, SW England,
formed behind a shingle bar. The wetland is
divided into two basins, the Higher Ley which is
mainly reedswamp, and the Lower Ley which is
a shallow freshwater lake (Table 1).

Table 1. Morphometric details of Slapton Ley (after Van
Vlymen, 1980)

Mean depth (z)	1.55 m
Maximum depth (z_{max})	2.80 m
Area (A)	77 ha
Volume (V)	$1.19 \times 10^6 \, \mathrm{m^3}$
Mean water residence time	17.5 days

Slapton Ley is a UK Grade 1 Site of Special
Scientific Interest, and a National Nature Reserve. It is a conservation site of European importance, and a significant factor in the economy of
the South Hams district of South Devon. In recent years there has been concern that the Ley is
becoming increasingly eutrophic. Consequently,
any information regarding its recent history is of
prime importance. This paper presents a summary of research into the paleolimnology of Slapton
Ley over the past ten years. It is also intended as
background to the paper by Heathwaite &
O'Sullivan (1991).

116

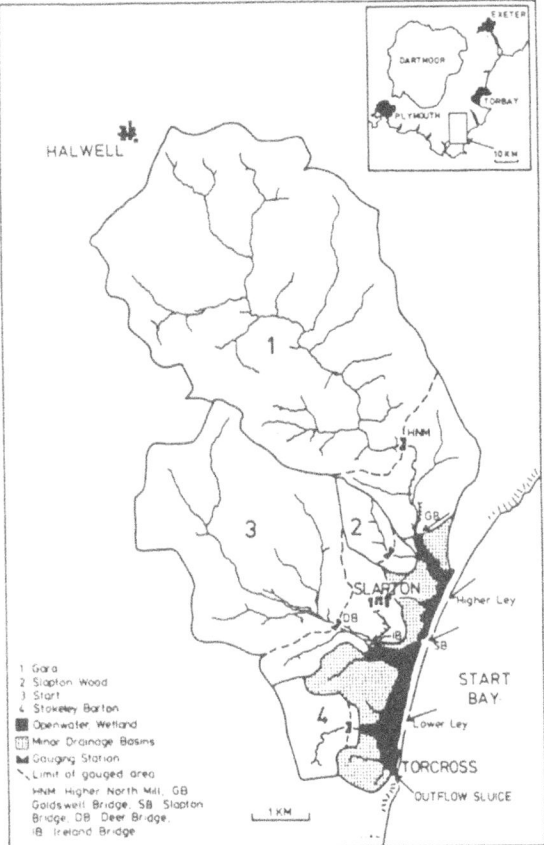

Fig. 1. Slapton Ley and its catchments.

Site description

The Lower Ley, Slapton (Fig. 2) is a long, narrow, shallow coastal lake orientated NNE/SSW, impounded behind a shingle barrier some 6 km in length. Nowhere is the Ley deeper than 2.8 m, despite its area of 77 ha, and it is fringed by extensive reedswamps and other habitats making up a Reserve of some 116 ha.

The catchment of the Ley is a deeply dissected plateau (maximum elevation > 200 m), developed over Lower Devonian slates which weather to shallow (0–0.5 m) silty loams (Trudgill, 1983) rich in haematite (αFe_2O_3). The main inflowing tributaries of the Ley are the River Gara (catchment area (D) = 27 km^2), which enters via the Higher Ley and Slapton Bridge, and the Start Stream (D = 13 km^2) which flows in through Ireland Bay

(Fig. 1). Other drainage totals 6 km^2. The Ley drains through a culvert at its extreme southern end, at the village of Torcross. This was opened in 1856 (Stanes, 1983).

The main land use in the catchment of the Ley is agriculture. Johnes & O'Sullivan (1989) predict that in 1985, losses of nitrogen and phosphorus from this catchment to the Ley were 160 t N, and 4.8 t P, of which ca 148 t N and 2.44 t P were from agriculture, and 8 t N and 2.3 t P from sewage. The total nitrogen load on the Ley may in fact be somewhat higher, and of the order of 260 t a^{-1} (O'Sullivan *et al.*, 1989). Historical studies (Acott, 1989) suggest that external N and

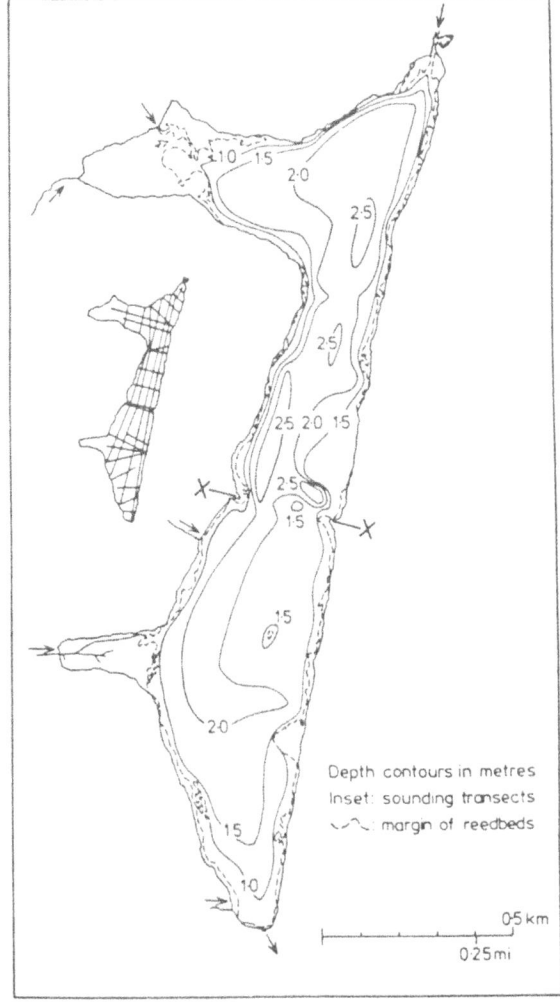

Fig. 2. Bathymetry of Slapton Ley (from Van Vlymen, 1979).

117

P loads have risen to present levels from 42 t N, 1.35 t P in 1905, and 58 t N, 2.3 t P in 1945.

Previous work

The sediments of Slapton Ley have previously been studied by Crabtree & Round (1967), and by Morey (1976). The first authors analysed a 1.2 m core from Ireland Bay (Fig. 1), and found that for most of its existence, the Ley has been a shallow, eutrophic lake whose dominant diatoms have been members of the genus *Fragilaria*. That the former Ley was clear is shown by the abundance of epiphytic and epipelic taxa. There is little evidence for prolonged marine incursions, but there is some indication of episodes of erosion of catchment material.

Morey (1976) studied the longer-term development of the Ley. He found that it was formed by onshore movement of shingle, and that below the present *gyttja* are peat and clay layers indicating succession from estuarine deposits to reedswamp and then open water. Radiocarbon dates from the uppermost peat showed that the present freshwater Ley cannot be older than 1800 years. Placing the Ley on a curve of Holocene sea level rise (Hails, 1975) gives an age of ca. 1000 years.

Stratigraphy

The majority of cores used in our investigations have been 1 m Mackereth cores. A typical stratigraphy would be: –

0–0.35/0.4 m Dark brown *gyttja*.
0.35/0.4–0.45/0.6 m Red brown clay.
below 0.45–0.6 m Dark grey clay-mud.

This is very similar to the top part of the core analysed by Crabtree & Round (1967), except that the uppermost unit is now much thicker.

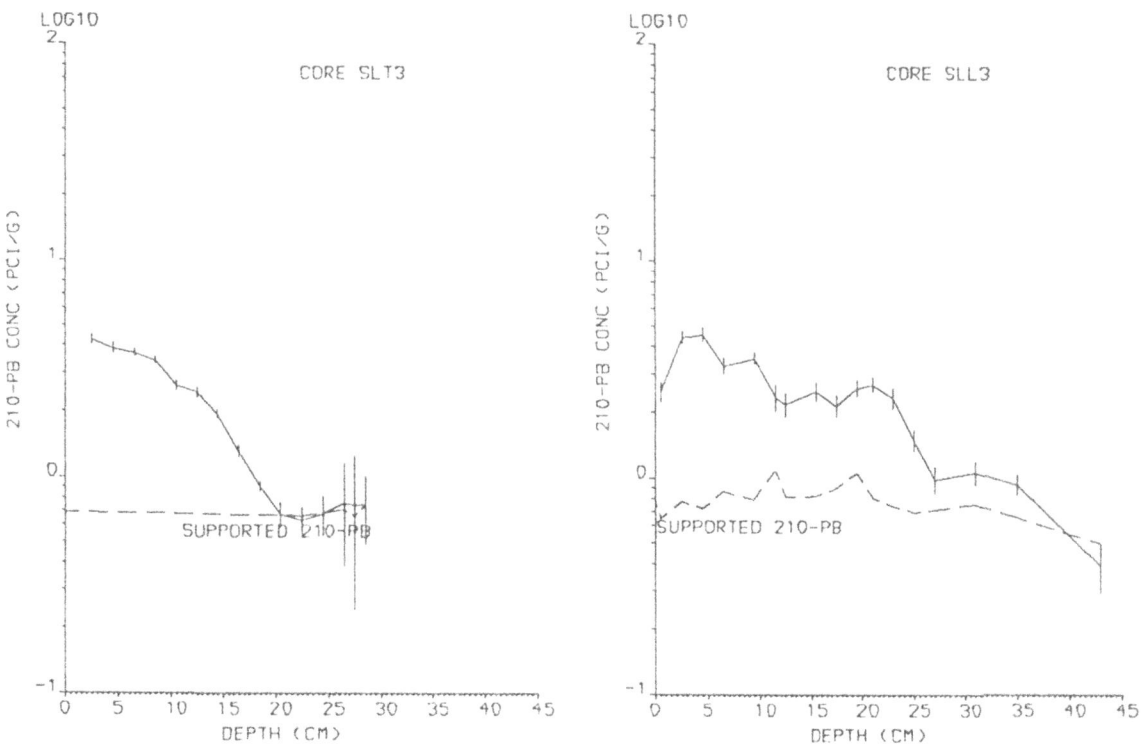

Fig. 3. Total Lead-210 concentration in cores SLT3 and SLL3 from Slapton Ley.

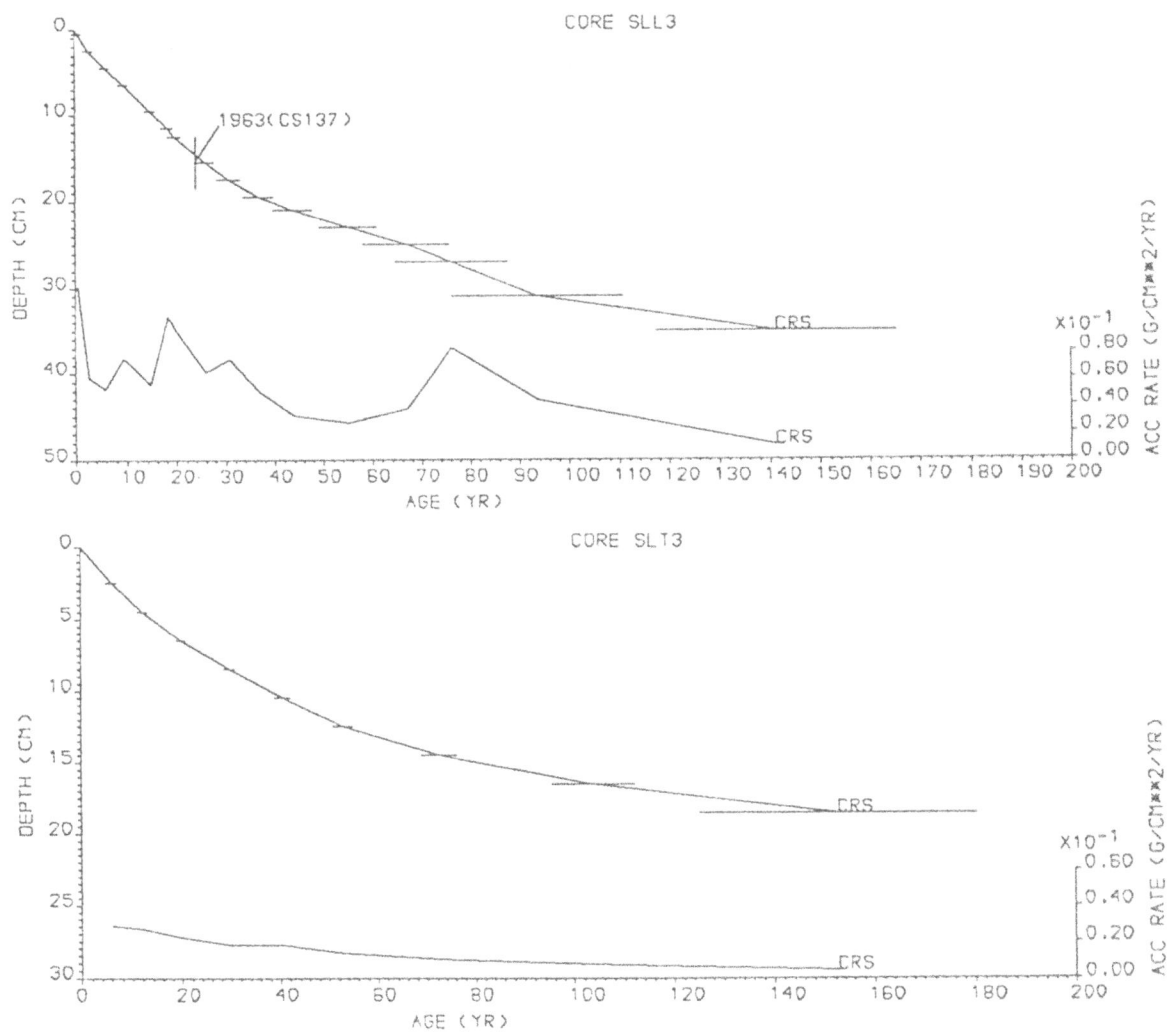

Fig. 4. Age-depth relationships for cores SLL3 and SLT3.

Chronology

Lead-210 profiles (Figs. 3–4) have been obtained from two cores from Ireland Bay (SLT3, SLL3). Variation in concentration of ^{210}Pb with depth for both profiles is shown in Fig. 3. The unsupported ^{210}Pb inventory of cores SLL3 was 7.73 pCi cm^{-2}, a constant influx of 0.24 pCi cm^{-2} a^{-1}. Owing to the non-monotonic nature of both unsupported ^{210}Pb profiles, it was only possible to derive a chronology using the CRS model (Appleby & Oldfield, 1978, Oldfield & Appleby, 1984). Time-depth relationships are shown in Fig. 4.

The results indicate that sediment influx has increased since ca. 1950, with a maximum in 1968. This is confirmed by ^{137}Cs determination (not shown here), in which the mean peak (1963) is located at 12.5 ± 3 cm depth. This is dated by the ^{210}Pb chronology at 1967 ± 6 years. The diffuse nature of the caesium peak is a reflection of (a) the fast accumulation rate, and (b) mixing caused by turbulence and/or bioturbation. Slapton Ley is a very turbulent lake, in which Langmuir cells incorporating the whole water column often form. Similarly, macrobenthos such as tubificids are abundant. Significant concentrations of ^{137}Cs were found below the level which by ^{210}Pb dating is 1954.

The ^{210}Pb chronology also indicates a faster period of sediment accumulation at 0.27 m depth, or in ca. 1910 AD. Unsupported ^{210}Pb activities at this level are low, owing to the fast accumulation rate, and the relatively low ^{210}Pb influx. The time-depth relationship for core SLT3 (Fig. 4b) records a much slower rate of accumulation (0.177 cm a^{-1}). Sediment accumulation is thus not particularly even, in Ireland Bay at least.

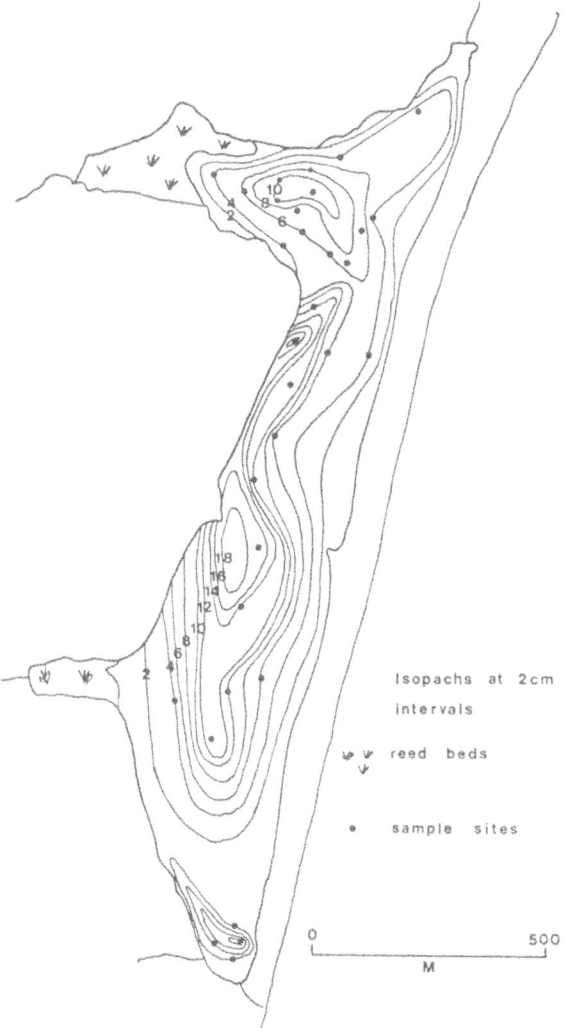

Fig. 5. Isopach map for the Lower Ley, Slapton, showing locations of cores used in estimating sediment thickness.

Sedimentation and erosion

Patterns of sedimentation across the bed of the Ley have been studied by means of the multiple-coring and whole core susceptibility (K) approach pioneered by Dearing (1983). Figure 5 illustrates the location of each core across the lake bed, and isopachs derived by determining the thickness of sediment above a peak in K found just below the present SWI in all cores, and dated in cores SLL3 and SLT3 at 10 years bp. In Fig. 6 is shown a montage of K profiles for the suite of ca. 20 cores on which the isopachs are based. The cores fall into three main sets: (1) Ireland Bay, (2) the narrow central 'neck' area of the Ley, and (3) Stokeley Bay/Torcross.

Results suggest that a 'tongue' of sedimentation is proceeding south along the lake, away from the main inflows which are in the north. Sedimentation rates are high in the narrow 'neck' area, on the landward side of the Ley (cf. Loe Pool, O'Sullivan, Coard & Pickering, 1982, and Merevale Lake, Foster *et al.*, 1985), but much slower in the exposed areas close to the shingle ridge. In the southern part of the Ley, north of Torcross, is an area where the bed is composed of shingle material rather than *gyttja*, which presumably results from a marine incursion. This pattern follows that derived by Morey (1976), except that he did not record the small, separate basin of sedimentation which is located offshore of Torcross, close to the main outflow.

Determination of the dry mass content of all cores indicates that some 8 t ha^{-1} of dry matter are deposited on the bed of the Ley annually. This is equivalent to an accumulation rate of 9 mm a^{-1}, or an erosion rate from the catchment of 610 t, or 13.4 t km^{-2} a^{-1}. This agrees quite well with the current monitored input of suspended sediment of 640 t a^{-1} (O'Sullivan *et al.*, 1989), but gives rather faster accumulation rates that the ^{210}Pb dated cores (see above). These, however, are from Ireland Bay, an area of the lake bed where accumulation rate is about half that of much of the lake.

120

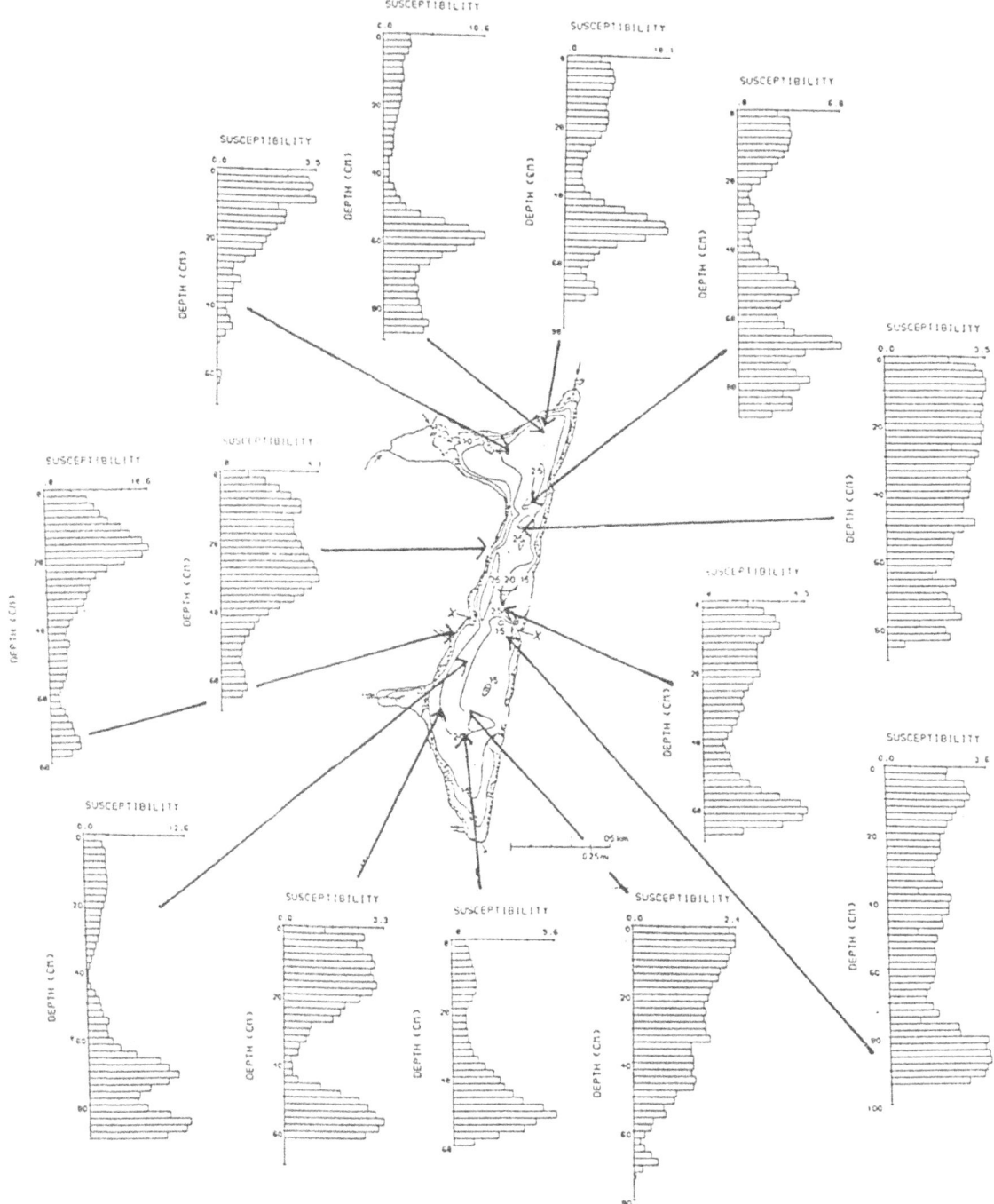

Fig. 6. Whole core susceptibility (K) profiles from Slapton Ley.

Sediment source

Crick (1985) investigated the magnetic characteristics of potential sediment sources in the catchment of the Ley. On the basis of determination of X, X_{fd}, SIRM, SIRM/X and coercivity ('S') of bedrock, subsoil, topsoil, stream bank material, and fine bedload, he concluded that the magnetic properties of the uppermost sediments of the Ley most closely resembled those of the topsoil of local arable fields, which are rich in haematite. A similar analysis of a core from Ireland Bay (Fig. 7) shows X and SIRM rising towards the sediment surface, and that SIRM/X increases from <500 below 0.2 m, to >500 above that level (dated by ^{210}Pb at ca. 1950). 'S' ratio suggests that the haematite component was not particularly abundant, and in fact is greater below 0.4 m (ca. 1800). Nevertheless it seems that in increasing amount of topsoil has been eroded into Slapton Ley in the post-war period. The increased sedimentation rate is confirmed by the ^{210}Pb chronology.

Eutrophication

As found by Crabtree & Round (1967), for most of its existence, Slapton Ley has been a shallow, clear, eutrophic lake, dominated by macrophytes. Since the 1960s, however, the Ley has been increasingly prone to producing substantial algal blooms (Benson-Evans *et al.*, 1967, Van Vlymen, 1980). From 1984 until 1989, the fish population of the Ley fell to very low levels indeed in comparison to previous years (K. Chell, C. Kennedy, pers. comm.).

Moscrop (1986) found that in the top 0.06 m of the sediment (Fig. 8), a major change in the diatom flora of the Ley is recorded. The abundant *Fragilaria* dominance found by Crabtree & Round (1967), and present also throughout most of Moscrop's core, is replaced in the top few cm by centric diatoms, primarily *Cyclostephanos dubius*, *Melosira varians*, *Aulacoseiva granulata*, and also by *Asterionella formosa*. These species are not only indicative of much more eutrophic conditions, and in the case of *C. dubius*, hypertrophy, but also

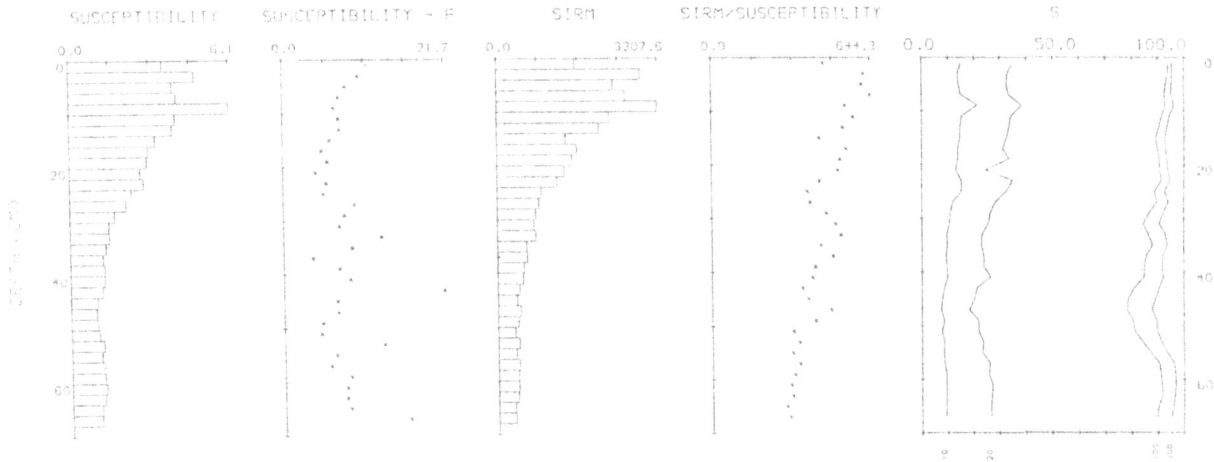

Fig. 7. Mineral magnetic properties of a core from Slapton Ley.

122

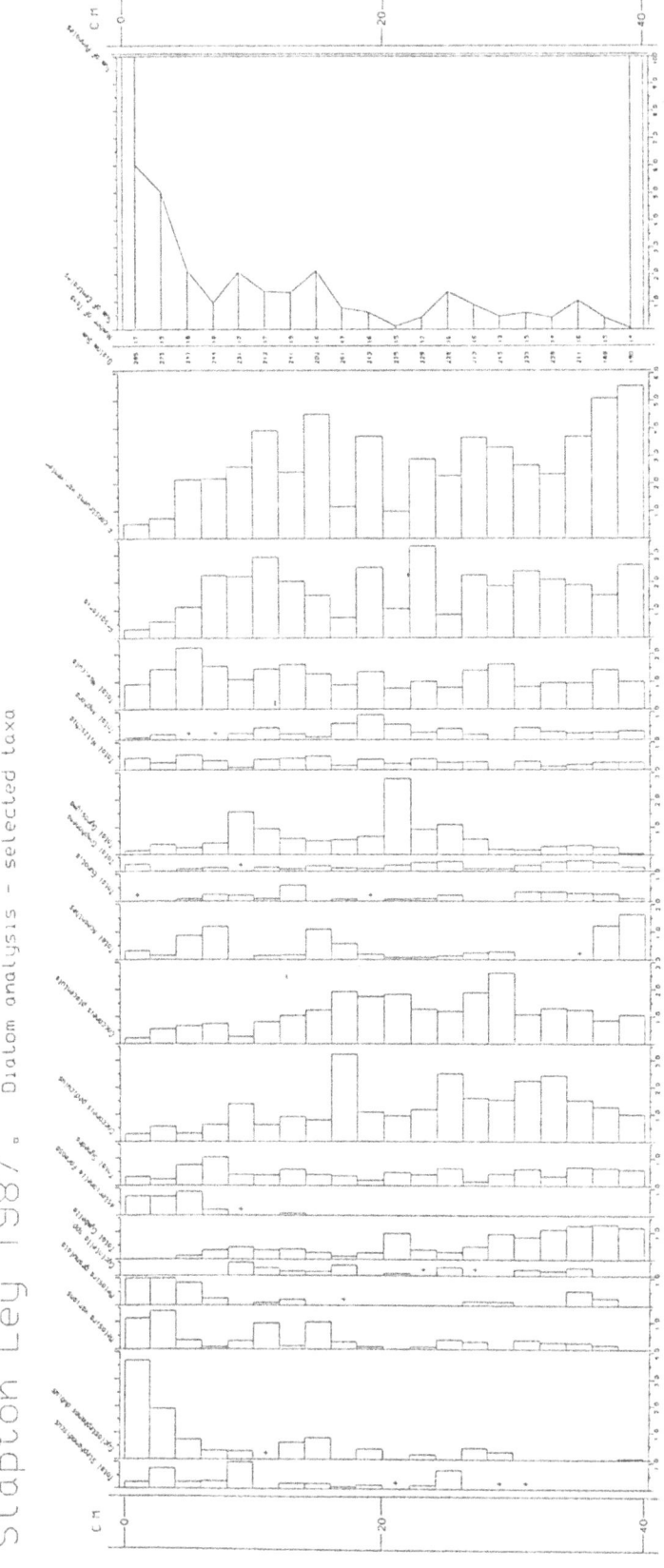

Fig. 8. Diatom profile (% total diatoms) of the uppermost sediments of Slapton Ley (from Moscrop, 1986).

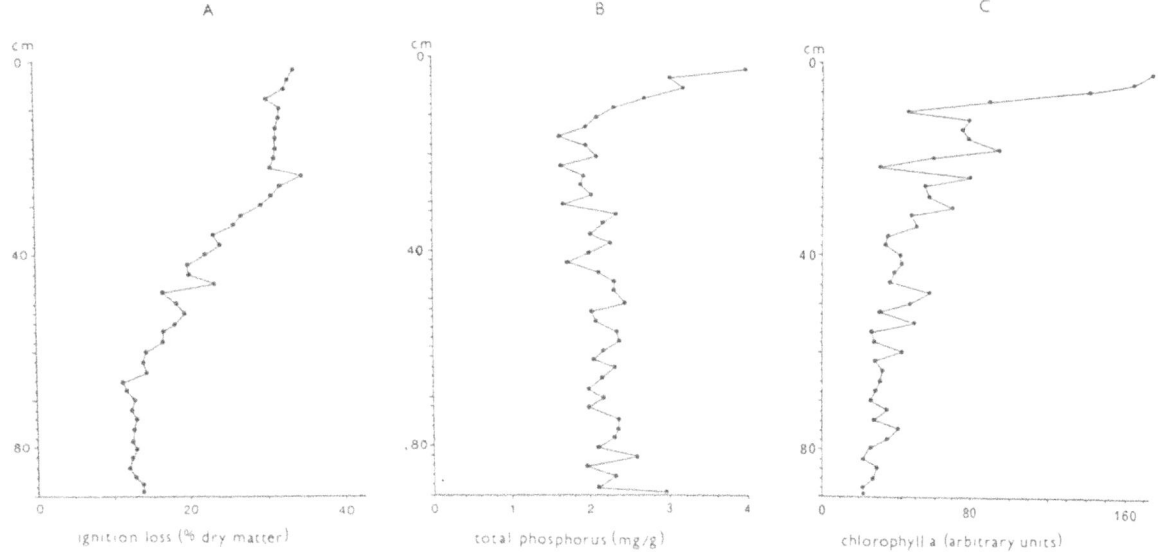

Fig. 9. Ignition loss (A), total available phosphorus (B), and chlorophyll a (C) from a core from Slapton Ley (from Brookfield, 1981).

they are planktonic forms, showing that the Ley has become generally much more turbid. A change from macrophyte to phytoplankton dominance, similar to that found in the Norfolk Broads (Moss, 1983), is suggested.

Brookfield (1981) studied aspects of the chemistry associated with the above changes. He found (Fig. 9) that organic matter, available phosphorus, and chlorophyll a derivatives increase towards the sediment surface, peaking just below the SWI (ie in the mid 1970s).

Conclusions

Slapton Ley, a shallow eutrophic lake in SW England, formed behind a shingle barrier ca. 1000 years ago. For most of its existence, Slapton Ley has been a clear water lake, with an abundant macrophyte and epiphytic diatom flora. Since 1950, erosion of catchment material, especially topsoil from arable fields, associated with the intensification of agriculture, has resulted in an increase in sedimentation within the Ley. Current sediment influx is calculated at 610 t a^{-1}, which

agrees well with the presently monitored value of 642 t. This equivalent to a sediment accumulation rate of 9 mm a^{-1}, or an erosion rate of 13.4 t km^{-2} a^{-1} from the catchment. Diatom and chemical analysis of the topmost sediments shows that in the mid-1970s, the Ley experienced a major shift in its trophic status, changing from a clear macrophyte lake to a turbid, plankton dominated hypertrophic system.

References

Acott, T. G., 1989. An investigation into nitrogen and phosphorus loadings on Slapton Ley, 1905–1985. unpublished BSc dissertation. Polytechnic South West.

Appleby, P. G. & F. Oldfield, 1978. The calculation of ^{210}Pb dates assuming a constant rate of supply of unsupported ^{210}Pb to the sediment. Catena 5: 1–8.

Benson-Evans, K., D. Fish, G. Pickup & P. Davies, 1967. The natural history of Slapton Ley Nature Reserve. II – Preliminary studies of the freshwater algae. Field Studies 2: 493–519.

Brookfield, D., 1981. Chemical analysis of lake sediment from Slapton Ley. unpublished BSc dissertation, Polytechnic South West.

Crabtree, K. & F. E. Round, 1967. Analysis of a core from Slapton Ley. New Phytol. 66: 255–270.

124

Crick, M. W., 1985. Investigation into the relationship between sediment accumulation in the Lower Ley (Slapton), and spatial patterns of erosion within its catchment. unpublished BSc dissertation, Polytechnic South West.

Dearing, J. A., 1983. Changing patterns of sediment accumulation in a small lake in Scania, Southern Sweden. In Paleolimnology (J. Meriläinen, P. Huttunen & R. W. Battarbee, eds), Dr W. Junk bv, The Hague.

Foster, I. D. L., J. A. Dearing, A. A. Simpson, A. D. Carter & P. G. Appleby, 1985. Lake-catchment-based studies of erosion and denudation in the Merevale catchment, Warwickshire, UK. Earth Surface Processes & Landforms. 10: 45–68.

Hails, J. R., 1975. Some aspects of the Quaternary history of Start Bay, Devon. Field Studies 4: 207–222.

Heathwaite, A. L. & P. E. O'Sullivan, 1990. Sequential inorganic chemical analysis of a core from Slapton Ley. Hydrobiologia, this volume.

Johnes, P. J. & P. E. O'Sullivan, 1989. The Natural History of Slapton Ley Nature Reserve. XVIII – Nitrogen and phosphorus losses from the catchment – an export coefficient approach. Field Studies 7: 285–309.

Morey, C. R., 1976. The Natural History of Slapton Ley Nature Reserve. IX – The morphology and history of the lake basins. Field Studies 4: 353–368.

Moscrop, C., 1986. A diatom profile from recent sediments in Slapton Ley. unpublished BSc dissertation, Polytechnic South West.

Moss, B., 1983. The Norfolk Broadland: experiments in the restoration of a complex wetland. Biol. Revs. 58: 521–561.

Oldfield, F. & P. G. Appleby, 1984. Empirical testing of ^{210}Pb dating models for lake sediments. – In Lake Sediments and Environmental History, E. Y. Haworth & J. W. G. Lund (eds) Butterworth: 93–124.

O'Sullivan, P. E., Coard, M. A. & D. A. Pickering, 1982. The use of laminated lake sediments in the estimation and calibration of erosion rates – in Recent developments in the explanation and prediction of erosion and sediment yield (Proceedings of the Exeter Symposium, July 1982) IASH publ. no. 137: 385–396.

O'Sullivan, P. E., A. L. Heathwaite, K. M. Farr & J. P. Smith, 1989. Southwest England and the Shropshire-Cheshire meres. Guide to Excursion A, Vth International Symposium on Paleolimnology, Ambleside, Cumbria, 1–6 September, 1989.

Stanes, R., 1983. A Fortunate Place: a history of Slapton in South Devon. Field Studies Council, London.

Trudgill, S. T., 1983. The Natural History of Slapton Ley Nature Reserve. XVI – The soils of Slapton Wood. Field Studies 5: 833-840.

Van Vlymen, C. D., 1980. The water balance, physico-chemical environment, and phytoplankton studies of Slapton Ley, Devon. PhD thesis, University of Exeter.

Hydrobiologia **214**: 125–135, 1991.
J. P. Smith, P. G. Appleby, R. W. Battarbee, J. A. Dearing, R. Flower, E. Y. Haworth, F. Oldfield & P. E. O'Sullivan (eds), 125
Environmental History and Palaeolimnology.
© 1991 *Kluwer Academic Publishers.*

Sequential inorganic chemical analysis of a core from Slapton Ley, Devon, UK

A.L. Heathwaite[1] & P.E. O'Sullivan[2]
[1]*Dept. Geography, University of Sheffield, Sheffield, S10 2TN, UK*; [2]*Dept. Environmental Sciences, Polytechnic South West, Plymouth PL4 8AA, UK*

Key words: Paleolimnology, sediment chemistry, sequential analysis, ^{210}Pb dating, eutrophication, lake level changes

Abstract

Analysis of the upper 40 cm of a sediment core from Slapton Ley, a coastal lake in SW England was based on the fractionation procedure devised by Engstrom (Engstrom & Wright, 1984). This allows separation of the sediment into authigenic, biogenic and allogenic components. Lead-210 dating of the same core enabled trends in both concentration (mg g^{-1} dry mass) and influx (mg cm^{-2} a^{-1}) to be evaluated, and to be compared with events over the past two hundred years in the Ley and its catchment. The results show that before ca. 1950 AD, Slapton Ley was a shallow, clear, eutrophic lake, into which, in the mid-C19th, calcareous material was introduced by a phase of lime-kiln operation, road construction and lake level control. Since 1950, erosion of detrital material from the catchment has increased, as has the input of both allogenic and authigenic phosphorus, and biogenic silica. These changes reflect the intensification of agriculture post-1945, and the construction of Slapton sewage treatment works in 1953. A major peak in authigenic nitrogen, 6 cm below the present sediment surface, is correlated with the severe 1976 drought in the UK. The ecosystem of the Ley appears to have been triggered by this event into its present hypertrophic state.

Introduction

Chemical analysis of freshwater lake sediments was pioneered by Mackereth (1965, 1966). He concentrated on bulk (or total) inorganic elemental determinands, and expressed his results in terms of the relative composition of the material (concentration in mg g^{-1} dry sediment or mg g^{-1} mineral matter). For twenty years, this was the model which all inorganic chemical analyses of lake sediments followed. Engstrom (Engstrom & Wright, 1984) devised a step-wise procedure for isolating the various fractions (authigenic, biogenic and allogenic) of a lake sediment. This

method allows a more precise identification of changing sediment source, and when allied with techniques for dating near-surface sediments (for example, ^{210}Pb or ^{137}Cs analysis, Oldfield, 1981; palaeomagnetism, Thompson, 1973, or varve counts, O'Sullivan, 1983) enables the expression of results in terms of sediment influx (mg cm^{-2} year^{-1}), and thus represents a much more powerful means of investigation, from an inorganic chemical perspective, of the recent history of a lake and its catchment.

The present investigation seeks to increase knowledge regarding the history of eutrophication of Slapton Ley and to apply Engstrom's sequential analytical procedure to this problem. Slapton Ley is a UK Grade 1 Site of Special Scientific interest, so that data concerning the history and causes of eutrophication are an important input into its conservation and management. As far as we know Engstrom's procedure has not so far been successfully applied to the investigation of very recent lake history. We have not adopted a multiple core approach as used by Engstrom & Swain (1986), but instead have concentrated on the detailed chemical fractionation of a single core. Slapton Ley, and the coring site indicated in

Fig. 1, are described more fully in O'Sullivan *et al.* (1991).

Methods

Five cores were collected in July 1987 from beneath 2 m of water in the area of the Lower Ley known as Ireland Bay (Fig. 1) using a 1 m 'mini'-Mackereth corer (Mackereth, 1969). The cores were stored intact at 5 °C until subsampling took place. The pattern of sedimentation in the lake at the coring site was evaluated by whole core susceptibility (K) using a Bartington MS1B susceptibility bridge and loop sensor designed specifically for use with Mackereth cores. Of the five cores taken from Slapton Ley, one (SLL3) was selected for chemical analysis.

Twenty 1 cm slices were extracted from the top 20 cm and 2 cm intervals from 20 cm downwards. The sediment below 40 cm was discarded because the ^{210}Pb profile for the adjacent core SLT3 indicated that below this depth material would be of an age beyond the range of our principal dating technique. The core slices were subsampled for the determination of wet volume (V), bulk density (D), dry matter content (DM), loss on ignition (LOI), mineral matter content (MM, reciprocal of LOI), and carbonate content (CO_3^{2-}) immediately after slicing (after Allen, 1989). The sequential chemical fractionation procedure used here followed that of Engstrom & Wright (1984) except in the respect that we began our analysis with 1 ml fresh wet sediment rather than 1 g dry sediment. This makes no difference to the calculation, provided the dry mass/wet volume relationship for the sediment has been determined.

All elements except K, N and P were determined using atomic absorption spectrometry (AAS) on a Pye-Unicam SP9. Potassium was measured by means of flame emission using the same instrument. Nitrogen and phosphorus were determined by autoanalysis using the modified methods of Crooke & Simpson (1970) and Murphy & Riley (1962) respectively.

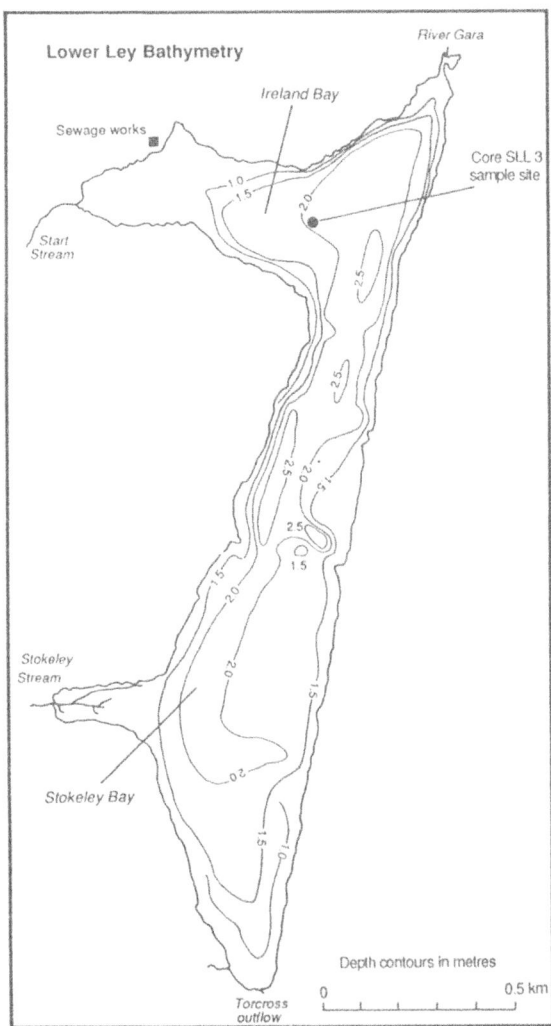

Fig. 1. The Lower Ley, Slapton, showing lake bathymetry, and location of coring site.

Results

Magnetic susceptibility

Whole core initial apparent reversible magnetic susceptibility (K) for core SLL3 is shown in Fig. 2. K rises from low values at the bottom of the core to a peak at 12–14 cm. There is also a minor peak at 62–64 cm. These results indicate that SLL3 is a typical 'Ireland Bay' core and easily correlated with others used in the investigation of the palaeolimnology of Slapton Ley (see O'Sullivan et al., this volume).

Stratigraphy

The stratigraphy of core SLL3 was as follows : -

0–11.5 cm mottled black and pale brown gyttja
11.5–23 cm pale brown gyttja
23–40 cm mottled black and pale brown gyttja, becoming sandier towards the base of the core.

Chronology

The ^{210}Pb chronology adopted for core SLL3 is that described by O'Sullivan et al. (1991).

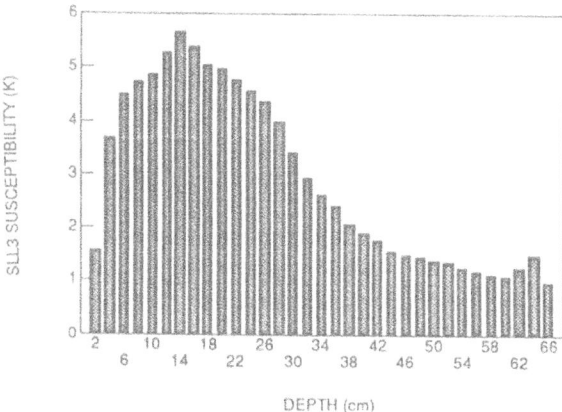

Fig. 2. Whole core initial apparent reversible magnetic susceptibility for core SLL3 from Slapton Ley.

Physical and chemical analyses

Bulk density, dry matter content, ignition loss, mineral matter and carbonate content, total influx

Figure 3 depicts total sediment influx, bulk density, dry matter content, ignition loss, mineral matter and carbonate content for core SLL3. Also added, is the ^{210}Pb chronology from the same core. A major change in the physical properties of the sediment takes place in the section 24–28 cm, where D, DM and MM fall, and LOI increases. The change is mirrored in the profile for CO_3^{2-} content, which exhibits a sharp decrease, and in the stratigraphy and the magnetic susceptibility of the core (see above). From the ^{210}Pb chronology, it may be seen that this change occurred in the period 1906–1926 AD. Peaks in total sediment influx are dated at 1822, 1908, 1968 and 1987 AD.

Chemical properties

The data presented here must be considered in terms both of concentration and influx, for as explained by Engstrom & Wright (1984), the two methods of expressing the results give different, but complementary information. In the diagrams (Figs. 4–5), the complete curves represent total concentration (Fig. 4) or total influx (Fig. 5) as appropriate, and the light portions, the authigenic fraction. The abundance of the allogenic component (the black portion of the curve) may therefore be assessed visually by subtracting the authigenic component from the total. The only exception to this rule is the curve for Si, where the paler shaded portion consists of the biogenic rather than the authigenic fraction.

Two elements (K, Al) are located mainly in the allogenic fraction. Their contribution to the sediment consists therefore mostly of mineral matter embedded in the crystal lattice, which only passes into solution during the last stage of the fractionation procedure. Six others (Mg, Ca, Mn, Fe, N and P) are attached mainly to the authigenic component, which means that, whatever their ultimate source, they are delivered to the sediment by processes of chemical fixation within the lake. Two of these (Ca, Mn) are almost exclusively

128

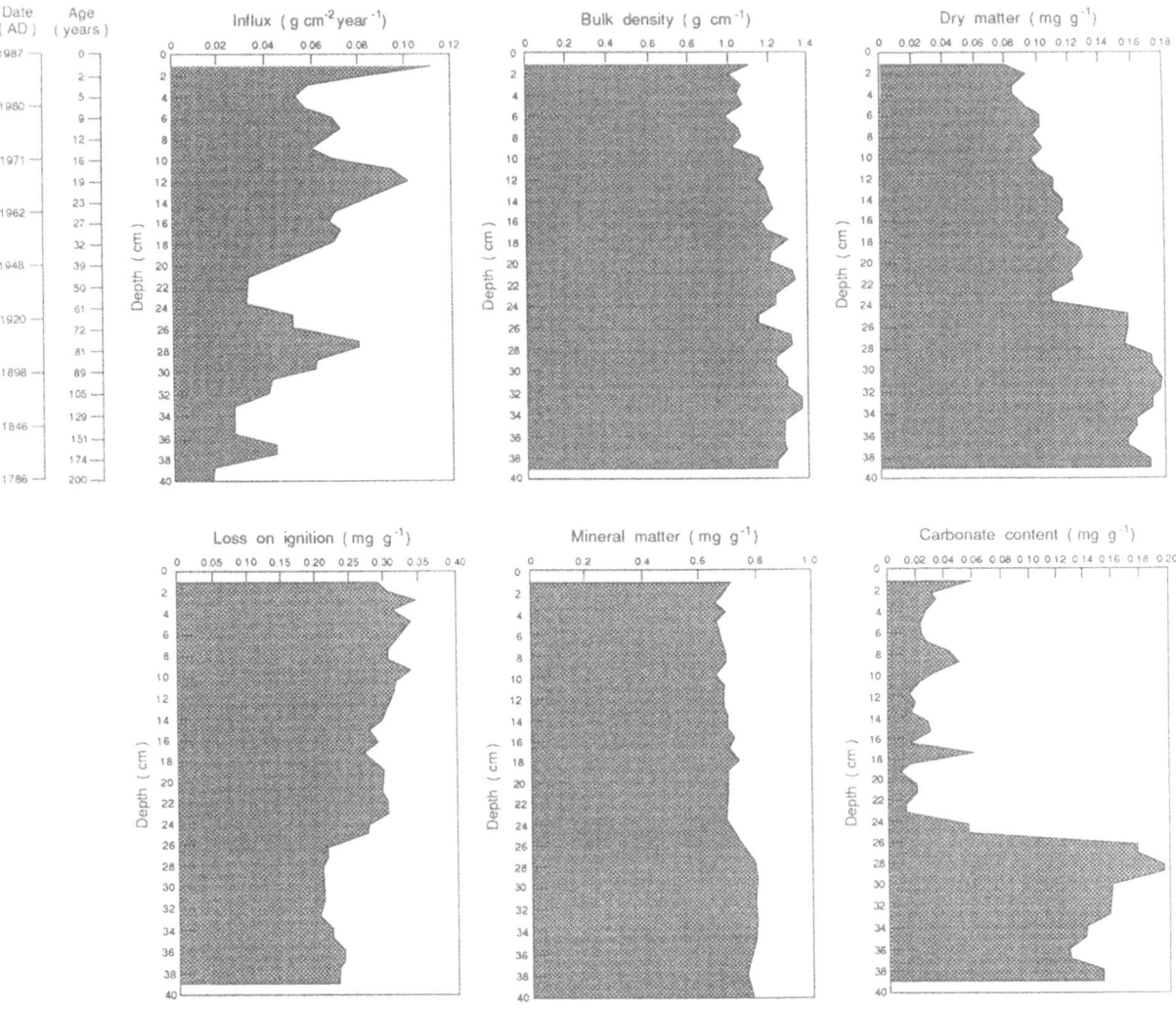

Fig. 3. Chronology, total sediment influx, bulk density, dry matter content, loss on ignition, mineral matter and carbonate content of core SLL3.

authigenic in origin, whereas the others possess small allogenic components. Of these, Fe, N and P follow each other quite closely, whilst the curve for Mg is similar to that of Fe in the top half of the core. In the lower sections, however, it mimics those for Ca and Mn.

Whereas, in the case of Fe, the proportion of allogenic material sedimented *decreases* towards the sediment water interface (SWI), for phosphorus, the opposite is the case. Above ca. 21 cm, the contribution of allogenic P to the total increases. As stated above, the profile for Si records the

presence of roughly equal proportions of allogenic and biogenic material, with an increase in influx of *both* components above 21 cm, and especially of the biogenic Si just below the SWI.

The results may therefore be described in terms of four categories of determined, these being (a) the exclusively 'allogenic' elements (K, Al), (b) those elements which are mainly authigenic in their provenance (Mg, Fe, N and P), (c) the exclusively 'authigenic' fraction (Ca, Mn, CO_3^{2-}), and (d) the 'allogenic/biogenic' element, Si.

In terms of concentration, the allogenic com-

129

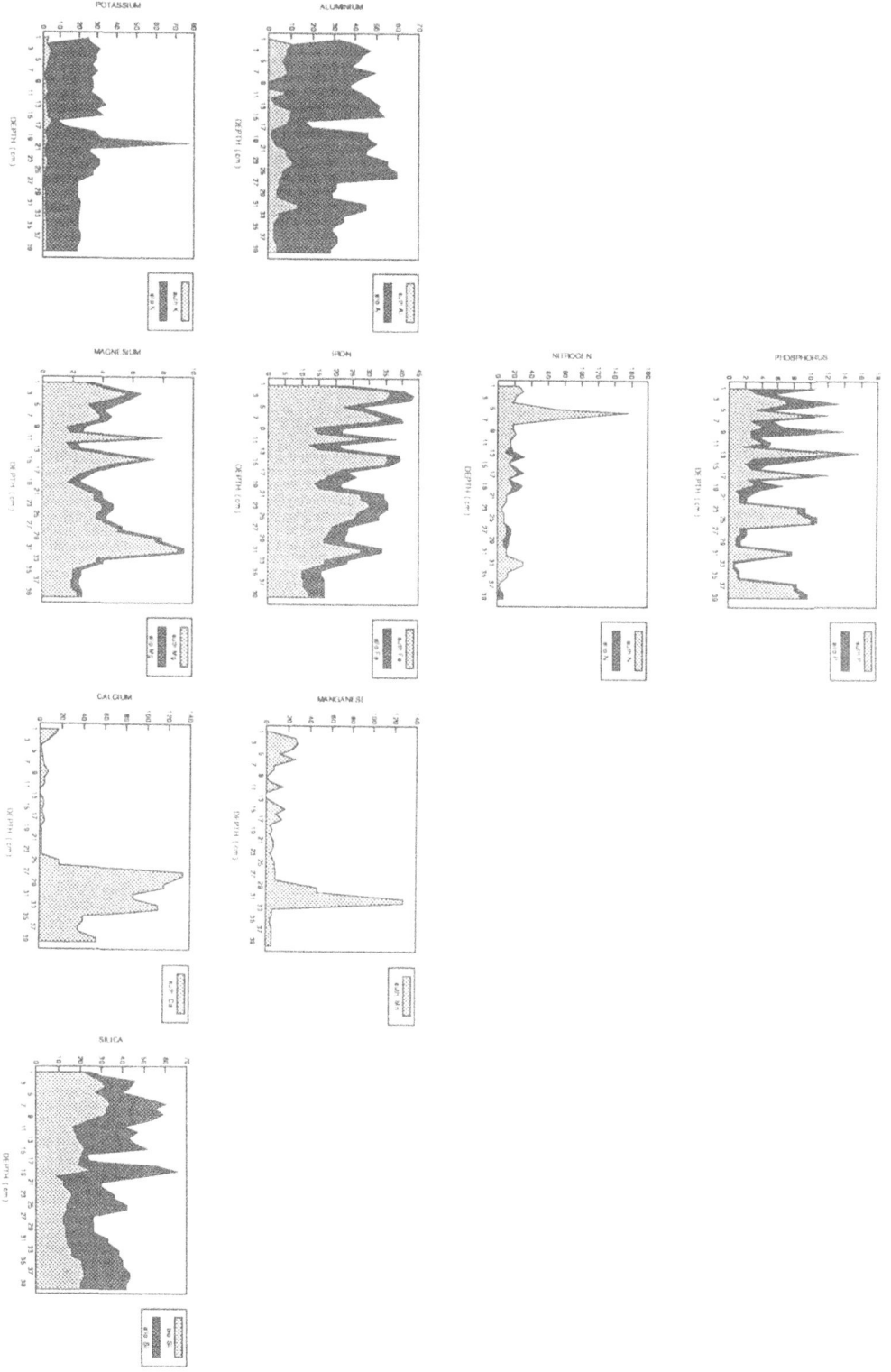

Fig. 4. Concentration (mg g⁻¹) of selected elements in core SLL3.

130

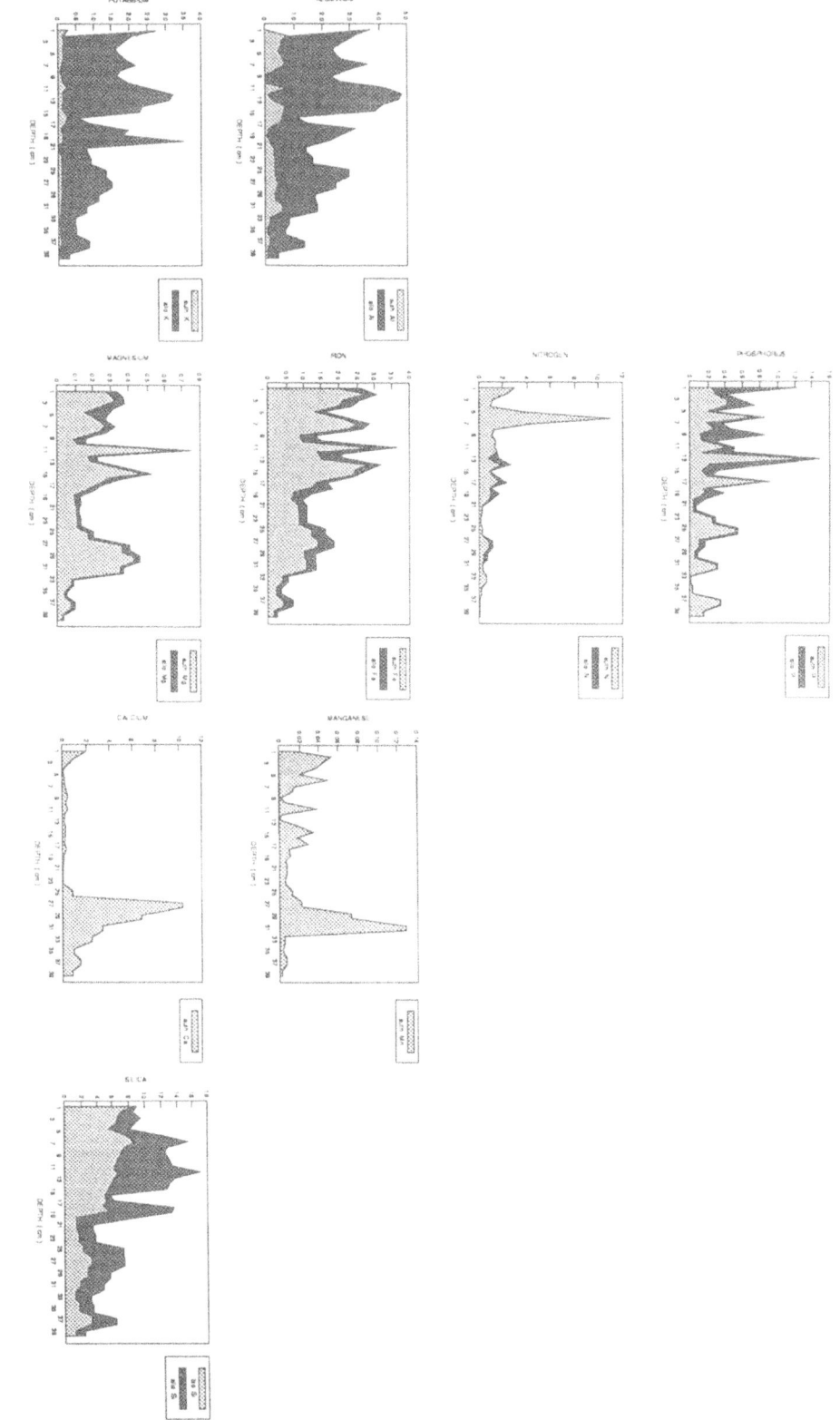

Fig. 5. Influx (mg cm^{-2} a^{-1}) of selected elements to core SLL3.

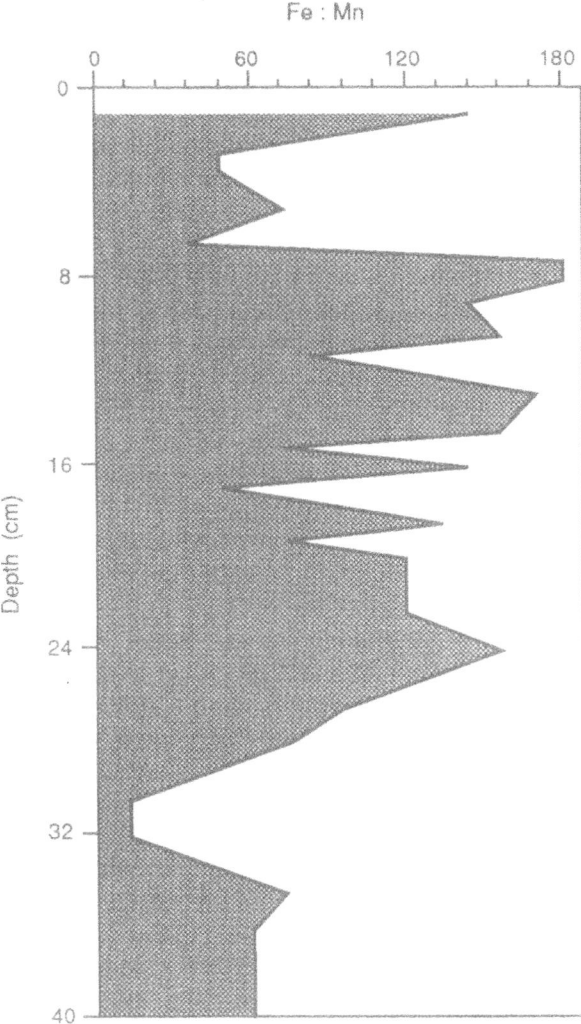

Fig. 6. Authigenic Fe:Mn ratio for core SLL3.

its curve follows much more closely those of Fe, N and P, varying erratically towards the SWI. Changes in concentration of biogenic and allogenic Si have already been described.

Expressing the results in terms of influx appears to clarify most of the above trends. Allogenic influx, mainly K, Al and Si, is very low below 32 cm, reaches a maximum at 26 cm, falls, and then exhibits a general increase above 21 cm. The influx profiles of the mainly authigenic determinands (Mg, Fe, N and P) record similar changes. The exclusive authigens (Ca and Mn) reach maxima at 28 and 32 cm respectively, and then fall very low, but recover just below the SWI. Biogenic Si as a proportion of total Si also increases above 5 cm, as does allogenic P.

Discussion

As may be seen from Fig. 3, in core SLL3 there is a general upward trend in total sediment influx, superimposed upon which are peaks in the periods ca. 1822 AD (36–38 cm), 1908 AD (27–28 cm), 1968 AD (12 cm), and at the SWI (1987 AD). Both the 1968 peak, and that for 1908 (1910 by ^{210}Pb dating), are recorded by the ^{210}Pb chronology (O'Sullivan *et al.*, this volume). Similarly, the section 28–20 cm (1906–1948 AD) represents a major change in sediment quality, with events below this datum being dominated by in-lake sedimentation of mainly authigenic species, and above, until very recent times, the record of increased allochthonous inputs. The results will be discussed under three headings : -

The interpretation of catchment processes from the sediment record

In terms of both concentration and influx (Figs. 4 and 5), the profiles for K, Al and Si are similar, with all three elements located mainly in the allogenic fraction. There is only a slight increase in concentration from the base to the top of the core, but the influx data exhibit a strong upward trend. On a concentration basis these elements show no

ponent (K and Al) remains constant throughout most of the core, with a general increase above ca. 25 cm, and fluctuating values around 16 cm, which are also recorded by allogenic Si. In contrast, the exclusively authigenic elements Ca and Mn are abundant in the lower parts of the core, especially between 35 and 26 cm, but fall to very low concentrations above 24 cm. Values rise again in the top few cm. Magnesium is also abundant in the lower sections, reaching a peak in concentration at 32 cm. Above 24 cm, however, correlation with mineral matter, which would originally have led us to reject Mackereth's (1966) hypothesis that they are indicators of erosion. However, influx exhibits a strong association with mineral matter (see Fig. 3), so like Engstrom & Wright (1984) we can conclude that this may be used to show that the amount of eroded material reaching Slapton Ley has gradually increased over the past two centuries, at least as far as this part of the lake is concerned. Multiple-core studies of other parts of the Ley (O'Sullivan *et al.*, this volume) confirm this conclusion for very recent times.

Mn concentrations in core SLL3 are low (< 3 mg g^{-1}) whereas those for Fe are generally high (> 40 mg g^{-1}). The Fe:Mn ratio of the sediments (Fig. 6) is, therefore, normally greater than 50:1, which is consistent with the origin of these elements in free-draining aerated soils (Trudgill, 1983), and their deposition in a highly-flushed, turbulent lake (Van Vlymen, 1979) in which oxygen concentration at the sediment surface rarely falls to levels at which remobilisation would occur (Engstrom & Wright, 1984; Mackereth, 1966).

There are however, two exceptions to this rule, which occur during (a) the pronounced peak in authigenic Mn, at 28–32 cm (ca. 1882 to 1906 AD), and the top 6 cm of the core (deposited since AD 1978). In each case, it would appear that rather than recording changes in redox conditions in the soils of the catchment, what these peaks constitute are signals of the variation of solubility of Mn within the lake itself (Engstrom & Wright, 1984).

The upper of the two may be explained in terms of increased productivity of the Ley (see below), which has recently affected oxygen availability at the SWI (A. Bark, unpublished). In such circumstances, the Mn:Fe ratio would be enhanced. Crabtree & Round (1967) also found that Mn values increased in the uppermost sediments of the Ley. The lower peak, however, is related to other factors, which are now discussed.

The impact of water level changes on the sediment record

Authigenic Ca, Mg and Mn reach maxima together at 27–33 cm. In this part of the core, for all three elements, the authigenic fraction is the major component of the total influx. This is in contrast to the findings of Engstrom & Wright (1984) who suggested that in lake sediments, Ca and Mg are principally found in the allogenic fraction. Furthermore, whereas there is little correlation to be seen between organic matter (Fig. 3) and either Ca concentration, or influx, the element is strongly correlated with CO_3^{2-} content. Most of the Ca in the sediments of Slapton Ley is therefore present as $CaCO_3$.

The period of enhanced sedimentation of these elements is dated, on the basis of the ^{210}Pb chronology (Fig. 3), to the period 1958–1908 AD. In 1856, the sluice draining the Ley, at the village of Torcross (Fig. 1) came into operation, and a road was constructed along the shingle bank. Mean lake level was also raised and subsequently artificially controlled (Morey, 1976; Stanes, 1983). The shingle ridge is partly local in origin, and made up of fragments of Devonian slates, and partly calcareous, this latter fraction being composed of flint and chert pebbles whose source is the bed of the English Channel (Hails, 1975). It would appear therefore that the construction of the road, and the raising of the Ley level, introduced a new source of $CaCO_3$ to the system.

An additional source of $CaCO_3$ was probably the lime kilns located at the now extinct hamlet of Slapton Cellars, near Slapton Bridge (Stanes, 1983). These were present in the C18th, however, so that it was the building of the sluice and the causeway, which *raised* the level of the Ley, and increased its water residence time, which led to enhanced authigenic sedimentation, presumably by increasing the effective mean annual concentrations of these elements in the lake waters.

It is not known precisely why, in 1908, high sedimentation of Ca, Mg and Mn ceased. Perhaps operation of the lime-kilns was discontinued. However, the coincidence of this event with the horizon which denotes a much more general change on lake sedimentation suggests that a more general cause must be sought. One working hypothesis would be that the raising of the lake level in 1856, created, for that period of about sixty years, a water body which was deep enough for the development, in summer, of conditions of oxygen deficiency in at least some parts. Whether this corresponded to a true hypolimnion, or was just temporary stagnation of a few deep parts of the Ley remains an open question. As sedimentation proceeded, eventually the Ley once more became too shallow, and hence, too turbulent for oxygen gradients to form, and sediments in which the Fe:Mn ratio exceeds 50:1 were once more deposited.

The evidence for eutrophication in the Slapton sediments

The main feature of the nitrogen profiles from core SLL3 is the large peak of authigenic N at 6 cm. Here, for example, concentration of this form of the element increases from <40 mg g^{-1} to ca 160 mg g^{-1}, and influx from an average of 1–2 mg cm^{-2} a^{-1} to over 11 mg cm^{-2} a^{-1}. According to the ^{210}Pb chronology, this peak coincides with the years 1977–1980. Water quality records for one of the catchments discharging into Slapton Ley (Burt *et al.*, 1988) suggest that, over the period 1970–1985, stream nitrate concentration exhibits a general upward trend (from 5 mg l^{-1} to 7 mg l^{-1}). A major peak in concentration (to 14 mg l^{-1}) was observed following the 1976 drought. It would appear from our results that this event may have triggered a major shift in the trophic status of the Ley.

Biogenic Si is often used as an index of primary productivity (Schelske *et al.*, 1987). In the uppermost sediments of our core (above 5 cm), biogenic Si forms up to 93% of total Si. Phosphorus here is located primarily in the allogenic fraction, and influx of this element also increases towards the sediment surface. Diatom analysis on an adjacent core (O'Sullivan *et al.*, this volume) indicates that from the mid-1970s onwards, the flora changes from one with abundant species of *Fragilaria*, to

one in which planktonic centric diatoms, especially *Cyclostephanos dubius*, were prominent.

We interpret these changes as follows. In the lower part of the core (below 21 cm, or before 1942 AD), authigenic P is more abundant than allogenic. With the other chemical and the microfossil evidence, this points to the existence at that time (pre-1950) of a shallow eutrophic lake, influenced by its catchment, but less strongly so than at present. In the recent sediments (1950 to the present), allogenic P is at least as important as authigenic, indicating an increase in a catchment, minerogenic source for phosphorus. The lake appears to have become more eutrophic at this time, as witnessed by the expansion of the biogenic silica influx.

Current research into nitrogen and phosphorus inputs from the catchment of Slapton Ley (Burt *et al.*, 1990; Heathwaite *et al.*, 1990) has shown that phosphorus adsorbed on to sediment particles forms a major component of the total external P load to the lake. The source of this material is the topsoil of fields in the catchment, especially those under grassland which is being heavily grazed, and those under arable cultivation. Historical studies by Acott (1989) have shown that agriculture has intensified in the Slapton catchment since 1945. Furthermore, O'Sullivan *et al.* (this volume) attribute increased influx of magnetically 'hard' mineral matter to the Ley after AD 1950 to erosion of topsoil from arable land. Enhanced influx of allogenic P to the Ley is therefore attributable to agricultural changes in the period since 1945.

The increase in both biogenic Si and authigenic P around 19 cm marks a further change in the composition of the sediments and suggests that another potential source of phosphorus for the Lower Ley may be important. This is a small sewage works on Slapton village stream which discharges into the Start Stream and ultimately into Ireland Bay (Fig. 1). It currently contributes up to 20% of the total annual P load to the Lower Ley (Heathwaite, unpublished; Johnes & O'Sullivan, 1989) and may, since its installation in 1953, have formed a significant proportion of the authigenic P in the sediment (primarily because of its proximity to the Lower Ley rather than to the magnitude of the input). Using the ^{210}Pb chronology, 1953 would correlate with a depth of 19 cm. Both authigenic P and biogenic Si influx increase significantly above this level.

There is no long term monitoring data on the magnitude of the P load from the Slapton sewage works, or from that at Blackawton in the north of the Gara catchment (Fig. 1), but Acott (1989) showed how the population of the Slapton catchment has increased from ca. 1600 persons in 1945, to ca. 2200 in 1981. Many of these live in dwellings which are probably not directly connected to the sewers, however, but this permanent population is increased each summer by an unknown number of tourists who must also add to the sewage phosphorus load.

Conclusions

Sequential inorganic chemical analysis of core SLL3 has revealed a number of points concerning the history of Slapton Ley and its catchment over the last 200 years. A major change in the sedimentation regime of the Ley occurred in the period following the construction, in 1856, of a road along the margin of the Ley, and the raising of the lake level by the building of an outflow sluice. These events resulted in an increase in the lake's hydraulic retention time, and considerable influx of $CaCO_3$, both from disturbed beach material, and probably from lime-kilns operating on the lake shore.

The record of sedimentation of Fe and Mn suggests that, until recently, apart from the period immediately following raising of the water level, the lake does not appear to have been one in which deoxygenation of its bottom waters has ever been common. Before 1950 Slapton Ley was a shallow, clear, eutrophic lake dominated by macrophytes and epiphytic (*Fragilaria*) diatoms. Post-1945, erosion of allogenic material into the Ley increased. Enhanced levels of sedimentary N and P and biogenic Si also indicate that in the period since 1945, Slapton Ley has become more eutrophic. Intensification of agriculture in the

post-war period and the installation of Slapton sewage works seem to account for both of these processes. A peak in sedimentary authigenic N is correlated with the 1976 drought. The lake appears to have been triggered by this event to transcend a threshold in its trophic status, and to become hypertrophic.

References

Acott, T., 1989. An investigation into nitrogen and phosphorus loading on Slapton Ley, 1905–1985. Unpublished BSc. dissertation, Polytechnic South West.

Allen, S. E., 1989. Chemical Analysis of Ecological Materials. Blackwell.

Burt, T. P., B. P. Arkell, S. T. Trudgill & D. E. Walling, 1988. Stream nitrate levels in a small catchment in southwest England over a period of 15 years (1970–1985). Hydrol. Proc. 2: 267–284.

Burt, T. P., A. L. Heathwaite, S. T. Trudgill, P. E. O'Sullivan, P. J. Johnes & K. Chell, 1990. Hydrological processes in the Slapton catchments, and their relationship to sediment and solute losses. Proc. Wageningen Symp. IAHS, June 1990.

Crabtree, K. & F. E. Round, 1967. Analysis of a core from Slapton Ley. New Phytol. 66: 255–270.

Crooke, W. M. & W. E. Simpson, 1971. Determination of ammonium in Kjeldahl digests of crops by an automated procedure. J. Sci. Fd. Agric. 22: 9–10.

Engstrom, D. R. & H. E. Wright, 1984. Chemical stratigraphy of lake sediments as a record of environmental change. In: Haworth, E. Y. & Lund, J. W. G. (eds), Lake sediments and environmental history. Leicester University Press, Leicester, UK: 11–69.

Engstrom, D. R. & E. B. Swain, 1986. The chemistry of lake sediments in time and space. Hydrobiologia 143: 37–44.

Hails, J. R., 1975. Some aspects of the Quaternary history of Start Bay, Devon. Field Studies 4: 207–222.

Heathwaite, A. L., T. P. Burt & S. T. Trudgill, 1990. The effect of land use on nitrogen, phosphorus and suspended sediment delivery to streams in a small catchment in southwest England. In Thornes, J. B. (ed.) Vegetation and Erosion, Wiley: 161–179.

Mackereth, F. J. H., 1965. Chemical investigations of lake sediments and their interpretation. Proc. Roy. Soc. Lond. B 161: 295–309.

Mackereth, F. J. H., 1966. Some chemical observations on post-glacial lake sediments. Phil. Trans. R. Soc. Lond., B 250: 167–213.

Mackereth. F. J. H., 1969. A short-core sampler for subaqueous deposits. Limnol. Oceanogr. 14: 145–151.

Morey, C. R., 1976. The natural history of Slapton Ley Nature Reserve IX: The morphology and history of the lake basins. Field Studies, 4: 353–368.

Murphy, J. & J. P. Riley, 1962. A modified single solution method for the determination of phosphate in natural waters. Anal. Chim. Acta., 27: 31–36.

Oldfield, F., 1981. Peats and lake sediments: formation, stratigraphy, description and nomenclature. In: Goudie, A. S. (ed.), Geomorphological Techniques. George Allen & Unwin: 306–326.

O'Sullivan, P. E., 1983. Annually laminated lake sediments and the study of Quaternary environmental changes – a review. Quat. Sci. Rev. 1: 245–313.

O'Sullivan, P. E., A. L. Heathwaite, P. G. Appleby, D. Brookfield, M. W. Crick, C. Moscrop, T. B. Mulder, N. J. Vernon & J. M. Wilmshurst, 1991. Palaeolimnology of Slapton Ley, Devon, UK. In J. P. Smith et al. (eds), Environmental History and Palaeolimnology. Kluwer Academic Publishers, Dordrecht: 115–124. Reprinted from Hydrobiologia 214.

Schelske, C., D. J. Conley, E. F. Stoermer, T. L. Newberry & C. D. Campbell, 1987. Biogenic silica & phosphorus accumulation in sediments as an index of eutrophication in the Laurentian Great Lakes. In Löffler, H. (ed.), Palaeolimnology IV. Dr W. Junk, The Hague.

Stanes, R., 1983. A fortunate place: the history of Slapton in South Devon. Field Studies Council, London.

Trudgill, S. T., 1983. The Natural history of Slapton Ley Nature reserve XV – The soils of Slapton Wood. Field Studies 5: 835–840.

Van Vlymen, C. D., 1970. The natural history of Slapton Ley Nature Reserve XIII – The water balance of Slapton Ley. Field Studies 5: 59–84.

Hydrobiologia **214**: 137–142, 1991.
J. P. Smith, P. G. Appleby, R. W. Battarbee, J. A. Dearing, R. Flower, E. Y. Haworth, F. Oldfield & P. E. O'Sullivan (eds),
Environmental History and Palaeolimnology.
© 1991 *Kluwer Academic Publishers.*

Sediment characteristics in relation to cultivation history in two varved lake sediments from East Finland

Elisabeth Grönlund
University of Joensuu, Karelian Institute, Section of Ecology P.O. Box 111, SF-80101 Joensuu, Finland

Key words: Canonical correspondence analysis, varved sediments, sediment chemistry, pollen analysis, cultivation history

Abstract

Changes in agricultural land-use patterns was studied with canonical correspondence analysis (CCA). The data sets consisted of sediment quality variables and pollen analysis data of two varved lake sediments in East Finland. The strong correlations between the sediment variables and pollen taxa reflect the simultaneous reactions to land-use changes of both vegetation and sediment quality; reactions which are reflected more drastically in the sediment quality of small Heinälampi than the larger Lake Suurjärvi.

Introduction

Palaeoecological research on cultivation history of East Finland has been carried out in the Karelian Institute since the 1970's (Vuorinen & K. Tolonen, 1975; Vuorinen, 1978; Simola *et al.*, 1985, 1987, 1988; Grönlund *et al.*, 1986). Special attention has been paid to the pollen-analytical and palaeolimnological tracing of slash-and-burn cultivation, an efficient means of environmental exploitation that was widely practiced in Finland from the Iron age to the early 20th century (e.g. M. Tolonen, 1985). The main problem in these studies is that the relationships between land use practices and sediment properties are usually obscured by many interacting biogeochemical processes, some of which are still poorly understood (Brugam, 1984).

A recent statistical tool to deal with large and multidimensional palaeoecological data sets is canonical correspondence analysis (CCA), available in the computer program package CANOCO (Ter Braak, 1987). This method provides a simul-taneous analysis of measured environmental variables and species data (Ter Braak, 1986; Ter Braak & Prentice, 1987). In the present study CCA was applied to establish and quantify relationships between some sediment variables and pollen-analytically detected human activity.

Sites and methods

The two lakes chosen for the study differ from each other in terms of catchment and lake areas: Lake Suurjärvi (62° 51′ N, 29° 05′ E), is a 32 m deep oligotrophic lake with an area of 8 km² and watershed of 48 km². It lies in the middle of the Kerimäki island which is surrounded by the waters of the Lake Saimaa complex in the province of Mikkeli. The topography around the lake is hilly, the tops of the till-covered hills above 105 m a.s.l. are supra-aquatic with unleached soils (Simola *et al.*, 1985). Lake Heinälampi (63° 07′ N, 27° 39′ E) is an 8 m deep lake in the Kuopio province with an area of 0.026 km² and drainage area of 3.6 km². The drainage area around the lake is relatively flat with fine silt and clay soils (Sandman *et al.*, 1989).

Lake Suurjärvi was freeze-cored in the winter 1981. The sample was photographed and the varve years counted from photographs. Some 3600 varves were observed, dating the bottom of the core (165 cm) to 1600 BC. The Lake Heinälampi sample was taken in 1988 with a long piston corer. The varves were counted from the fresh core, the oldest date being 650 BC at 185 cm. The topmost 10 cm of the Heinälampi core was omitted from the present analysis due to anomalously high inorganic sediment deposition caused by extensive forest ditching in 1980 in the catchment (Sandman *et al.*, 1989).

The sediment samples for pollen and chemical analyses were treated essentially according to Bengtsson & Enell (1986) and Berglund (1986).

In the canonical correspondence analysis the species data set contained percentages ($> 2\%$) of terrestrial pollen taxa, *Pteridium* spores and *Isoëtes* microspores. The environmental data set of the lakes had three sediment quality variables in common: sediment accumulation rate (cm yr^{-1}), loss-on-ignition (%) reflecting organic matter in the sediment and concentration of charcoal particles (cm³). Additionally, five elements were used as environmental variables in the Suurjärvi core: Fe, Mg, Ca, K, and total P measured on a dry weight basis. Also, arithmetic means, coefficients of variations and variations

ranges were calculated for all sediment quality parameters.

Results

In Lake Heinälampi the mean sediment accumulation rate is somewhat higher than in Lake Suurjärvi, but the coefficients of variation of sediment accumulation rates do not significantly differ between the lakes. However, the variation of organic matter and charcoal concentration were significantly different between the lakes (Table 1).

Canonical correspondence analysis of sediment variables and pollen data of both lakes reveal a pattern of land-use change through time. In a plot of the first two CCA axes, the coefficient of species-environment correlation is highly significant in both lakes (Table 2). Also, the sediment accumulation rate has the strongest correlation in both lakes with the first CCA-axis.

In the CCA space of Lake Suurjärvi data the stratigraphic samples ordinate loosely into three groups (Fig. 1):

Group 1 (1600 BC–1000 AD) samples are characterized by low accumulation rate and high and stable organic content; they represent the natural type of sediment deposited prior to human occupation. In the ordination these levels show affinity to the mixed deciduous forest trees (QM) as well as to spruce. Some samples, dated to 1625 BC, 400–425 BC and 660–775 AD shift out

Table 1. Arithmetic means (x), coefficients of variation (cv) and ranges of sediment properties of sthe study lakes (*** = $p < 0.001$, * = $p < 0.05$).

Variable	Lake Suurjärvi			Lake Heinälampi		
	x	cv	range	x	cv	range
Sed. rate cm yr^{-1}	0.056	0.441	0.02–0.125	0.0793	0.466	0.045–0.25
L.O.I. %	30.645	0.078***	23.26–34.50	31.095	0.367***	6.19–60.22
Char. cons. exx. cm³	53269	0.490*	3656–103940	38908	0.814*	7470–102976
Ca mg g^{-1}	2.261	0.137	1.66–3.59			
Fe mg g^{-1}	47.171	0.229	26.20–86.67			
K mg g^{-1}	1.083	0.909	0.49–1.62			
Mg mg g^{-1}	1.391	0.221	0.77–2.08			
P mg g^{-1}	2.861	0.247	1.37–3.48			

Table 2. Significant correlation coefficients of species-environment axes and species axes vs. sediment variables, (*** = $p < 0.001$, ** = $p < 0.01$).

	Lake Suurjärvi		Lake Heinälampi	
Env. axis 1	0.789***		0.732***	
Env. axis 2		0.662***		0.841***
Sed. rate cm yr⁻¹	0.725***		0.521***	− 0.481***
L.O.I. %	− 0.304***			− 0.781***
Ca mg g⁻¹	0.432***			
Fe mg g⁻¹		− 0.310***		
Mg mg g⁻¹		− 0.423***		
P mg g⁻¹		0.352***		
	Spec. axis 1	Spec. axis 2	Spec. axis 1	Spec. axis 2

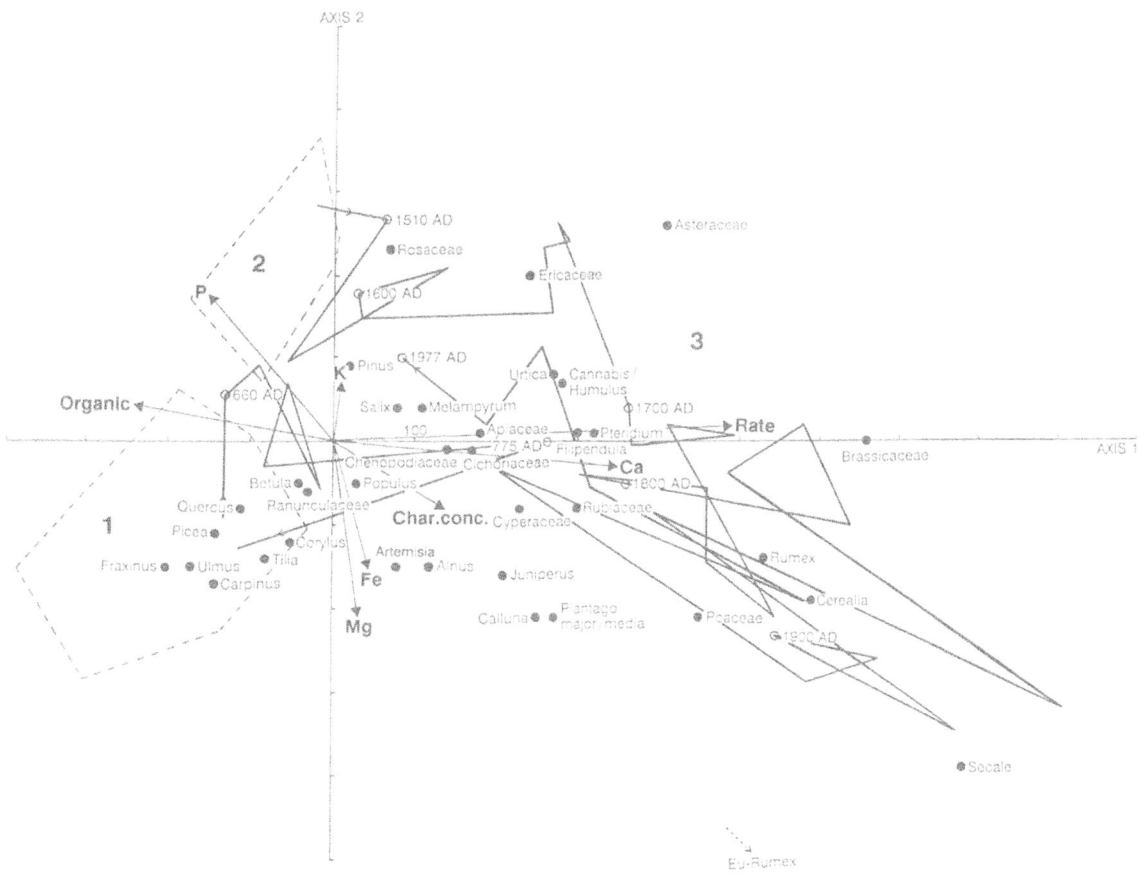

Fig. 1. CCA-plot of Lake Suurjärvi, axis 1 vs. 2. Arrows represent the sediment variables, black dots the pollen species. Biplot scores of sediment variables multiplied by three. Sample points are separated in three groups (see text): In groups 1 and 2 the sample points lie within the framed area. In group 3 selected sample points are represented as open circles with their dates. Eigenvalues: axis 1 = 0.048, axis 2 = 0.012.

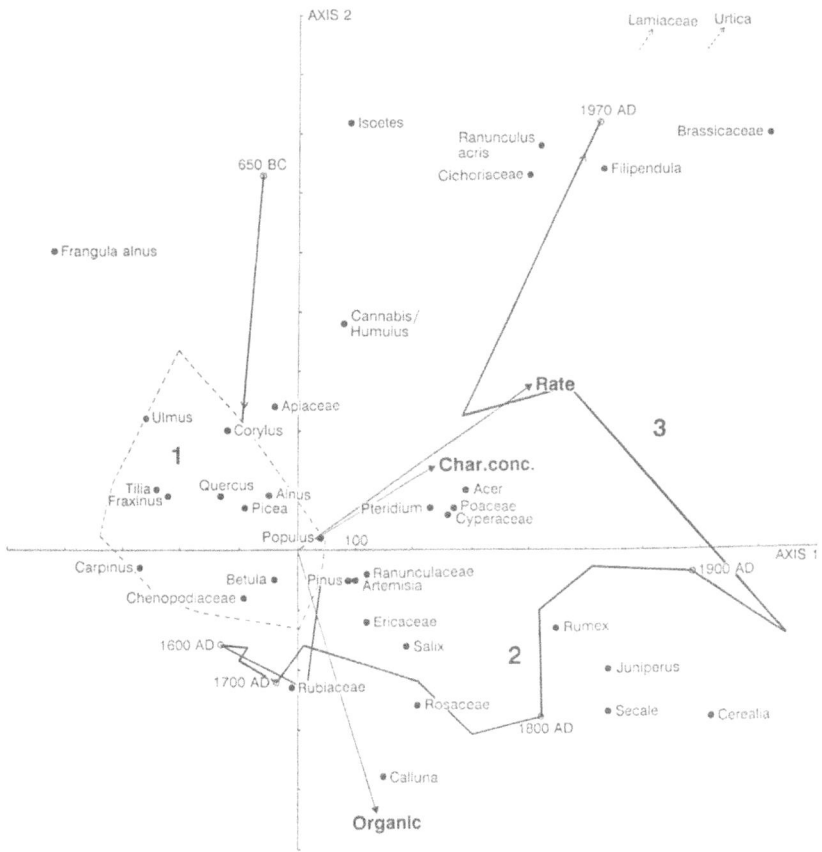

Fig. 2. CCA-plot of Lake Heinälampi, axis 1 vs. 2. Legend as in Fig. 1. Eigenvalues: axis 1 = 0.037, axis 2 = 0.024.

from the group and are ordinated in the vicinity of much younger samples; in the pollen analysis the first signs of cereal cultivation were found from the 660–775 AD samples.

Group 2 (1100–1500 AD) samples are characterized by high organic matter and phosphorus content and low Fe and Mg content of the sediment. The pollen analysis of these levels reveal signs of weak but continuous and gradually intensifying cultivation.

Group 3 (1500–1977 AD) samples spread out in the plot along the first axis mainly in accordance with increasing sediment accumulation rate, calcium content and charcoal concentration. The shift along the second axis (samples 1800–1900) is affected by increases in sedimentation and iron and manganese concentrations. The samples dated to 1700–1900 AD are ordinated close to the

indicator species of slash- and burn cultivation such as cerealia, rye, juniper and *Rumex*.

Three sample groups can be distinguished also in the Lake Heinälampi CCA-plot (Fig. 2):

Group 1 (600 BC–1500 AD) samples do not show affinity to any measured environmental variables due to the uniformly low values of the variables. The samples are associated with the pollen species of QM-forest trees, spruce and birch.

Group 2 (1500–1900 AD) samples move down along the 2nd axis as organic matter of the sediment increases. From 1700 onwards, the sediment accumulation rate and charcoal concentration increases and the samples move rapidly along the first axis. There is a close relationship between the pre-1900 samples and the pollen indicator species of slash-and-burn cultivation.

Group 3 (1900–1970 AD) samples are ordinated parallel to the second axis as organic matter content decreases and sediment accumulation and charcoal concentration increase rapidly. This shift is associated with *Isoëtes* microspores and pollen species indicating pasture and permanent settlement e.g. *Ranunculus acris*, *Filipendula ulmaria* and *Urtica*.

Discussion

Earlier pollen analytical studies on cultivation history in East-Finland have revealed a general pattern in land-use practices (Simola *et al.*, 1987, 1988). A similar pattern was also observed in the present study of lakes Suurjärvi and Heinälampi. Stages from pre-agriculture through phases of weak, sporadic and, in time, rapidly expanding slash-and-burn cultivation characterize the pollen assemblage results up till the 19th century. Since that time the cultivation signal gets weaker indicating a shift from extensive slash-and-burning to permanent fields. The shift to modern agriculture with plough fields and manure fertilization was forced by legislation and partly by increasing commercial value of wood.

In Lake Suurjärvi the earliest signs of slash-and-burn cultivation (660–775 AD) are associated with high phosphorus and relatively high organic matter content of the sediment. The supra-aquatic hilltops around the lake are the most suitable lands for cultivation and probably the first ones reclaimed; these small areas are quite far from the lake. Thus the first reaction of Suurjärvi to cultivation (sample group 2 in Fig. 1) could indeed be mediated through an increase in dissolved phosphorus and consequent organic matter production in the lake.

The expansion of cultivation to land at lower altitudes and the shortening of slash-and-burn rotation cycles between 1600–1900 AD (group 3) led to high inputs of eroded inorganic material to the lake. In the CCA-plot the length, and thus the importance, of the vectors of sediment accumulation rate, calcium and charcoal concentration can be interpreted on the basis of these phenomena.

The corresponding reaction of sediment iron and manganese may be connected either to their increased proportion in the sediment or to a change in redox conditions of the sediment.

In Lake Heinälampi the organic content of the sediment increased during the earliest period of slash-and-burn cultivation (group 2). However, in 1800–1970 AD the rapidly sedimenting material became highly inorganic, which is plausible considering the small catchment area and its fine grained soils.

In the CCA-plot the species point of spruce is positioned on the opposite side of the 'cultural axis' origo (axis 1) than the vector of charcoal concentration in both lakes. This shows clearly again the role os spruce as a negative cultivation indicator for the slash-and burn period in Finland as discussed earlier by e.g. M. Tolonen (1978) and Huttunen (1980).

In the present study canonical correspondence analysis proved to be a powerful method of distinguishing the changes in sediment properties that reflect the different periods of land use. However, in order to gain a fuller picture of the complicated relationships between land use and sediment quality we need more fine resolution pollen analyses and more detailed analysis of sediment chemistry.

Acknowledgements

I wish to thank Ms. Leena Kivinen, she drew the figures and let me use her unpublished pollen data of Lake Heinälampi. Dr. Heikki Simola's kind comments were invaluable. I also recall thankfully the stimulating discussions with Prof. John Birks in Bergen. The skillful staff of Karelian Institute prepared the samples. This study has been financially supported by the Academy of Finland (project 1003-6: 'Palaeoecological study of settlement and cultivation history in East Finland').

References

Berglund, B. E. & M. Ralska-Jasiewiczowa, 1986. Pollen analysis and pollen diagrams. In: Berglund, B. E. (ed.),

Handbook of Holocene Palaeoecology and Palaeohydrology. John Wiley. New York: 455–484.

Bengtsson & M. Enell, 1986. Chemical analysis. In: Berglund, B. E. (ed.) Handbook of Holocene palaeoecology and palaeohydrology. John Wiley. New York: 423–451.

Brugam, R. E., 1984. Holocene paleolimnology. In Wright, H. E. Jr. (ed.) Late quaternary environments of the United States. University of Minnesota Press. Minneapolis: 208–221.

Grönlund, E., H. Simola & P. Huttunen, 1986. Paleolimnological reflections of fiberplant retting in the sediment of a small clear-water lake. Hydrobiologia 143: 425–431.

Huttunen, P., 1980. Early land use, especially the slash-and-burn cultivation in the commune of Lammi, southern Finland, interpreted mainly using pollen and charcoal analyses. Acta Bot. Fenn. 113: 1–45.

Sandman, O., A. Liehu & H. Simola, 1989. Paleolimnology of two Finnish lakes with recent field- and forest-ditch erosion sediments, dated by 210 Pb and 137 Cs. (manuscript).

Ter Braak, C. J. F., 1986. Canonical correspondence analysis: a new eigenvector method for multivariate direct gradient analysis. Ecology 67: 1167–1179.

Ter Braak, C. J. F., 1987. CANOCO – a FORTRAN program for Canonical Community Ordination by (Partial) (Detrended) (Canonical) Correspondence Analysis, Principal Components Analysis and Recundancy Analysis (Version 2.1). TNO Institute of Applied Computer Science, Wageningen.

Ter Braak, C. J. F. & I. C. Prentice, 1987. A theory of gradient analysis. Advances in Ecol. Res. 18: 271–317.

Tolonen, M., 1978. Palaeoecology of annually-laminated sediments in Lake Ahvenainen, South Finland. I: Pollen and charcoal analyses and their relation to human impact. Ann. Bot. Fenn. 15: 177–208.

Tolonen, M., 1985. Cereal cultivation with particular reference to rye: some aspect on pollen analytical records from SW Finland. Fennoscandia Archaeologica II: 85–89.

Simola, H., E. Grönlund, P. Uimonen-Simola & P. Huttunen, 1985. Pollen analytical evidence for iron age origin of cupstones in the Kerimäki area. Iskos 5: 527–531.

Simola, H., E. Grönlund & P. Uimonen-Simola, 1987. Conquest of wilderness: early agriculture in the eastern Finnish Lake District. Manuscript. (Fourth Nordic Conference on the Application of Scientific Methods in Archaeology, Haugesund, Norway, Oct. 1987).

Simola, H., E. Grönlund & P. Uimonen, Simola, 1988. Paleoecological investigation of the history of agriculture in the province of South Savo, Finland. University of Joensuu, Publications of Karelian Institute. 55 p. (In Finnish, English abstract).

Vuorinen, J., 1978. The influence of prior land use on the sediments of a small lake. Pol. Arch. Hydrobiol. 25: 453–451.

Vuorinen, J. & K. Tolonen, 1977. Flandrian pollen deposition in Lake Pappilanlampi, Eastern Finland. Publications of University of Joensuu BII 3: 1–12.

Hydrobiologia **214**: 143–147, 1991.
J. P. Smith, P. G. Appleby, R. W. Battarbee, J. A. Dearing, R. Flower, E. Y. Haworth, F. Oldfield & P. E. O'Sullivan (eds),   143
Environmental History and Palaeolimnology.
© 1991 *Kluwer Academic Publishers.*

The influence of land use on the sedimentation of the river delta in the Kyrönjoki drainage basin

Raimo Heikkilä
University of Helsinki, Department of Geography, Hallituskatu 11, SF-00100 Helsinki, Finland

Key words: river delta, sedimentation, drainage basin processes, land use

Abstract

Sediment surface samples (0–2 cm) from 66 sites, and longer cores (up to 540 cm) from 9 sites in the estuary of the river Kyrönjoki, Western Finland, were analysed for organic content, P, Fe, Mn, Pb, Cd, Cu and Zn. One core was dated on the basis of annual laminations.

Chemical analyses of the cores showed that organic matter and heavy metal content have increased in recent decades. The heavy metal content was clearly lower than in areas polluted by industry. The sedimentation rate in the delta increased between the 1930s and 1950. It decreased in the 1960s, and has been below 1930s levels since 1970. The sedimentation rates of organic matter were fairly stable all through the period measured, even though the organic content increased.

The increase in organic matter and the heavy metal content of the sediment in recent decades is evidently due to the increased intensity of agriculture, forestry and peat harvesting in the drainage basin. Drainage of peatlands in particular has increased erosion and the organic sediment load of the river. Reservoir building after 1970 has decreased the sedimentation rates in the delta.

Introduction

Human activities have brought about massive changes in the hydrology and sediment load of rivers in Western Finland in recent decades. Agriculture, forestry drainage of peatlands, peat harvesting and watercourse works have had the biggest influence on sedimentation in the drainage basin of the river Kyrönjoki. Human influence on water quality and sediment transport in the river estuary has been intensively studied recently by Meriläinen (1984, 1985, 1986), Mäkelä (1986), Pitkänen (1986) & Heikkilä (1986a, 1986b). Mansikkaniemi (1985) examined the sedimentation in the Ilmajoki flood plain. The aim of the present study is to find out the effects of human activities in the drainage basin on the sedimen-

tation rates and sediment quality in the delta of the river kyrönjoki.

Study area

The drainage basin of the river Kyrönjoki is situated in western Finland, southeast of the city of Vaasa (Fig. 1). It covers 4920 km^2, of which 47% is made up of forests, 26% of peatlands, 24% of arable fields, 2% of built-up areas and 1% of lakes. Agriculture is exceptionally intensive in the basin. Only 8% of the whole of Finland is arable fields. About 100 000 people live in the basin, but there is very little industry.

The dominating soils in the river valleys are clay and silt. Peat and glacial till cover much of the

144

Fig. 1. The location of the river Kyrönjoki drainage basin.

Fig. 2. The river system of the Kyrönjoki and the names of the localities in the drainage basin. Reservoirs are shaded with horizontal lines.

watershed areas. Glaciofluvial sand and gravel exist only in limited patches near the watersheds. Most of the fields lie along the riversides in the area between Ilmajoki and Vähäkyrö (Fig. 2). The isostatic land upheaval in the delta is 8 mm a^{-1}, and in the upstream area 7 mm a^{-1} (Kääriäinen, 1975). The slopes in the drainage basin are very gentle, and the highest elevation in the southernmost part of the basin is 200 metres a.s.l.

The river discharge fluctuates over a wide range (Minimum : Mean : Maximum = 1 : 43 : 528 m^3/s). Spring, summer and autumn floods are frequent.

The river channel has been cleared many times since the 1600s to prevent floods and to make it agriculturally viable. The most extensive dredging work was done during the 1930s, 1950s and 1970s in the downstream area. Each time, c. 1 million m^3 of land mass was moved. Four reservoirs and four hydroelectric power stations (with daily regulation) have been built since 1963.

The water of the river Kyrönjoki contains a lot of humic substances due to drainage from exten-

sive peatland areas, mostly drained for forestry or peat harvesting (Kenttämies, 1981). About 15% of the basin is drained for forestry. 1% of it is used for peat harvesting. Peat harvesting areas have significant localized effects on water quality and sedimentation, especially in small brooks and lakes where the water flows from the ditches (Sallantaus, 1984). Furthermore, the river water is very acid, especially during periods of flood, due to the cultivation of the sulphide-bearing Litorina clays in the downstream area along the riversides, from Ilmajoki to Hemfjärden. In some cases the pH falls below 4.0, and it is usually about 5.5 (Eronen, 1974; Alasaarela, 1982).

Materials and methods

The material used in this study consists of data about land use in the drainage basin of the river

Kyrönjoki, obtained by field and map survey, from official statistics and sediment samples from the estuary of the river. Samples of the sediment surface (0–2 cm) from 66 sites and longer cores (up to 540 cm) from 9 sites in the estuary of the river were taken. One sediment core from Nabbviken (site 22, Heikkilä, 1986b) was studied in detail. The uppermost part (65 cm) of the core was dated using annual laminae (See Simola, 1983), up to the year 1931. The core was 430 cm long, covering approximately 800 years. The uppermost part of the core was taken using the core freezing technique (Renberg, 1981), and the lower parts using a Russian peat sampler. Another core from Getlaxfjärden was dated according to the depth of severe change in sediment chemistry due to damming in 1957.

The sediment samples were analysed for organic content, P, Fe, Mn, Pb, Cd, Cu and Zn using standard methods. Organic content was determined as loss-on-ignition, P was determined spectrophotometrically using the molybdenum blue method. Heavy metals were analysed by AAS.

Sedimentation rates over the period 1931–1983 were calculated as grams of dry matter per m^2 per year on the basis of thickness of the varves and the dry weight of the sediment. Sedimentation rates were compared with historical data of watercourse works in the basin and with hydrological data.

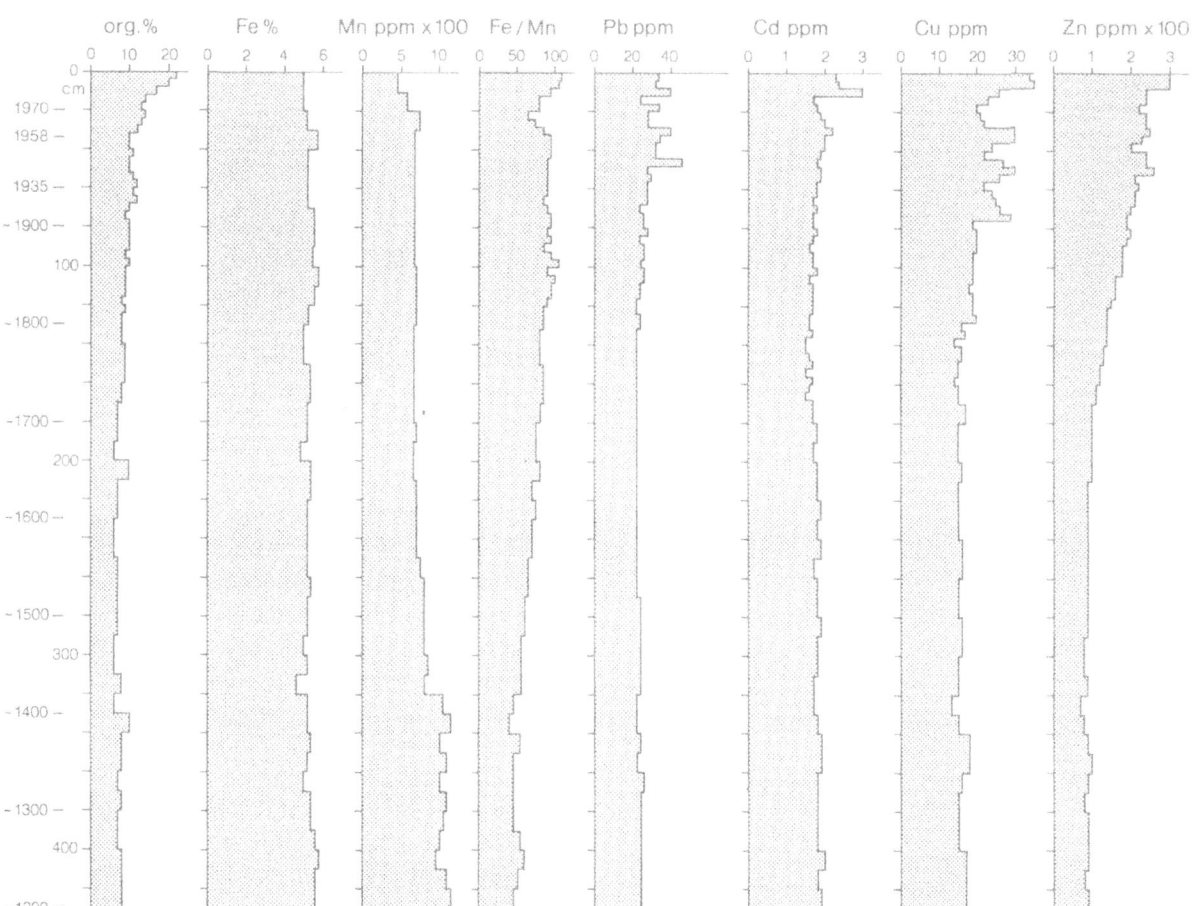

Fig. 3. Chemostratigraphy of the sediment core from Nabbviken in the delta of the river Kyrönjoki.

146

Results

Chemical analyses of the cores showed that the organic matter and heavy metal content have increased in recent decades, compared with the natural background in the lower parts of the core (Fig. 3). The organic content in particular has increased during the last 20 years. The heavy metal content, however, was in all cases clearly lower than in areas polluted by industry (See e.g. Förstner, 1976). The zinc content rose continuously from the latter half of the 18th century up until the 1930s. Since then, it has been around 250 ppm. The highest lead content value occurred in the early 1940s (Fig. 3). The same peak could be seen in other cores, too. The phosphorus con-

tent of the sediment was very high in many cases, up to 20 mg/g (cf. e.g. Simola, 1983), but only a slight concentration in the sediment surface could be observed.

The sedimentation rates over the period 1931–1983 vary between 1000 and 4500 g m^{-2} a^{-1}. The mean annual rate was 2500 ± 800 g m^{-2} a^{-1}. Until 1970, the rates were mostly above the mean value, and annual variation was wide. There were increases from the 1930s up to the middle of the 1950s. The highest rate occurred in 1953, which was a year of very high floods. In the 1950s, the rates were generally high, lower in the 1960s and at a stable, low level, clearly lower than in the 1930s, from 1970 on. The sedimentation rate of organic matter was fairly stable throughout the whole period studied (Fig. 4).

Discussion

The increase of organic matter and heavy metal content in the sediment over recent decades is evidently due to the increased intensity of agriculture, forestry and peat harvesting in the drainage basin. Drainage of peatlands in particular has increased erosion and the organic sediment load of the river. The increase in the content of zinc in the dated core fits very well with the increase in the cultivation of sulphide-bearing Litorina clays along the lower course of the river between Ilmajoki and Vähäkyrö (See Fig. 2). It is evident that zinc has been diluted in the acidic waters (pH as low as 3.0) flowing from the cultivated clay areas. Mixing with neutral brackish water in the delta causes zinc precipitation (Heikkilä, 1986b). It is not yet clear why the lead content peaked around 1943. The same peak can be seen in the stratigraphy of most cores, and it could possibly be used for dating them. Waste water from towns and villages has evidently caused an increase in the amount of phosphorus.

Reservoir building since 1970 has clearly decreased the sedimentation rates in the delta, because most of the suspended solids eroded in the upstream areas have been deposited in the reservoirs. The use of the reservoirs to prevent

Fig. 4. Sedimentation rates in the Nabbviken core. A = Annual sedimentation rates. The portion of organic matter is shaded black. The thickness of the bar for each year corresponds with the thickness of each varve. B = The annual amount of dredging of the river channel under the water in the downstream area. Data for the 1930s was not available. C = Wet years (Mean discharge > 50 m³/s). D = Flood years; one dot: Maximum discharge > 350 m³/s, two dots: Maximum discharge > 400 m³/s.

floods has also caused a decrease in the number of flood years (Fig. 4), which probably in turn has decreased the sedimentation rates. The peak in sedimentation rates in the 1950s was evidently caused by intensive draining of arable fields, especially in the downstream area of the river. The rates fell in the 1960s as a result of increased sub-drainage (see Seuna & Kauppi, 1981). The effects of other watercourse works (dredging of the channel and building of artificial embankments) are not clearly visible in the sediment stratigraphy or in the sedimentation rates. This is because the river channel has been dredged during periods of low discharge, and the dredged land masses have been used to build embankments along the riverside.

Acknowledgements

I am very grateful to the following persons: Professors Matti Seppälä, Toive Aartolahti and Kimmo Tolonen, and Mr. Pertti Sevola for their advice during the work; Hanna Heikkilä & Kati Heikkonen for the laboratory analyses; Jukka Päivärinta & Leena Heiskanen for technical assistance and Joan Nordlund for language revision. I am also grateful to the National Board of Waters and the Ministry of the Environment for financing the study.

References

Alasaarela, E., 1982. Acidity problems caused by flood control works of the river Kyrönjoki. Publ. Wat. Res. Inst. 49: 3–16.

Eronen, M., 1974. The history of the Litorina Sea and associated holocene events. Comm. Physico-Math. 44: 79–195.

Förstner, U., 1976. Metal concentrations in freshwater sediments – natural background and cultural effects. In H. L. Golterman (ed.), Interactions between sediments and fresh water. Dr W. Junk Publishers, The Hague: 94–103.

Heikkilä, R., 1986a. Recent sedimentation in the delta of the river Kyrönjoki, western Finland. Publ. Wat. Res. Inst. 66: 24–28.

Heikkilä, R., 1986b. Recent sedimentological conditions in the delta of the river Kyrönjoki, Western Finland. Hydrobiologia 143: 371–377.

Kääriäinen, E., 1975. Land uplift in Finland on the basis of sea level recordings. Rep. Finnish. Geod. Inst. 75(5): 1–14.

Kenttämies, K., 1981. The effects on water quality of forest drainage and fertilization in peatlands. Publ. Wat. Res. Inst. 43: 24–31.

Mansikkaniemi, H., 1985. Sedimentation and water quality in the flood basin of the river Kyrönjoki in Finland. Fennia 163: 155–194.

Meriläinen, J. J., 1984. Zonation of the macrozoobenthos in the Kyrönjoki estuary in the Bothnian Bay, Finland. Ann. Zool. Fenn. 21: 89–104.

Meriläinen, J. J., 1985. Spreading of river waters and behaviour of different fractions of particulate matter and dissolved organic matter in the non-tidal Kyrönjoki estuary, Bothnian Bay. Aqua Fennica 15: 53–64.

Meriläinen, J. J., 1986. Loads of nutrients, organic matter and suspended solids discharged by the river Kyrönjoki into the Bothnian Bay. Publ. Wat. Res. Inst. 66: 67–71.

Mäkelä, K., 1986. Variations in dry matter, phosphorus and organic carbon in two Bothnian Bay sediment cores in relation to hydraulic engineering works. Publ. Wat. Res. Inst. 66: 36–39.

Pitkänen, H., 1986. Discharges of nutrients and organic matter to the Gulf of Bothnia by Finnish rivers in 1968–1983. Publ. Wat. Res. Inst. 66: 72–83.

Renberg, I., 1981. Improved methods for sampling, photographing and varve-counting of varved lake sediments. Boreas 10: 255–258.

Sallantaus, T., 1984. Quality of runoff water from Finnish fuel peat mining areas. Aqua Fennica 14: 223–233.

Seuna, P. & L.Kauppi, 1981. Influence of sub-drainage on water quantity and quality in a cultivated area in Finland. Publ. Wat. Res. Inst. 43: 32–47.

Simola, H., 1983. Limnological effects of peatland drainage and fertilization as reflected in the varved sediment of a deep lake. Hydrobiologia 106: 43–57.

Hydrobiologia **214**: 149–154, 1991.
J. P. Smith, P. G. Appleby, R. W. Battarbee, J. A. Dearing, R. Flower, E. Y. Haworth, F. Oldfield & P. E. O'Sullivan (eds). 149
Environmental History and Palaeolimnology.
© 1991 *Kluwer Academic Publishers.*

Heavy metals (Cu and Zn) in recent sediments of Llangorse Lake, Wales: non-ferrous smelting, Napoleon and the price of wheat – a palaeoecological study

Roger Jones[1], Frank M. Chambers[2] & Kathryn Benson-Evans[3]
[1]*Department of Biology, Trent University, Peterborough, Ontario, Canada K9J 7B8*; [2]*Environmental Research Unit, University of Keele, Staffs ST5 5BG, UK*; [3]*University of Wales College of Cardiff, P.O. Box 915, Cardiff CF1 3TL, UK*

Key words: palaeolimnology, sediments, heavy metal pollution, Cu and Zn, land-use, crops, Industrial Revolution, Napoleonic Wars, Wales

Abstract

Elevated concentrations of Cu and Zn have been found in the upper part of three sediment cores collected from Llangorse Lake, in south Wales. Palaeomagnetic evidence from one of the cores and ^{210}Pb analysis of another, suggests that the increase in sediment Cu and Zn concentrations began during the eighteenth century. A sharp increase in the concentrations of these metals in the sediment profile appears to have occurred during the latter part of the eighteenth century and these concentrations remained high until the mid to late nineteenth century.

The absence of known ore deposits and industry around the lake suggests that the lake and catchment soils were increasingly contaminated by long-range aerial transport of emissions from the expanding activity of Cu and Zn smelters located some 80 km upwind in the Swansea area during the Industrial Revolution. Evidence from agricultural crop returns indicates a significant increase in the amount of land devoted to tillage in the catchment, particularly to cereal production, during the late eighteenth and the first half of the nineteenth century which included the Napoleonic Wars. This agricultural shift appears to coincide with increased concentrations of Cu and Zn in the lake sediments. It is suggested that newly ploughed soils, contaminated with metals for many years by long-range aerial transport from the Swansea area, eroded, and were carried into the lake by catchment run-off and added to the sediment burden of Cu and Zn. A subsequent decline of Cu and Zn emissions due to the collapse of the non-ferrous smelting industry and reduced soil erosion because of a 50% reduction of tillage due to an agricultural depression in the second half of the 19th century may explain the fall in Cu and Zn concentrations in the upper part of the sediment profile. The most recent sediments (20th century) show the increase in heavy metals characteristic of many lakes around the world.

Introduction

Elevated concentrations of Cu and Zn were found in the upper parts of three cores collected for palaeoecological studies of Llangorse Lake and its catchment (Jones *et al.*, 1978; Jones *et al.*, 1985). Jones (1984), using dating evidence and cross-correlation of metal profiles of two cores, suggested that the increase in Cu and Zn concentrations began during the 18th century. Historical

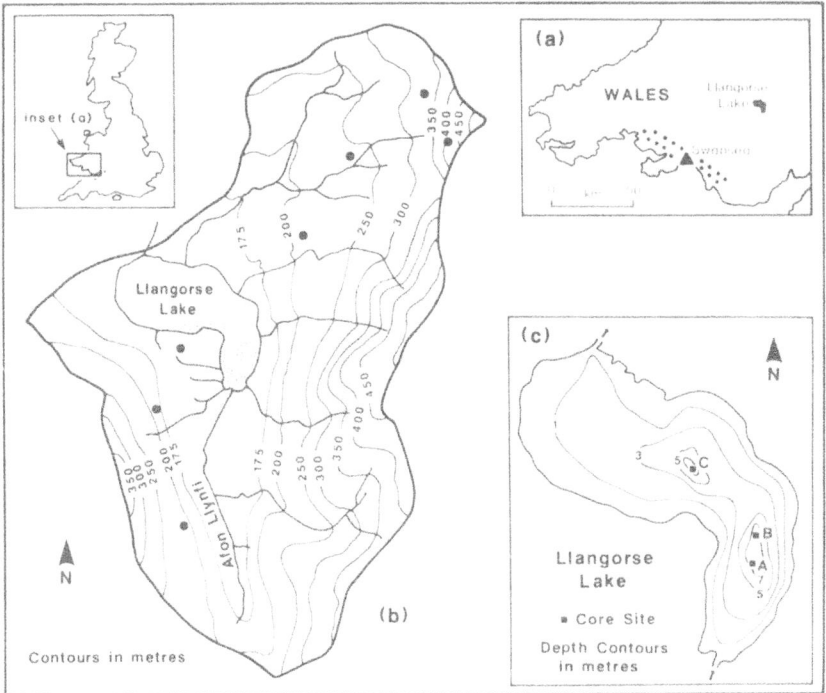

Fig. 1. Llangorse Lake and its catchment. (a) relationship of the lake to the non-ferrous smelting industry (●) in the Swansea area. (b) catchment showing sites (•) where soil samples were collected. (c) core sites in the lake.

records of the activity of the non-ferrous smelting industry located up-wind from the lake and availability of agricultural crop data combined with sediment pollen spectra allowed us to investigate the hypothesis that the pattern of Cu and Zn in the upper sediments in Llangorse Lake is related to long-range aerial transport of these metals prior to the 20th century and to temporal shifts of catchment land-use.

Site description

Llangorse Lake is situated in a rural area in south-central Wales (Fig. 1). Mixed farming abounds in the catchment up to an elevation of ca. 300 m. Heathland and grassland used for rough grazing occur at higher elevations. Limnological studies of the lake are described in Jones & Benson-Evans 1974; Tai 1975; and Cragg *et al.*, 1980.

Materials and methods

This paper deals mainly with a core collected in 1980 from site A (Fig. 1) with a Livingstone piston cover. Details of the coring procedure, sub-sampling and chemical analytical methods are given in Jones (1984) and Jones *et al.* (1985). Samples for pollen determination were prepared at intervals down the core and at closer intervals (2 or 4 cm apart) in the section from 32 to 132 cm depth, using the procedure given in Chambers (1985) and Jones *et al.* (1985).

Soil samples were taken from the top 10 cm of the soil profile at 7 locations in the catchment (Fig. 1). After air drying, subsamples were oven-dried and treated in the same way for chemical analysis as the lake sediments (Jones, 1984).

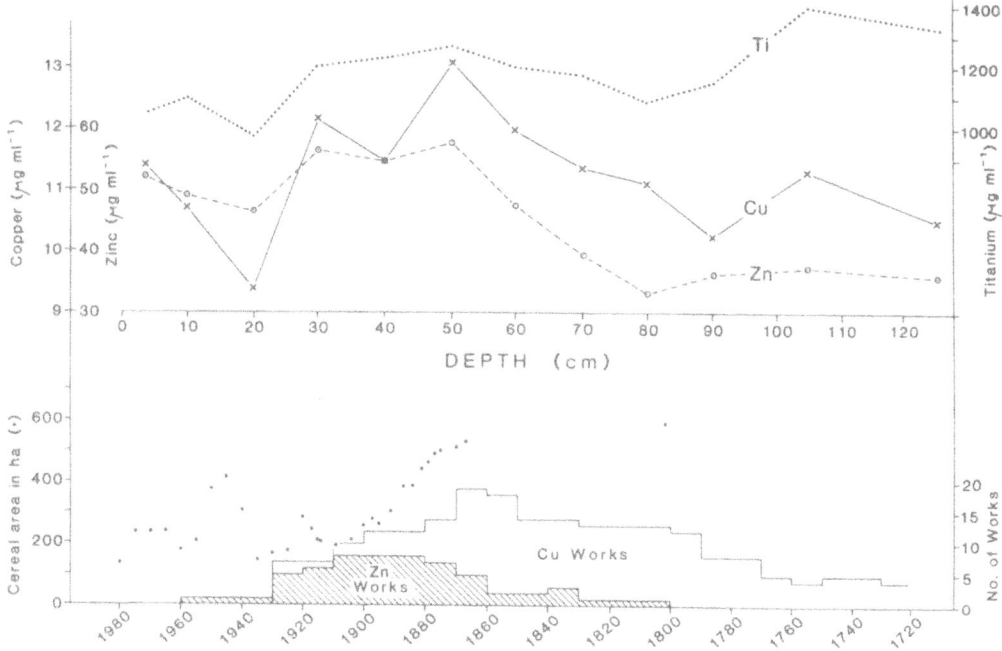

Fig. 2. Top: Distribution of Cu, Zn and Ti in the upper 120 cm of a core from Llangorse Lake, Wales. Bottom: Number of Cu and Zn works per decade in the Swansea area of south Wales and the area (ha) of Llangorse catchment under cereals for particular years (MAF 68). 1801 data from Williams (1950–52).

Results and discussion

The upper 2 m of sediments discussed in this paper did not exhibit any visible signs of stratigraphy. The increase of Cu of interest in this study began at a depth of ca. 90 cm, while the increase of Zn occurs further up the profile at ca. 80 cm (Fig. 2). A ^{210}Pb date of 1841 ± 12 yr A.D. (Jones, 1984), using a constant rate of supply (crs) model (Oldfield & Appleby, 1984), was obtained for sediments at a depth of 65 cm in this core. A core collected from site B (Fig. 1) in 1972 and subjected to palaeomagnetic study, yielded a tentative date of 1820 A.D. for sediments at a depth of 47 cm (Seddon, unpublished data) in Jones *et al.* (1978). Cross-correlation of sediment profiles of Cu, Zn and Pb in the cores from sites A and B (Jones, 1984) and the compatibility of the above dates suggest that the increases in concentrations of Cu and Zn occurred during the latter part of the 18th century.

The absence of known ore deposits and industry near the lake implies aerial transport of these metals to the catchment with the source apparently being the non-ferrous smelting industry in the Swansea area (Fig. 1), some 80 km upwind of the lake. This industry began in the 17th century and grew to such an extent that it became the world centre for Cu smelting during the late 18th and early 19th centuries (Davies, 1971; Roberts, 1980). Pollution controls were minimal and deposition of metals onto the lake and its catchment, and hence flux to the sediments, would have increased as the metal burden of the prevailing south-westerly air masses increased with expansion of the smelting industry during the Industrial Revolution.

Data on the amount of ore smelted or metals produced are incomplete (Roberts, 1980), so an estimate of smelting activity was made by determining the number of active works per decade (Fig. 2) from Roberts (1980). While the rate of sediment accumulation is not precisely known, Cu in the sediment profile appears to follow the

152

Table 1. Summary characteristics and ages* (A.D.) of pollen diagram phases.

30–57 cm	1915–1858*	Phase 5	High Gramineae. Sub-zone 5a shows reduced *Triticum* representation; sub-zone 5b has lower *Corylus* and higher *Pteridium*
57–73 cm	1858–1823*	Phase 4	Higher Gramineae; reduced *Corylus*; presence of *Triticum*
73–77 cm	1823–1815*	Phase 3	Higher *Corylus*; reduced Gramineae
77–97 cm	1815–1770*	Phase 2	Reduced *Corylus*; higher Gramineae
97–132 cm	1770–1700*	Phase 1	High *Corylus*; low Gramineae

* Ages are approximate only. Dates have been interpolated assuming a linear accumulation rate of 0.47 mm y^{-1}, based on the ^{210}Pb date of 1841 $\pm$ 12 yr A.D. at 65 cm depth.

pattern of increasing numbers of Cu works per decade (Fig. 2), beginning with the increase in the late 18th to a maximum in the mid-19th century. The peak of Zn smelting came towards the end of the 19th century.

During the Industrial Revolution, burgeoning urban populations led to increased national food production (Chambers & Mingay, 1966) and it seems likely on the local level too, that food production increased around Llangorse Lake since catchment soils are inherently fertile (Davies, 1960). A pollen study of the sediments from site A showed, for example, a decline in AP : NAP (arboreal : non-arboreal) pollen ratios and increasing Gramineae and cereal pollen in the depth interval 100 to 96 cm. Detailed pollen data (Chambers *et al.*, in prep) are summarised in Table 1 to show characteristics of catchment vegetation and approximate ages.

During the Napoleonic Wars the price of wheat and other cereals rose substantially because of poor harvests caused by inclement weather and the isolation of Britain from continental cereal supplies (Ernle, 1961). Farmers took advantage of high cereal prices, particularly of wheat, by extending cultivation to marginal lands (Howells, 1977). It is hypothesised that increased cultivation in the catchment, probably by ploughing grassland on lower hillslopes resulted in a concommitant rise in minerogenic sediment burden to the lake, owing to erosion from the expanded acreage of exposed arable soils. It seems unlikely that low-lying water meadows around the lake and along the inflowing Afon Llynfi were ploughed since these areas would have been liable

to flooding and wheat prefers well-drained soils. The Afon Llynfi outlet from the lake was in fact straightened and deepened in the mid-19th century in an attempt to alleviate flooding. The acreage of cereals reported for the three parishes which comprise the lake catchment in the 1801 Agricultural Returns (591.3 ha) (Williams, 1950–1952) is higher than the acreage (531.9 ha) reported for the catchment in 1867 (MAF 68). In 1867, the percentage of catchment land below the 300 m contour interval under cultivation was ca. 50% so it seems reasonable to assume a similar percentage was cultivated in 1801. Of the cultivated land in 1867, ca. 15% was fallow and since fallowing was a common agricultural practice in earlier periods then at least a similar area of fallow may be speculated in 1801. An increase in concentration of Ti, suggested as an indicator of accumulation of clastic materials in some lakes (Engstrom & Wright, 1984), at ca. 80 cm depth in Llangorse Lake sediments, may reflect expansion of crop cultivation during the Industrial Revolution and Napoleonic Wars Period (Fig. 2). While data are not presented, the increased concentrations of Ti were accompanied by concommitant increases of K and Mg, elements which have also been suggested to indicate erosional input into lakes (Mackereth, 1966; Rippey *et al.*, 1981). Prior to this period, it is possible that more of the catchment below 300 m was pastureland since farmers in the adjacent Vale of Usk had found it more profitable to increase the size of their flocks of sheep and grow less corn (Emery, 1984).

Copper and Zn concentrations in the sediments (Fig. 2) apparently increased at ca. the same depth as Ti in the profile. It is possible that suspended matter in run-off from newly ploughed areas added to the metal burden of the lake because catchment soils had been receiving for many years aerial deposition from smelters in the Swansea area (Fig. 2).

In the case of Cu, at least, enrichment of topsoil by aerial deposition probably had occurred since the 17th century and increased during the 18th century as the smelting industry in the Swansea area expanded.

On the other hand, preliminary studies of mineral magnetic measurements by Dr. J. Dearing attempting to identify shifts in sediment sources in the recent history of Llangorse Lake proved contradictory (Chambers, Dearing & Jones, 1989). Indications of increased influx of soil material to the lake during the period of agricultural intensification and decreased influx during the agricultural depression in the last quarter of the 19th century (referred to below) are not readily apparent in mineral magnetic data. It is hoped that further studies will help clarify the situation.

The decline of the respective smelting industries (Fig. 2) appears to correlate with falling concentrations of Cu and Zn in the lake sediments (Fig. 2). Also, the last quarter of the 19th century was a time of a great depression in British Agriculture (Ernle, 1961) and was reflected around Llangorse Lake by decreased cereal production, particularly of wheat (MAF 68). During the last 20 years of the 19th century the percentage of cultivated land in the catchment below 300 m elevation declined from 40% in 1880 to 22% by 1900. Arable land taken out of production (both cereal and other crops) reverted ('tumbled down') to pasture and this appears to be shown in pollen spectra (Table 1) by high Gramineae, higher *Pteridium* (bracken), and by reduced *Triticum*, reflecting the declining acreage of wheat during this period (MAF 68). If indeed there was declining erosional input of soil into the lake as the acreage of grassland increased then this may explain the falling Ti (Fig. 2) and also K and Mg in the sediments. Furthermore, less suspended solids in runoff would mean a reduction of the Cu

and Zn burden to the lake associated with top soil as arable was converted to pasture.

The increase of Cu and Zn in more recent sediments (20th century) appears to be similar to that found in lakes in many parts of the world (Christensen & Chien, 1981; Hamilton-Taylor, 1979; Jones *et al.*, 1984; Rippey *et al.*, 1981). The levels of Cu and Zn in these sediments are much higher than concentrations in catchment soils ($24.9 \pm 4.6\ \mu g\ g^{-1}$ and $89.5 \pm 15.4\ \mu g\ g^{-1}$ respectively). These higher sediment concentrations may, in part, be related to decomposition of phytoplankton capable of accumulating metals (Stokes, 1979; Trollope & Evans, 1976) in this highly eutrophic lake subject to annual algal blooms (Jones & Benson-Evans, 1976). Metals may also have entered the lake in sewage effluent which ran directly into the lake from the village of Llangorse until 1981 when the discharge was redirected into the Afon Llynfi outlet.

Acknowledgements

The authors thank Dr H.J.B. Birks for the use of a Livingstone corer and floating platform and Dr. K.D. Bennett & Dr. P. Kerslake for operating the corer. Mr S. Gardiner and Mr A. Lawrence kindly prepared the figures. Research funds were provided by Trent University and NRC, Ottawa, Canada.

References

Chambers, F. M., 1985. Flandrian environmental history of the Llynfi catchment, South Wales. Ecol. Mediterr 11: 73–80.

Chambers, F. M., J. Dearing & R. Jones, 1989. Palaeolimnology of Llangorse Lake: Eutrophication and Catchment Links. In Excursion Guide B – Lowland Lake Environments. V[TH] International Symposium on Palaeolimnology. Cumbria U.K., September 1989.

Chambers, J. D. & G. E. Mingay, 1966. The Agricultural Revolution 1750–1880. B.T. Batsford, London.

Christensen, E. R. & Nan Kwang Chien, 1981. Fluxes of arsenic, lead, zinc and cadmium to Green Bay and Lake Michigan sediments. Envir. Sci. Technol. 15: 553–558.

Cragg, B., J. Fry, Z. Bacchus & S. S. Thurley, 1980. The

aquatic vegetation of Llangorse Lake, Wales. Aquat. Bot. 8: 187–196.

Davies, H. R. J., 1971. The Industrial Revolution. In: Balchin, W. G. V. (ed.), Swansea and its Region, Univ. College of Swansea: 163–178.

Davies, H. T., 1960. The soils of Brecknock. Brycheiniog 6: 51–66.

Emery, F., 1984. Wales. In Thirsk, J. (ed.), The Agrarian History of England and Wales Vol. V 1640–1750. I. Regional Farming Systems. Cambridge University Press, U.K.

Engstrom, D. R. & H. E. Wright, 1984. Chemical stratigraphy of lake sediments as a record of environmental change. In Haworth, E. Y. & J. W. G. Lund (eds), Lake Sediments and Environmental History. University Press, Leicester: 11–67.

Ernle, Lord, 1961. English Farming Past and Present. Heinemann, London.

Hamilton-Taylor, J., 1979. Enrichments of zinc, lead and copper in recent sediments of Windermere, England. Envir. Sci. Technol. 13: 693–697.

Howells, B., 1977. Modern History. In Thomas, D. (ed.), Wales – A New Study. David & Charles, London: 94–120.

Jones, R., 1984. Heavy metals in the sediments of Llangorse Lake, Wales, since Celtic-Roman times. Verh. Int. Ver. Limnol. 22: 1377–1382.

Jones, R. & K. Benson-Evans, 1974. Nutrient and phytoplankton studies of Llangorse Lake, a eutrophic lake in the Brecon Beacons National Park, Wales. Field Studies 4: 61–75.

Jones, R., K. Benson-Evans & F. M. Chambers, 1985. Human influence upon sedimentation in Llangorse Lake, Wales. Earth Surf. Proc. Landf. 10: 227–235.

Jones, R., K. Benson-Evans, F. M. Chambers, B. A. Seddon & Y. C. Tai, 1978. Biological and chemical studies of sedi-ments from Llangorse Lake, Wales. Verh. Int. Ver. Limnol. 20: 642–648.

Jones, R., M. Dickman, R. D. Mott & M. Ouellet, 1984. Late Quaternary diatom and chemical profiles from a meromictic lake in Quebec, Canada. Chem. Geol. 44: 267–286.

Mackereth, F. J. H., 1966. Some chemical observations on postglacial lake sediments. Phil. Trans. R. Soc. (London) B 250: 165–213.

MAF 68. Agricultural Statistics: Parish Summaries. Public Record Office, Kew, London.

Oldfield, F. & P. G. Appleby, 1984. Empirical dating of ^{210}Pb dating models for lake sediments. In Haworth, E. Y. & J. W. G. Lund (eds), Lake Sediments and Environmental History. University Press, Leicester: 93–124.

Roberts, R. O., 1980. The smelting of non-ferrous metals since 1750. In John, A. H. (ed.), Glamorgan County History. Industrial Glamorgan, Cardiff: 47–95.

Rippey, B., R. J. Murphy & S. W. Kyle, 1981. Anthropogenically derived changes in sedimentary flux of Mn, Cr, Ni, Zn, Hg, Pb and P in Lough Neagh, Northern Ireland. Envir. Sci. Technol. 16: 23–30.

Stokes, P. M., 1979. Copper accumulations in freshwater biota. In Nriagu, J. O. (ed.), Copper in the Environment Part 1: Ecological Cycling. John Wiley & Sons, New York: 358–381.

Tai, Y. C., 1975. Phytoplankton Studies of Llangorse Lake, Breconshire. Unpublished Ph.D. thesis, University of Wales.

Trollope, D. R. & B. Evans, 1976. Concentrations of copper, iron, lead, nickel and zinc in freshwater algal blooms. Envir. Pollut. 11: 109–116.

Williams, D., 1950–52. The Acreage Returns of 1801 for Wales. Bull. Bd Celtic Stud 14: 54–68.

Hydrobiologia **214**: 155–162, 1991.
J. P. Smith, P. G. Appleby, R. W. Battarbee, J. A. Dearing, R. Flower, E. Y. Haworth, F. Oldfield & P. E. O'Sullivan (eds),   155
Environmental History and Palaeolimnology.
© 1991 *Kluwer Academic Publishers.*

A comparative study of heavy metal contamination and pollution in four reservoirs in the English Midlands, UK

I.D.L. Foster, S.M. Charlesworth & D.H. Keen
Centre For Environmental Science Research And Consultancy, Department of Geography, Coventry Polytechnic, Priory St., Coventry, UK

Key words: urbanisation, heavy metals, phosphorus, macrofossils, lake and fluvial sediments, lake rehabilitation

Abstract

Chemical and palaeoecological analysis of lake and fluvial sediments reveals a range of human impact on the sediment chemistry of four reservoirs in the English Midlands. Atmospheric pollution is recorded in both inner city and rural sites over the last 150 years. Catchment derived heavy metals at one urban site reveals high contamination factors for Pb, Cu, Ni, Zn and Cd. From the phosphorus record and from reconstructions based on macrofossil remains, eutrophication is recorded at all sites. Recent attempts to evaluate rehabilitation programmes for the inner city sites have proved problematical owing to two major problems. First, their shallow nature results in a high cost of desilting and, secondly, heavy metal contamination makes treatment and disposal of the sediment expensive.

Introduction

Heavy metal contamination of lake and reservoir sediments derives from both atmospheric and catchment inputs. The former is generally diffuse and represents short, medium and long distance atmospheric transport (Müller & Barsch, 1980; Nriagu, 1979; Renburg, 1986; Foster & Dearing, 1987). Catchment inputs usually consist of point sources, including reworked mine waste and spoil, accidental spillage, industrial and sewage works discharges, and leakage from land-fill sites. (Förstner & Wittmann, 1979; Christenson & Chien, 1981; Dearing *et al.*, 1981; Håkanson & Jansson, 1983). Diffuse catchment inputs may come from urban storm runoff, particularly when the storm sewer capacity is exceeded by high runoff.

Interpretation of lake sediment heavy metal content is made difficult, however, when several sources interact. For example, lakes and reservoirs in heavily industrialised regions will receive both a direct and an indirect atmospheric input: the latter deriving from eroded catchment soils which may either be enriched with atmospherically-derived heavy metals, or diluted by sources depleted of metals, such as channel bank erosion (Foster & Dearing, 1987).

A further dimension to the problem of identifying heavy metal sources is revealed through an examination of sediments stored within the fluvial system. For example, in the English Midlands, high heavy metal concentrations are to be found in the sediments of river channel substrates (Thoms 1987a and b). Fluvial sediments may therefore contribute to present and future contamination of reservoir sediments despite attempts to treat point source inputs or rehabilitate a contaminated lake.

This brief communication reviews the trends in heavy metal concentrations and pollution in four Midland catchments, and attempts to identify the main sources of contamination. Part of this work was undertaken in association with Coventry City Council in order to evaluate particular problems associated with the rehabilitation of inner city reservoirs. These issues are discussed after a consideration of some of the pollution problems encountered.

The catchments

The four catchments are located in the English Midlands. Seeswood Pool and Merevale Lake are situated in the rural area of North Warwickshire, and Wyken Slough and the Swanswell Pool in the urban area of the City of Coventry. Summary site data are contained in Table 1 and further details of these sites have been published elsewhere (Foster *et al.*, 1985, 1986, 1988a and b). All reservoirs, except for the Swanswell Pool, currently receive channelled inflows; the Springfield Brook was diverted from the Swanswell Pool in the mid-19th century, and since that time the level of water has been maintained artificially either from controlled springs or, more recently, from mains water supplies.

Sediment thicknesses in these reservoirs varies significantly. Maxima of ca 1.2 m have been found in the rural lakes, but at Wyken Slough and Swanswell Pool, thicknesses reach 1.7 and 2.8 m respectively. Sedimentation in the last records a dramatic change from mineral-rich material to organic sediments above 70 cm. This dramatic change probably relates to the isolation of the reservoir from its catchment in the mid 19th century.

Results of chemical analysis of water samples demonstrate wide variations in composition

Table 1. Catchment Characteristics

Site	Merevale lake	Seeswood pool	Swanswell pool	Wyken slough
Grid Reference	SP300970	SP327905	SP335795	SP363833
Reservoir Area (ha)	6.5	6.7	0.73	2.25
Catchment Area (ha)	201.0	238.0	220.0[1]	4500.0
Reservoir: Catchment ratio	1:30	1:33	1:301[1]	1:2000
Max. Altitude (m)	175	160	105[1]	112
Mean Reservoir Level (m)	118	125	70	85
Relative Relief (m)	57	35	35[1]	27
Max. Reservoir Depth (m)	8.0	3.5	1.3	0.8
Mean Reservoir Depth (m)	4.0	1.5	0.9	0.6
Volume (m^3)	154219	103241	6389	13050
Date Impounded	1838	1765	1265	c.1850

Contemporary Land Use of Catchment Areas (%)

Deciduous Woodland	78.4	4.0	0.0	0.0
Coniferous Plantation	7.5	0.0	0.0	0.0
Permanent Pasture	14.1	55.0	0.0	60.0
Grass Ley	0.0	11.0	0.0	0.0
Arable	0.0	30.0	0.0	0.0
Urban	0.0	0.0	100.0	40.0

[1] In *ca* 1855, the catchment area was reduced to less than 1 ha, the reservoir: catchment ratio to *ca* 1:1, and the maximum altitude to less than 2 m above water level.

between the four reservoirs. Electrical conductance values are lowest at Merevale Lake ($466.0\ \mu S\ cm^{-2}$) although the Swanswell Pool also possesses low values similar to those of the mains water supplies from which it is fed ($520.0\ \mu S\ cm^{-2}$). In contrast, both Seeswood Pool and Wyken Slough exhibit higher conductance values of 628.4 and $1022.5\ \mu S\ cm^{-2}$ respectively. Average pH values at Merevale Lake, Seeswood Pool, Swanswell Pool and Wyken Slough are 7.42, 7.46, 8.70 and 8.10 respectively. Phosphorus levels exceed $0.1\ mg\ l^{-1}$ in Swanswell Pool and Wyken Slough and, at the same sites, NO_3 concentrations are above $5.0\ mg\ l^{-1}$.

The four catchments receive heavy metals from a range of sources and possess rather different catchment and sedimentological characteristics.

1. Merevale Lake: Little variation in sedimentation rate since 1838 (Foster et al., 1985). The only source of heavy metal pollution is from the atmosphere.

2. Seeswood Pool: Rapid increase in sedimentation after 1960 (Foster et al., 1986). The only source of heavy metal pollution is from the atmosphere.

3. Wyken Slough: No data on changing sedimentation rate. Average accumulation ca 1.0 m since 1850. Heavy metals derived from atmospheric inputs, discharge from metal plating works, leachate from a toxic waste tip, runoff from the M6 and urban storm runoff.

4. Swanswell Pool: No data on changing sedimentation rate. Average accumulation rate ca 70 cm since 1855 despite no channelled inflow. Heavy metal pollution from inner city atmospheric inputs, limited urban storm runoff from its small catchment, and possibly from occasional spillages from a nearby hospital.

Table 2a. Minimum and maximum concentrations of heavy metals in lake and fluvial sediments ($\mu g\ g^{-1}$)

Site	Pb		Cu		Ni		Zn		Cd	
	min	max	min	max	min	max	min	max	min	max
Merevale	37.7	77.7	64.0	114.3	61.8	143.5	52.5	785.7	–	–
Seeswood	18.4	61.7	33.3	80.0	41.9	62.8	124.0	531.4	–	–
Wyken	51.2	475.9	27.5	490.1	26.2	163.4	52.0	1000.0	3.2	29.0
Swanswell	66.0	311.5	25.5	292.2	75.0	165.3	60.0	1800.0	6.1	16.3
River Sherbourne Samples[1]										
Upstream	219.1		146.8		217.5		510.8		1.0	
Urban Area	684.0		612.2		256.0		2325.7		19.5	
Downstream	442.0		223.0		142.0		1093.0		5.0	

Table 2b. Contamination Factors[2] in Lake[3] and Fluvial Sediments[4]

	Pb	Cu	Ni	Zn	Cd
Merevale	2.06	1.79	2.32	5.15	–
Seeswood	3.35	2.40	1.50	4.29	–
Wyken	9.29	17.82	6.24	19.23	9.31
Swanswell	4.72	11.46	2.20	30.00	2.67
River Sherbourne Samples[1]					
Urban Area	3.12	4.17	1.17	4.55	19.50
Downstream	2.01	1.52	0.65	2.14	5.00

[1] From Data in Thoms, 1987.

[2] Based on the method of Håkanson and Jansson 1983.

[3] Based on comparisons with minimum concentrations in basal sediment.

[4] Based on comparison with upstream unpolluted samples.

158

Heavy metal content

The heavy metal content of the lake and some fluvial sediments has been determined by both total acid digestion (Foster *et al.*, 1987) and by a selective extraction procedure (Tessier *et al.*, 1979) with the total acid digest used to determine residual concentrations. Summary data for all four sites are presented in Table 2 and sediment based Zn and Pb profiles in Fig. 1. The profiles from the Merevale sediments are remarkably similar to the record of anthropogenic emissions, with generally increasing concentrations since 1900 and greatest increases post 1945. Since sedimentation rates have varied little over the last 150 years, this profile most probably represents a record of atmospheric pollution. For example, the present loading of Zn to the sediments is around $0.2 \text{ g m}^{-2} \text{ yr}^{-1}$. The Seeswood Pool concentration profiles contain a sharp decrease in the period post 1960 which correlates with a dramatic increase in sedimentation since this time. The present day loading of Zn is around $0.6 \text{ g m}^{-2} \text{ yr}^{-1}$. This value is higher than that of Merevale lake, and the excess probably represents the erosion and transport of metal enriched topsoil to the lake. The decline in concentration, however, is probably caused by a greater loading of metal-deficient channel marginal sediments. This interpretation is confirmed by independent studies of sediment sources (Foster *et al.*, 1990).

Both Swanswell Pool and Wyken Slough possess higher heavy metal concentrations. The former is derived from atmospheric sources alone, and the *average* loading of Zn at this inner city site over the last 132 years exceeds $2.5 \text{ g m}^{-2} \text{ yr}^{-1}$. One of the main contributory factors is the position of the site relative to the city centre, with high traffic volumes and considerable local industrial pollution. The much lower ratio of lead to zinc at this inner city site probably reflects localised pollution from car exhausts.

The profile from Wyken Slough represents a much more complex interaction of processes. With a larger catchment to lake ratio, and a stream inflow, sedimentation rates are much higher than at the Swanswell Pool, yet concentrations for Pb, Cu and Cd are also much higher. Taking Zn again as the reference element, the average loading to the site for the last 137 years is ca $1.6 \text{ g m}^{-2} \text{ yr}^{-1}$, somewhat lower than at Swanswell Pool. The lower loading of Zn may in part relate to its solubility at high pH. On occasions, Severn Trent Water Authority have reported pH values for sections of the inflowing river of 8.4 and Zn concentrations in filtered water samples as high as $560 \text{ } \mu\text{g l}^{-1}$. Ratios of filtered to unfiltered concentrations for Pb, Cu, Ni, Zn and Cd are 3.69, 1.67, 1.03, 1.25 and 1.93. A lower value indicates a greater proportion in solution which may, therefore, not be recorded in the sediments. The heavy metal record at these

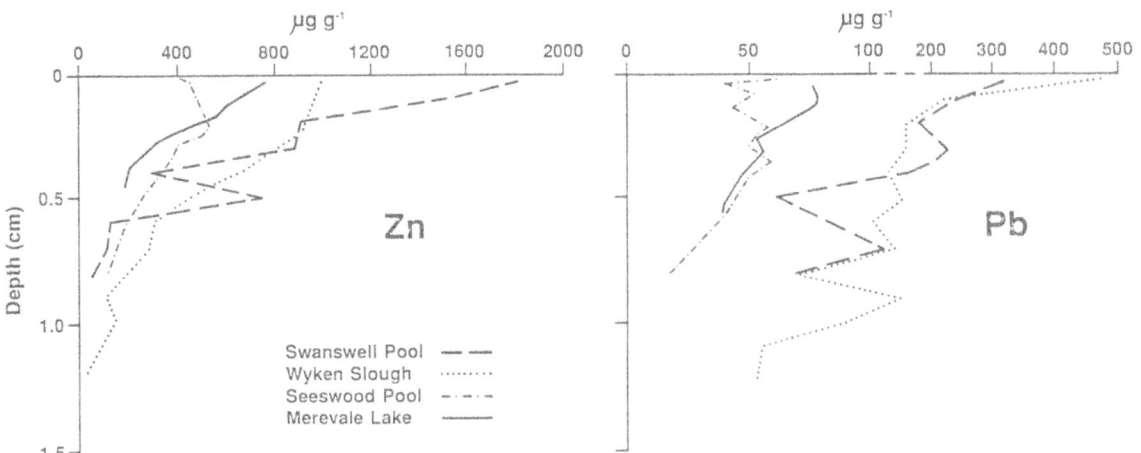

Fig. 1. Zn and Pb concentrations for the four Midland Lakes. (Note the change of scale on the Pb concentration axis)

159

sites, thus, probably under-represents the historical throughput to a varying degree. Indeed, much of the Zn is known to derive from leachates from a toxic waste landfill site, where concentrations have exceeded 21 mg l^{-1}, and from a metal plating works where the levels have reached

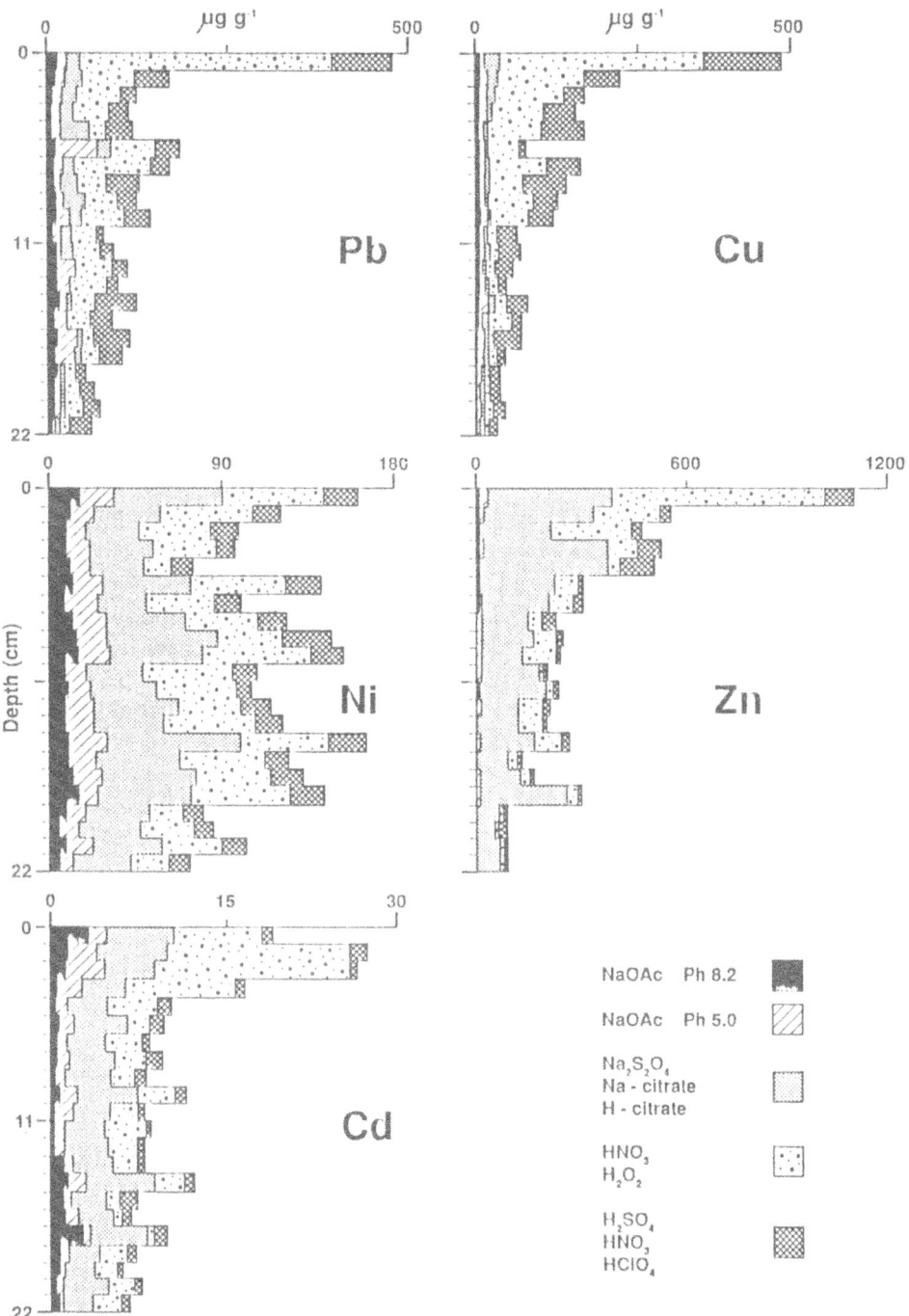

Fig. 2. Fractionation of heavy metals in the upper 22 cm of the sediments of the Wyken Slough. (Method based on that of Tessier *et al.*, 1979).

Table 3. Phosphorus and organic matter in lake sediments.

	Phosphorus $\mu g\ g^{-1}$		Organic Matter %	
	min	max	min	max
Merevale	482.9	982.3	2.0	6.0
Seeswood	602.4	1760.6	7.0	12.0
Wyken	123.8	876.0	1.4	8.2
Swanswell	120.0	999.5	5.4	22.4

43 mg l^{-1}. Eighty and 40% of the Zn from the two sources respectively was found to be in solution.

In addition to the heavy metals retained in the lake sediments are those trapped in the substrates of fluvial systems. The sediments of the river Sherbourne, which runs through the City of Coventry, have been analysed by Thoms (1987a) and summary results are given in Table 2. Concentrations in the fraction smaller than 63 μm frequently exceeds those in the urban lake sediments, and contamination factors are similar to those in lake sediments. Preliminary analysis of heavy metal concentrations in the substrate of the stream flowing into the Wyken Slough exhibits concentrations of Pb, Zn and Cd of 225, 200 and 11.8 μg g^{-1} upstream and 60, 100 and 3.0 μg g^{-1} downstream of the reservoir.

An attempt has also been made to divide the sediments into five fractions by the sequential dissolution method of Tessier *et al.* (1979). These are the exchangeable, carbonate bound, iron and manganese bound, organic and residual fractions (Fig. 2). The results agree with those of Thoms (1987a), in that the largest fraction for most heavy metals is bound to organic matter.

Eutrophication

In most lake sediments, phosphorus and organic matter concentrations increase towards the sediment surface (Table 3). At Seeswood Pool, phosphorus concentrations exceed 1700 μg g^{-1}. Organic matter concentrations are highest in the Swanswell Pool sediments, reaching a maximum of 22.4%. Evidence for a deterioration in water quality at the latter site is supported by analysis of molluscan faunal remains (Fig. 3). Changes in the taxa present at the base of the upper organic-rich sediments suggest the establishment of a weed-rich, still-water community. This is followed by evidence of bottom disturbance consistent with the use of the reservoir as a boating lake in the 1920's and 30's. The pond now supports a limited fauna dominated by the somewhat pollution-tolerant species *Lymnaea peregra* (Müller). The same species dominates the upper sediments of the Wyken Slough.

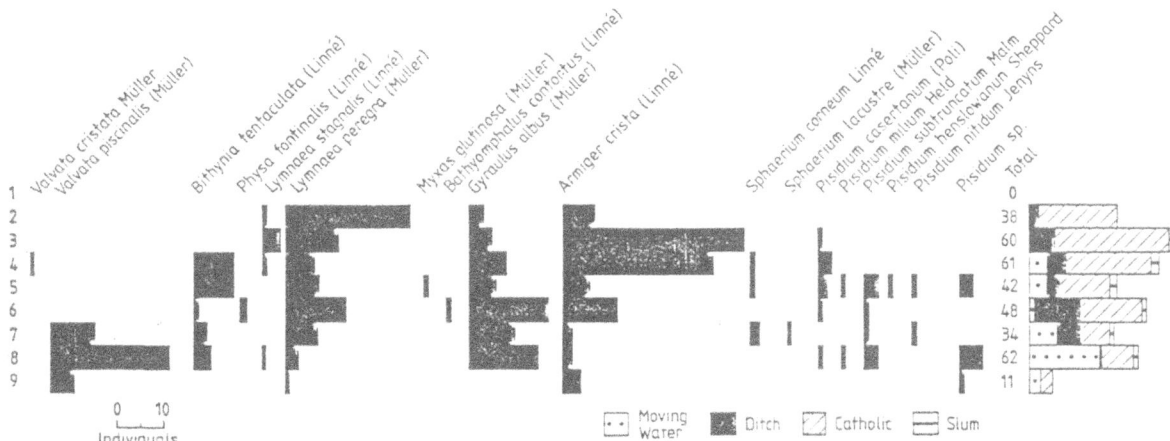

Fig. 3. Molluscan Fauna in the basal sediments of Swanswell Pool. (Samples 1 to 9 on the Y axis are from 2 cm thick subsamples taken at 10 cm intervals from the sediment surface).

Rehabilitation

Part of the present research programme concerns an evaluation of methods for rehabilitating urban lakes in order for them to be used as centres of water-based recreational activity. In the two lakes considered here, common problems include shallow water depths and metal, phosphorus and organic matter contaminated sediments. It is desirable to increase depths by 0.75 m in the Swanswell Pool and remove all lake sediment from the Wyken Slough so as to provide a solid substrate. The following specific limitations have been encountered:-

1 No suitable sites on which to allow extracted sediments partially to dry before disposal exist in close proximity to the reservoirs.
2 Their high liquid content precludes disposal in controlled land-fill sites.
3 The close proximity of residential areas, health hazards and environmental nuisance caused by odours and noise, restrict the methods of removal and transport which may be employed.
4 The high metal content (ca 9.5 and 3.7 t of Pb, Cu, Ni, Zn and Cd in Wyken and Swanswell respectively) precludes the disposal of these sediments on agricultural land.

The rehabilitation scheme for the Wyken Slough also requires management of the contributing river systems in order to prevent further influx of heavy metals from point sources, and from reworking of enriched bottom sediments. This may involve use of lagoons upstream of the lake, diversion of river water to a treatment works, or controlled filtering through expansion of an area of marsh upstream of the inflow point.

Solutions to these problems are expensive. For example, the rehabilitation scheme currently under consideration for the Swanswell Pool involves:

1 Removal of sediment by airflow dredging equipment.
2 Transfer of sediment to sealed tankers for transport.
3 Transport to the nearest treatment plant (In this case ca 80 km round trip).

4 Dewatering and disposal of sediment in a land-fill site.

Highest costs are associated with stages 3 and 4, which together add 400% to the cost of stages 1 and 2.

Conclusion

On the basis of the above results, the following conclusions may be drawn.

1 The sediments of Merevale Lake contain a record of atmospheric heavy metal loading to rural environments in the English Midlands which correlates closely with historical atmospheric emission data. At Seeswood Pool, changing sedimentation rates and sources of heavy metals through time make interpretation more difficult.
2 The Swanswell Pool provides valuable information on inner city atmospheric pollution, clearly demonstrating enrichment of Pb in particular.
3 Interpretation of the history of metal loading at the Wyken Slough is more complex owing to the variety of sources, the impact of metal storage in the fluvial system, and the variable contribution of dissolved and particulate loading to the reservoir.
4 All Midland lakes show some evidence of recent eutrophication via the phosphorus, organic matter and, at Swanswell Pool, molluscan records.
5 Rehabilitation of the inner city lakes is made complex by both high heavy metal content and their shallow water depth.
6 A rehabilitation programme for these sites involves significant operational and disposal constraints leading to high cost.

Acknowledgements

We are grateful for the cooperation and financial assistance of several organisations and individuals. Coventry City Council and Coventry Poly-

162

technic funded the research. Particular thanks go to Geoff Hall, Robin Ashby & Don Reed (Coventry City Council) and John Batty (Severn Trent Water Authority) for advice, comment, field assistance and the provision of data. A number of undergraduate students have contributed significantly to the research programme at all sites through undergraduate project work. We would especially like to thank Kayzi Ambridge, Liam Kelly, Andy Love, Ceri Owen & Trevor Price. Shirley Addleton drew the diagrams to the usual high standard.

References

Christensen, E. R. & N. K. Chien, 1981. Fluxes of arsenic, lead, zinc and cadmium to Green Bay and Lake Michigan sediments. Envir. Sci. Technol.15: 553–558.

Dearing, J. A., J. J. Elner & C. M. Happey-Wood, 1981. Recent sediment flux and erosional processes in a Welsh upland lake-catchment based on magnetic susceptibility measurements.. Quat. Res. 16: 356–372.

Förstner, U. & G. T. W. Wittmann, 1979. Metal Pollution in the aquatic environment. Springer-Verlag. Berlin. 486 pp.

Foster, I. D. L. & J. A. Dearing, 1987. Quantification of long term trends in atmospheric pollution and agricultural eutrophication: A lake-watershed approach. IAHS Publication 168: 173–189.

Foster, I. D. L., S. M. C. Charlesworth & D. H. Keen, 1988a. An evaluation of the quality and character of the Wyken Slough and of its bottom sediments. Sediments and Water Group Report. Geography Department, Coventry Polytechnic. 46 pp.

Foster, I. D. L., S. M. C. Charlesworth & D. H. Keen, 1988b. The sedimentology and water balance of Swanswell Pool, Coventry. Sediments and Water Group Report. Geography Department, Coventry Polytechnic. 41 pp.

Foster, I. D. L., J. A. Dearing, A. D. Simpson, A. D. Carter & P. G. Appleby, 1985. Lake catchment based studies of erosion and denudation in the Merevale catchment, Warwickshire, U.K. Earth Surface Processes and Landforms., 10: 45–68.

Foster, I. D. L., J. A. Dearing & P. G. Appleby, 1986. Historical trends in catchment sediment yields: a case study in reconstruction from lake sediment records in Warwickshire, U.K. Hydrol. Sci. J., 31: 427–443.

Foster, I. D. L., J. A. Dearing, S. M. C. Charlesworth & L. A. Kelly, 1987. Paired catchment studies: a framework for investigating chemical fluxes in small drainage basins. Appl. Geogr.7: 115–133.

Foster, I. D. L., R. Grew & J. Dearing, 1990. Magnitude and frequency of sediment transport in Agricultural Catchments: a paired lake catchment study in Midland England. In J. Boardman, I. D. L. Foster & J. A. Dearing (eds). Soil Erosion on Agricultural Land. Wiley, Chichester: 153–171.

Håkanson, L. & M. Jansson, 1983. Principles of Lake Sedimentology. Springer-Verlag. Berlin. 316 pp.

Müller, G. & D. Barsch, 1980. Anthropogenic lead accumulation in the sediments of a high arctic lake, Ooblayouh bay, N. Ellesmere Island. N.W.T. Canada. Envir. Technol. Lett. 1: 131–140.

Nriagu, J. O., 1979. Global inventory of natural and anthropogenic emissions of trace metals to the atmosphere. Nature 279: 409–411.

Renburg, I., 1986. Concentration and annual accumulation of heavy metals in lake sediments: Their significance in studies of the history of heavy metal pollution. Hydrobiologia 143: 379–385.

Tessier, A., P. G. C. Campbell & M. Bisson, 1979. Sequential extraction procedure for the speciation of particulate trace metals. Analyt. Chem. 51: 844–851.

Thoms, M. T., 1987a. Channel Sedimentation within Urban Gravel-bed rivers. Unpublished Ph.D. Dissertation. Loughborough University of Technology. 222 pp.

Thoms, M. T., 1987b. Channel sedimentation within the Urbanised River Tame, U.K. Regulated Rivers: Research and Management 1: 229–246.

Hydrobiologia **214**: 163–169, 1991.
J. P. Smith, P. G. Appleby, R. W. Battarbee, J. A. Dearing, R. Flower, E. Y. Haworth, F. Oldfield & P. E. O'Sullivan (eds),   163
Environmental History and Palaeolimnology.
© *1991 Kluwer Academic Publishers.*

Sedimentary diatom concentrations and accumulation rates as predictors of lake trophic state

Thomas J. Whitmore [1]
*Florida Museum of Natural History, University of Florida, Gainesville, FL 32611, USA; [1]present
address: Department of Fisheries and Aquaculture, University of Florida, Gainesville, FL 32606, USA*

Key words: diatoms, trophic state, paleolimnology, Florida

Abstract

Diatom concentrations in surface sediments are positively correlated with limnetic chlorophyll *a* concentrations in Florida (USA) lakes. Using this relationship, I examine models that provide quantitative inferences for trophic state in historical applications.

The best model predicts chlorophyll *a* trophic state index (TSI) values from log-transformed diatom concentrations and explains approximately half the variance in the dependent variable. Diatom accumulation rates are not better than sedimentary diatom concentrations as predictors of TSI. The entire diatom assemblage is as sensitive an indicator of TSI as are the planktonic diatoms alone. A model that considers the ecological preferences of specific taxa was found to be a better predictor than the model based on total diatom concentration.

The sedimentary diatom concentration model provides a useful method for assessing historical changes in primary productivity, except in lakes where factors (e.g., silica limitation, blue-green bacterial inhibition) limited diatom production, or post-depositional changes removed sedimentary diatoms. TSI inferences are presented for sediment cores from two Florida lakes, one of which demonstrates a problematic application, and the other of which does not.

Introduction

Chlorophyll *a* is strongly correlated with algal biomass in Florida lakes (Canfield *et al.*, 1985), and it is less expensive and labor-intensive to measure than estimates of algal biomass. Canfield *et al.* (1985) have shown that chlorophyll *a* yields higher coefficients of determination than phytoplankton biomass when regressed in empirical models with nutrient measures. Because chlorophyll *a* is often the preferred proxy measure of algal production, historical data for chlorophyll *a* values are available for many Florida lakes,

whereas phytoplankton biomass data are not. Chlorophyll *a* is therefore a logical dependent variable in empirical models used to infer historical algal production.

Lake trophic state is often assessed from algal standing crop and its effect on water transparency, and it is quantitatively described by trophic state indices (TSI). For Florida lakes, I prefer the TSI defined by Huber *et al.* (1982) that is based on chlorophyll *a*. This index (TSI(Chl *a*)) takes into account the fact that relationships between chlorophyll *a* and subindex variables (P, N, Secchi depth) in Florida lakes are different from

164

those in north-temperate lakes (Baker *et al.*, 1981) on which other trophic indices (e.g., Carlson, 1977) have been based.

The strong relationship between chlorophyll *a* and algal biomass suggests a logical method for inferring historical TSI(Chl *a*) values. Diatoms are an important component of algal production in lakes, and their siliceous valves are usually well-preserved in lake sediments. Sedimentary diatom concentrations should therefore reflect past levels of algal standing crop and limnetic chlorophyll *a*. Diatom accumulation rates might be expected to relate more strongly than sedimentary diatom concentrations to TSI(Chl *a*) because they minimize variance caused by differences in inorganic sedimentation rates. Because chlorophyll *a* values are measured on limnetic water samples, it is also necessary to investigate whether planktonic diatoms are more highly correlated than the entire diatom assemblage with TSI(Chl *a*).

Several factors may cause sedimentary diatoms to reflect past trophic states inaccurately. Silica

limitation of planktonic diatom populations has been observed in Lakes Ontario (Stoermer *et al.*, 1985) and Michigan (Schelske *et al.*, 1983; Schelske, 1988) because phosphorus loading increased the production of diatoms and the loss of biogenic silica to sediments. Sediment cores from these lakes show recent reductions in sedimentary diatom concentrations and accumulation rates. Silica dissolution within sediments also has been implicated in reducing sedimentary diatom concentrations (Parker & Edgington, 1976). Blue-green bacteria produce allelopathic substances that have been shown to limit diatom production under conditions of advanced eutrophy (Keating, 1978).

The objectives of this study are to develop predictive models for TSI based on sedimentary diatom concentrations or accumulation rates and to identify factors that might limit the historical application of such models.

Methods

Sediment samples were collected from the sediment-water interface of 30 Florida lakes (Fig. 1) with TSI(Chl *a*) values ranging from 19.4 to 86.0, and pH ranging from 4.8 to 9.0. Samples were taken using either an Ekman dredge or a 4-cm diameter plastic-barrel piston corer. The top 2 cm of sediment were removed with a pipette and transferred to water-tight containers. Subsamples were removed for diatom analyses and to estimate bulk sediment accumulation rates by ^{210}Pb assay.

One cm^3 subsamples of wet sediment were cleaned for diatom analyses with hydrogen peroxide and potassium dichromate (Van der Werff, 1956). Cleaned samples were settled onto coverslips in evaporation trays that permitted estimation of diatom numbers in the subsample (Battarbee, 1973). Coverslips were then mounted with Hyrax mounting medium. A minimum of 500 diatom valves was counted in each sample by medium-phase microscopy with an oil immersion objective. Diatoms were identified using standard floras including Hustedt (1930, 1930–1966) and Patrick & Reimer (1966–1975). Specimens that

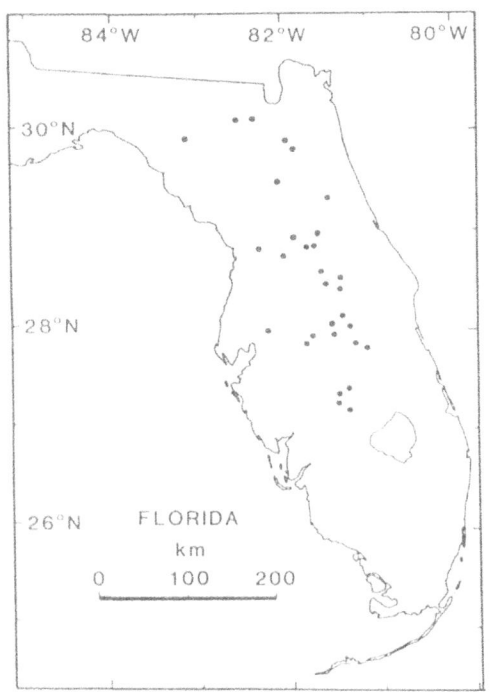

Fig. 1. Geographical distribution of lakes in surface sediment survey.

were taxonomically difficult were studied with an Hitachi S-145A scanning electron microscope.

Bulk density values (g dry cm^{-3} wet) were determined by weighing volumetric subsamples that were dried at 110 °C. Sedimentary concentrations of diatoms (CONC), expressed as valves g^{-1} dry wt, were estimated by dividing the concentration of diatoms in the wet subsamples (valves cm^{-3} wet) by the bulk density values for each sample. Bulk sediment accumulation rates (g cm^{-2} yr^{-1}) were obtained by the ^{210}Pb dilution tracer method of Binford & Brenner (1986). This ^{210}Pb method yields estimates of the contemporary mean lakewide surface sediment accumulation rates. Diatom accumulation rates (ACCUM), expressed as valves cm^{-2} yr^{-1}, were then calculated as the product of sedimentary diatom concentrations and bulk sediment accumulation rates.

I calculated the planktonic proportion of each assemblage using autecological data presented by Lowe (1974) and supplemented by information from other sources (Patrick & Reimer, 1966–1975; Hustedt, 1930, 1930–1966). The percentage of the diatom assemblage in Lowe's euplanktonic category was calculated by summing the percentages of individual taxa. Tychoplanktonic taxa were assumed to be $\frac{1}{3}$ euplanktonic. The percentage of taxa represented by more than one habitat category was divided between those categories.

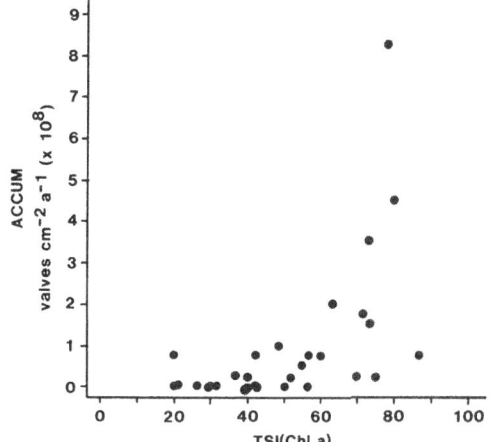

Fig. 3. Diatom accumulation rate vs. Huber *et al.* (1982) TSI(Chl *a*) for survey lakes.

gories. The sedimentary concentrations (PL-CONC) and accumulation rates of planktonic diatoms (PL-ACCUM) were then calculated, respectively, by multiplying the planktonic proportion of the diatom assemblages by the total concentrations (CONC) and annual accumulation rates (ACCUM) of sedimentary diatoms.

Because TSI values have curvilinear relationships with sedimentary diatom concentrations and diatom accumulation rates (e.g. Figs. 2 & 3), the diatom variables were log-transformed to construct the predictive models for trophic state. Median TSI values for each lake were obtained from the Florida Lakes Data Base, a large data set maintained by the Water Resources Research Center at the University of Florida. Pearson product-moment correlation coefficients were calculated between log-transformed diatom variables and TSI values using the SAS PROC CORR procedure (SAS Institute, Inc., 1985). Predictive models were selected based upon the strength of the correlations, and regression equations for these models were constructed using the SAS PROC GLM procedure (SAS Institute, Inc., 1985).

Results

TSI variables are more strongly correlated with log-transformed sedimentary diatom concentra-

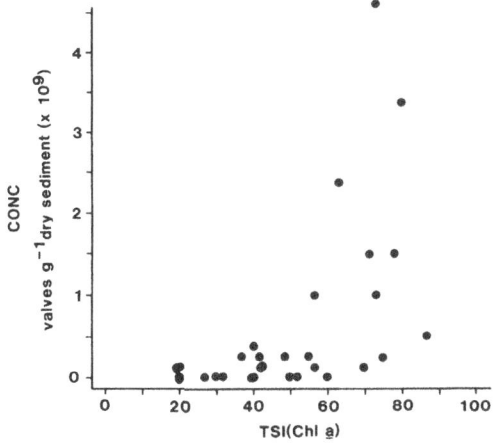

Fig. 2. Total diatom concentration in surface sediments vs. Huber *et al.* (1982) TSI(Chl *a*) for 30 lakes in survey.

166

Table 1. Correlation coefficients between trophic state indices and log-transformed diatom concentration and accumulation rate variables. TSI(TP) is a subindex based on total P, and TSI(AVE) is an averaged subindex based on Secchi depth, chlorophyll *a*, and total N or total P depending upon which is limiting. Sample size is shown in parentheses.

	Trophic state indices		
	TSI(TP)	TSI(AVE)	TSI(Chl *a*)
log_{10}(CONC)	0.38 (28)	0.61 (30)*	†0.71 (30)**
log_{10}(ACCUM)	0.30 (28)	0.54 (30)*	0.65 (30)**
log_{10}(PL-CONC)	0.47 (28)	0.64 (30)**	0.70 (30)**
log_{10}(PL-ACCUM)	0.40 (28)	0.59 (30)*	0.66 (30)**

* $p < 0.01$.
** $p < 0.001$.
† Relationship selected for predictive model.

tions than with log-transformed diatom accumulation rates (Table 1). Correlations between planktonic diatom variables and TSI are not significantly better than correlations based on the entire diatom assemblage. Two lakes were removed from the data set prior to correlations with the TSI for total phosphorus (TSI(TP)) because they were nitrogen limited. Their diatom standing crops did not reflect the high epilimnetic concentrations of phosphorus.

The strongest correlations in Table 1 were

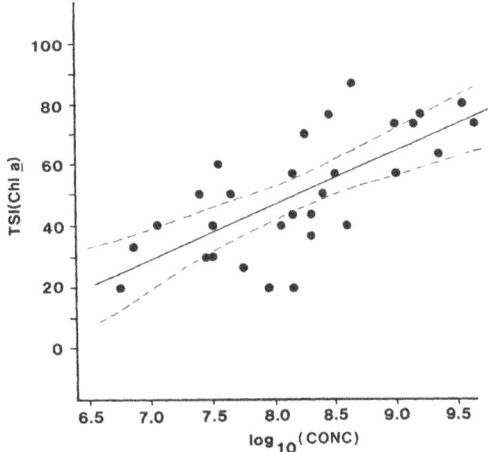

Fig. 4. Huber *et al.* (1982) TSI(Chl *a*) vs. log-transformed sedimentary diatom concentration. Regression line is shown with 95% confidence intervals.

between log-transformed sedimentary diatom concentrations (log_{10}(CONC)) and the TSI variable for chlorophyll *a*. The equation (Fig. 4) predicting Huber *et al.*'s (1982) TSI for Chl *a* was:

$$TSI(Chl\ a) = -95.54 + 17.78 \times log_{10}(CONC)$$
$$R^2 = 0.50, n = 30, \sqrt{MSE} = 14.11.$$

The predictive model for Huber *et al.*'s TSI(Chl *a*) was applied to fossil diatom assemblages in the sediment cores of two Florida lakes. Diatom assemblages were analyzed in a 90 cm sediment core from Lake Francis in Highlands County. The assemblages show that Lake Francis has generally maintained a uniform trophic state, and has been dominated by *Aulacoseira ambigua* (Grun.) Simonsen var. *ambigua* ($\bar{x} = 47\%$) and *Staurosirella pinnata* (Ehr.) Williams & Round ($\bar{x} = 11\%$). Trophic state may have declined recently, as suggested by *Cyclotella stelligera* Cl. u. Grun. var. *stelligera*, which increased from an average of 4% throughout most of the core to 27% in the surface sample. Diminished productivity may be a consequence of runoff and subsurface inputs of herbicides from citrus groves adjacent to Lake Francis.

The second sediment core was from Lake Parker in Polk County. Lake Parker has been influenced in recent decades by extensive phosphate mining and urbanization in its watershed. The lake has undergone considerable eutrophication, as indicated by diatom predictive models, and increased rates of phosphorus and organic matter sedimentation (Whitmore, in prep.). The diatom assemblage at the 60 cm level (^{210}Pb date of A.D. 1875) indicates a lower water level. Most valves in this sample showed evidence of mechanical breakage. The lake then proceeded through a period of moderate to high trophic state until the 1960s, and its diatom assemblage was dominated by phytoplankton including *Aulacoseira ambigua*, and several species of *Synedra* and *Nitzschia*. Since the 1960s, planktonic taxa declined while periphytic species of *Staurosirella* and *Pseudostaurosira* increased.

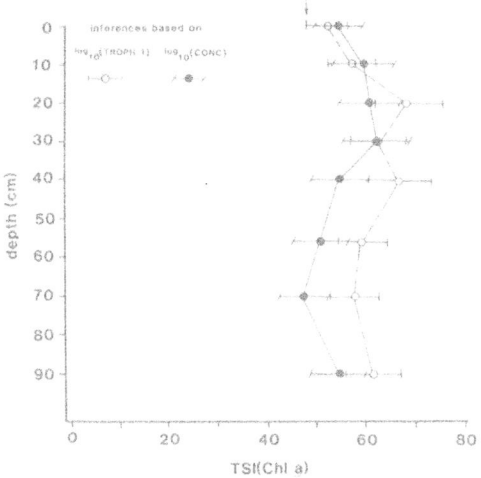

Fig. 5. Historical TSI(Chl *a*) inferences for 90 cm sediment core from Lake Francis, Highlands Co. Inferences based on sedimentary diatom concentration and TROPH 1 index (Whitmore, 1989) are shown with 95% confidence intervals. Arrow indicates median of recent TSI(Chl *a*) values from Florida Lakes Data Base.

Figures 5 & 6 show historical TSI(Chl *a*) reconstructions based on the concentration of fossil diatom valves in Lakes Francis and Parker. Confidence intervals are shown rather than prediction intervals because I wish to infer mean rather than actual TSI(CHL *a*) values. In both figures, inferences are compared with TSI(Chl *a*) values obtained from a predictive model utilizing autecological preferences of specific diatom taxa. The diatom index in this latter model, TROPH 1, is essentially a ratio of the percentage of diatoms indicating high trophic state to those indicating low trophic state conditions. TROPH 1 was defined and used in other TSI predictive models by Whitmore (1989). The linear regression equation (Fig. 7) for the present application of the TROPH 1 index is:

$$TSI(Chl\ a) = 40.98 + 40.68 \times \log_{10}(TROPH\ 1)$$

$$R^2 = 0.66,\ n = 30,\ \sqrt{MSE} = 11.61.$$

The TSI(Chl *a*) inferences for Lake Francis (Fig. 5) show close agreement between the model based on sedimentary diatom concentrations and the model using the TROPH 1 index. 95% confidence intervals for inferences obtained from the two models overlap at all but the 70 cm sample.

In contrast, several discrepancies appear between inferences obtained from these models in the Lake Parker reconstruction (Fig. 6). Confidence intervals overlap for most samples, but the inference based on sedimentary diatom concen-

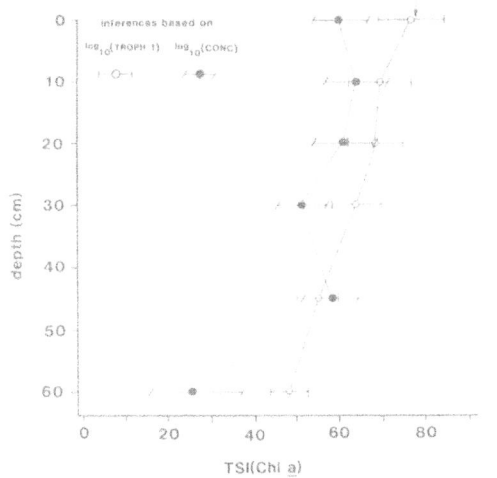

Fig. 6. Historical TSI(Chl *a*) inferences for sediment core from Lake Parker, Polk Co. Inferences based on sedimentary diatom concentrations and TROPH 1 index are shown with 95% confidence intervals. Arrow indicate median of recent TSI(Chl *a*) values from Florida Lakes Data Base.

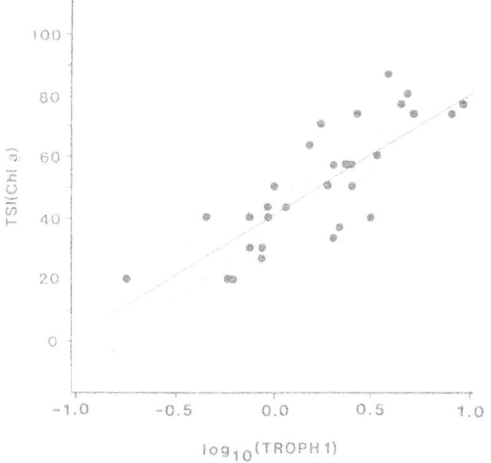

Fig. 7. Huber *et al.* (1982) TSI(Chl *a*) vs. log-transformed TROPH 1 index (Whitmore, 1989) that is based on autecological nutrient preferences. Regression line is shown with 95% confidence intervals.

trations is significantly lower in the 60 cm sample. Mechanical breakage during the lower water level probably reduced sedimentary diatom concentrations. The modern (0 cm) inference based on the TROPH 1 index is very close to the measured TSI(Chl a) value, but the inference based on the diatom concentration is significantly lower, possibly due to mechanisms limiting diatom production.

Discussion

When correlated with TSI, log-transformed planktonic diatom variables did not show larger correlation coefficients than variables based on the entire diatom assemblage. Stronger correlations were anticipated because TSI(Chl a) is measured from planktonic algal biomass in the limnetic zone, whereas periphytic diatoms, which comprised 60–90% of the diatom assemblages in several lakes, do not contribute to the measured chlorophyll a values. Increases in nutrient availability may affect the planktonic and periphytic communities equally, so that higher periphyton production is also associated with higher values of water-column chlorophyll a. It is therefore more expedient to base historical TSI inferences on the total diatom assemblage without separating individual diatoms according to habitat preference.

Contrary to expectation, diatom concentrations are correlated as strongly with TSI as are diatom accumulation rates. Sedimentary diatom concentrations increase over 3 orders of magnitude with increasing trophic state. Bulk sediment accumulation rates, however, are relatively constant in Florida lakes, and are not correlated with trophic state. The main determinant of diatom accumulation rate is sedimentary diatom concentration, and it is highly correlated with diatom accumulation rates ($r = 0.68$, $p = 0.001$, $n = 30$). Sedimentary diatom concentration, therefore, is the preferred variable for TSI predictive models in this lake district because it predicts as well as the diatom accumulation rate and it is easier to measure.

Even if diatom accumulation rates were better indicators of TSI(Chl a), downcore application of such a predictive model might have proven problematic because of the way that bulk sediment accumulation rates were assessed. The ^{210}Pb method I employed (Binford & Brenner, 1986) yields an estimate of the contemporary mean lakewide sediment accumulation rate, whereas core-derived accumulation rates measure local sedimentation that may vary spatially within a lake due to sediment focusing (Binford & Brenner, 1988). A model based on Binford & Brenner's method could be applied to a single sediment core only when it is reasonable to assume that the sediment accumulation rate at the core site is representative of the accumulation rate throughout the lake. If sedimentation rate were to vary within a lake basin, it would be necessary to sample several sediment cores and estimate the mean lakewide diatom accumulation rate in order to apply a TSI predictive model based on diatom accumulation rates. Substantially larger sampling regimes would also be necessary to construct a model relating TSI to diatom accumulation rates as determined by the whole-core method of dating.

The Lake Parker reconstruction illustrated problems inherent in trophic reconstructions from sedimentary diatom concentrations. The paucity of diatoms at the 60 cm level resulted in TSI inferences significantly lower than inferences obtained from the diatom index. Mechanical breakage of diatoms during the lower water level reduced the number of countable diatoms in the sediment, and this breakage may have in turn promoted frustule dissolution (Parker & Edgington, 1976). The surface sediment sample suggested another inherent problem; Lake Parker is highly eutrophic and sustains blue-green bacterial populations. It appears that diatom production may currently be limited by dissolved silica or by blue-green allelopathic substances. TSI(Chl a) values for the lake, nevertheless, are high because of chlorophyll a from non-diatom phytoplankton components. Inferences based on the diatom index still accurately predict TSI in Lake Parker because of qualitative differences in the diatom assemblage that indicate high nutrient conditions.

Predictive models based on sedimentary

diatom concentrations should be applied cautiously when factors may have limited diatom production, or when diagenetic mechanisms may have affected preservation. Inferences based solely on concentrations are convenient, but appear to be less informative than methods based on autecological information of specific taxa (Whitmore, 1989).

Acknowledgements

I thank M. W. Binford, M. Brenner and E. Fisher for collecting surface sediment samples. I am grateful to M. Binford for calculating bulk sediment accumulation rates, and to M. Brenner for discussions and his thoughtful review of the manuscript. This work was supported in part by NSF grant DAR 79-2418 to E. S. Deevey, Jr., P. L. Brezonik and T. L. Crisman.

References

Baker, L. A., P. L. Brezonik & C. R. Kratzer, 1981. Nutrient-loading trophic state relationships in Florida lakes. Water Resour. Res. Center, Univ. Florida. Publ. 56.

Battarbee, R. W., 1973. A new method for the estimation of absolute microfossil numbers, with reference especially to diatoms. Limnol. Oceanogr. 18: 647–653.

Binford, M. W. & M. Brenner, 1986. Dilution of ^{210}Pb by organic sedimentation in lakes of different trophic states, and application to studies of sediment-water interactions. Limnol. Oceanogr. 31: 584–595.

Binford, M. W. & M. Brenner, 1988. Reply to comment by Benoit & Hemond. Limnol. Oceanogr. 33: 304–310.

Canfield, D. E., S. B. Linda & L. M. Hodgson, 1985. Chlorophyll-biomass-nutrient relationships for natural assemblages of Florida phytoplankton. Wat. Res. Bull. 21: 381–391.

Carlson, R. E., 1977. A trophic state index for lakes. Limnol. Oceanogr. 22: 361–369.

Huber, W. C., P. L. Brezonik, J. P. Heany, R. E. Dickinson, S. D. Preston, D. S. Dwornik & M. A. DeMaio, 1982. A classification of Florida lakes, Vol. 1–2. Florida Dep. Environ. Regulation, Rep. ENV-05-82-1. Tallahassee.

Hustedt, F., 1930. Bacillariophyta (Diatomeae). Heft 10. In: A. Pascher (ed.), Die Süsswasser-Flora Mitteleuropas. G Fischer, Jena.

Hustedt, F. 1930–1966. Die Kieselalgen Deutschlands, Österreichs und der Schweiz. Teil 1–3. In: L. Rabenhorst (ed.), Kryptogamen-Flora von Deutschland, Österreich und der Schweiz Band 7, AVG, Leipzig.

Keating, K. I., 1978. Blue-green algal inhibition of diatom growth: transition from mesotrophic to eutrophic community structure. Science 199: 971–973.

Lowe, R. L., 1974. Environmental requirements and pollution tolerances of freshwater diatoms. U.S. EPA 670/4-74-005. Cincinnati, Ohio.

Parker, J. I. & D. N. Edgington, 1976. Concentrations of diatom frustules in Lake Michigan sediment cores. Limnol. Oceanogr. 21: 887–893.

Patrick, R. & C. W. Reimer, 1966–1975. The diatoms of the United States. Monogr. Acad. Nat. Sci. Phila., No. 13, Part 1, Vol. 1–2.

SAS Institute, Inc., 1985. SAS user's guide: Statistics. SAS Institute, Inc. Cary, NC.

Schelske, C. L., 1988. Historic trends in Lake Michigan silica concentrations. Int. Revue ges. Hydrobiol. 73: 559–591.

Schelske, C. L., E. F. Stoermer, D. J. Conley, J. A. Robbins & R. M. Glover, 1983. Early eutrophication in the lower Great Lakes: new evidence from biogenic silica in sediments. Science 222: 320–322.

Stoermer, E. F., J. A. Wolin, C. L. Schelske & D. J. Conley, 1985. An assessment of ecological changes during the recent history of Lake Ontario based on siliceous algal microfossils preserved in the sediments. J. Phycol. 21: 257–276.

Van der Werff, A., 1956. A new method of concentrating and cleaning diatoms and other organisms. Int. Ver. Theor. Angew. Limnol. 12: 276–277.

Whitmore, T. J., 1989. Florida diatom assemblages as indicators of trophic state and pH. Limnol. Oceanogr. 34: 884–897.

Hydrobiologia **214**: 171–180, 1991.
J. P. Smith, P. G. Appleby, R. W. Battarbee, J. A. Dearing, R. Flower, E. Y. Haworth, F. Oldfield & P. E. O'Sullivan (eds),   171
Environmental History and Palaeolimnology.
© 1991 Kluwer Academic Publishers.

The sediment column as a record of trophic status: examples from Bosherston Lakes, SW Wales

A. W. G. Rees[1], G. C. F. Hinton[2], F. G. Johnson[1] & P. E. O'Sullivan[1]
[1] Department of Environmental Sciences, Polytechnic South West, Drake Circus, Plymouth
PL4 8AA, UK; [2] Nature Conservancy Council (East Anglia Region), 60, Bracondale, Norwich
NR1 2BE, UK

Key words: sediment geochemistry, ^{210}Pb dating, diatom analysis, eutrophication, phosphorus, Chara, marl lakes

Abstract

Bosherston Lakes are a series of interconnected, mesotrophic to hypereutrophic, artificially-created coastal marl lakes in Dyfed, South West Wales. Progressive eutrophication of the lake system has been produced by a high external phosphorus loading which includes phosphorus-rich effluent from a sewage treatment works (STW) in the catchment of the Lakes.

Cores were taken from four sites of varying eutrophic status within the Lakes. In the surface sediment layer, organic C, N and P concentrations generally correlate directly with trophic status and reflect distance from the source of P input. At one site, sediment stratigraphy records a clear transition at 20–15 cm depth, marked by a sharp upward increase in porosity, organic C, N, and P, and 'iron-associated'-P; decreases in organic matter C/N, C/P and N/P ratios; a sharp decrease in carbonate, and a change in the subfossil diatom assemblage. Lead-210 dating indicates that this change occurred in the period 1919 to 1938.

The diatom stratigraphy and sediment geochemistry suggest that this transition reflects an increase in trophic status at this site, probably as a result of the influx of nutrient-rich water. This took place when the management of the Stackpole estate surrounding the lake system, fell into decline during the period 1919–1938.

Area of investigation

Bosherston Lakes, South West Dyfed, form three interconnected arms, each of which is an artificially drowned stream valley that once drained directly to the sea (Fig. 1). The Eastern and Central Arms were dammed first in ca. 1780 AD. The Western Arm and Central Lake (which forms the body of water where the three main arms converge, Fig. 1) were created in ca. 1850. The entire lake system was designed as a source of

recreation, and as a fishery for the surrounding Stackpole Estate, under the ownership of the House of Cawdor.

The lakes are shallow (maximum mean depths <2.2 m; Table 1). The Western and Central Arms are spring fed with the whole of their inferred catchment area lying within Carboniferous Limestone. The Eastern Arm is supplied by two tributaries – the Merrion Stream (Fig. 1) which flows predominantly over Carboniferous Limestone, and the Stackpole Stream which runs

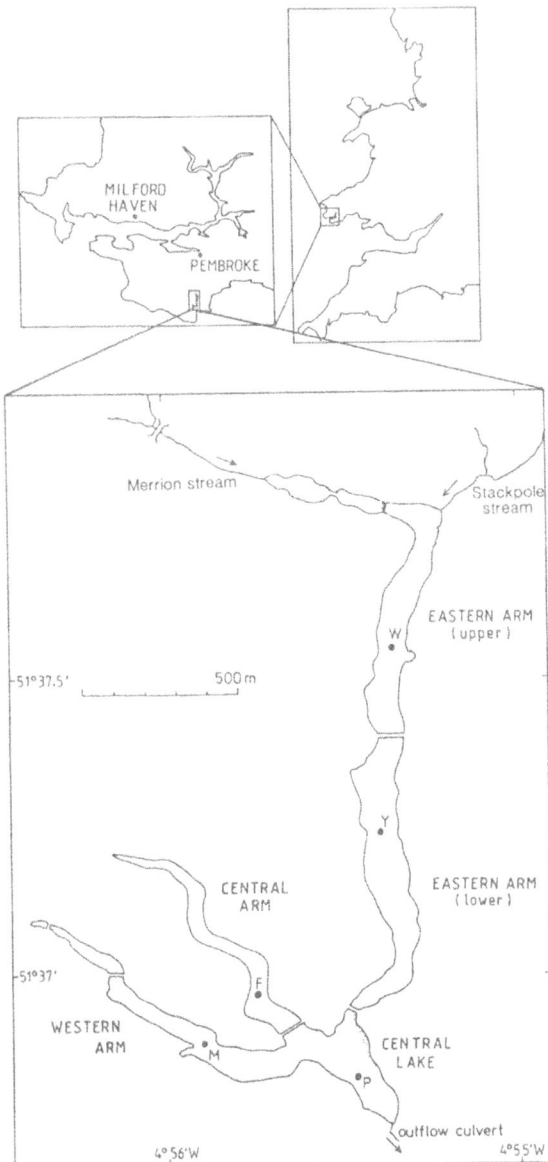

Fig. 1. Location of Bosherston Lakes and core sampling sites (●).

sediments are dominated by detrital and biogenically and bio-induced calcium carbonate ($>60\%$ $CaCO_3$; Rees, ms).

Fluctuations in water level in each arm are highly seasonal. In the past, the level of each was controlled by weirs and sluices. Most of these are still operational, but an important exception is the structure at the southernmost end of the Eastern Arm, which originally operated as a dam, with excess water flowing out via a side overspill weir. A tunnel (ca. 1 m diameter) at the base of the dam was sealed with a wooden sluice board. Examination by divers has revealed that sometime in the past the board disintegrated. Anecdotal evidence (Haycock, pers comm.) indicates that this probably occurred in the period between the two World Wars, when the management of the Stackpole Estate, and hence the Lakes, fell into decline.

The Eastern Arm has probably been eutrophic since its creation (Rees, ms), in response to agricultural practices within its catchment, and to point and non-point sources of domestic sewage. Part of the latter once consisted of a continual discharge of treated sewage effluent into the Stackpole Stream which was in operation from 1967 until redirection (straight out to sea) in 1984. The very high loading of nutrients created hypereutrophic conditions in the Eastern Arm, with maximum total phosphorus and chlorophyll concentrations of 0.76 mg l^{-1} and 301 μg l^{-1} (Table 1) respectively.

As a consequence of the low retention by the sediments of the Eastern Arm, there is currently a high phosphorus loading on the Central Lake (Rees, ms). Originally, the outflow of water from the Eastern Arm via the side-overspill weir would have been of relatively nutrient-depleted surface water. However, the disintegration of the sluice board at the base of the dam separating the two arms allowed the influx of more nutrient-rich bottom-water into the Central Lake.

The influx of phosphorus enriched water on the Central Lake has caused progressive eutrophication. This has been marked by an increase in plankton productivity, and the growth of nutrient-tolerant macrophyte species, and by a decline in the growth of the calcareous macroalga *Chara*

mainly over red Devonian sandstones and marls. The Central Lake, which receives waters from all three main arms, discharges at its southwesternmost point from an outflow above the maximum high tide mark, into the outer part of the Bristol Channel (Fig. 1). The waters of the entire lake system are highly calcareous (maximum hardness <295 mg l^{-1} $CaCO_3$; Table 1). Similarly, the

Table 1. Physical and chemical characteristics of each arm of Bosherston Lakes.

	Characteristics of each arm of Bosherston Lakes				
	U. East. Arm (site W)	L. East. Arm (site Y)	Central Lake (site P)	Western Arm (site M)	Central Arm (site F)
Area (m^2)	56 900	96 800	52 300	61 900	43 300
Max. volume (m^3)	82 300	214 000	108 400	105 400	63 600
Max. mean depth (m)	1.45	2.21	2.07	1.70	1.47
pH	7.7–9.8	7.5–9.7	7.9–9.8	7.9–9.2	7.5–9.4
Hardness (mg/l CaCO$_3$)	63–277	34–265	47–265	51–295	112–295
Chlorophyll (μg/l)	5–301	4–144	3–55	1–21	1–8.5
Nitrate (mg/l)	0–7	0–6	0–4.5	0–4.2	0–4.5
Total P (mg/l)	0–0.5	0–0.76	0–0.18	0–0.36	0–0.04
Phosphate (mg/l)	0–0.3	0–0.56	0–0.05	0–0.03	0–0.02
Trophic status	HEu	HEu	Eu	Meso	Meso

(HEu = hypereutrophic; Eu = eutrophic; Meso = mesotrophic.)
(Chemical data for 10/83–10/84.)

Table 2. Comparison of the characteristics of the sediments from two distinct layers in the Central Lake core with sediments from the Central and Lower East Arms.

Location	Depth (cm)	Porosity (%)	Carbonate (%)	Org. C (%)	Org. N (mg/g)	Org. P (mg/g)	C/N ratio	C/P ratio	N/P ratio
L. East Arm	0–30	80–96	34–71	4.6–7.1	4.7–10.2	0.25–0.84	8.2–11.0	219–586	27–58
Central Lake	0–15	88–95	73–74	4.7–5.5	7.2– 9.2	0.48–0.72	6.1– 7.8	196–266	28–42
	20–31	77–81	82–88	1.6–3.0	2.2– 3.9	0.07–0.16	8.2–10.4	470–814	56–83
Central Arm	0–10	83–88	89–92	1.4–1.9	2.9– 3.4	0.06–0.08	5.7– 6.4	566–667	54–68

(Hinton, ms). The eutrophic conditions in the Central Lake are characterised by high total phosphorus and chlorophyll concentrations (maximum 0.18 mg l^{-1} and 55 μg l^{-1} respectively). *Chara* does not tolerate the high nutrient and low light penetration conditions produced by eutrophication.

The Western Arm is of mesotrophic/eutrophic status (Table 1) and is suffering from the gradual encroachment of nutrient-rich waters from the Central Lake. This is having an adverse effect on the once abundant *Chara* (Hinton, ms). The Central Arm is the least enriched of the Bosherston Lakes and is of mesotrophic status (Table 1) with a high *Chara* productivity.

Methods

Sediment cores for geochemistry 10 to 40 cm in length were taken in Nov 1984 from the sites marked in Fig. 1, using a hand-operated coring device. Each core was logged and then sectioned at 1 cm intervals (2–3 cm for site F) within 6 hours of collection. Each section was weighed, dried to a constant weight at 60 °C and then reweighed. Water content and porosity were calculated from weight measurements (using a measured average sediment density of 2.37 g cm^{-3}).

A representative portion of each sediment section was ground to < 120 Mesh and was used for all subsequent determinations. All analyses

174

were carried out in duplicate. Organic carbon was estimated by means of the chromic acid digestion technique of Gaudette et al. (1974), with a precision of 2.5%. Total carbonate was determined by the acidification method of Black (1965). Total-N and total-P were analysed simultaneously by micro-Keldahl digestion (Lennox & Flanagan, 1982) and subsequent determination of ammonium and phosphate in digests by autoanalysis (Howland, pers. comm.; Petts, 1979). Precision was 5.3% for total-N and 4.2% for total-P. In soils and lake sediments, organic N usually constitutes 95–98% of the total-N present (Bremner, 1965; Keeney et al., 1970). Hence the total-N values determined in this study will be referred to as organic N.

Inorganic P was determined by the 1N HCl digestion method of Aspila et al. (1976) with a precision of 2.0%. An estimate of the 'iron-associated' P was obtained by a similar technique using 1N NaOH. Subtraction of iron-associated P from inorganic P gives an estimation of 'calcium-associated' P. The 1N HCl digestate was further analysed for acid-soluble iron using A.A.S.

Eight regularly spaced sediment sections were analysed by AERE (Harwell) for total and supported ^{210}Pb. Sediment sections were dated by the 'constant rate of supply' model of Appleby and Oldfield (1978). Sedimentation rates (in dry mass per year and depth per year) were calculated from the dating and cumulative weight measurements.

A further core was obtained from site P in the Central Lake (Fig. 1) in October 1988. The core was subsectioned at 2 cm intervals, and measured volumes of wet sediment removed from each slice for determination of water content (by drying at 110 °C for 24 hours) and for analysis of diatom remains. For the latter, portions of wet sediment were treated with hydrogen peroxide and dilute HCl to remove organic matter and carbonates respectively. A portion of the material remaining was then mounted on a microscope slide for subsequent diatom identification and counting. Diatom influx has so far not been calculated.

Results

Surface sediment geochemistry

Figure 2 shows the concentrations of organic carbon, organic nitrogen and organic and inorganic phosphorus in the surface sediments at each of the four sampling sites. There is a clear trend in decreasing sediment nutrients away from the main sources of high nutrient loading (the Stackpole and Merrion Streams which enter in the north) and a distinct correlation with the trophic status (Table 1) of each arm.

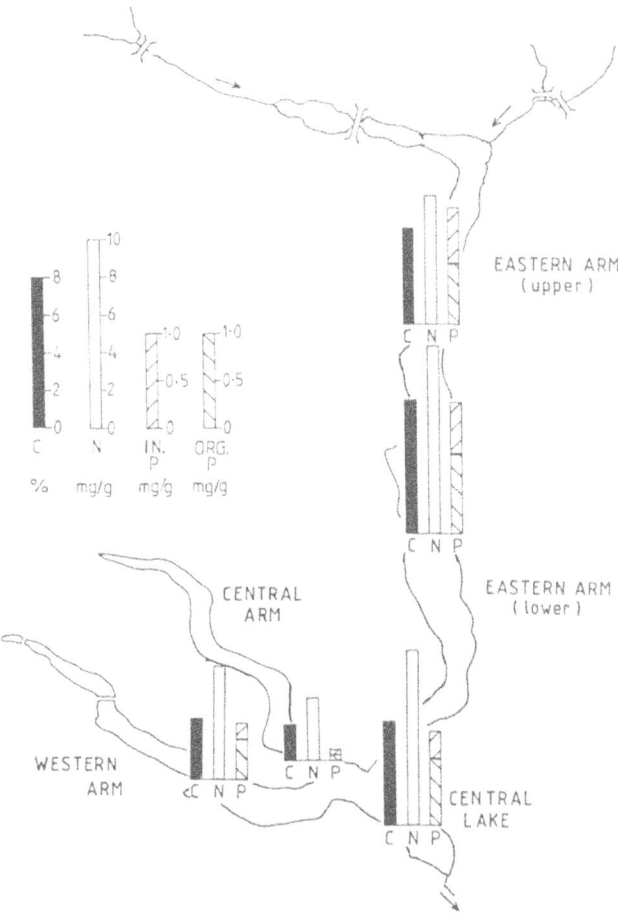

Fig. 2. Concentrations of organic carbon, nitrogen and organic and inorganic phosphorus in surface (0–1 cm) sediments of Bosherston Lakes.

Central Lake (Site P) cores

Stratigraphy

The stratigraphy consisted of a 20 cm thick surface layer of grey/brown mud, followed by 11 cm of cream/brown sediment at the base of which is a 4 cm thick section of coarse grey sand (Fig. 3a). A similar stratigraphy was also found in the core used for diatom analysis (see below), and is therefore employed as the basis of inter-core correlation. Each layer possesses a unique range of water content values (which decrease with sediment depth). The water content profile records a transition at 15–20 cm depth (ie at the base of the grey/brown layer). The coarse grey sand deposit at the base of the core is assumed to pre-date the formation of the Central Lake (in ca. 1850) when the area was open to the sea.

Lead-210 dating

Dating by ^{210}Pb gives a current (1984) sedimentation rate in the Central Lake of 1.5 kg m^{-2} a^{-1} by weight. The sedimentation rate appears to increase above 30 cm, and then remain approximately constant from 22–13 cm depth, and then to increase again near the sediment surface (Fig. 3a). The transition from estuarine to freshwater conditions (ca. 1850 AD) cannot be dated accurately (Fig. 3a) using this method as ^{210}Pb dating is inaccurate at such an age. The boundary at 20 cm depth is dated ca. 1919.

Sediment geochemistry

Organic C, organic N and organic P are all enriched in the upper layer and decrease sharply in concentration in the transition zone from 15–20 cm depth (Fig. 3b). By contrast, carbonate content increases from the upper to lower layer. In the upper layer, nutrient concentrations generally increase steadily towards the sediment-water interface. Fluctuations in the profile of organic C may be due to analytical problems experienced in the analysis of the sections from 3–5 cm (because of the very limited amount of sample available).

Organic matter C/N, C/P and N/P ratios generally increase gradually with depth in the upper layer, and then rise sharply in the transition zone (Fig. 3c). Total inorganic P, calcium-associated P and iron-associated P concentration are all significantly greater in the upper than in the lower layer (Fig. 3d).

Diatom analysis

The main trend found in the diatom record is that of a major increase and abundance of *Fragilaria* (Fig. 4). This genus accounts for approximately 50% of all diatoms recorded in almost half the samples. For the purpose of description, the diatom profile is divided into three zones – I, II and III.

Zone I (28–24 cm) contains several abundant taxa, notably *Cymbella microcephala*, *Epithemia sorex*, *E. zebra*, *Fragilaria brevistriata* and *F. pinnata*, *Gomphonema intricatum*, *Navicula microcephala* and *Nitzschia denticula*. Some of these occur in the basal sample (28–26 cm) and not in the one above. The taxon referred to here as *cf. Cyclotella* appears in the upper sample.

Zone II (24–12 cm) is characterised by abundant *cf. Cyclotella* and *C. kuetzingiana*. A peak of *Amphora ovalis*, mainly var. *pediculus* occurs at 20–22 cm. *Epithemia sorex*, *Cymbella microcephala* and *Gomphonema intricatum* decline, whereas all *Fragilaria* species expand, as does *Navicula* species A. *Stephanodiscus astrea* var. *minutula* appears.

Zone III (12–0 cm) is rich in species of *Fragilaria*, notably *F. brevistriata*, *F. construens* and *F. pinnata*. Also abundant are *Navicula* species A. and sp. B. At the top of this zone there is an increase in *Cocconeis diminuta* and in *Stephanodiscus hantzschii*, *S. parvus* and *S. tenuis*.

Throughout the profile there is clearly an abundance of pennate diatoms (Pennales). Centric diatoms (Centrales) reach a maximum in Zone II and also increase just below the sediment surface. The expansion in Zone II can be attributed to a large *Cyclotella* peak, and that in Zone III, to *Stephanodiscus* species.

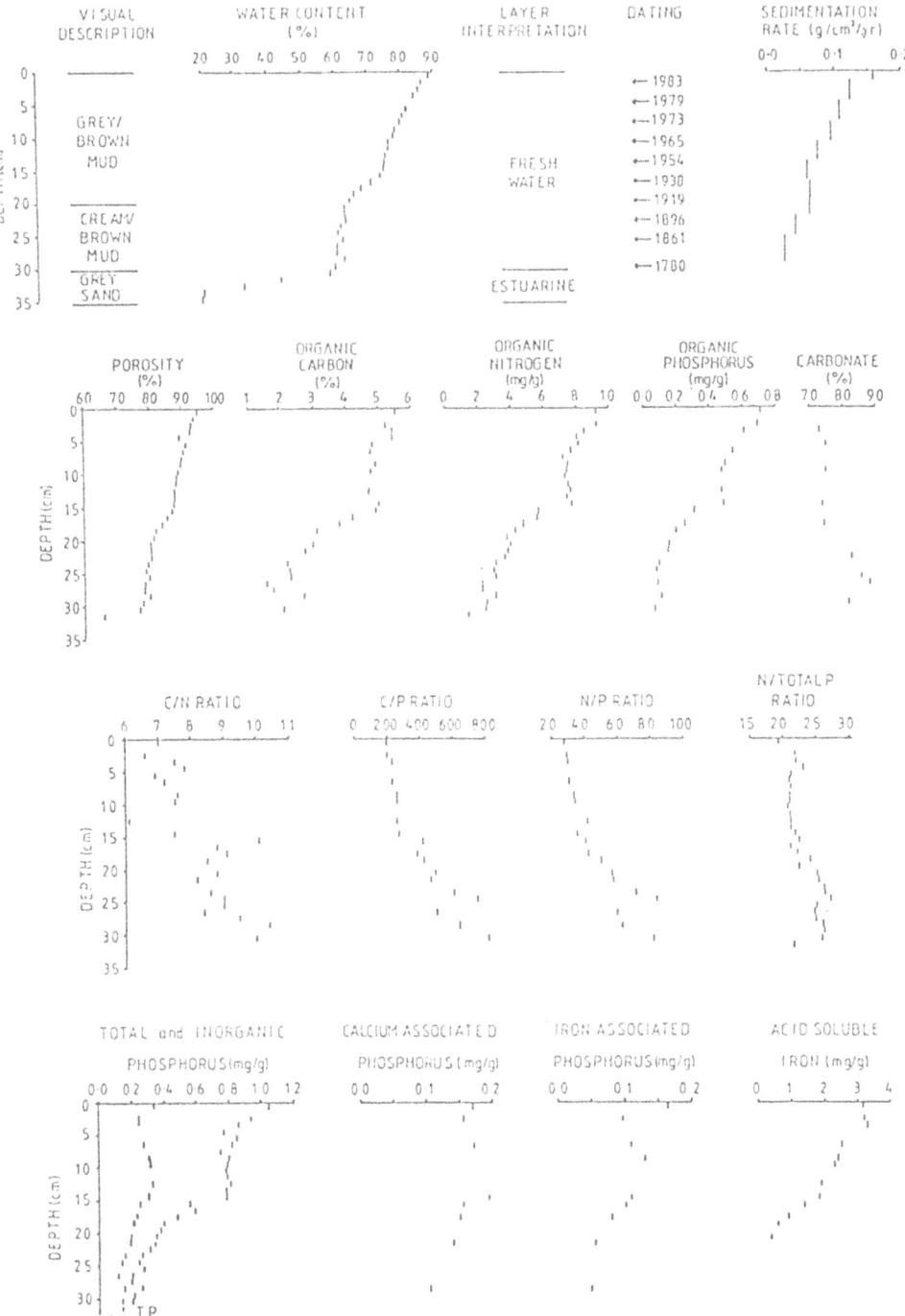

Fig. 3. Vertical variability in geochemistry of a sediment core taken at site P in the Central Lake, Bosherston Lakes.
a. Sediment appearance, water content, interpretation, dating and sedimentation rate.
b. Porosity, organic carbon, nitrogen and phosphorus and calcium carbonate.
c. Organic matter nutrient ratios (atomic).
d. Phosphorus fractions and acid-soluble iron.

177

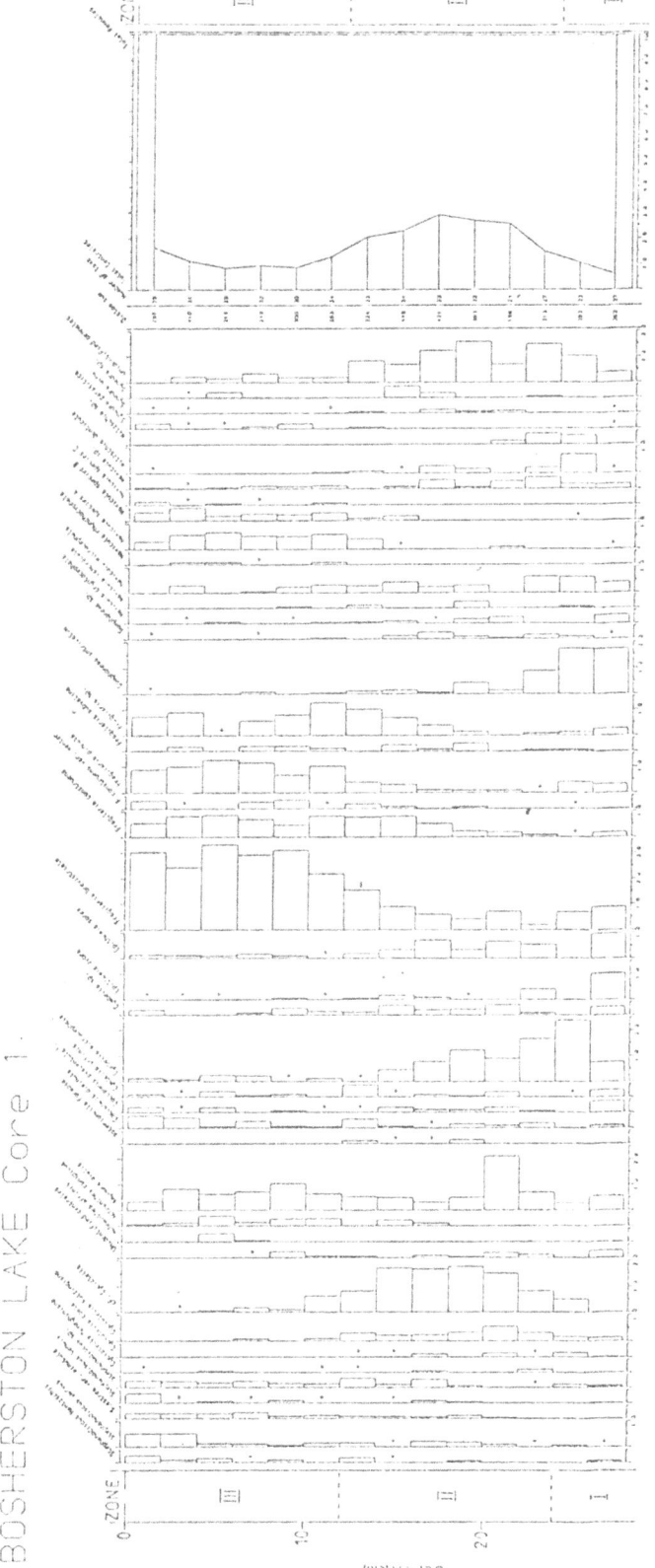

Fig. 4. Selected diatom frequencies from a core from site P, Central Lake, Bosherston Lakes.

178

Discussion

Geochemistry

It is notable that the geochemistry of the surface sediments in each of the sampling sites shows a clear correlation of sediment nutrient levels with trophic status (Fig. 2, Table 1). A clear indication of a change in trophic status in the Central Lake is recorded in the geochemical results at a transition zone between 20–15 cm depth. This is dated by ^{210}Pb to between 1919–1938. The sediment deposited below the transition is characterised (Table 2) by a high calcium carbonate content (81–88%), a relatively low porosity (78–80%), low organic carbon concentration (2.2–3.0%), low nitrogen (2.2–3.9 µg N g^{-1}), low organic phosphorus (0.08–0.16 µg P g^{-1}) and a high C/P ratio (470–750). These characteristics most closely resemble those of the sediment currently being deposited in the Central Arm (Table 2).

The cream colour of the lower sediment layer also closely resembles that of the *Chara* detritus-dominated sediment of both the Central and Western Arms. The above evidence strongly suggests that in the Central Lake initial sedimentation after the creation of freshwater conditions was overwhelmingly dominated by *Chara* detritus. *Chara* is well known as an aggressive coloniser of newly created freshwater habitats.

The sediment deposited above the transition zone (15 cm depth) possesses characteristics more akin to those of the top 15 cm of the sediments of the Eastern Arm (Table 2). However, while porosity, organic C, organic N, organic P and nutrient ratios in the top 15 cm are similar, the concentration of inorganic P and iron-associated P are lower and the concentration of calcium carbonate higher. These differences reflect the lower inorganic detrital component (particularly clays) of the sediments of the Central Lake. The similarity between the upper 15 cm of the sediments of the Central Lake, and those of the Eastern Arm must be due to the dominant influence of nutrient-rich waters from the Eastern Arm on the sedimentation processes occurring in the Central Lake.

The sediments of the Eastern Arm are devoid of *Chara* detritus. *Chara* was the dominant macrophyte species in the Central Lake in 1977–1978, although it was recorded as completely absent by 1981 (Hinton, ms). Significant quantities of aquatic angiosperms were present in the Central Lake as far back as 1967. It is thus likely that the *Chara* productivity here has been substantially lower than that of the Western and Central Arms since then. The C/P ratios of the material in the uppermost 15 cm of Central Lake sediment closely resemble those of the top 5 cm of the sediments in the Western Arm (Rees, ms). These are mainly dominated by relatively P rich *Chara* detritus with some contribution from *Myriophyllum spicatum* and *Potamogeton pectinatus* (Hinton, ms). Analysis of phosphorus concentrations in living *Chara* specimens from various sites in the Western and Central Arms shows that when exposed to higher P concentrations, it tends to incorporate more P into its biomass (Hinton & Rees, ms).

The enhanced concentrations of organic nutrients in the topmost 0–15 cm of sediment in the Central Lake must reflect the input of nutrients from the Eastern Arm. An increase in nutrient concentrations may raise the productivity of phytoplankton and nutrient-tolerant macrophyte species, which may in turn inhibit the productivity of *Chara*. The loss of *Chara* is likely to be accompanied by a shift in the mode of calcium carbonate production from biogenic to bio-induced. Sediments from the Central Arm, typical of biogenic carbonate production by *Chara*, possess a calcium carbonate content of up to 92%. At the other extreme, the sediments of the Eastern Arm, typical of bio-induced carbonate production (Rees, ms), contain around 64% carbonate. The decline in *Chara* productivity in the Central Lake is recorded by the decrease in calcium carbonate concentration at the transition zone in the sediment column.

The change in the sediments of the Central Lake is consistent with the idea of a minor influence of nutrients from the Eastern Arm pre 1919, and a major influx post 1938. Although it is likely that nutrient concentrations in the waters of the Eastern Arm will have been increasing steadily during this period (as a result of local population increase), in order to account for the observed sedimentation pattern a more abrupt change is required. The sediments of the Eastern Arm do not show comparable changes within this period (Rees, ms), thus indicating that activities in the catchment did not directly influence the Central Lake at this time. The conclusion is that before ca. 1919–1938, the tunnel at the base of the dam in the Eastern Arm was permanently closed by the sluice board, and that between 1919 and 1938 the board disintegrated. Thus, before ca. 1919, relatively nutrient-poor surface water left the Eastern Arm via a slide sluice, whereas after ca. 1938, nutrient-rich bottom water from the Eastern Arm entered the Central Lake and caused a significant increase in its trophic status.

A comparable example of preservation in the geochemical record of features indicating an increase in external nutrient loading has been found in Lake Mendota (Bortleson & Lee, 1972; Brock, 1985). The sediment of this lake is composed of a surface layer (max. 60 cm) of black sludge (*gyttja*), overlying a layer of buff marl. The *gyttja* is low in carbonate ($<35\%$) and high in organic C while the marl layer is high in carbonate ($>55\%$) and low in organic matter. The upper layer is enriched in P and iron compared to the lower layer. The change in sedimentation in Lake Mendota is closely correlated with human activity.

Diatom analysis

The results of the diatom analyses suggest that the sedimentation history of the Central Lake may be divided into at least three periods. These are represented by the three zones in the diatom diagram (Fig. 4). Full interpretation of the diatom data is however, constrained by the fact that the results show only relative, not absolute, differences, in the abundance of each taxon.

Zone I displays a mixed, complex diatom assemblage with some variability in trophic and salinity preferences. Of the major species present,

Epithemia zebra, E. sorex, N. cryptocephala, Gomphonema intricatum and *Cocconeis placentula*, are tolerant of fresh to brackish waters. *Cymbella microcephala* and *Nitzschia denticula* are, however, freshwater species. These become more abundant in the top half of this zone. The presence of abundant *Nitzschia denticula* and *Gomphonema intricatum* indicates relatively oligotrophic conditions.

Zone II is distinguished by the relative abundance of *Cyclotella* and *Cymbella microcephala*. Zone II appears to be one of transition with notable decreases in the relative abundance of a variety of taxa, and increases in others. Conditions appear to be favourable to the planktonic *Cyclotella* which rise to a peak and decline as epiphytic *Fragilaria* become more abundant. The rapid decline in *Gomphonema intricatum* and in *Cymbella microcephala* also indicates changing conditions. The decrease in *G. intricatum* suggests an increase in trophic status although the high relative abundance of *Cyclotella* indicates that conditions were still fairly mesotrophic. The rise in abundance of species of the epiphytic *Fragilaria* denotes the existence of a lake whose waters were still relatively clear, and which therefore supported substantial numbers of macrophytes (Crabtree & Round, 1967, Moss, 1983).

In Zone III, *Fragilaria* remains the most abundant taxon, but members of the Centric genus *Stephanodiscus* become much more prominent in the record towards the sediment surface. This, and the increase in abundance of other alkalibiontic taxa, and those associated with increases in nutrient status, indicates the onset of eutrophication, and the change in the Central Lake from conditions most conductive to the growth of *Chara* to those which favour phytoplankton. This event, which coincides with major stratigraphic and geochemical changes, took place, according to the ^{210}Pb dates, in the period 1919 to 1938, when it is thought that installations controlling the water level in the Lakes, and essentially separating each water body from the others, fell into disrepair (see above).

Conclusions

Part of Bosherston Lakes contains a sedimentary record which demonstrates the geochemical and ecological changes which accompany eutrophication. With the onset of this process, sediment quality changes from a marl, rich only in refractory organic matter, to a *gyttja* in which algal detritus and other fresh, labile organic material are abundant. The mechanism for this change is an increase in external nutrient loading. The sedimentary diatom record contains complementary changes, from a mixed flora deposited in the early history of the system, to one in which epiphytic species of *Fragilaria* are abundant. These probably grew initially on *Chara* and then on the more nutrient tolerant macrophytes that have progressively replaced it. Just below the sediment surface, members of the genus *Stephanodiscus* increase in frequency, indicating the onset of advanced eutrophication. Lead-210 dating confirms that this process began in the interval between the two World Wars, when the intensity of management of the estate on which the Lakes are located, declined.

Acknowledgements

The authors would like to thank the Stackpole Nature Conservancy Council warden R. Haycock for his invaluable help in this study, Welsh Water Authority for carrying out water chemistry analyses, and R. Davies, G. Owrid, K. Nolan, T. Denton and T. Patel for help in the field and the laboratory. The work described herein forms part of the Bosherston Lakes Study, funded by the Chief Scientist's Directorate of the Nature Conservancy Council (contract No. HF3/03/276), and awarded to the former Department of Oceanography, University College of Wales, Swansea.

References

Appleby, P. G. & F. Oldfield, 1978. The calculation of lead-210 dates assuming a constant rate of supply of unsupported ^{210}Pb to the sediment. Catena 5: 1–8.

Aspila, K. I., H. Agemian & A. S. Y. Chau, 1976. A semi-automated method for the determination of inorganic, organic and total phosphate in sediments. Analyst 101: 187–197.

Black, C. A., 1965. Methods of soil analysis. Am. Soc. Agronomy. Monograph No. 9, Madison, Wisc., USA. 1572 pp.

Bortleson, G. C. & G. F. Lee, 1972. Recent sedimentary history of Lake Mendota. Envir. Sci. Technol. 6: 799–808.

Bremner, J. M., 1965. Inorganic forms of nitrogen. In C. A. Black, (*op. cit.*): 1179–1237.

Brock, T. D., 1985. A eutrophic lake, Lake Mendota, Wisconsin. Ecological Studies 55, Springer Verlag.

Crabtree, K. J. & F. E. Round, 1967. Analysis of a core from Slapton Ley. New Phytologist 66: 255–270.

Keeney, D. R., J. G. Konrad & G. Chesters, 1970. Nitrogen distribution in some Wisconsin Lake sediments. J. Water Poll. Contr. Fed. 42: 411–417.

Lennox, L. J. & M. J. Flanagan, 1982. An automated procedure for the determination of total Kjeldahl nitrogen. Wat. Res. 16: 1127–1133.

Moss, B., 1983. The Norfolk Broadland: experiments in the restoration of a complex wetland. Biol. Rev. 58: 521–561.

Petts, K. W., 1979. The determination of ammonia in estuarine water by autoanalyser. Water Research Centre Technical Report TR 119, 24 pp.

Hydrobiologia **214**: 181–186, 1991.
J. P. Smith, P. G. Appleby, R. W. Battarbee, J. A. Dearing, R. Flower, E. Y. Haworth, F. Oldfield & P. E. O'Sullivan (eds), 181
Environmental History and Palaeolimnology.
© 1991 *Kluwer Academic Publishers.*

Recent changes to upland tarns in the English Lake District

Elizabeth Y. Haworth & Jean P. Lishman
Institute of Freshwater Ecology, Ambleside, Cumbria, UK

Key words: palaeolimnology, diatoms, acidification, trace metals

Abstract

Studies of diatom assemblages in the sediments of three lakes with small catchments in west Cumbria show recent increases in proportions of acid-tolerant taxa. These changes are correlated with the steep increases in atmospheric pollutants. In other sites with better catchment soils, diatom assemblages are less changed.

Introduction

In her diatom study of Lake District tarns, Knudson (1954) found that Scoat Tarn was one of only two lakes in the area where the acidic species – *Tabellaria binalis* – occurred. This was the first British record of this species (Knudson, 1952). We have therefore examined the sedimentary record in order to find out whether the lake has always had an acidic flora and if not, when it changed. We have also examined sediment cores from two nearby lakes, to see whether recent palaeolimnological records are similar and, since the catchments appear little changed over the last few centuries, if this is related to atmospheric pollution as inferred from other lakes in Scandinavia and Scotland (Renberg & Hellberg, 1982; Battarbee *et al.*, 1985).

The three small lakes near Wastwater, on the western side of Cumbria, occupy adjacent catchments within 10 miles of the coast (Fig. 1). They lie on Borrowdale Volcanic rocks of Ordovician age, which are acidic andesite lavas and are covered by shallow, acid, peaty soils which now support a treeless rough grassland. Scoat Tarn is a typical corrie lake, while Greendale and Low Tarns occupy deepened moraine hollows. All

have pH of c. 5.0 and negative alkalinity (Sutcliffe & Carrick, 1988).

Methods

In 1984, 1 m cores of sediment were collected from Scoat, Low and Greendale Tarns in the Wastwater catchment (Nat. Grid. Ref. NY150090). This material was sliced into 0.5 cm or 1 cm slides for diatom, geochemical and other analyses. According to ^{14}C dating of material from the bottom of the Scoat Tarn profile (S.R.R., East Kilbride dates:

  50–52 cm = 650 AD ± 70,
  73–75 cm = 910 AD ± 70,
  78–80 cm = 950 AD ± 70,
  82–84 cm = 930 AD ± 80)

the c. 85 cm includes the environmental record of the last c. 1000 years (Haworth *et al.*, 1987). The more recent sediments have been dated by AERE, Harwell and Liverpool University (Fig. 2) using ^{210}Pb and ^{137}Cs analyses. The geochemical analyses have included trace metals as well as major elements and these concentrations were measured using an Erba elemental analyser (Haworth *et al.*, 1988). Biogenic silica and

182

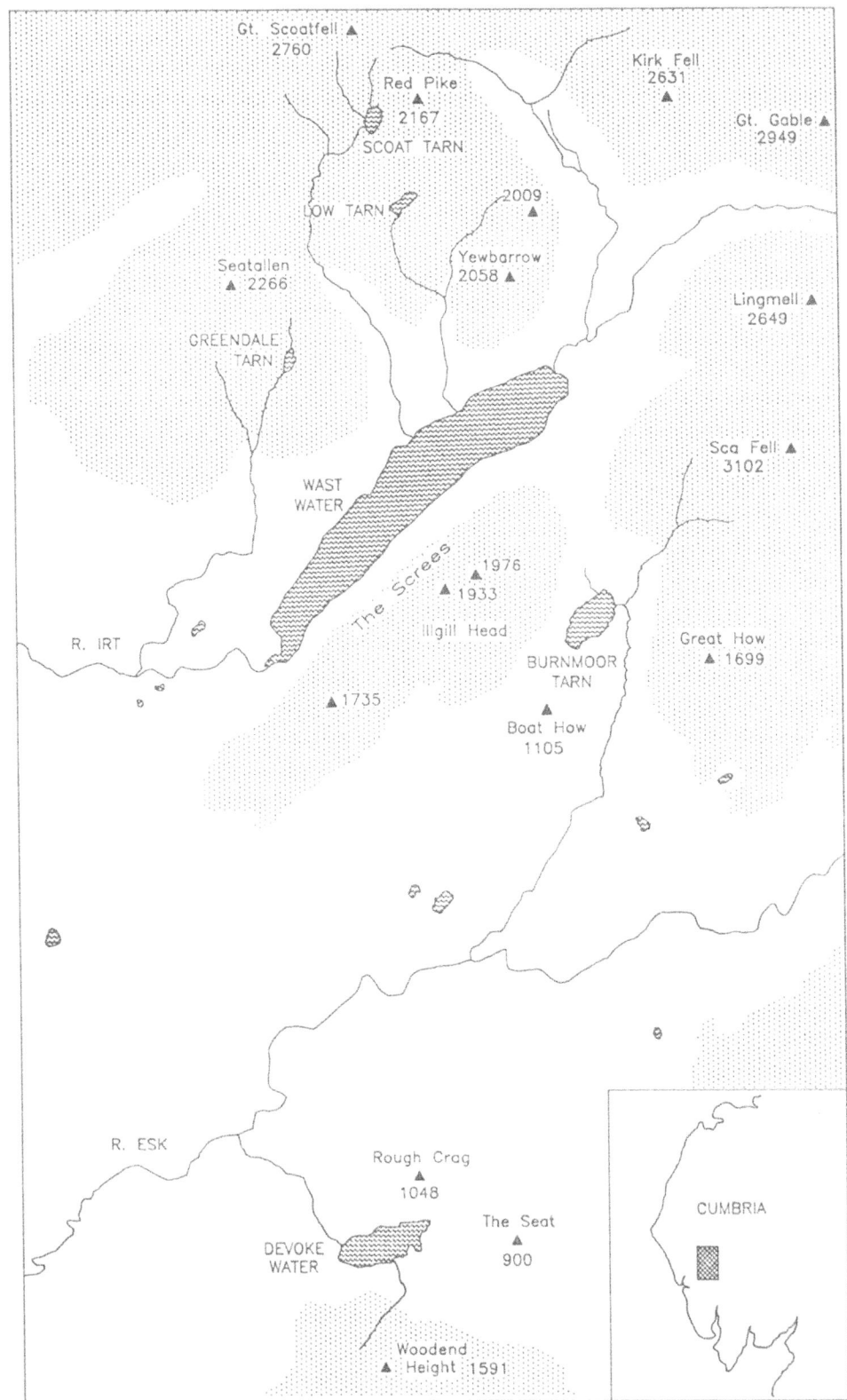

Fig. 1. Map of Wastwater and Eskdale catchments showing locations of the sites in Cumbria (inset).

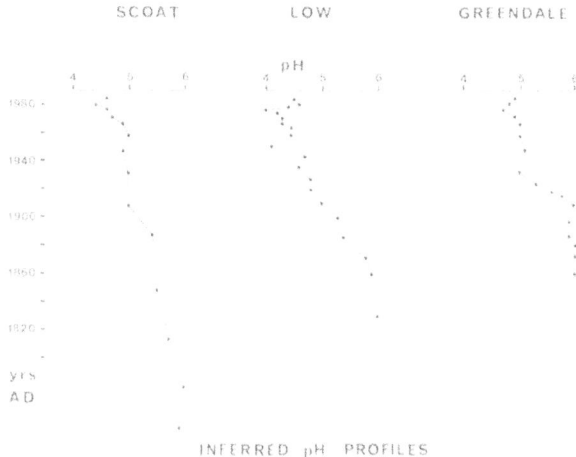

Fig. 2. Inferred pH profiles from Wasdale lake sediments.

mineral magnetics have also been studied, as have the pollen and macroscopic plant remains (Haworth *et al.*, 1987, 1988, & in prep.). Percentage diatom analyses were made on prepared microscope slides containing a measured volume of acid-cleaned sediment, following the method in Haworth (1984). Past changes in pH were calculated from the diatom assemblages using the Index B equation of Renberg & Hellberg (1982).

Results

Analysis of the diatom assemblages show that, although each tarn has had a slightly different flora, there is an overall similarity in the pattern of change within the last c. 1000 years. Acidophilous taxa have clearly increased (Fig. 3), both

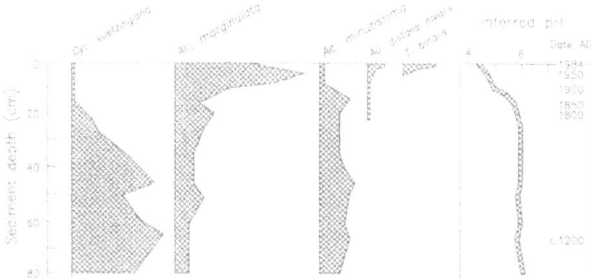

Fig. 3. Diatom profiles showing the increase in the acidic taxa in Scoat Tarn.

in percentage of each sample and in absolute terms, over the last c. 200 years (Haworth *et al.*, 1987). These include *Frustulia rhomboides*, *Aulacoseira distans* var. *nivalis*, *Navicula leptostriata*, *Tabellaria binalis*, *T. quadriseptata*, *Achnanthes marginulata*, *Cymbella aequalis*.

In the deepest lake, Scoat Tarn, there was a dominance of *Cyclotella kuetzingiana* plankton in the lower part of the core but this is not found in the shallower sites. Commonly, and specifically in local lakes, significant populations of this species only occur where the pH is >6.0 and alkalinity >30 μeq l^{-1}. In the Wasdale tarns, the upper, post-1800's sediment horizons include a marked decline of this taxon, together with *Cymbella lunata*, *Achanthes minutissima* and *Fragilaria virescens* var. *exigua*, all circumneutral taxa.

The answer to the initial question of the timing of any change in diatom assemblages is that the sediment record shows that *Tabellaria binalis* had only just become a frequent component of diatom assemblages in Scoat Tarn when Knudson (1954) made her survey in 1950. The taxon is now also present in Greendale and Low Tarns, as well as in several other lakes, having arrived more recently (Haworth *et al.*, 1987, 1988).

The profile of inferred pH shows a clear decline in the pH of all three lakes (Fig. 2) of >1.5 pH units since c. 1850 AD. There is some gradation in the onset and severity of this acidification (Haworth *et al.*, 1988), as the Scoat Tarn pH appears to have declined very early in the 1800's, while the change did not occur in Greendale until c. 1900 and the shallow Low Tarn, with an area of peatbog on its catchment, appears to have become more acidic than the others.

A few remnant sediment samples from a core of the whole post-glacial, collected by W. Tutin in 1964, show that Low Tarn, at least, never passed through the *Fragilaria*-dominated, alkaliphilous diatom assemblages typical of the early post-glacial period in more lowland lakes (Evans, 1961; Haworth, 1985).

Chemical analysis emphasizes the peaty nature of the catchment soils, with >20% carbon in these sediments. The profiles show that there has been little change in soil constituents, except for

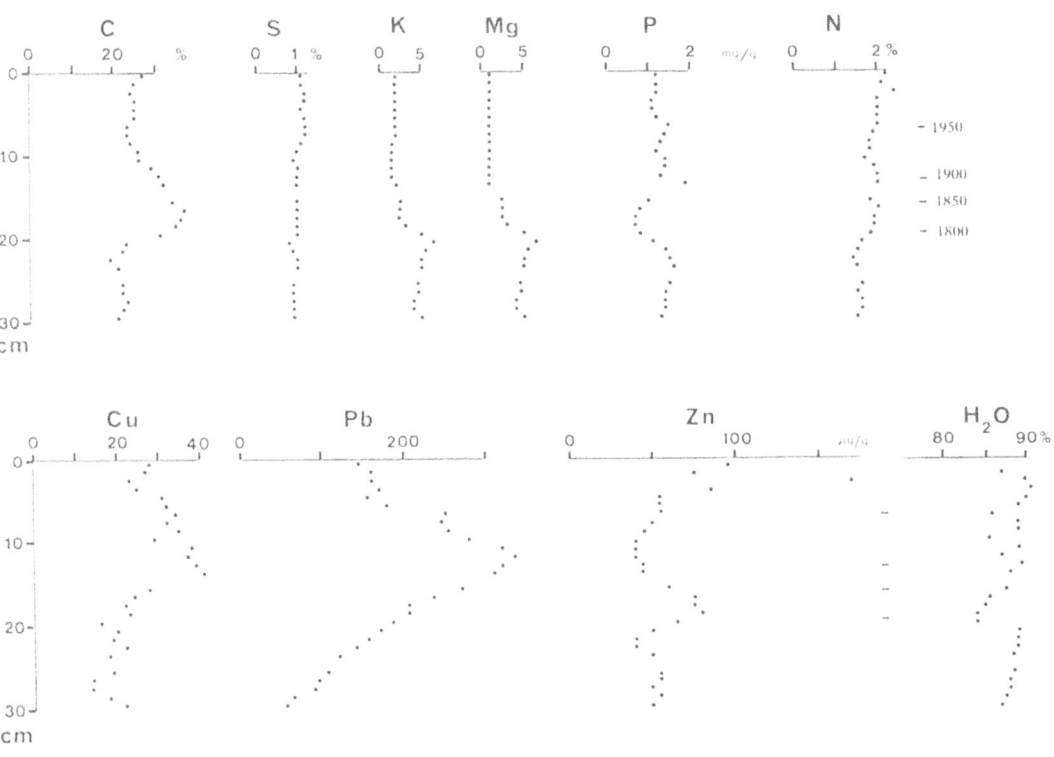

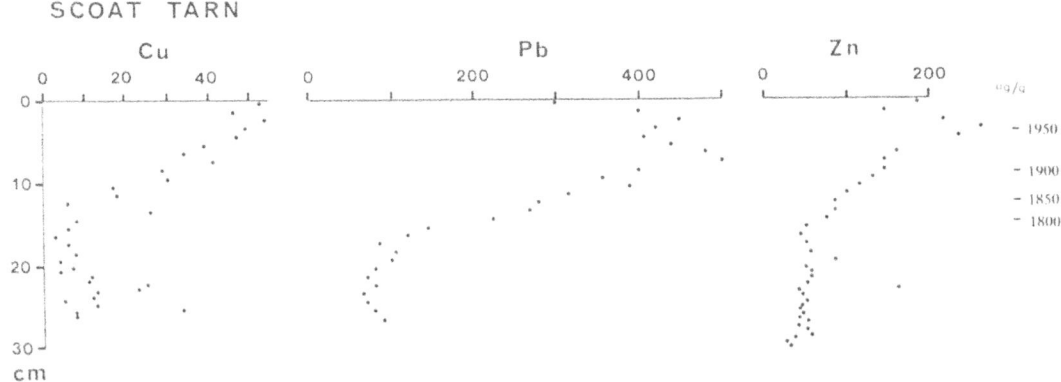

Fig. 4. Trace elements in the recent sediments.

an horizon of increased minerogenic potassium and magnesium between 55 and 65 cm in Scoat Tarn in which the ^{14}C date is older than expected (650 AD $\pm$ 70), when the pollen profile suggests a date c. 1300 AD (Pennington, pers. comm.). In the upper sediments there are clear increases of the concentrations of copper (Cu), lead (Pb) and zinc (Zn) (Fig. 4) from the background levels of the pre-1850 material. Pb concentrations have increased greatly since 1800, with the highest concentration in the deepest site, Scoat Tarn. The subsequent decreases are the consequence of recent increased sedimentation rates, especially in Low Tarn, and this dilution emphasizes the atmospheric source of this input. Cu increased from 1900 onward, while the Zn profile shows a similar rise in Scoat and Greendale and a less obvious and later increase in Low Tarn, however Zn is known to be more readily remobilized from sediments in more acid water (Rippey, 1990) and this profile has to be interpreted in relation to the pH decline (Fig. 2) which suggests an initial rise c. 1800 when the pH of the lake water was still c 6.0, with a decline suggesting remobilization when the pH dropped below 5.0. It is more obvious here than in the other sites because of the greater accumulation of sediment and earlier decline in pH.

In these small, montane lakes there is little evidence for any catchment changes to account for the decreased pH and alkalinity that caused such changes in diatom assemblages. Records show that sheep and cattle have been pastured in the area since the 12th century AD (Tallantire, in Haworth *et al.*, 1988), although it is probable that there has been some over-grazing during the last few decades, for *Calluna* has clearly declined in the Wastwater catchment, as elsewhere in the Lake District. Pennington (pers. comm.) has found a change from *Calluna* to Gramineae dominance in the pollen of the upper 10 cm and Tallantire (pers. comm.) records a similar decline in the amount of *Calluna* macroscopic remains. This vegetation change is found in uplands elsewhere, eg. Galloway (Battarbee *et al.*, 1985).

These three catchments are clearly more susceptible to acidification by virtue of the low buffering capacity of their local soils. Two lakes in the nearby Eskdale catchment area, Burnmoor Tarn and Devoke Water are not only larger and occur at a lower altitude but they also have better soils on a greater amount of glacial drift. The diatom assemblages therefore show little or no change through the last 1000 years with *Cyclotella* spp. dominant. Only in Devoke is there any recent increase in acidophilous taxa (Atkinson & Haworth, 1990). Here too there is evidence of the increase in atmospheric pollutants in the form of higher concentrations of Cu, Pb and Zn and a change in magnetic minerals (Hilton, pers. comm.).

All these lakes are within 5 miles of each other and there have been no significant land-use changes in the area. Over the past two hundred years the Cumbrian coast has been an important industrial area – shipbuilding in local ports, especially in Barrow-in-Furness, iron-mining at Broughton and Cleator, coal and steel near Workington (Marshall & Davies-Shiel, 1977), as well as nuclear power generation at Sellafield – all have contributed to the general atmospheric pollution borne on the prevailing SW and NW winds, in addition to material carried from industrial areas of the UK out into the Atlantic on easterly anticyclonic winds to return with cyclonic westerlies.

There are, however, further signs of change, with a slight increase in the diatom inferred pH in the most recent sediments of all these Wasdale tarns (Fig. 2). This change is still within the limits of statistical error and only subsequent monitoring will prove positive.

Acknowledgements

Grateful thanks are due to the landowners (the National Trust and Muncaster Estate); RAF Boulmer and the coring team of P.R. Cubby, P.V. Allen and others; J.D. Eakins & P.G. Appleby (for isotope analysis); Prof. W. Pennington, Dr P. Tallantire and J. Hilton. This research was funded by DoE, the Royal Society S.W.A.P. and N.E.R.C.

186

References

Atkinson, K. M. & E. Y. Haworth, 1990. Devoke Water and Loch Sionascaig: the post-glacial overview. Phil. Trans. R. Soc. B. 327: 341–355.

Battarbee, R. W., R. J. Flower, A. C. Stevenson & B. Rippey, 1985. Lake acidification in Galloway: a palaeoecological list of competing hypotheses. Nature 314: 350–352.

Evans, G. H., 1961. A study of the diatoms in a core from the sediments of Devoke Water. M. Sc. Thesis, Univ. Coll. Wales, Aberystwyth,

Haworth, E. Y., 1984. Stratigraphic changes in algal remains (diatoms and chrysophytes) in the recent sediments of Blelham Tarn, English Lake District. In Lake Sediments and Environmental History, Haworth, E. Y. & J. W. G. Lund, (eds) Leicester Univ. Press: 165–190.

Haworth, E. Y., 1985. 'The highly nervous system of the English Lakes': aquatic ecosystem sensitivity to external changes, as demonstrated by diatoms. Rep. Freshwat. Biol. Ass. 53: 60–79.

Haworth, E. Y., K. M. Atkinson & E. M. Riley, 1987. Acidification in Cumbrian waters: past and present distribution of diatoms in local lakes and tarns. Report to Dept. of the Environment, pp. 107.

Haworth, E. Y., J. P. Lishman & P. Tallantire, 1988. A further report on the acidification of three tarns in Wasdale, Cumbria, north-west England. Report to Dept. of the Environment. pp. 34.

Knudson, B., 1952. The diatom genus *Tabellaria*. I. Taxonomy and morphology. Ann. Bot. NS 16: 421–440.

Knudson, B., 1954. The ecology of the diatom genus *Tabellaria* in the English Lake District. J. Ecol. 42: 345–358.

Marshall, J. D. & M. Davies-Shiel, 1977. Industrial Archaeology of the Lake Counties. M. Moon, Beckermet, (2nd edition).

Renberg, I. & T. Hellberg, 1982. The pH history of lakes in southwestern Sweden, as calculated from the subfossil diatom flora of the sediments. Ambio 11: 30–33.

Rippey, B., 1990. Sediment chemistry and atmospheric contamination. Phil. Trans. R. Soc. B. 327: 311–317.

Sutcliffe, D. W. & T. R. Carrick, 1988. Alkalinity and pH of tarns and streams in the English Lake District (Cumbria). Freshwat. Biol. 19: 179–189.

Hydrobiologia **214**: 187–190, 1991.
J. P. Smith, P. G. Appleby, R. W. Battarbee, J. A. Dearing, R. Flower, E. Y. Haworth, F. Oldfield & P. E. O'Sullivan (eds), 187
Environmental History and Palaeolimnology.
© 1991 *Kluwer Academic Publishers.*

Palaeolimnological study of an environmental monitoring area, or, Are there pristine lakes in Finland?

Heikki Simola, Pertti Huttunen, Jukka Rönkkö & Pirjo Uimonen-Simola
Univ. Joensuu, Karelian Inst., Sect Ecology and Dept Biology, P.O. Box 111, SF-80101 Finland

Key words: palaeolimnology, integrated monitoring, heavy metals, ^{210}Pb, diatoms, Cladocera

Abstract

A palaeolimnological study of Lake Iso-Hietajärvi in Patvinsuo National Park, Lieksa, East Finland was conducted. The drainage area of the lake is one of four Integrated Monitoring areas established in Finland. Lead-210 dating reveals a period of increased sedimentation in the lake from 1920 to 1950. Increased atmospheric burden of several heavy metals in the order Pb > Cu > Zn > Ti = Al > Cr = Ni = V is recorded. The first to expand during the 19th century is Pb, whilst V increases after 1950. Sedimentary chlorophyll derivatives expand in the early part of the 20th century. Assemblages of diatoms and Cladocera were also changed somewhat during this time, but water quality seems not to have varied much: e.g. the diatom-inferred pH has remained in the range 6.4–6.8 (with a slight decrease) throughout the period of study.

Introduction

Integrated Monitoring is a newly established international programme aimed at observing atmospheric pollution and its effects on natural ecosystems (WIM II, 1989). The Monitoring Area Network is composed of small natural drainage basins, in which atmospheric deposition, hydrology, and many ecosystem parameters will be monitored according to an internationally consented protocol. Four such monitoring areas are being founded in Finland by the Ministry of the Environment. All of these include a lake, and a palaeolimnological survey is planned for each site. The present study is the first one completed.

The main aims of the palaeolimnological study are twofold:

1. To determine whether the area is pristine enough to qualify for integrated monitoring. If the stratigraphic sediment analyses revealed some major, hitherto unknown changes, e.g. considerable local pollution, the suitability of the area might be questioned.

2. To establish the history and magnitude of environmental change in the area. Many of the changes that the Integrated programmes aim to monitor are subtle, began decades or even centuries ago. Stratigraphic analyses of lake deposits may be the only source of information on the background concentrations of pollutants, and of the early phases of change.

Study site, material and methods

The Hietajärvi Integrated Monitoring Area contains the 3.3 km² drainage area of Lake Hietajärvi (63° 10′ N, 30° 42′ E) and lies entirely within the boundaries of Patvinsuo National

Park, which was founded in 1985. The area is unique for southern Finland in that it forms a reasonably sized paludified drainage area in which no peatland ditching has taken place; some 80% of southern Finnish peatlands have been ditched during this century.

Hietajärvi is an oligotrophic mesohumic lake (total P 5–15 μg l^{-1}, colour 30 mg Pt l^{-1}, pH 6.5–6.8) with a surface area of 0.9 km^2 and maximum depth of 7 m. The present study focuses on the recent palaeolimnology of Iso-Hietajärvi, with special interest on the lake's development during the industrial era.

Samples of sediment were collected in May 1987 at the deepest part of the lake using a large-diameter Kajak-corer. The cores were extruded and sampled in the field: 0–5 cm in 0.5 cm slices, 5–15 cm in 1.0 cm slices and 15–30 cm in either 5 cm slices (chemistry) or 1 cm slices at every 5 cm (diatoms and Cladocera; the lowermost level for these was 27–28 cm). Standard chemical procedures were applied for the element and pigment analyses (Bengtsson & Enell, 1986). Lead-210 analysis was carried out at the Technical Research Center of Finland (Häsänen, 1977); the dates were derived from the CRS model (Appleby & Oldfield, 1978). Diatom slides were made by HNO$_3$ and H$_2$SO$_4$ treatment. At least 200 frustules were counted in randomly chosen fields at each level. Cladocera were prepared by heating the samples in KOH, sieving them with 25 μm mesh and strewing the residue onto slides in glycerol. Usually several slides at each level were screened for Cladocera, until at least 100 Chydorids had been recorded, which normally brings the total count, with Bosminids etc., up to some hundreds.

Results

Sediment chemistry of Iso-Hietajärvi is presented in Fig. 1, which includes the ^{210}Pb dates, stratigraphies of various elements (concentrations in dry sediment) and the pigment concentrations.

Within the lower half of the profile, 30–15 cm, ignition residue and element concentrations gradually decline, but from 15 cm upwards concentrations of many of the determinands increase. A marked rise in the sediment accumulation rate takes place in the interval 8–4 cm or 1920–1950. The conspicuous surface maxima of P, Fe and Mn are a dynamic feature, connected with the oxic sediment surface; increase of K indicates erosion. The expansion of most heavy metals, beginning during the 1800's (Pb) or later, may be attributed to atmospheric fallout. Increase of chlorophyll derivatives could indicate

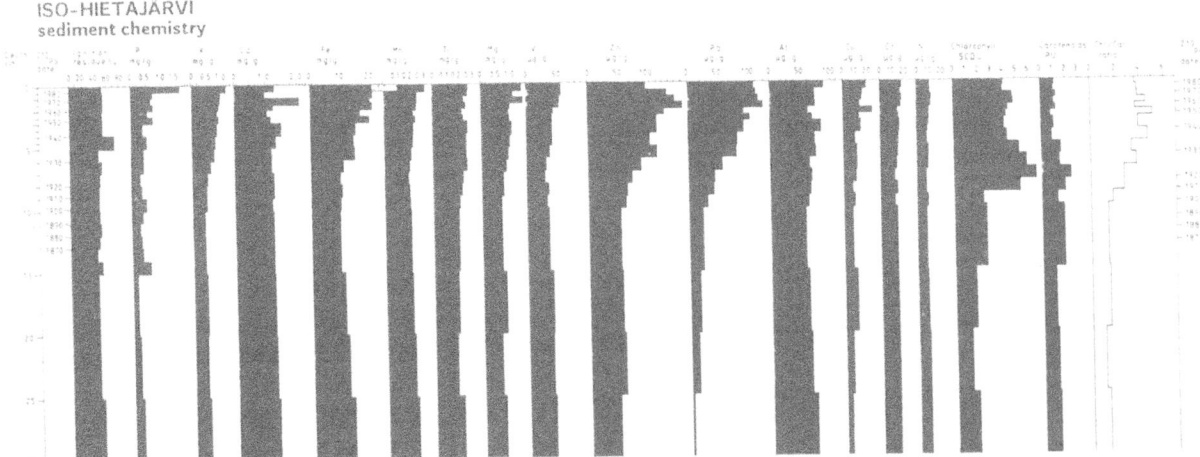

ISO-HIETAJARVI
sediment chemistry

Fig. 1. Sediment dating (CRS), stratigraphy of some elements and chlorophyll and carotenoid units as well as chlorophyll/carotenoid-ratio within the uppermost 30 cm of sediment of Lake Iso-Hietajärvi. Interpretation in the text.

Table 1. Atmospheric fluxes of heavy metals into the sediment of Iso-Hietajärvi. For each metal, a background flux was established from the pre-1800 concentrations and estimated dry matter deposition (45 g m^{-2} yr^{-1}). An excess flux, taken as atmospheric fallout was then calculated for the dates 1900, 1950 and 1980, by subtracting the background (pre-1800 concentration times sedimentation rate at each level) from the total flux at the corresponding levels. In the Table, the left-hand column for each element gives the background flux and the excess fluxes (if any) for the given times as mg m^{-2} yr^{-1}, and the right-hand column gives the excess as percentage of the total flux at that level.

	Date	Pb		Cu		Zn		Ti	
		Flux	%	Flux	%	Flux	%	Flux	%
Excess flux	1980	3.2	84	0.36	57	1.9	49	3.9	17
	1950	4.9	82	0.47	47	3.6	49	5.9	17
	1900	0.5	39	0.05	12	–	–	1.8	0.8
Background	pre-1800	0.7		0.35		2.6		9.1	

	Date	Al		Cr		Ni		V	
		Flux	%	Flux	%	Flux	%	Flux	%
Excess flux	1980	0.6	21	0.12	19	0.07	18	0.08	24
	1950	1.3	23	0.14	13	–	–	–	–
	1900	–	–	–	–	–	–	–	–
Background	pre-1800	3.0		0.65		0.42		1.9	

eutrophication; constantly low levels of carotenoids show throughout the studied sequence good oxygen conditions.

Lead-210 dating allows transformation of concentrations into yearly fluxes, and thus estimation of the history and present rates of atmospheric fallout to the lake (Table 1). The general pattern is consistent with that observed in other oligotrophic headwater lakes in Finland (Verta *et al.*, 1989).

Both diatoms and Cladocera show stratigraphic changes around the middle of this century that indicate some slight disturbance of the biotic communities (Figs. 2 and 3). We attempted reconstruction of pH and water colour from both diatom and Cladoceran assemblages with CCA (Canonical Correspondence Analysis) using a data base collected from 60 Eastern Finnish lakes (Huttunen & Meriläinen, 1986). The pH-inferences by both groups merge together (octave-scaling of species proportions), indicating very slight upcore decline of pH, with lowest values

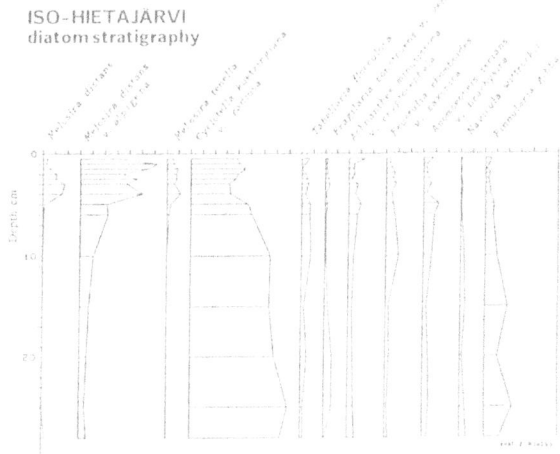

ISO-HIETAJÄRVI
diatom stratigraphy

Fig. 2. Diatom stratigraphy of Iso-Hietajärvi. Selected taxa only; percentage diagram. Substitution of *Cyclotella kuetzingiana* by *Melosira* spp. may indicate slight eutrophication and/or water level lowering. There is no indication of recent acidification. Complete species list is given in Simola (1987).

190

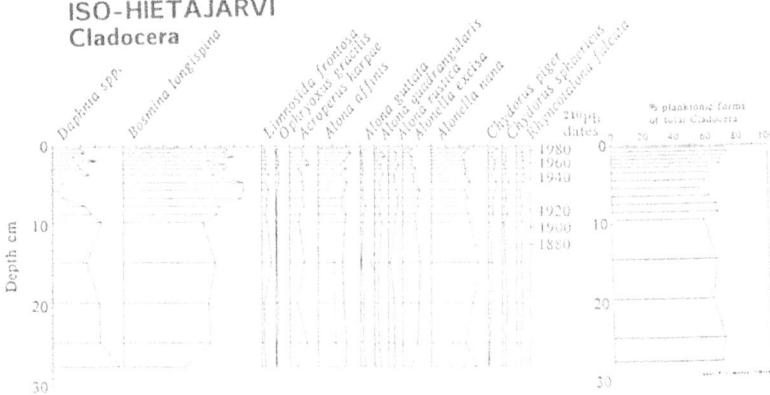

Fig. 3. Stratigraphy of Cladocera (water-fleas) in the recent sediment of Iso-Hietajärvi. Percentage diagrams: A: selected taxa; B: proportion of plankton forms. A peak of *Bosmina* coincides with the major change in diatoms; the littoral assemblages have remained stable throughout the studied interval.

6.4–6.6 between 10 and 3 cm. As to water colour, however, the two groups gave contradictory inferences: according to diatoms the colour would have risen within the core while Cladocera indicated the opposite. Possibly the Cladocera have not reacted to changing colour, but to lowered water level (i.e. more extensively illuminated lake bottom), which can be postulated from other analyses.

Conclusions

The following conclusions may be drawn:

1. The lake has been, and still is, a clearwater oligotrophic lake. There is some indication of slight eutrophication (chlorophyll increase around 1910) and possibly dystrophication (diatom-colour inference) around 1930, which could be related to water level lowering.

2. From the 1920's to late 1940's the sediment accumulation rate was high. It is possible that the lake level in that period was regulated to facilitate timber floating in the stream system downstream.

3. Concentrations of heavy metals, especially Pb, Zn and Cu, clearly increase towards the sediment surface. Dating of the sediment (^{210}Pb; CRS) made it possible to estimate background fluxes, as well as recent increases that can be attributed to atmospheric fallout (Table 1).

4. Although chosen for a monitoring site as an undisturbed drainage basin, the area has clearly suffered from human influence, both locally (possible water-level lowering) and generally, by atmospheric pollution. It may indeed be impossible to find truly pristine watershed ecosystems in Finland.

Acknowledgements

This study has been funded by the Ministry of Environment, Finland.

References

Bengtsson, L. & M. Enell, 1986. Chemical analysis. In B. Berglund (ed.), Handbook of Holocene palaeoecology and palaeohydrology. John Wiley & Sons, Chichester: 423–451.

Huttunen, P. & J. Meriläinen, 1986. Applications of multivariate techniques to infer limnological conditions from diatom assemblages. Dev. Hydrobiol. 29: 201–211.

Simola, H., 1988. Paleolimnological study of Hietajärvi area of integrated monitoring (in Finnish). Ministry of Environment, Finland, Publications of Environmental Protection Department D46: 1–19.

Verta, M., K. Tolonen & H. Simola, 1989. History of heavy metal pollution in Finland as recorded by lake sediments. Science of the Total Environment 87/88: 1–18.

WIM 2, 1988. Second Workshop on Integrated Monitoring, 5–8 October 1988, Finland. Ministry of Environment, Finland, Publications of Environmental Protection Department D55: 1–84.

Hydrobiologia **214**: 191–199, 1990.
J. P. Smith, P. G. Appleby, R. W. Battarbee, J. A. Dearing, R. Flower, E. Y. Haworth, F. Oldfield & P. E. O'Sullivan (eds), 191
Environmental History and Palaeolimnology.
© 1990 *Kluwer Academic Publishers.*

The eutrophication history of Lake Särkinen, Finland and the effects of lake aeration

Olavi Sandman[1], Kristiina Eskonen[2] & Anita Liehu[3]
[1]*Mikkeli Water and Environment District, Box 77, SF-50101 Mikkeli, Finland;* [2]*Kylätie 10 B 8,*
SF-82300 Rääkkylä, Finland; [3]*Technical Research Center of Finland, Reactor Laboratory, Otakaari 3*
A, SF-02150 Espoo, Finland

Key words: paleolimnology, lake aeration, eutrophication

Abstract

Lake Särkinen is a small lake in the parish of Sotkamo, Finland. The lake has been strongly enriched since the middle of the 1960's. The nutrient load was greatly reduced in 1969 and aeration was started in 1980.

According to ^{210}Pb dating sediment accumulation rates are lowest (ca 9 mg cm^{-2} yr^{-1}) between about 1920 and 1960. Thereafter they rise to the present level (22 mg cm^{-2} yr^{-1}).

The diatom flora indicates rising eutrophy from the beginning of the 20th century and again in the 1950-60's period. The surface sample, which represents the 1980's, shows a change in diatom flora indicating lake recovery. Changes in nutrient concentrations and in the solubility of phosphorus in the sediments indicate signs of oxygen depletion.

Introduction

Lake sediments can record the evolution of a lake through time. Sediments can also increase nutrient supply, following eutrophy, by releasing stored nutrients. Anoxia of the hypolimnion and sediment surface is often the factor responsible for phosphorus dissolution.

In this paper we examine Lake Särkinen in the parish of Sotkamo, northeastern Finland. The lake has been aerated since 1980 by a Mixox device (Lappalainen, 1982) but there is no systematic knowledge of the recent evolution of lake Särkinen. The aim of this study is to reveal the eutrophication history of the lake and to gather information about the effects of lake aeration.

Study site, material and methods

Lake Särkinen lies on an esker. Its area is 0,45 km^2, mean depth is 5,5 m, greatest depth is 18,0 m, and the arithmethic mean residence time is about 2 yr. Its drainage area is 3,8 km^2, field percentage being 8%.

The Nurmes-Sotkamo railway and the Vuokatti-Valtimo road by-pass the lake; both were built between 1924–28. Other human activities which may have increased soil erosion are the building of the Vuokatti Sporting Centre in 1945 and its enlargement in 1972. The State Railways holiday centre was constructed in 1952 and enlarged in 1956 and 1959.

Eutrophication of the lake is greatly influenced by the sewage load resulting from about 45 inhabitants between 1956–59 and about 70 inhabitants between 1959–69. The first public awareness of increasing eutrophication occurred in the 1960's. Despite reducing sewage input, anoxia and plankton blooms increased. This led to lake aeration in 1980.

Between 1971–77 phosphorus in the surface water increased from 2 to 90 μg P l^{-1} (mean 15 μg l^{-1}) and in the hypolimnion from 10 to 1550 μg P l^{-1}. The mean phosphorus in the hypolimnion varied from year to year; in 1972 it was 45 μg P l^{-1}, 1976 627 μg P l^{-1}. During a major bloom of blue-green algae in August 1976 plankton biomass was 600 g km^{-2}. However, the nutrient concentration in the water is now decreasing (Wahlgren & Lappalainen, 1989).

Sediment sampling was done on 24. – 25.3.1986 and 14. – 15.6.1986 at about 13 m depth (Fig. 1). It was thought that the aeration could have mixed the deeper sediment.

Sediment sampling was done by gravity coring (Axelsson & Håkanson, 1978) and cores were sliced at 1 cm or 2 cm intervals. The visual characterisation was made from a frozen core (Huttunen & Meriläinen, 1978). The preliminary results, ignition loss and dating by ^{210}Pb analysis, showed that the sediments from sites A and B were nearly identical and the latter was chosen for further investigations.

The loss-on-ignition, total P, total N and metals: Fe, Mn, Zn were analyzed using standard methods (Zink-Nielsen, 1975; Erkomaa et al., 1977). The crude amount of base soluble phosphorus was analyzed from 10 ml of sediment, which was mixed with 1 l of distilled water. The pH was raised to pH 11 with KOH and samples were restirred after 60 min before allowing to settle. Phosphorus concentration was measured in the supernatant. Carbonaceous particles

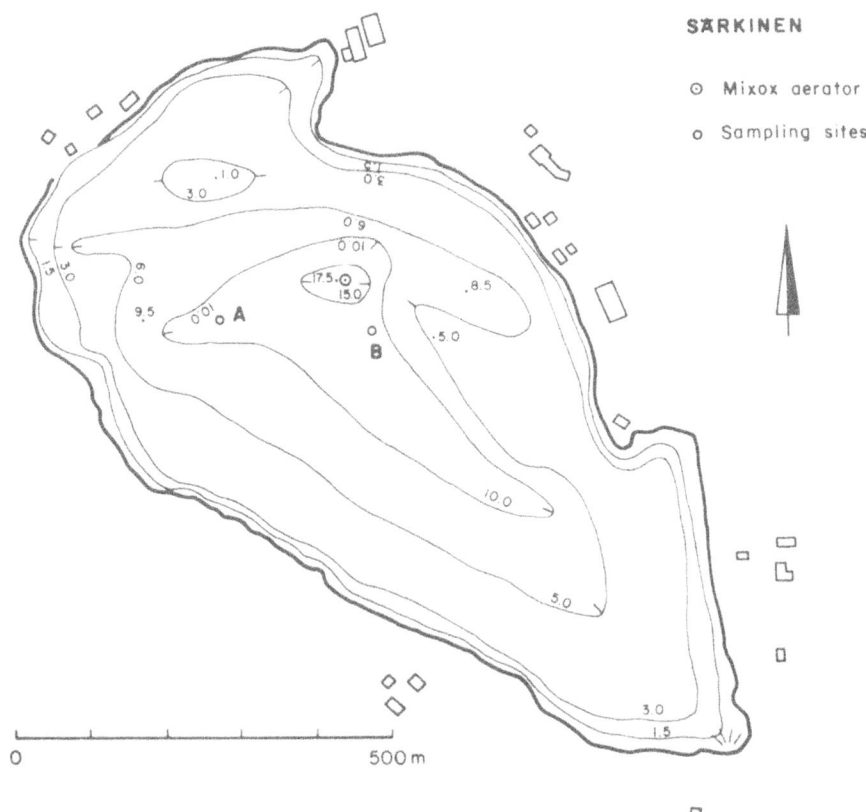

Fig. 1. Bathymetric map of Lake Särkinen, Finland.

analysis was done after Renberg & Wik (1984).

The diatom analysis at 1 cm intervals of the core top 15 cm was carried out. A small amount of sediment was diluted with pure alcohol and carefully homogenized with a glass rod. Sub-samples were then sonicated to disaggregate the particles. Mounting was done with Naphrax. The efficiency of this rapid method was controlled by the traditional H_2O_2- and HNO_3-oxidations. The number of diatom frustules counted was 300–400 per subsample. Diatom identification was done after Mölder & Tynni (1967–73), and Hustedt (1930, 1937–66).

The diatom pH was calculated using index α (Meriläinen, 1967) and index B (Renberg & Hellberg, 1982). To express the development of diatom taxa the DCA (Detrended Correspondence Analysis) -program of Cornell University was used.

The isotope datings from two profiles (A and B) were made in the Reactor Laboratory of the Technical Research Centre of Finland. ^{210}Pb was analyzed by the aid of its alpha emitting grand-daughter ^{210}Po (Häsänen, 1977). Dates were calculated using the C.R.S. model (Appleby & Oldfield, 1978).

Results and discussion

At the top of the sediment core there was a brown, oxidised 1,5–2,0 cm layer followed by a dark layer with lighter streaks which continued for about 5–7 cm. There was no apparent difference in the appearence of the top sediment between summer and winter. The layer from 7 to 13 cm was more grey and under that the sediment colour became medium brown. At the depth of 20–22 cm was a dark zone.

Figure 2 shows the sediment accumulation

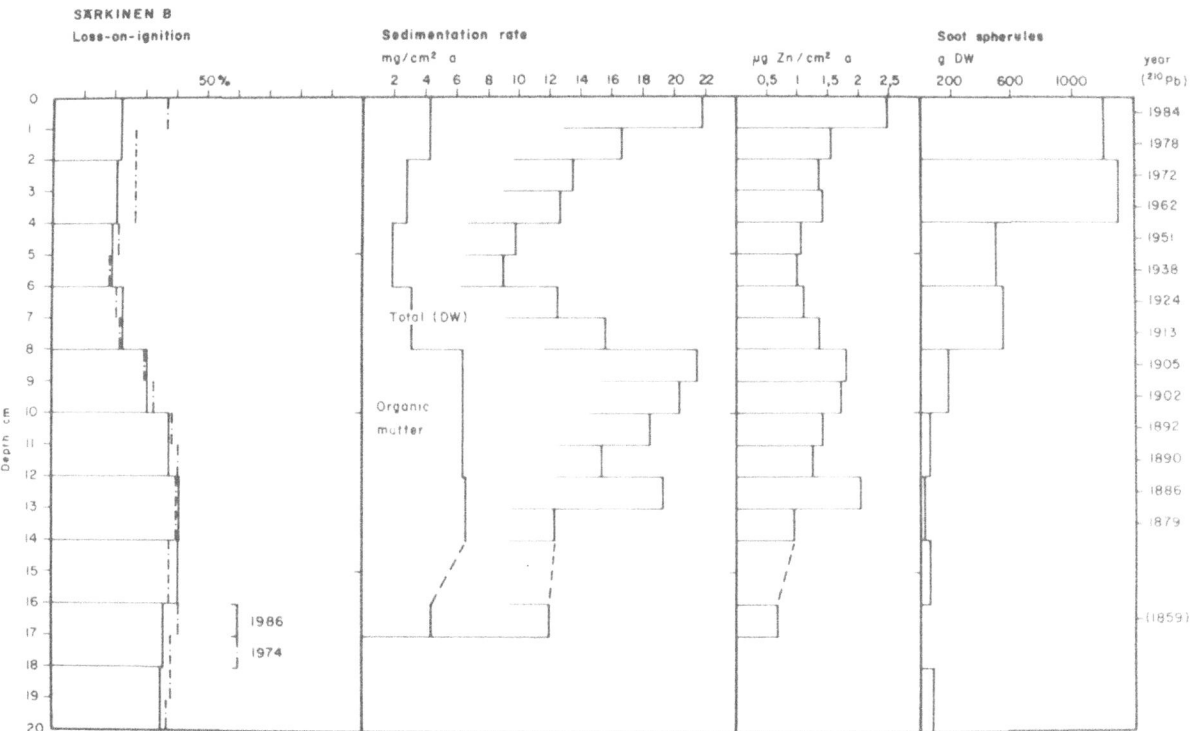

Fig. 2. Ignition loss of Lake Särkinen sediments compared with values from a core taken in the year 1974 (Sandman, 1982), sedimentation rates of dry weight (DW), organic matter and Zn. The number of carbonaceous spherules are given per gramme dry matter.

194

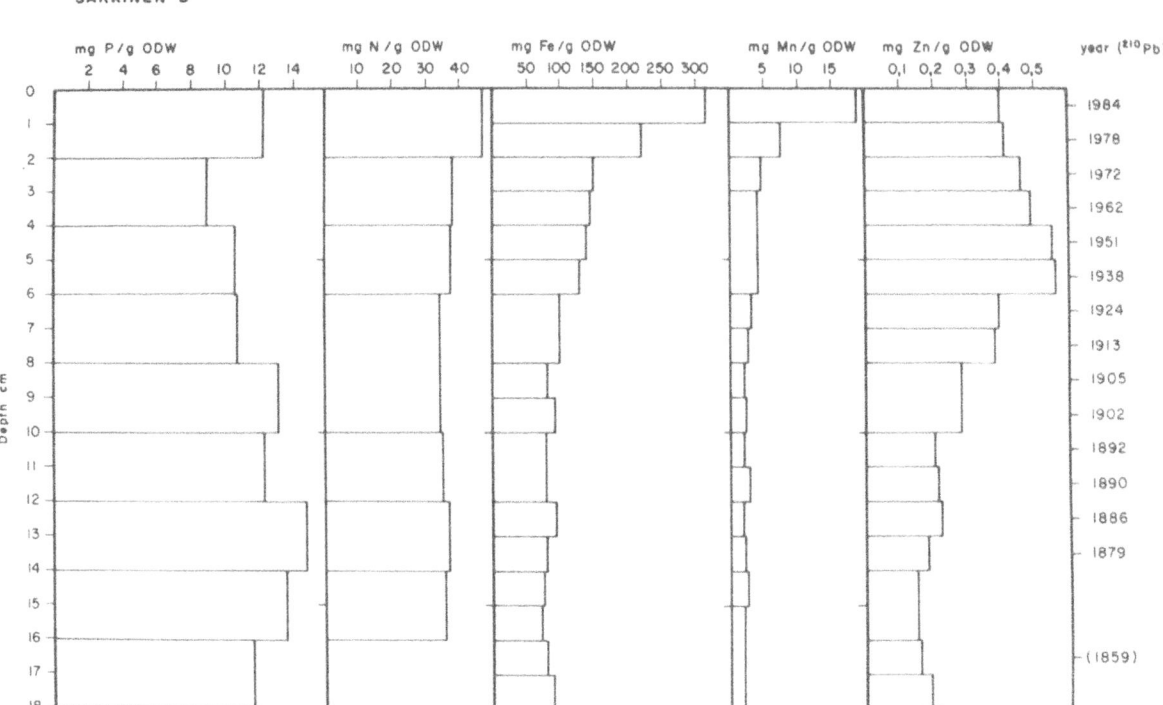

SÄRKINEN B

Fig. 3. The concentrations of total P, N, Fe, Mn and Zn in the sediment of Lake Särkinen. The amounts are expressed as per gramme dry organic matter.

rates of dry matter and mineral matter for core B. The rate maxima for both cores are at the sediment surface, at 9–8 cm (1906–1914) and at 13–12 cm (1885–1890); the minimum is at 6–4 cm (1939–1962).

The maximum zinc concentration is at the sediment depth of 6–4 cm (1939–1962) (Fig. 2, Fig. 3). The amount of contamination is quite small: the background value is ca 60 μg g^{-1} and the maximum is about 110 μg Zn g^{-1} DW. Kansanen & Jaakkola (1985) found a background content of 260 μg in Lake Vanajavesi while in Lake Iso-Kontunen it was about 10 μg Zn g^{-1} DW (Sandman & Simola, 1983).

The phosphorus maximum (Fig. 3, Fig. 4) is at 14–12 cm depth in the sediment (1879–90), which corresponds with the minimum dry matter sediment accumulation rate. The total P and total N curves (Fig. 3) resemble each other. However

there is an anomaly in phosphorus at 4–2 cm (1962–1979). The curves of Fe and Mn are also similar. Iron has a maximum at 14–13 cm (1979–85) and there is a minimum at 8–6 cm (Fe and Mn) (1914–39). The values tend to increase towards the sediment surface.

The diatoms give considerable information about the history of the lake ecosystem (Fig. 5). The deepest subsamples from 15–12 cm (1874–1890) reflect a stable phase with the planktonic forms *Cyclotella comta* and *Cyclotella kützingiana* dominating. Littoral taxa represent 30–40% of all counted individuals and *Achnanthes-* and *Fragilaria* genera dominate.

According to the diatom flora, Lake Särkinen was probably a lake with clear water but perhaps not strictly oligotrophic (Huttunen & Meriläinen, 1986). The fraction of organic matter in sediment is high and there is a maximum of nutrients, too.

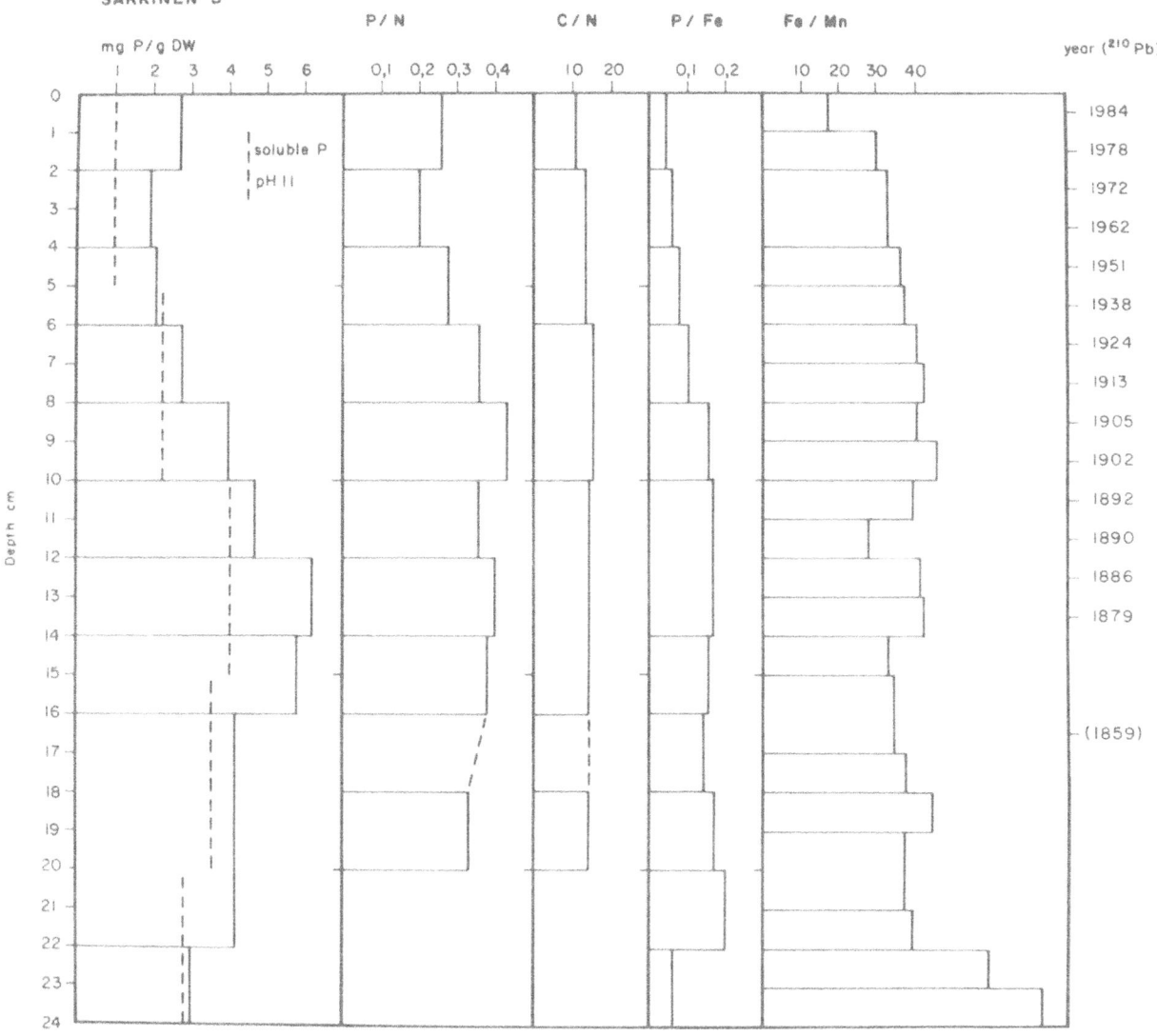

Fig. 4. The concentrations of phosphorus in the sediment of Lake Särkinen compared with base soluble P and P/N, C/N, P/Fe, Fe/Mn ratios.

The phenomenon is associated with the first maximum of dry matter in the years 1885–90 (Fig. 2). The nitrogen and C/N-values (calculated by estimating C to be half the loss-on-ignition) are constant.

The next peak in sediment accumulation rate is at 11–8 cm (1896–1914) and here sediment is more minerogenic. The chemically inactive silts minimizes the g^{-1} dry weight computed for other constituents. There is a change in diatom flora, which is shown in the DCA ordination (Fig. 6).

At 12–11 cm depth (1880–96) frequencies of epiphytic and especially benthic genera increase. The total number of taxa is low at 10–9 cm and the amount of fractured valves is great.

It seems that eutrophication increased (the maxima of *Melosira granulata* and *M. ambigua*) to 9–8 cm after which control the primary productivity in open water shifted to non-diatom algae taxa: e.g. blue-green and green algae. The chryso-

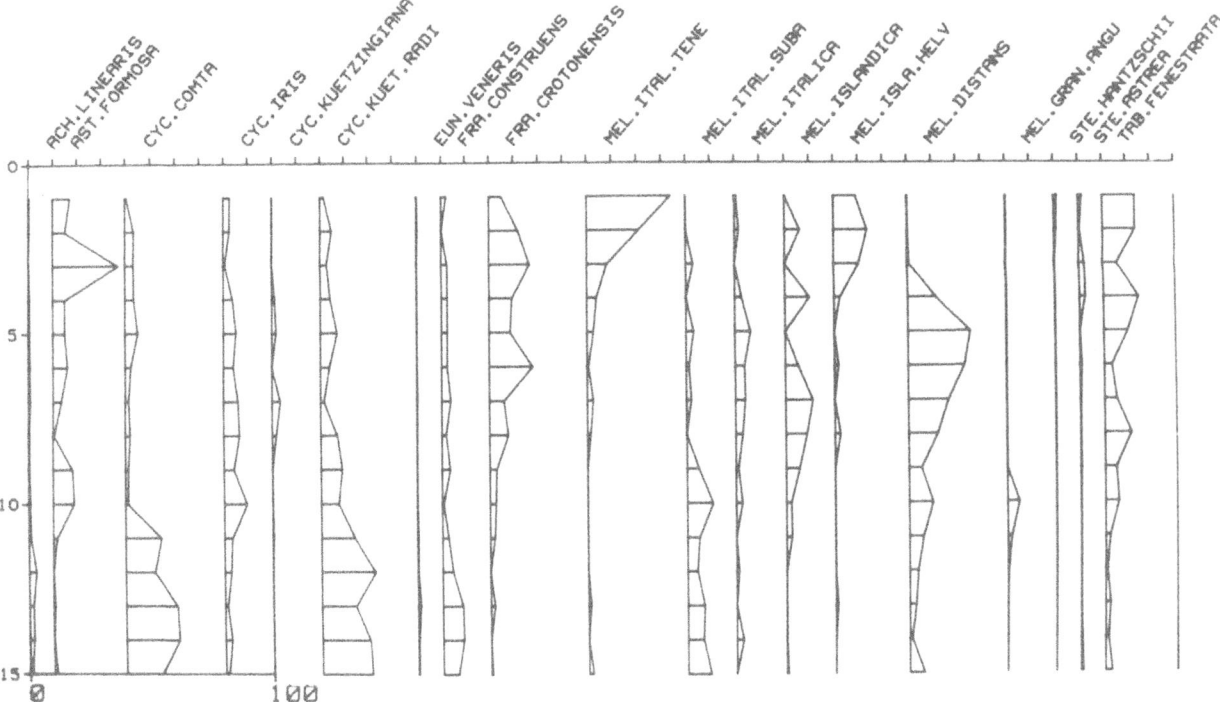

Fig. 5. Diatom stratigraphy for selected taxa from the sediment core of Lake Särkinen.

phyte cyst/diatom-ratio shows that frequencies of cyst decreases at 9–8 cm. Because these fossils can preserve better than diatom valves, it may be expected that their frequency would also increase. Cysts have, however, a clear minimum at this level. It seems that both these nutrient rivals had decreased at the same time but benthic diatoms derived benefit from the environmental changes which took place.

Such large changes can not be caused by landuse changes such as field clearing since all the fields lie near the outflow and there was no known slash and burn cultivation at that time. There is a possible allochtonous inwash of a pond drained at the beginning of the 1900's. This seems to be the cause of the high sediment accumulation rate, the large proportion of in organic matter, fractured and corroded diatom valves, and the major changes in diatom taxa. The fluctuation of Fe/Mn-ratio (Fig. 4) at 14–9 cm could reflect redox-potential changes in the hypolimnion (Mackereth, 1966).

In the next phase 8–4 cm depth (1914–1962) the diatom/chrysophyte cyst ratio increases permanently. According to Smol (1985) increasing eutrophy causes diatoms to supercede chrysophytes. Simultaneously the *Cyclotella* taxa are reduced and *Melosira ssp* (especially *M. distans var. distans*), *Fragilaria crotonensis* and *Tabellaria fenestrata* dominate. These changes may arise from increasing nutrient status or perhaps from the change in the ratio of photic zone to epilimnion caused by water colour increasing after forest drainage (see Harris, 1986). The decrease of littoral diatoms may also arise from diminishing transparency.

The minimum in sediment accumulation rate occurs at 8–4 cm depth (1914–1964) (Fig. 2). This is surprising as many events occur at this time which should increase soil erosion: the construction of the road and the railway, field clearing and building works. The clear minimum in accumulation rate at 6–5 cm (1939–52) however occurs in the war years, but eutrophication in Lake

Eig. v.

Ax I 0.256

Ax 2 0.146

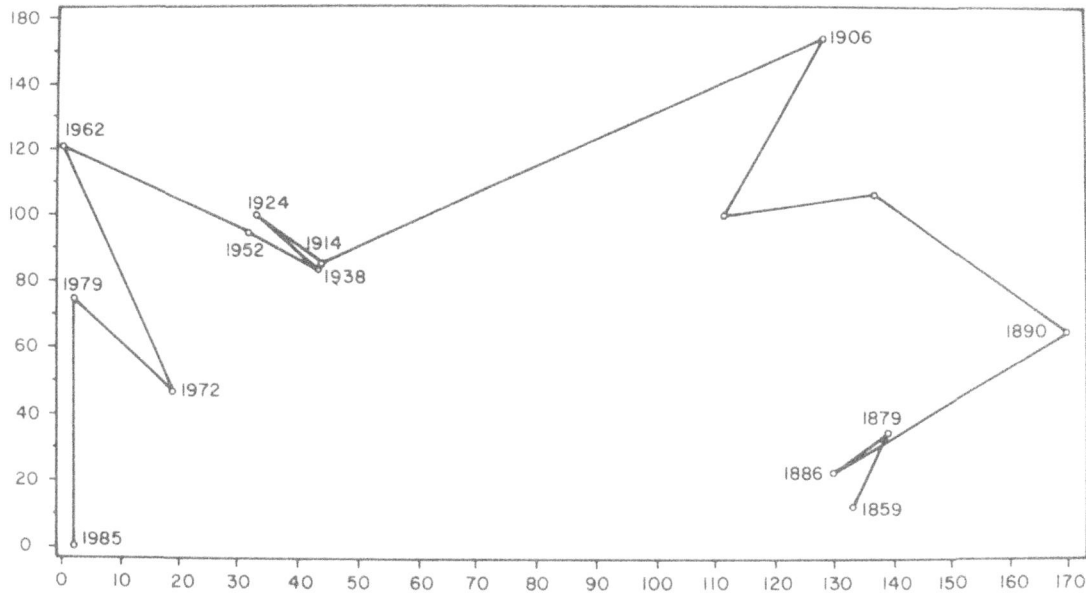

Fig. 6. Ordination of samples from Lake Särkinen in DCA-space.

Särkinen continued since *Asterionella formosa,* *Cyclotella pseudostelligera, Fragilaria construens* are common in this period.

The top 4 cm record an increasing rate in sediment accumulation eutrophication maximum and the effects of lake aeration. The sediment depth of 4–3 cm dates to the beginning of the 1960's and it thus signals the known second enrichment phase. Above this depth the number and diversity of benthic diatoms clearly increase. In the years 1972–79 (sediment depth 3–2 cm) *Asterionella formosa* fragments reaches a maximum. At a depth of 3–2 cm (1972–79) the diatom pH rises suddenly from about pH 6.5 to pH 7.2 (index B) being due to the further expansion of alkaliphilic *Asterionella formosa* and *Melosira islandica*. In the top 2 cm (from the year 1979) indifferent *Melosira italica* ssp *tenuissima* clearly increases. Ilmavirta (1980) considers this species typical of very eutrophic waters but Mölder & Tynni (1963) consider it as a littoral species in dystrophic and oligotrophic waters. From our experience it is an

indicator of meso-eutrophy, especially when abundant in the plankton. In the light of these data, we believe the effects of enrichment are declining.

There is a phosphorus minimum at 4–2 cm (1962–72) when we relate the concentration to organic matter (ODW), (Fig. 3). The same feature occurs in the phosphorus/nitrogen ratio (P/N), (Fig. 4) and this may reflect anoxia before aeration. Water quality data show that after 1965 much phosphorus was released. In late winter and in the summer of 1986 the oxidised top layer of sediment was about 2 cm thick and contained an increased concentration of Fe, Mn and P (Fig. 3).

Several researchers have studied the release of phosphorus from sediments in anoxic conditions but conclusions concerning rates vary (Nürnberg, 1988). However, a general concept is that base-soluble phosphorus, bound with iron, would be a good measure of the phosphorus which escapes from sediment in anoxic conditions (e.g. Messer, 1984).

The proportion of alkali-soluble phosphorus in total phosphorus is large but the lowest values occur at the sediment surface. This may be the consequence of anoxia. Phosphorus is however mobile in sediment (e.g. Carignan & Flett, 1981) and thus the oxidized state of the deep of Lake Särkinen and the high iron values should help keep phosphorus in the sediment (Williams et al., 1979).

In Fig. 2 one can see the ignition loss values for the years 1974 (Sandman, 1982) and 1986. The accumulation rate is about the same in both years, but the loss-on-ignition values has dropped from 30% to 12%. The decline in loss-on-ignition begins at the sediment depth of 5–4 cm.

Lake aeration seems to have had a positive effect on the lake whereas reducing the nutrient load had no effect. The aeration is thus not only artificial respiration, but it binds the nutrients to sediment and makes mineralization more effective, see Fig. 2 (Fillos, 1977). It is, however, worth pointing out that for lasting recovery the nutrient load should be reduced as well as aerating the lake (see Gächter, 1986).

Acknowledgements

Research funding from the National Board of Waters and Environment (to OS), The Foundation for Research of Natural Resources in Finland (to KE) and the Technical Research Centre (to AL) are gratefully acknowledged.

References

Appleby, P. G. & F. Oldfield, 1978. The calculation of lead-210 dates assuming a constant rate of supply of unsupported ^{210}Pb to the sediment. Catena 5: 1–8.

Axelsson, V. & L. Håkanson, 1978. A gravity corer with a simple valve system. J. Sed. Pet. 48,2: 630–633.

Carignan, R. & R. J. Flett, 1981. Post-depositional mobility of phosphorus in lake sediments. Limnol. Oceanogr. 26: 361–366.

Erkomaa, K., I. Mäkinen & O. Sandman, 1977. Vesiviranomaisen ja julkisen valvonnan alaisten vesitutkimuslaitosten fysikaaliset ja kemialliset analyysimenetelmät (Methods of water analyses used by authorised and water authority laboratories). National Board of Waters, Report 121. 54 pp. (in Finnish).

Fillos, J., 1977. Effects of sediment on the quality of the overlying water. In Golterman (ed.), Interactions between sediment and freshwater. Dr W. Junk. The Hague: 266–271.

Gächter, R., 1987. Lake restoration. Why oxygenation and artificial mixing cannot substitute for a decrease in the external phosphorus loading. Schweiz. Z. Hydrol. 49/2: 172–185.

Harris, G. P., 1986. Phytoplankton ecology. Structure, Function and Fluctuation. Chapman and Hall Ltd. London: 239–247.

Hustedt, F., 1927–1966. Die Kieselalgen Deutschlands, Österreichs und der Schweiz unter Berücksichtigung der übrigen Länder Europas sowie angrenzenden Meeresgebiete. In: Dr. L. Rabenhorsts Kryptogamen-Flora von Deutschland, Österreich und der Schweiz. VII 1: 1–920, 2: 1–845, 3: 1–816.

Hustedt, F., 1930. Bacillariophyta (Diatomae). In: Pacher, A. (ed) Die Süsswasser-flora Mitteleuropas 10: 1–466. Jena.

Hustedt, F., 1937–39. Systematische und ökologische Untersuchungen Über die Diatomeen-Flora von Java, Bali und Sumatra nach dem Material der Deutschen limnologischen Sunda Expedition. Arch. Hydrobiol. Suppl. 15: 131–177, 187–295, 393–506, 638–790, Suppl. 16: 1–155, 274–394.

Huttunen, P. & J. Meriläinen, 1978. New freezing device providing large unmixed sediment samples from lakes. Ann. Bot. fenn. 15: 128–130.

Huttunen, P. & J. Meriläinen, 1986. Diatom response to pH and humic matter of the water. Publ. Karelian Inst. Univ. Joensuu. 79: 47–54.

Häsänen, E., 1977. Dating of sediments, based on ^{210}Po measurements. Radiochem. Radioanal. Lett. 31: 207.

Ilmavirta, V., 1980. Phytoplankton in 35 Finnish brownwater lakes of different trophic status. Devel. Hydrobiol. 3: 121–130.

Kansanen, P. & T. Jaakkola, 1985. Assessment of pollution history from recent sediments in Lake Vanajavesi, Southern Finland. I. Selection of representative profiles, their dating and chemostratigraphy. Ann. Zool. fenn. 22: 27–69.

Lappalainen, K. M., 1982. Convection in bottom sediments and its role in material exchange between water and sediments. Hydrobiologia 86: 105–108.

Mackereth, F. J. H., 1966. Some chemical observations on postglacial lake sediments. Phil. Trans. r. Soc. (B) 250: 165–213.

Meriläinen, J., 1967. The diatom flora and the hydrogen-ion concentration of the water. Ann. Bot. fenn. 4: 51–58.

Messer, J. J., J. M. Ihnat & D. L. Wegner, 1984. Phosphorus release from the sediments of Flaming Gorge Reservoir, Wyoming, USA. Verh. Int. Ver. Limnol. 22: 1457–1464.

Mölder, K. & R. Tynni, 1967–73. Über Finnlands rezente und subfossile Diatomeen I–VII. Bull. Geol. Soc. Finland 39: 199–217 (1967), 40: 151–170 (1968), 41: 235–251 (1969), 42: 129–144 (1970), 43: 203–220 (1971), 44: 141–149 (1972), 45: 159–179 (1973).

Nürnberg, G., 1988. Prediction of phosphorus release rates from total and reductant-soluble phosphorus in anoxic lake sediments. Can. J. Fish. aquat. Sci. 45: 453–462.

Renberg, J. & T. Hellberg, 1982. The pH history of lakes in southwestern Sweden, as calculated from the subfossil diatom flora of the sediment. Ambio 11: 30–33.

Renberg, J. & M. Vik, 1984. Dating recent lake sediments by soot particle counting. Verh. Int. Ver. Limnol. 22: 712–718.

Sandman, O., 1982. The eutrophication of some pelotrophic lakes; a palaeolimnological study. Publ. Wat. Res. Inst. Nat. Bd. Wat., Finland. 49: 76–96.

Sandman, O. & H. Simola, 1983. Juvan Iso-Kontusen tutkimus (Abstract: Recent paleolimnology of Lake Iso-Kontunen, East Finland). Univ. Joensuu, Publ. Karelian Inst. 55: 63–77.

Smol, J. P., 1985. The ratio of diatom frustules of chrysophycean statospores. A useful paleolimnological index. Hydrobiologia 123: 199–208.

Williams, J. D. H., J.-M. Jaquet & R. L. Thomas, 1976. Forms of phosphorus in the surficial sediments of Lake Erie. J. Fish. Res. Bd. Can. 33: 413–429.

Zink-Nielsen, I., 1975. Interkalibrering af sedimentkemiske analysemetoder II (Intercalibration of chemical sediment analyses). Nordforsk Miljövårdssekretariatet. 6: 1–19.

Hydrobiologia **214**: 201–206, 1990.
J. P. Smith, P. G. Appleby, R. W. Battarbee, J. A. Dearing, R. Flower, E. Y. Haworth, F. Oldfield & P. E. O'Sullivan (eds),
Environmental History and Palaeolimnology.
© 1990 *Kluwer Academic Publishers.*

Are we building enough bridges between paleolimnology and aquatic ecology?*

John P. Smol

Dept. Biology, Queen's University, Kingston, Ontario, K7L 3N6, Canada

Abstract. No.

Few paleolimnologists would argue against the fact that our science has made substantial advances over the last decade. A simple perusal of the development in procedures and applications, conveniently documented at about four to five year intervals in previous symposium volumes (Frey, 1969; Klekowski, 1978; Meriläinen *et al.*, 1983; Löffler, 1986), should convince any sceptic. The papers presented at this symposium further demonstrate that paleolimnology continues to gain momentum.

Yet, despite these advances, it is my opinion that we have been slow, or perhaps reluctant, to use our new and powerful approaches to test a wealth of hypotheses being generated by rapidly expanding ecological theory. The many advantages of having a temporal record providing information on time-scales ranging from seasonal events to thousands (or in some cases millions) of years of community interactions has much to offer aquatic ecology. Nonetheless, I believe that many potential bridges have not been built.

Throughout this commentary, I use the word 'neolimnologist' to designate limnologists working with present-day aquatic systems. Among the practitioners of the synthetic science of 'limnology', I prefer not to differentiate between the more commonly used headings of 'limnologists' and 'paleolimnologists', as both groups of researchers are 'limnologists'. One might argue that this is a minor point of semantics – perhaps it is – but I think it adds to an artificial isolation of two interdependent sub-disciplines.

Are we ready to build these bridges?

I think the answer to this first question is a resounding 'Yes!'. Our techniques have advanced to the point where we can accept challenges posed by neolimnologists and begin evaluating ecological hypotheses that otherwise remain untested.

Part of my mandate for this introductory talk is to review our recent developments. This is a difficult task, for few areas of science have enjoyed such leaps in progress. About a year ago (1988) I was offered a similar indulgence, when I was asked to offer my opinions on paleolimnology's recent advances and future challenges (Smol, in press). To briefly reiterate my comments made then (and I acknowledge these represent my parochial view of ecologically-based paleolimnology, heavily modified by research interests in my laboratory), I believe that our most significant advances can be divided into three interdependent categories.

I would first identify refinements made to sampling techniques, and the resulting temporal resolution we can attain, as primary hurdles that we have cleared. As in most scientific endeavors, technology drives progress, and the continued refinements to coring apparatus (although it is hard to improve on some of the original designs!) and sediment handling protocols (including, but certainly not restricted to, the development of freeze coring, close interval sediment extrusion, tape-peel adhesions, and resin impregnated sedi-

_____

* Opening keynote lecture to the 'Aquatic Ecology and Palaeolimnological Interpretation' session at the Vth International Symposium on Palaeolimnology, held at Ambleside, U.K., Sept. 1, 1989.

ments) have done much to improve our resolution. These developments go hand-in-hand with advances made by scientists working on dating techniques, upon which all ecologically-based paleolimnologists are dependent. Very often, paleolimnologists can select the time scale (e.g. centuries, decades, years, and in some cases seasons) – a distinct advantage over workers restricted to neolimnological protocols.

Secondly, the quantity and quality of information we have learned to tease from the sedimentary record continues to escalate; in fact we have now shown that most aquatic groups leave some sort of morphological or chemical marker. As with sampling procedures, technological advances, such as the application of high pressure liquid chromatography (HPLC) to the study of fossil pigments, have done much to foster the quantity and quality of our primary paleolimnological data. Researchers have repeatedly shown that each additional group of indicators *adds* information, and does not simply confirm information gleaned from other groups. The whole is greater than the sum of the individual parts.

Thirdly, and very importantly, tremendous progress has been made on the ways we are learning to use our indicators – how we have learned to calibrate and quantify their ecological optima and tolerances along environmental gradients. Over a short time, we have progressed from simply using qualitative inferences based primarily on literature citations of ecological distributions, to the use of various indices, to the development of surface sediment calibration sets. With the latter came a variety of new statistical approaches, now culminating in what seems to be our new Rosetta Stone – weighted averaging regression, calibration, ordination, and constrained ordination (e.g. canonical correspondence analysis) as developed by Cajo ter Braak (1987). The power of surface sediment calibrations is hard to over-estimate. Equally important has been the development of error estimates, validation techniques, and quality assurance/quality control protocols, so vital to any science. For several groups of organisms, it is the paleolimnologist, not the neolimnologist, who has gathered the primary ecological data.

Why are bridges so necessary?

Paleolimnologists are dependent on present-day lake processes to interpret paleoecological data and therefore we obviously need the information generated by neolimnologists. My main thesis, however, is that neolimnologists also need our data and the long-term perspective we provide in order to constrain their theories, to add realism to their models, and to test their hypotheses. My friend, Patrick DeDeckker, uses an analogy to make this point: neolimnologists are like people taking photographs of a fast train with the latest photographic equipment. They perhaps take several photographs of the train, but they don't know what the train looked like when it left the station, what trajectory the train followed before reaching cruising speed, and what it will look like on arrival. Neolimnologists tend to only look at a particular portion of time, yet a much longer record of community change is archived in a lake's sediments. I believe that it is this message that we should try to develop and communicate. A few examples might strengthen these arguments.

In 1986, the ecologist Patrick Weatherhead asked the simple but powerful question 'How unusual are unusual events?'. Weatherhead's goal was to document the extent that ecological researchers invoke 'unusual' events to interpret their data. He wondered whether these were really unusual events, or were instead artifacts of the observational approach, or perhaps more precisely, a result that researchers expected to find, given their perspective on the science. He looked at a sample of 380 papers and found that about 11% of ecological studies (averaging about 2.5 years in duration) invoked 'unusual' events to explain the outcome of their observations. These data are especially striking because none of the above studies were published specifically to report the unusual event. Interestingly, almost all of these 'unusual events' were abiotically, and especially climatically, controlled.

It seems that 'unusual events' in ecology are not so 'unusual' after all. Clearly, the temporal perspective and background of ecologists (and other scientists) are dominant operating factors influencing their interpretations.

This latter point is reinforced by another interesting outcome of Weatherhead's analysis. Apparently, the relationship between study duration and the probability of an unusual event is not as clear-cut as one might expect. In general, the frequency of unusual events increased with time in studies lasting up to about 6 years, but for longer studies (up to 15 years), unusual events were less common. Weatherhead observes that the likely reason for this apparent paradox is that we cannot treat these events as statistical phenomena because the criterion used to designate an 'unusual event' was the author's perception that something was unusual (based on the data he or she had at hand). In short, '... we tend to overestimate the importance of some unusual events when we lack the benefit of the perspective provided by a longer study' (Weatherhead, 1986; p. 154).

If we superimpose Weatherhead's finding onto a second study, we see the serious dilemma ecologists presently have with time scales. David Tilman (1989) documented the duration of field studies published in the journal *Ecology* for a randomly selected subset of issues for 1977–1987: a total of 623 observational and experimental studies. Of the observational studies, about 40% were less than one year in duration, and fully 90% were less than 3 years long. The curve for experimental studies is even more skewed towards short-term studies. The 'long-term perspective' is not generally available to most ecologists.

I think there is ample evidence that ecologists increasingly recognize the need for long term data; certainly there is a groundswell of interest and at least moral (if not yet financial!) support for the collection and archiving of long term data (e.g. Likens, 1989). Yet, I feel strongly that many neolimnologists do not fully recognize paleolimnological data as an important source of this information. Many neolimnologists still seem to equate 'long term ecological data collection' exclusively with repeated field sampling, even if this approach is not usually feasible. In my opinion, this epitomizes the lack of sufficient

bridges between these interdependent sections of 'limnology'.

Like many people at this symposium, I have recently spent much time as a speaker or discussant at many university, government, and industry sponsored workshops and colloquia, relating to a variety of ecological and environmental problems. The fact that we, as paleolimnologists, are even invited to these meetings should be encouraging, as neolimnologists at least suspect that we have something to offer. However, I have been equally dismayed when, for example, I talk about some of the simplest paleolimnological techniques, such as surface sediment calibrations, a portcullis seems to fall down – once sediments are involved in any way, this all becomes foreign and largely inapplicable territory. Many neolimnologists wrongly believe that our data are somewhere in the realm of fossils, or geology, but certainly not 'limnology' or aquatic ecology.

In general, neolimnologists are unaware of the wealth of excellent ecological data that we can gather and/or that we already hold in our data bases. Even if they are aware of the quantity and quality of these data, many seem oblivious or at least unsure of how they can apply these data to their own research needs. In addition, their perspectives on paleolimnology are frequently rife with misconceptions, not recognizing that many of these at least potential problems have been investigated, quantified, and in many cases, resolved. Yes, we need to build more bridges.

It is, of course, a two-way street. Neolimnologists desperately need long-term data. Paleolimnologists, however, can also greatly benefit from the wealth of ecological theory currently accumulating. Much of this theory, in my opinion, is still poorly developed and tested. By embracing this theory, we can further develop our science and integrate it into a broader ecological framework. We have much to gain from these interactions.

The future opportunities for these interactions are especially bright, as paleolimnologists are presently faced with tremendous research challenges. Below, I identify a few that I am more familiar with, but I appreciate that there are many more – several of which have been discussed at this symposium.

Paleolimnological approaches have been widely used to study the long-term effects of damage to ecosystems. I begin with acidification: partly because it is a subject with which I am familiar, and partly because it has important legacies that we can exploit. Many of the new techniques and protocols, as well as the sizeable data bases that this research is generating, can now be transferred to other research endeavours. In addition, there is still much we can learn from ongoing and newly proposed acidification studies. For example, lakes can be randomly chosen from a region. Pre-industrial age pH levels can then be inferred, and compared to the present lakewater pH. The resulting 'change-in-pH', or for that matter change in other variables (such as dissolved organic carbon or selected metals), can be mapped and population based estimates of lakewater changes can be generated. The study of lake recovery (if in fact it occurs) can also be documented by fine-interval sediment analyses. In addition, paleolimnological studies provide one of the only avenues for model validation.

Long-Term Monitoring Programs (LTMPs) continue to gain popularity with many lake managers, as well as other scientists. The significant recognition that organisms are important adjuncts to 'snap-shot' water chemistry measurements is equally gaining momentum. All of the indicators we use in paleolimnology should be prime candidates for inclusion in LTMPs, not only because we already hold a wealth of high quality ecological data on species distributions that could easily be applied to LTMPs, but also because these organisms respond quickly (in a predictable and quantifiable manner) to even subtle environmental changes. Equally, paleolimnological studies should be included in all LTMPs, for without these data the background variability (which any new environmental changes must be measured against) cannot be assessed. It is impossible to extrapolate reliable ecological models without first determining if organisms are responding to an unusual phenomenon (e.g. a culturally induced environmental change) or the result of a long-term trend.

Lake trophic dynamics continue to be a main focus in neolimnological studies, and one to which paleolimnologists can make an important contribution. Quantification of trophic variables (e.g. nutrient concentrations) to species distributions is under way in a number of paleolimnology laboratories, often applying similar calibration techniques to those developed in acidification studies. Once calibration is available, paleolimnologists should be able to quantify past lake trophic status, from which a plethora of hypotheses could be tested and developed (e.g. the terrestrial/aquatic linkage, anthropogenic effects on lake systems, etc.). In addition, these studies will be vital to help evaluate the rapidly developing theory associated with topics such as cascading lake trophic dynamics (e.g. Carpenter, 1988). More process-oriented paleolimnological techniques (e.g. biogenic silica, chemical accumulation rates, etc.) may be especially applicable to these types of studies.

Just as acidification was a major environmental focus in the 1980's, global climate change appears to be rapidly becoming the major concern of the 1990's and beyond. Virtually every technique paleolimnologists use can be applied to both hindcasting climatic change (absolutely vital for model validation) and studying the limnological effects of past climatic changes. Although palynology will provide much important proxy data, paleolimnological approaches have much to offer, and in fact, have many advantages over more traditional techniques. Saline lake systems and high polar regions will likely become important research foci for these studies.

What obstacles do we face?

Given these research opportunities and the recognized need for long-term data in ecological studies, what obstacles do we face in building bridges?

Aquatic ecology is presently rich in ecological theory, whereas paleolimnology is, in general, theory-poor. We will always be plagued by critics who see this as a major drawback to making paleolimnology an 'exciting' science. This, however, did not bother nor hinder such eminent scholars as E.S. Deevey (1984). Also, because we are an historical science, we cannot undertake 'true experimentation'. I do not see these as serious problems. To use my favourite example of acidification and its ensuing problems, paleolimnology did not provide many new elegant theories about the acidification process and its effects (in some ways, I am thankful for this, given the large numbers of untested theories already in the literature). Nor did paleolimnology (technically, at least) 'prove' anything in the acid rain debate – but it certainly presented a very powerful case! When it came to even the most basic questions, such as: 'Have lakes acidified and, if so, when, and by how much and how fast?', it was mainly paleolimnology that provided these answers. Paleolimnology provided data that were vital to falsify many theoretical proposals.

Nonetheless, several obstacles remain. I think most of these center on the misconceptions and communication problems that I alluded to earlier. I think communication is the only approach that will overcome these barriers. We must continue to attend neolimnological and ecological workshops and symposia and explain how our approaches are applicable to the research interests of neolimnologists. Publication in a broad spectrum of journals, in terms that are useful and applicable to neolimnologists, is also vital. Large advances have occurred in neolimnology, large advances have occurred in paleolimnology – it is time we start talking to each other. Although paleolimnology makes its way more and more into the 'psyche' of neolimnologists, I strongly feel that we, as paleolimnologists, must take the initiative in showing how important, applicable, and necessary our approaches are. A good start might be to begin framing our research projects more in tune with current ecological thinking. We should keep astride of these new developments, and use our procedures to test these theories.

Finally, cross-disciplinary research initiatives might be especially helpful in forming conduits of communication and interest between neo- and paleolimnologists. For example, data obtained from sediment trap studies bridge the usual boundaries of neo- and paleolimnological research. Moreover, these data provide the opportunity to fine-tune the temporal scale and quality of our calibration data, as well as provide important information on taphonomic processes.

Conclusion

Paleolimnologists share with neolimnologists many of the same problems and complexities. I believe we should foster a new activism, one in which paleolimnological research becomes more integrated with theoretical and applied limnology. We are at an important juncture in time, especially with the almost insatiable appetite ecologists now have for long-term data. We should seize the opportunity while it is before us. Paleolimnology can accommodate the scrutiny of neolimnologists and other scientists, and paleolimnology can benefit from the infusion of fresh ideas and criticisms that would inevitably come from these synergisms.

Ecologically-based paleolimnology is still a young science, but even a cursory look at its development documents the tremendous advances that we have made. A scan of the papers presented at this symposium is further evidence of our progress. I can't wait to see what we will accomplish in the next four years.

Acknowledgements

I would very much like to thank Dr. E. Y. Haworth and the Paleolimnology V organizing committee for inviting me to present this lecture, and for providing such a stimulating and enjoyable symposium. The quality of this manuscript was greatly improved by the comments provided by the 18 scientists in my lab, as well as Drs. Livingstone, Birks, Schelske, Stoermer, Binford, DeDeckker, Engstrom, Haworth, Charles, Brenner, Leavitt, and Siver.

The day I presented this paper was 10 years to the day that I arrived at Queen's University and began to work on my doctoral research with Prof.

S. R. (Ted) Brown. I would like to dedicate this paper to him.

References

Carpenter, S. R. (ed.), 1988. Complex lake interactions. Springer-Verlag, N.Y., 283 pp.

Deevey, E. S. Jr., 1984. Stress, strain, and stability of lacustrine ecosystems. In E. Y. Haworth & J. W. G. Lund (eds), Lake sediments and environmental history. University of Minnesota Press, Minneapolis: 203–229.

Frey, D. G. (ed.), 1969. Symposium on paleolimnology. Mitt. int. Ver. Limnol. 12: 1–448.

Klekowski, R. A., 1978. Second international symposium on paleolimnology. Pol. Arch. Hydrobiol. 25: 1–498.

Likens, G. E. (ed.), 1989. Long term studies in ecology: Approaches and alternatives. Springer-Verlag, N.Y., 214 pp.

Löffler, H. (ed.), 1986. Paleolimnology IV. Hydrobiologia 143: 1–431.

Meriläinen, J., P. Huttunen & R. W. Batterbee (eds), 1983. Paleolimnology. Hydrobiologia 103: 1–318.

Smol, J. P., (in press). Paleolimnology – recent advances and future challenges. In R. De Bernardi, G. Giussani & L. Barbanti (eds), Scientific Perspectives in Theoretical and Applied Limnology. Mem. Ist. Ital. Idrobiol. 47 (in press).

ter Braak, C. J. F., 1987. Unimodal models to relate species to environments. Thesis, Wageningen.

Tilman, D., 1989. Ecological experimentation: Strengths and conceptual problems. In G. E. Likens (ed.), Long term studies in ecology: approaches and alternatives. Springer-Verlag, N. Y.: 136–157.

Weatherhead, P. J., 1986. How unusual are unusual events? Am. Nat. 128: 150–154.

Hydrobiologia **214**: 207–211, 1990.
J. P. Smith, P. G. Appleby, R. W. Battarbee, J. A. Dearing, R. Flower, E. Y. Haworth, F. Oldfield & P. E. O'Sullivan (eds), 207
Environmental History and Palaeolimnology.
© 1990 *Kluwer Academic Publishers.*

Weichselian chironomid and cladoceran assemblages from maar lakes

Wolfgang Hofmann
Max-Planck-Institut für Limnologie, Postfach 165, D-2320 Plön, Germany

Key words: Pleniglacial, maar lakes, biostratigraphy, Chironomidae, Cladocera

Abstract

Sediments from maar lakes in the periglacial area were analysed to obtain information about the limnetic fauna of Pleniglacial lakes. On the basis of abundance and species composition of the subfossil chironomid and cladoceran assemblages, three major zones were biostratigraphically separated: the Upper Pleniglacial, the Middle Pleniglacial, and the Eem Interglacial.

Introduction

Biostratigraphical studies on Cladocera and Chironomidae almost exclusively deal with Late-Glacial/Holocene sediments (Hofmann, 1987, 1988; Walker, 1987). Sediments from periglacial maar lakes provide information about the fully glacial period. The aim of this study is to assess whether these animal groups can be used for a biostratigraphy of the Pleniglacial.

Material and methods

Sediment samples from three eutrophic Eifel maars; Meerfelder Maar, Holzmaar, Schalkenmehrener Maar (Lorenz & Büchel, 1980; Irion & Negendank, 1984) and from Lac du Bouchet, in the Massif Central, France (Bonifay, 1987) were kindly provided by Prof. Negendank (Trier). Figure 1 shows core lengths, the position of the volcanic tuff layer of the Laach eruption (11 200 BP (Zolitschka, 1989)), and datings referring to Negendank (1989; in press) indicating that the sediments reach far into Weichselian period.

For preparation of the sediment samples and identification of the remains see Hofmann (in press). The sample fraction $> 200 \mu$m was used for chironomid analysis and $> 100 \mu$m for Cladocera.

Results and discussion

The Upper Pleniglacial sediments of the Holzmaar were characterized by extremely low abundances of cladoceran remains (Fig. 2). More than 10 000 specimens per g DW occurred in the Holocene layers and more than 1 000 in the Late-Glacial. Below the tuff layer concentrations rapidly decreased. In many samples no remains were found at all although the subsample size was mostly > 1 g wet sediment. In the section from 12.22 m to 31.80 m sediment depth only 6 specimens were found in a total subsample volume of 27.5 g WW (one sample with higher abundance not included). Mean abundance was 0.2 remains per g WW (Table 1). Likewise, in the corresponding layers from Schalkenmehrener Maar and Lac du Bouchet mean abundance was < 1 specimen per g WW. The value obtained in

208

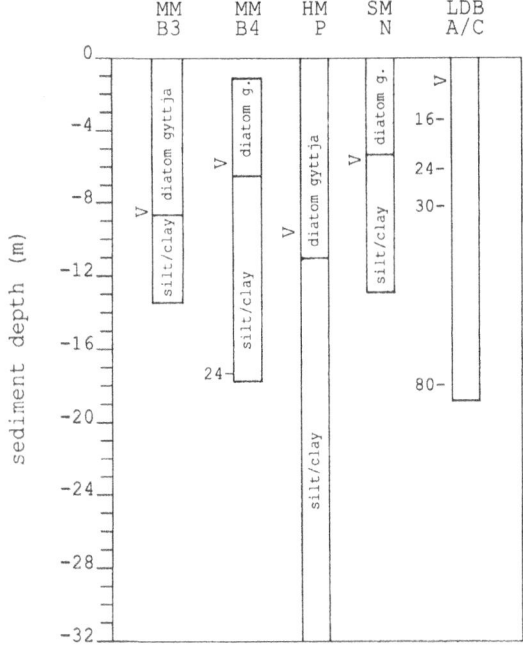

Fig. 1. Meerfelder Maar (MM), profiles B3, B4; Holzmaar (HM), profile P; Schalkenmehrener Maar (SM), profile N; Lac du Bouchet (LDB), profiles A/C: length of the profiles, position of the Laach volcanic tuff layer (11 200 BP) (arrows) and datings (10^3 years BP; classification of maar sediments and datings after Negendank (1989; in press).

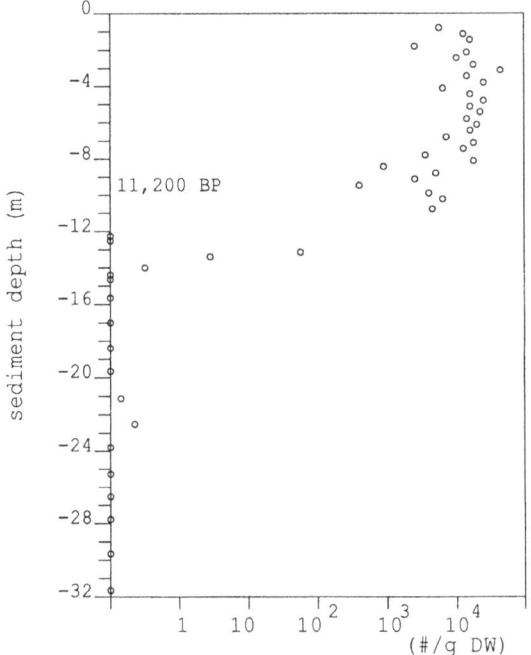

Fig. 2. Holzmaar P, Cladocera: abundance (numbers/g dry sediment).

Meerfelder Maar was slightly higher but is related to sediment dry weight.

These specimens are insufficient for any characterization of the cladoceran assemblages with respect to species composition. Hence, this zone can only be separated from the Late-Glacial by its extremely low concentrations of cladoceran remains. Similarly, chironomid remains were extremely rare during this period (Table 2). In all the maars studied abundance was less than 1 specimen per g WW/DW. The material was particularly scarce in the lower section of Holzmaar and Schalkenmehrener Maar.

However, in contrast to the Cladocera the chironomid assemblage of this zone exhibited a particular species composition and was thus well separated from the Late-Glacial. The most abundant taxa were *Diamesa*, *Protanypus*, *Paracladopelma*, and *Micropsectra* (Table 3). Species of Diamesinae were found in all the maars and

were predominating in some cases. Species of the genus *Diamesa* are cold-stenothermous inhabiting flowing water, in particular running waters of mountain regions and less frequently shallow still waters (Oliver, 1983). In the lakes under discussion *Diamesa* was confined to this zone. The other three taxa are typical elements of the profundal zone of oligotrophic lakes under temperate conditions. In arctic/subarctic regions, they preferably occur in the littoral zone (Brundin, 1949).

Below this horizon, the data from Lac du Bouchet indicated a significant change in the cladoceran and chironomid assemblages with respect to both abundance and species composition (Fig. 3). Below 7 m concentrations of cladoceran remains increased again with maximum values of > 100 specimens/g. But abundances varied by an order of magnitude over small depths. The development of the chydorid assemblage of this period was characterized by repeated alternations between two predominating species, *Alona quadrangularis* and *Chydorus sphaericus* (Hofmann, in press) (Table 4).

Table 1. Abundance of cladoceran remains in the sediment section below the Late-Glacial in three Eifel maars and the Lac du Bouchet.

	sediment depth (cm)	g sediment analysed	Nos.	Nos./g sediment
Meerfelder Maar B3	1010–1299	8.2 DW	37	4.5
Holzmaar P	1222–3180	27.5 WW	6	0.2
Schalkenm. Maar N	693–1315	26.9 WW	21	0.8
Lac du Bouchet A/C	125–530	87.4 WW	56	0.6

Table 2. Abundance of chironomid remains in the sediment section below the Late-Glacial in three Eifel maars and the Lac du Bouchet.

	sediment depth (cm)	g sediment analysed	Nos.	Nos./g sediment
Meerfelder Maar B3	1010–1299	59.8 DW	44	0.74
Meerfelder Maar B4	630–1800	170.5 DW	55	0.32
Holzmaar P	1222–1703	107.4 WW	51	0.47
	1736–3205	295.6 WW	3	0.01
Schalkenm. Maar N	693–1062	60.3 WW	30	0.50
	1095–1315	50.5 WW	0	0.00
Lac du Bouchet A/C	125–660	967.9 WW	21	0.02

Table 3. Occurrence of four chironomid taxa in the sediment section below the Late-Glacial in three Eifel maars (MM: Meerfelder Maar, HM: Holzmaar, SM: Schalkenmehrener Maar) and Lac du Bouchet (LDB) (** – predominance).

Lake core sediment depth (cm)	MM B3 1010 – 1299	MM B4 734 – 1800	HM P 1222 – 3205	SM N 693 – 1062	LDB A/C 125 – 660
Diamesa	**	**	*	*	**
Protanypus			*	*	*
Paracladopelma	*	**	*	*	
Micropsectra		*	**	**	*

210

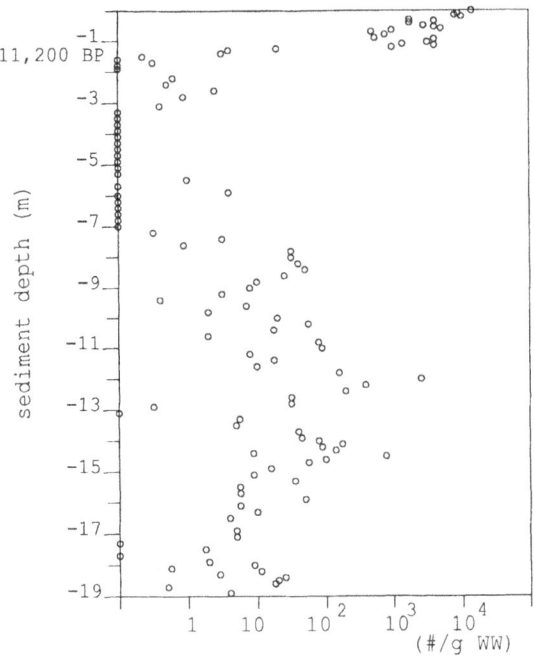

LAC DU BOUCHET A/C: CLADOCERA

Fig. 3. Lac du Bouchet, profiles A/C, Cladocera: abundance (numbers/g wet sediment); (from: Hofmann, in press).

Similarly, there were very distinct alternations in the chironomids, between the groups *Micropsectra/Tanytarus* and *Paracladius/Paratanytarsus* (Hofmann, in press).

In the lowermost section of the Lac du Bouchet core, a third stratigraphic unit was discernible by an increase in the number of chydorid species and appearance of taxa which prefer temperate climatic conditions: *Alona costata, Disparalona rostrata, Monospilus dispar* (Table 4). The restricted chironomid material from this zone also gave the impression of a species succession (Hofmann, in press).

In view of the dating of the Lac du Bouchet profile (3.20 m: 16 000 yrs.; 6.10 m: 24 000 yrs; 8.10 m: 30 000 yrs; 18.10 m: 80 000 yrs. Negendank, in press) and the chronostratigraphy of the last glaciation (Nilsson, 1982), the section of the profile under discussion represents the period from the Upper Pleniglacial to the Eem Interglacial.

Furthermore, the three biostratigraphic units presented correspond with the (1) Eem Interglacial, (2) the Middle/Lower Pleniglacial and Early Glacial, and (3) the Upper Pleniglacial.

The extremely low concentration of animal remains and the particular composition of the chironomid assemblage in the Upper Pleniglacial is obviously related to the temperature minimum and the glacial expansion occurring during this period (Nilsson, 1982).

In the Middle Pleniglacial abundance of remains was higher and the species composition

Table 4. Lac du Bouchet, profiles A/C, sediment depth: 550–1877, Chydoridae: mean percentage of predominant species in seven sections; N – numbers counted (from: Hofmann, in press).

ACO	*Alona costata*	MDI	*Monospilus dispar*
ARU	*Alona rustica*	DRO	*Disparalona rostrata*
AQU	*Alona quadrangularis*	ANA	*Alonella nana*
AAF	*Alona affinis*	CPI	*Chydorus piger*
GTE	*Graptoleberis testudinaria*	CSP	*Chydorus sphaericus*

sediment depth (cm)	N	ACO	ARU	AQU	AAF	GTE	MDI	DRO	ANA	CPI	CSP
550–800	139										100
820–860	220			100							
880–1020	287			9.8							89.5
1040–1369	3607			<0.1							>99.9
1389–1529	2415		1.8	1.0							97.0
1549–1749	168			94.6							3.0
1789–1877	113	7.1		15.0	7.1	5.3	28.3	13.3	8.0	15.0	0.9

of the cladoceran and chironomid fauna was different indicating that the climatic conditions were not so extreme as in the following period. However, low species diversity reflected the influence of the cold climate.

Increasing species diversity and occurrence of faunal elements of the temperate zone in the lowermost sediments indicates a warmer climate. These sediments therefore apparently originated from the Eem Interglacial (or Early-Glacial).

The results obtained from the Lac du Bouchet profile suggest that climatic changes during the Pleniglacial period significantly affected the abundance and species composition of the limnetic fauna of periglacial lakes. Hence, the remains of these organisms can be used for the purpose of a biostratigraphy.

The outline of the stratigraphy presented has to be considered as a first attempt. The general validity of the distribution pattern found in Lac du Bouchet will have to be checked by additional case studies. Furthermore, the conclusions deduced from these data will have to be discussed in connection with the results of geological and climatological studies.

The application of the stratigraphy proposed for the Lac du Bouchet profile to the cores from the Eifel maars leads to the conclusion that these profiles represent the Upper Pleniglacial only, because the section characterized by particular species composition and species alternations was missing. Thus, it would be desirable to see if, in older sediments from Eifel maars, assemblages occur which resemble the Middle Pleniglacial fauna of Lac du Bouchet. This would also answer the question whether the Pleniglacial biostratigraphy of Lac du Bouchet represents a general pattern of succession in lakes in periglacial regions.

Acknowledgements

I would like to thank Professor Negendank (Trier) who kindly provided the sediment samples and stimulated the study. I also wish to thank Bernd Zolitschka (Trier) and Bernt Haverkamp (Münster) for cooperation.

References

Bonifay, E. (ed.), 1987: Travaux francais en Paleolimnologie. Doc. C.E.R.L.A.T. 1: 1–275.

Brundin, L., 1949. Chironomiden und andere Bodentiere der südschwedischen Urgebirgsseen. Rep. Inst. Freshw. Res. Drottningholm 30: 1–914.

Hofmann, W., 1987. Cladocera in space and time: Analysis of lake sediments. Hydrobiologia 145: 315–321.

Hofmann, W., 1988. The significance of chironomid analysis (Insecta: Diptera) for paleolimnological research. Palaeogeogr., Palaeoclimatol., Palaeoecol. 62: 501–509.

Hofmann, W., in press: Stratigraphy of Chironomidae (Insecta: Diptera) and Cladocera (Crustacea) in Holocene and Wurm sediments from Lac du Bouchet (Haute Loire, France). Doc. C.E.R.L.A.T.

Irion, G. & J. F. W. Negendank (eds.), 1984. Das Meerfelder Maar – Untersuchungen zur Entwicklungsgeschichte eines Eifelmaares. Cour. Forsch.-Inst. Senckenberg 65: 1–101.

Lorenz, V. & G. Büchel, 1980. Zur Vulkanologie der Maare und Schlackenkegel der Westeifel. Mitt. Pollichia 68: 29–100.

Negendank, J. F. W., 1989. Pleistozäne und holozäne Maarsedimente der Eifel. Z. dt. geol. Ges. 140: 13–24.

Negendank, J. F. W., in press. Doc. C.E.R.L.A.T.

Nilsson, T., 1982. The Pleistocene. Reidel: Dordrecht, Boston, London.

Oliver, D. R., 1983. The larvae of Diamesinae (Diptera: Chironomidae) of the Holarctic region – Keys and diagnoses. Ent. scand. Suppl. 19: 115–138.

Walker, I. R., 1987. Chironomidae (Diptera) in paleoecology. Quat. Sci. Rev. 6: 29–40.

Zolitschka, B., 1989. Jahreszeitlich geschichtete Seesedimente aus dem Holzmaar und dem Meerfelder Maar. Z. dt. geol. Ges. 140: 25–33.

Hydrobiologia **214**: 213–221, 1990.
J. P. Smith, P. G. Appleby, R. W. Battarbee, J. A. Dearing, R. Flower, E. Y. Haworth, F. Oldfield & P. E. O'Sullivan (eds), 213
Environmental History and Palaeolimnology.

Stratigraphy of the fossil Chironomidae (Diptera) from Lake Grasmere, South Island, New Zealand, during the last 6000 years

Barbara Schakau
Department of Zoology, University of Canterbury, Christchurch, New Zealand

Key words: Chironomidae, Palaeolimnology, New Zealand

Abstract

The fossil chironomid fauna of a 3.26 m long sediment core from Lake Grasmere has been analysed. The fossil chironomid taxa belong mainly to the subgroups Tanytarsini, Orthocladiinae, Chironomini, and Tanyphodinae. Heptagyini and Podonominae were not common. Tanytarsini were the dominant component of the fauna with *Corynocera* sp. as the most abundant species during pre-Polynesian times (before 1000 yr BP). The abundance and the composition of the fossil chironomid taxa have fluctuated markedly over the last 6000 years. These fluctuations could be partly correlated with changes in the stratigraphy of the sediments in the core. Layers of highly minerogenic sediment contained the lowest numbers of remains whereas high abundances were found in the sections of the core with the greatest proportion of organic matter. It is suggested that major shifts in the structure of the chironomid community have been mainly caused by changes in the hydrology and inflows of the lake, and the rate and type of sedimentation, in addition to variations in lake productivity.

Introduction

The analysis of fossil Chironomidae has been of considerable interest to palaeolimnologists especially in the Northern Hemisphere, as emphasised by Hofmann (1988) and Walker (1987). The composition and successional changes of the fossil chironomid fauna have been used mainly to reconstruct past environmental conditions during lake ontogeny. But this type of analysis can also be employed to follow the development of certain taxa and gain information about the structure of chironomid communities over time. This aspect of the study is important for the general understanding of the chironomid communities in New Zealand lakes.

Study area

Lake Grasmere is located in the Cass Basin in the montane region of the Waimakariri River catchment, North Canterbury. The Waimakariri River drains a section of the Southern Alps and the topography of its catchment is the direct result of glacial and post-glacial activity (Hayward, 1967). Lake Grasmere (43° 05′ South, 171° 45′ East) is a small, mesotrophic lake, situated at about 600 m A.S.L. (Stout, 1972). The area of the lake is 0.63 km² and its maximum depth is 15 m (Fig. 1). There is one surface inlet stream, and in addition, the waters of Ribbonwood Creek reach the lake as springs after flowing underground through the New Ribbonwood Creek Fan. The recent chironomid fauna is composed of 17 species (Stark, 1981a; Timms, 1982, 1983). This

214

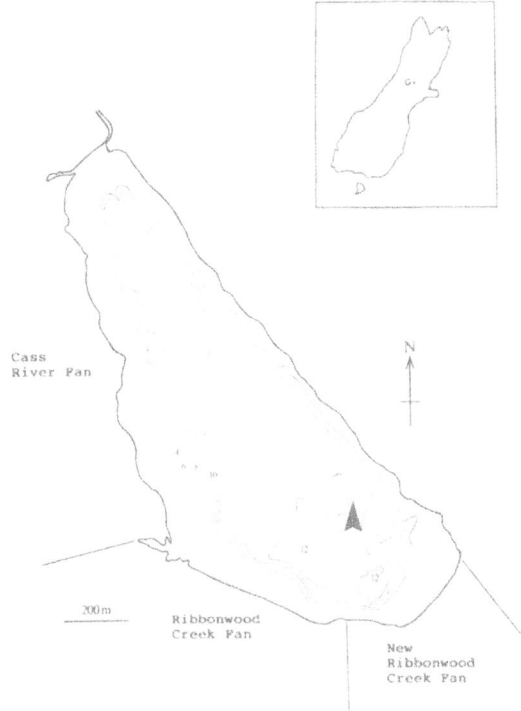

Fig. 1. Bathymetric map of Lake Grasmere (Contour Interval: 2 m). The black triangle indicates the sampling site of the sediment core. The inlet shows the South Island of New Zealand with the location of the lake (G).

is the highest number of species recorded from any New Zealand lake.

Lake Grasmere was formed partly by glacial activity, and partly by alluvial aggradation (Gage, 1959). The ice of the Waimakariri glacier, formed during the last advance of the Otira glaciation, withdrew possibly about 13500 years ago from the Cass area (Burrows, 1978) and Lake Grasmere originated after the ice retreat (Gage, 1959). The lake was partly dammed by ice-eroded rock. The greater part of the lake perimeter is formed by three extensive alluvial fans. The ages of the Cass Fan and the Old Ribbonwood Fan are uncertain, but both might have been deposited during late-glacial times (Soons, 1977). Deposition has still continued in the New Ribbonwood fan in historic times, decreasing the area of the lake. Originally, the lake probably extended eastwards (Gage, 1959).

The development of vegetation after the end of the Otira glaciation in the South Island followed a general pattern from open grassland-shrubland through shrubland to forest (Moar, 1971). Two forest zones could be distinguished in a pollen diagram from Kettlehole Bog, near Lake Grasmere (Lintott & Burrows, 1973). The original forest consisted of podocarps, indicating a climate with mild temperatures and reliable precipitation levels. The present climate at Cass does not allow the growth of these taxa. *Podocarpus* forest remained dominant in the area until at least 7500 yr BP (Moar, 1971). At some time after that, the vegetation changed to a *Nothofagus* (Southern Beech) forest. This transition may represent a change in climate to lower minimum temperatures, greater temperature extremes, lower rainfall and the occurrence of periodic drought (Burrows, 1979).

Pollen analysis of samples of the Lake Grasmere core, carried out by Dr. M. McGlone, showed that *Nothofagus* was already well established by 6000 yr BP in the catchment area of the lake. Most samples were dominated by *Nothofagus fusca* type pollen. An increase of grass pollen at the 30 cm level recorded the arrival of the Polynesians around 1000 years ago (McGlone, 1989). Burning of the forest cover was used by Polynesian hunters in this area especially between 500 and 1000 yr BP (Molloy, 1977). The presence of European settlers was shown by *Pinus* and *Rumex* pollen in the sediments of the core from 6 cm upwards. This means that the last 6 cm of the core were deposited after 1857 when the area was first explored and settled by Europeans (McLeod & Burrows, 1977).

Methods

The sediment core was obtained from Lake Grasmere using a modified hand-operating Livingstone piston corer (Green, 1979). The coring site was located in the southeastern part of the lake at a water depth of 10.5 m (Fig. 1). A total of 326 cm of sediment were recovered. The bottom 6 cm were comprised of medium grey clay and silt. This layer prevented more sediment from entering the coring tube. In the laboratory, the core was cut in half. One half is stored undisturbed for reference in the Zoology Department, University of Canterbury. The second half was sampled mostly at 10 cm intervals, each sample consisting of 1 ml of wet sediment. Smaller sampling intervals were chosen where changes in the stratigraphy of the sediments were visible. Parallel samples were taken for the determination of dry weight and loss-on-ignition. For the analysis of chironomid remains, the samples were treated with a slightly modified method as described by Boubee (1983) and Walker & Mathewes (1989). New Zealand (Forsyth, 1971; Stark, 1981b), as well as overseas keys (Simpson & Bode, 1980; Wiederholm, 1983) were used to identify the chironomid remains. The interpretation of the results was based on the relative abundance of the chironomid taxa in the samples.

Fig. 2. Summary figure illustrating major features of the sediment core (LG3), the age of the sediments, the results of the loss-on-ignition analysis, and the results of the preliminary pollen analysis.

Results

Sediment record

The Lake Grasmere core consisted mostly of clay gyttja of varying shades of olive grey. A layer between 196 and 204 cm contained sand particles sized from 0.02–2 mm. The bottom 6 cm were dominated by clay and silt (Fig. 2). Results of loss-on-ignition revealed a relatively low content of organic matter in the core with values ranging from 2.5 to 17.7% organic matter dry weight. The organic content throughout most of the core remained relatively stable, the only fluctuations being a minimum of 2.5% at the base and two pronounced maxima at the 315 and 25 cm horizons.

Radiocarbon dates are presented in Table 1. The date from the oldest sediments (at 318 cm) revealed that the sediment core from Lake Grasmere spans nearly 6000 years and might not cover the entire history of the lake. The clay/silt layer at the base of the core might have been deposited during a period of naturally accelerated erosion (Grant, 1985). Erosion and sedimentation

occurring in the mountainous regions of the South Island are very high and sediment yields from some basins in the Southern Alps are amongst the highest in the world (Griffiths, 1981, Whitehouse, 1984). Possible causes for increased rates of sediment transport include increased flood flows produced by a greater frequency and/or magnitude of rainstorms (Beschta, 1983; Grant, 1985) possibly in connection with damage to vegetation by fire. The fine sediment at the base of the Lake Grasmere core might have been trasported into the lake via Ribbonwood Creek or the Cass River. The Cass River could have been an inflow of Lake

Table 1. Radiocarbon dates for samples taken from the Lake Grasmere core.

Core depth (cm)	Laboratory number	Date
310–318	NZ–7568	5648 ± 200 yr BP
202–210	NZA–353	3410 ± 120 yr BP
42– 50	NZA–349	1228 ± 69 yr BP

The radiocarbon-dating was carried out by the Institute of Nuclear Science, Wellington, New Zealand.

Grasmere 6000 years ago as the form of its fan shows. A possible explanation of the origin of the clay/silt layer is that a series of frequent large rainstorms in the Craigieburn Ranges, where the headwaters of Ribbonwood Creek and the Cass River are located, caused severe hillslope erosion. Subsequently, sediment masses were transported downstream, the coarser sediments were deposited in the fans and at the lake margin, and the fine sediments finally in the deeper parts of the lake. The strongly increased sediment influx had a pronounced influence on the chironomid fauna as will be shown later.

Chironomid stratigraphy

The chironomid analysis was based on 56 subsamples, each consisting of 1 ml of sediment. For the analysis, absolute numbers of chironomid head capsules per volume unit were counted. A total of 5290 head capsules was retrieved averaging 96 remains per ml of sediment. The abundance of chironomid remains in the core fluctuated considerably (Fig. 3). These fluctuations could be partly correlated with changes in the stratigraphy of the core. Layers of highly minerogenic sediment contained the lowest numbers of remains whereas two peaks of higher abundance were found in the sections of the core with the highest proportion of organic matter. The greatest maximum abundance at the 150 cm level was not connected to any special sediment features.

Five chironomid subfamilies occurred in Lake Grasmere. Head capsules belonging to the Heptagyini and Podonominae were rare. Tanytarsini were the most abundant subgroup with a relative abundance of 37.3%, Chironomini were the sub-dominant group with 25.5%, Tanypodinae occurred with 17.9%, and Orthocladiinae with 15.8% relative abundance.

Comparing the relative abundances of the four main chironomid subgroups in Fig. 3, it is clearly shown that Tanytarsini represented the dominant form of the fossil chironomid fauna during most of the lake's developmental history under study.

Orthocladiinae showed a peak in dominance in the younger sediments between 10 and 30 cm. The Chironomini were important throughout the core but were especially dominant in the upper-

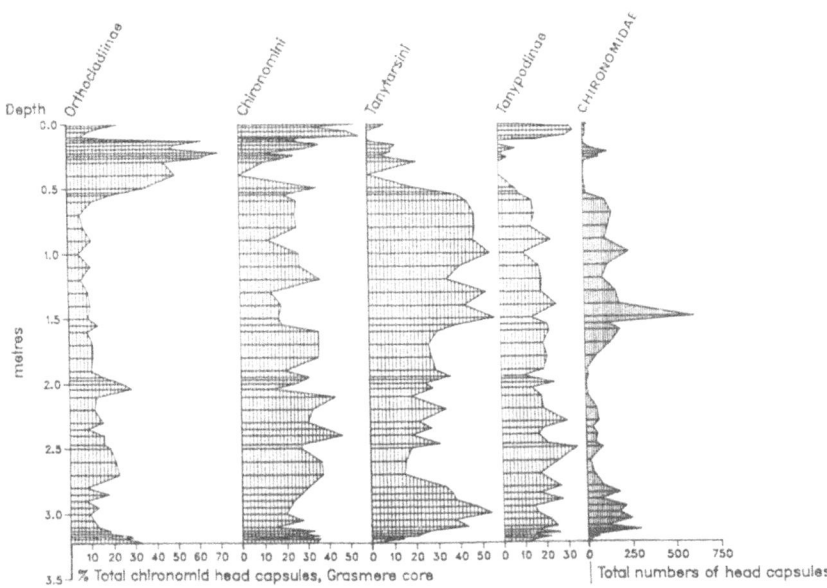

Fig. 3. Lake Grasmere: Total number of chironomid remains and the relative abundance (% of sum at each level) of the main chironomid subgroups in the samples of the core.

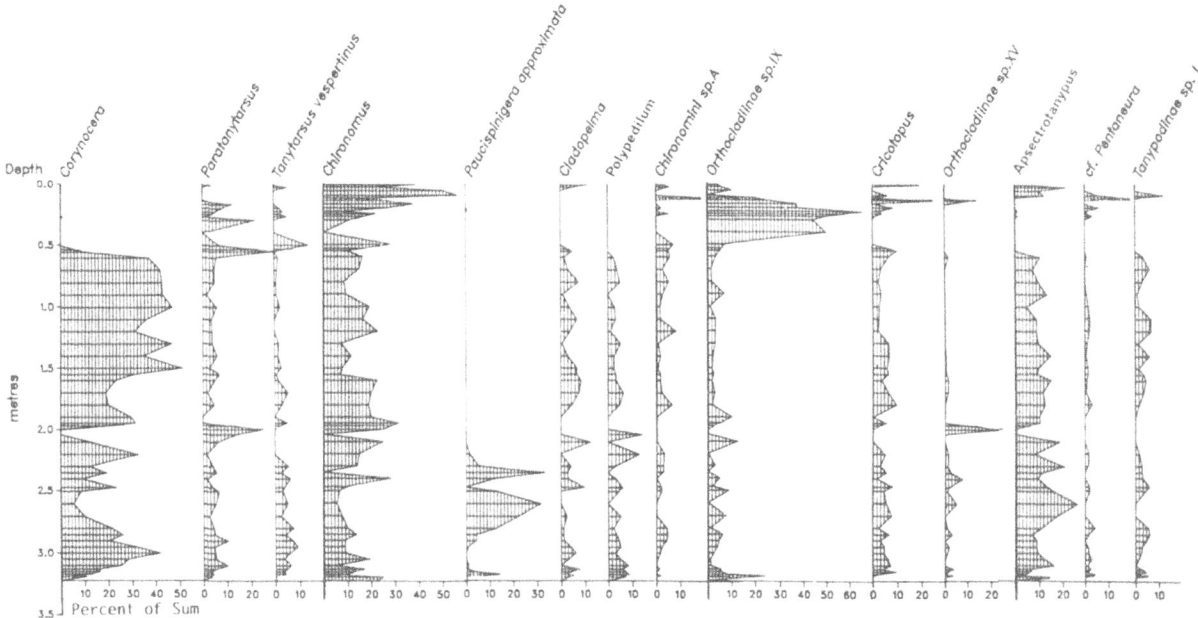

Fig. 4. Percentage diagram of the most abundant chironomid taxa for Lake Grasmere (percentages are calculated as proportion of all chironomid head capsules counted in each sample).

most samples. The relative abundance of Tanypodinae remained stable in most samples, except for a minimum in the more recent sediments.

The fossil chironomid remains from Lake Grasmere could be distinguished in terms of 35 taxa, but only 14 occurred with a relative abundance higher than 1% of the total number (Fig. 4). Chironomini and Orthocladiinae were the most diverse subgroups with 11 taxa in each. *Chironomus* (13.7%) was the most important Chironomini taxon. The number of undescribed taxa (6) in the Orthocladiinae was high owing to the poor taxonomic knowledge of this group in New Zealand. Orthocladiinae sp. IX (7.1%) was the most abundant taxon in the Orthocladiinae. In the Tanytarsini, only four taxa could be distinguished. *Corynocera* constituted the most dominant, with 28.8% relative abundance. The most abundant Tanypodinae head capsules belonged to *Apsectrotanypus* (9.3% relative abundance). Six Tanypodinae taxa were separated.

The temporal distribution of the chironomid taxa in the Lake Grasmere core showed some significant trends (Fig. 4). The dominant chironomid, *Corynocera*, appeared in low abundance in the top sample of the basal clay/silt layer, and became the dominant form in most samples from 310 cm up to the 55 cm level. In the youngest sediments (50 cm to the top of the core), and in two samples of the sandy layer (200–204 cm), *Corynocera* was not found. This taxon occurred in low numbers at the 260 cm horizon. The abundance of *Chironomus* fluctuated strongly in the core but the taxon was an important part of the fauna in nearly every sample. A pronounced peak in abundance of *Chironomus* occurred in the youngest sediments of the core. Orthocladiinae sp. IX occurred during the entire period covered by the core in relatively low abundances, but became the dominant chironomid taxon in the sediments between 12 and 40 cm. In the sample at the 23 cm level, 65.5% of all head capsules found belonged to this taxon.

Discussion

The sediment core from Lake Grasmere covered the last 6000 years of the lake's developmental history. All sediments in the core were deposited during the *Nothofagus* phase of the Holocene. During this period, no major climatic change occurred. Therefore, it is unlikely that influences of macroclimate (e.g. a major change in temperature) would have been the cause of changes in the faunal composition of the fossil chironomids of Lake Grasmere.

Factors to which variations in the chironomid fauna might be attributed include the trophic state or the productivity of the lake, the oxygen concentration of the hypolimnion (Brundin, 1951; Hofmann, 1978), and the quality and availability of food (Sæther, 1979; Kansanen *et al.*, 1984). Warwick (1980) emphasized the significance of sedimentation processes.

The interpretation of the observed changes in the subfossil chironomid associations of Lake Grasmere presents some difficulties. Two factors restrict a detailed analysis of the faunal succession and its ecological significance: the relatively high number of undescribed or previously unknown taxa (especially in the Orthocladiinae and Tanypodinae), and the rather sparse ecological information available for many chironomid species in New Zealand. However, the existing information about various fossil chironomids from Lake Grasmere makes it possible to reconstruct some aspects of the lake's history.

It is suggested that some pronounced shifts in the structure of the chironomid community have been mainly caused by changes in the hydrology and the inflow of the lake and by variations in the rate and type of sedimentation. Changes in lake productivity as well had an influence on the fossil chironomid fauna.

The clay/silt layer at the base of the core represented a major environmental disturbance for the bottom fauna of the lake. The massive input of fine mineral sediments was carried into the lake either by Ribbonwood Creek (in which case the waters of the creek must have entered the lake as a surface inlet), or the Cass River, and must have caused high turbidity. The rapid mineral sediment accumulation appears to have eliminated the profundal fauna. The sample at the bottom of the core consisted almost entirely of clay and silt particles.

The first chironomid remains were found at 322 cm. Initial numbers were low and included only five taxa. The change in sediment consistency to organic clay mud at the 318 cm level indicated the termination of the high sediment input, followed by clearing lake waters. Numbers and diversity of chironomids increased rapidly in correlation with a possible rise in lake productivity about 5500 years ago, as indicated by the relatively high organic content of the sediments.

Corynocera, the dominant subfossil chironomid in Lake Grasmere, can be used as an indicator of a distinct substrate type. Larvae of *Corynocera* were found in lakes of the Waikato region (Boubee, 1983) which had clear water with a maximum water depth of 1 m. The sediments consisted of a flocculent organic layer with a high concentration of algae (mainly diatoms). Boubee (1983) considered *Corynocera* to be an indicator for low water levels and Deevey (1955) found exceptionally high numbers (1500 hc/ml sediment) of this genus in Pyramid Valley lake which was shallow throughout its history. But the consistency of the substrate appears to be a more important factor than the water level influencing the distribution of *Corynocera*. Timms (1983) recorded *Corynocera* from Lake Letitia, near Cass, where larvae of this type were abundant down to a water depth of 12.5 m. Timms (1983) noted the special consistency of the mud which was 'brown, very soft, and composed largely of small aggregates of flocculent material' (p. 46). In the sediments of the Lake Grasmere core, *Corynocera* occurred in high abundances when sedimentation processes remained stable.

Visible changes in the sediment of the core were correlated with low abundances or the absence of *Corynocera*. At the 260 cm level at 1 cm wide dark lamination was noted. In this sample, the abundance of *Corynocera* declined to a minimum comparable to that in the older sediments of the core. The nature of the sediment changed considerably at 204 cm. The following 8 cm contained a relatively large amount of sand particles. Such sand deposits can accumulate as a result of unusual storms which produce sufficient water currents and wave action to move material out from the shore (Davies & Moeller, 1985). This is the most probable cause for a high country lake in the South Island. Grant (1989) suggested that in New Zealand periods of increased storminess, erosion and alluvation, caused by changes of atmospheric circulation, alternated with tranquil intervals when erosion and sediment transport declined and soil formed. Head capsules of *Corynocera* were not found in the samples derived from the sandy layer of the core but this taxon occurred again in a relatively high abundance at 195 cm immediately after the deposition of sand had ceased. Many of the other taxa of the chironomid community responded negatively to the disturbance, and numbers of head capsules and taxa declined considerably in this sediment layer.

For about the next 1500–1800 years (195–60 cm) *Corynocera* occurred in high abundances indicating a stable environment. The composition of the entire fauna remained relatively constant.

In the most recently deposited sediments of the core (55 cm-surface), a major shift in faunal composition was recorded. *Corynocera* disappeared entirely, suggesting a sedimentation – related disturbance. At 55 cm, numbers of chironomid remains as well as taxa, declined a suddenly although no apparent change in the sediments of the core was discernible. A colour change and an increase of organic matter in the core occurred between 30 and 22 cm. The succeeding sediments consisted of a sequence of alternating pale and light olive grey, implying a rapid change in sedimentation pattern during the last 1000 years.

At this stage of the study, it is not possible to determine the exact mechanism which governed the sudden change in the faunal composition at the 55 cm horizon. A possible explanation could be a change in the lake area caused by encroachment of the New Ribbonwood Fan. As mentioned earlier, deposition in the fan still continued in historic times (Gage, 1959). Another factor which

added to the instability of the environmental conditions of the lake during the last 1000 years was the appearance of the first Polynesians, in particular their influence on the landscape of the Cass Basin through use of fire (McGlone, 1989; McSaveney & Whitehouse, 1989). The increase of grass pollen at the 30 cm level in the core correlates with the maximum in organic matter and the pronounced peak in the abundance of Orthocladiinae sp. IX. This high abundance might indicate a rise in lake productivity as well as increased input of allochthonous material caused by the frequent fires in the early period of Polynesian settlement. Undescribed larvae of this genus have been found also in shallow ponds in the Botanic Gardens, Christchurch (pers. obs.) where they occur in large numbers. These ponds have a high density of macrophytes and also receive a high influx of allochthonous organic matter.

Towards the top of the core, the abundance of Orthocladiinae sp.IX decreases and *Chironomus* becomes the dominant chironomid. The top most 6 cm of the core cover the time of European settlement in the area and the dominance of *Chironomus* indicates a slight shift towards mesotrophy (Graham, 1976; Forsyth, 1986; Schakau, 1986).

In summary, the conditions in Lake Grasmere have changed considerably during the last 6000 years. Long periods with a relatively stable environment, reflected by high abundances of certain chironomid taxa and a comparatively unchanging faunal composition, were disrupted by short-term but high magnitude disturbances resulting in a decline of chironomid diversity and abundances and shifts in community structure. The subfossil chironomids from Lake Grasmere are good indicators of the general changes which have occurred in this geologically active environment of the Southern Alps of New Zealand.

Acknowledgements

This research was supported with a postgraduate scholarship to the author by the University Grants Committee, New Zealand. I wish to express my gratitude to my supervisor, Dr. V. M. Stout, for her valuable advice and support. I thank the technical staff of the Zoology Department, University of Canterbury, for their help during the study, Dr. M. McGlone for the results of the pollen analysis, Dr. M. Harper for plotting the stratigraphic results, Ms J. Matheson for her help with the graphical work, and Ms J. McKenzie for critically reading the manuscript.

References

Beschta, R. L., 1983. Channel changes following storm-induced hillslope erosion in the upper Kowhai Basin, Torlesse Range, N.Z. J. Hydrol. N.Z. 22: 93–111.

Boubee, J. A. T., 1983. Past and Present Fauna of Lake Maratoto with Special Reference to the Chironomidae. Ph.D. thesis, University of Waikato, New Zealand.

Brundin, L., 1951. The relation of O_2-Microstratification at the Mud Surface to the Ecology of the Profundal Bottom Fauna. Rep. Inst. Freshwat. Res. Drottningholm 32: 32–42.

Burrows, C. J., 1978. The Quaternary ice ages in New Zealand: A framework for biologists. Mauri Ora 6: 69–96.

Burrows, C. J., 1979. A chronology for cool climate episodes in the Southern Hemisphere 12000–1000 yr BP. Palaeogeogr., Palaeoclimatol., Palaeoecol. 27: 287–347.

Davies, M. B. & R. E. Moeller, 1985. Paleolimnology. Sedimentation. In: Likens, G. E. (Ed.). An Ecosystem Approach to Aquatic Ecology. Mirror Lake and Its Environment. Springer Verlag: 345–366.

Deevey, E. S., 1955. Paleolimnology of the Upper Swamp deposit, Pyramid Valley. Rec. Canterbury Mus. 6: 291–344.

Forsyth, D. J., 1971. Some New Zealand Chironomidae (Diptera). J. R. Soc. N.Z. 1: 113–144.

Forsyth, D. J., 1986. Distribution and production of *Chironomus* in eutrophic Lake Ngapouri. N.Z. J. mar. Freshwat. Res. 20: 47–54.

Gage, M., 1959. On the origin of some lakes in Canterbury. N.Z. Geographer 15: 69–75.

Grant, P. J., 1985. Major periods of erosion and alluvial sedimentation in New Zealand during the late Holocene. J. r. Soc. N.Z. 15: 67–121.

Grant, P. J., 1989. Effects on New Zealand vegetation of late Holocene erosion and alluvial sedimentation. N.Z. J. Ecol. 12 (Suppl.): 131–144.

Graham, A. A., 1976. Ecology and production of *Chironomus zealandicus* in Lake Hayes. M.Sc. thesis, University of Otago, N.Z., 96 p.

Green, J. D., 1979. Palaeolimnological studies on Lake Maratoto, North Island, New Zealand. In: Horie, S. (ed.), Palaeolimnology of Lake Biwa and the Japanese pleistocene 7: 416–438.

Griffiths, G. A., 1981. Some suspended sediment yields from South Island catchments, New Zealand. Wat. Resour. Bull. 17: 662–671.

Hayward, J. A., 1967. The Waimakariri Catchment. Tussock Grasslands & Mountain Lands Inst., Spec. Publ. 5, 288 p.

Hofmann, W., 1978. Analysis of animal microfossils from the Großer Segeberger See (F.R.G.). Arch. Hydrobiol. 82: 316–346.

Hofmann, W., 1988. The significance of chironomid analysis (Insecta: Diptera) for paleolimnological research. Palaeogeogr., Palaeoclimatol., Palaeoecol. 62: 501–510.

Kansanen, P. H., Aho, J. & L. Paasivirta, 1984. Testing the benthic lake type concept based on chironomid associations in some Finnish lakes using multivariate statistical methods. Ann. Zool. Fenn. 21: 55–76.

Lintott, W. H. & C. J. Burrows, 1973. A pollen diagram and macrofossils from Kettlehole Bog, Cass, South Island, New Zealand. N.Z. J. Bot. 11: 269–282.

McGlone, M. S., 1989. The Polynesian settlement of New Zealand in relation to environmental and biotic changes. N.Z. J. Ecol. 12 (Suppl.): 115–130.

McSaveney, M. J. & I. E. Whitehouse, 1989. Anthropogenic erosion of mountain land in Canterbury. N.Z. J. Ecol. 12 (Suppl.): 151–164.

McLeod, D. & C. J. Burrows, 1977. History of the Cass district. In: Burrows, C. J. (Ed.). Cass: history and science in the Cass district, Canterbury, New Zealand. Dep. of Botany, University of Canterbury: 23–36.

Moar, N. T., 1971. Contributions to the Quaternary history of the New Zealand flora. 6. Aranuian pollen diagrams from Canterbury, Nelson and North Westland, South Island. N.Z. J. Bot. 9, 80–145.

Molloy, B. P. J., 1977. The Fire History. In: Burrows, C. J. (Ed.). Cass: history and science in the Cass district, Canterbury, New Zealand. Dep. of Botany, University of Canterbury, New Zealand: 157–172.

Sæther, O. A., 1979. Chironomid communities as water quality indicators. Holarctic Ecol. 2: 65–74.

Schakau, B., 1986. Preliminary study of the development of the subfossil chironomid fauna (Diptera) of Lake Taylor, South Island, New Zealand, during the younger Holocene. Hydrobiologia 143: 287–291.

Simpson, K. W. & R. W. Bode, 1980. Common larvae of Chironomidae (Diptera) from New York State streams and rivers. N.Y. State Mus. Bull. 439, 105 p.

Soons, J. M., 1977. The Geomorphology of the Cass District. In: Burrows, C. J. (Ed.). Cass: history and science in the Cass district, Canterbury, New Zealand. Dep. of Botany, University of Canterbury, New Zealand: 79–92.

Stark, J. D., 1981a. Trophic Interrelationships, Life-Histories and Taxonomy of some Invertebrates associated with Aquatic Macrophytes in Lake Grasmere. Ph. D. thesis, University of Canterbury, Christchurch, N.Z., 256 p.

Stark, J. D., 1981b. Chironomidae (non-biting midges). In: Winterbourn, M. J. & L. D. Gregson. A Guide to the Aquatic Insects of New Zealand. Bull. Ent. Soc. N.Z. 5: 60–67.

Stout, V. M., 1972. Plankton composition in relation to nutrient inflow in a small New Zealand lake. Verh. Int. Ver. Limnol. 18: 605–612.

Timms, B. V., 1982. A study of the benthic communities of 20 lakes in the South Island, N.Z. Freshwat. Biol. 12: 123–138.

Timms, B. V., 1983. Benthic macroinvertebrates of seven lakes near Cass, Canterbury high country, New Zealand. N.Z. J. mar. Freshwat. Res. 17: 37–49.

Walker, I. R., 1987. Chironomidae (Diptera) in Palaeo-ecology. Quat. Sci. Rev. 6: 29–40.

Walker, I. R. & R. W. Mathewes, 1989. Chironomidae (Diptera) remains in surficial lake sediments from the Canadian Cordillera: analysis of the fauna across an altitudinal gradient. J. Paleolimnol. 2: 61–80.

Warwick, W. F., 1980. Palaeolimnology of the Bay of Quinte, Lake Ontario: 2800 years of cultural influence. Can. Bull. Fish. Aquat. Sci. 206, 117 p.

Wiederholm, T., 1983. Chironomidae of the Holarctic region. Keys and diagnosis. Part 1. Larvae. Ent. Scand. Suppl. 19, 457 p.

Whitehouse, I. E., 1984. Erosion in the eastern South Island high country – a changing perspective. J. Tussock Grasslands & Mountain Lands Inst., Rev. 42: 3–23.

Hydrobiologia **214**: 223–227, 1990.
J. P. Smith, P. G. Appleby, R. W. Battarbee, J. A. Dearing, R. Flower, E. Y. Haworth, F. Oldfield & P. E. O'Sullivan (eds), 223
Environmental History and Palaeolimnology.
© 1990 *Kluwer Academic Publishers.*

Modern assemblages of arctic and alpine Chironomidae as analogues for late-glacial communities

Ian R. Walker

Biology Dept., Queen's Univ., Kingston, Ontario K7L 3N6, Canada

Key words: arctic, alpine, Chironomidae, palaeoclimate, palaeolimnology

Abstract

Surficial sediment data, illustrating the differences between arctic and temperate chironomid faunas, are presented and briefly discussed.

When interpreting late-glacial assemblages of Chironomidae, palaeoecologists have relied principally upon ecological observations derived from studies of temperate lakes. Since arctic or subarctic climates are believed to have been prevalent at temperate latitudes during the late-glacial, it would seem more appropriate to discuss late-glacial chironomid palaeoecology on the basis of modern observations in lakes situated within these cold climatic regions. Although knowledge of arctic, subarctic, and alpine Chironomidae is broadly scattered through the literature, enough information exists for recent authors (e.g. Danks, 1981; Walker & Mathewes, 1989) to identify key characteristics of the fauna. Nevertheless, little reference to this literature can be found in chironomid palaeoecological discussions.

To illustrate the major differences between arctic and temperate faunas, and to facilitate future chironomid palaeoecological interpretation, I have compiled data on present chironomid distributions, as represented in the surficial sediments of lakes. These data are compiled from diverse sources (Table 1), including my unpublished data, and recent literature (Johnson & McNeil, 1988; Uutala, 1986; Walker & Mathewes, 1989; Walker, 1982; Warwick, 1982).

The data provide observations on the fauna throughout Canada, including grassland, parkland, and major forest regions in addition to arctic, subarctic, and alpine sites (Fig. 1 & 2). For the sake of brevity, little interpretation is offered here. These data supplement those supplied by Walker & Mathewes (1989). Readers are referred to Walker & Mathewes (1989) for a discussion of those environmental factors which regulate species composition in arctic/alpine regions.

The arctic/alpine fauna (Fig. 2), includes a distinctive element. *Parakiefferiella* sp.A is a common species in alpine lakes of the Canadian Rocky Mountains. *Parakiefferiella* sp.A also occurs in the high arctic, but I have not examined surficial sediments from any arctic lakes where it is common. *Paracladius* is characteristic of high arctic and alpine lakes. *Pseudodiamesa* occurs in alpine and upper subalpine lakes (e.g. Walker & Mathewes, 1989) as well as many arctic lakes. *Stictochironomus* is more abundant in arctic/alpine waters than elsewhere.

In lakes of the western Canadian low arctic, *Corynocera ambigua* is common, but has never been collected in the Canadian high arctic or in the eastern arctic. It is possible that temperature limitations prevent *Corynocera* from extending its

Table 1. Location and depth of lakes sampled for surficial sediments. [1-data from Walker & Mathewes, 1989; 2-sediments provided courtesy of R. King & A. Wolfe; 3-sediments provided courtesy of G. M. MacDonald; 4-sediments provided courtesy of D. R. Engstrom & S. C. Fritz; 5-data from Johnson & McNeil, 1988 (average for 14 cm surface core); 6-data from Uutala, 1986 (average for 20 to 40 cm long surface core); 7-data from Walker (1982, unpublished data); 8-data for Pasqua Lake from Warwick (1982), sediments from other lakes provided courtesy of R. E. Vance)].

Region	Lake Name	Latitude	Longitude	Altitude (m)	Depth (m)
Alpine (B.C. & Alberta)[1]	Opabin	51°20′N°°°	116°19′W	2270	6.0
	Ptarmigan	51 29	116 04	2330	8.5
	Hungabee	51 19	116 19	2240	2.5
High Arctic (Devon Island)[2]	Fish	75 39	84 32	30	4.5
	Immerk	75 41	84 34	30	6.8
	M. Beschel	75 39	84 28	30	8.2
	Phalarope	75 39	84 37	30	5.0
Western low arctic[3]	A2b	63 37	97 27	130	2.4
	B5b	63 15	97 20	110	2.3
	D2b?	63 35	97 21	115	4.8
Eastern low arctic[4]	Lab. 43	57 08	63 05	500	50.0
	Lab. 50	56 39	64 32	510	9.0
	Lab. 54	56 17	63 57	480	3.0
Western subarctic[3]	SS75	62 33	113 55	200	2.0
	SS33	62 16	109 33	360	6.5
Labrador[4]	Lab. 1	51 27	57 12	43	14.0
	Lab. 3	51 30	57 14	43	23.0
	Lab. 14	52 22	57 43	380	1.0
Ontario[5]	U. Batchawana	47 04	84 24	495	11.3
	Wishart	47 03	84 24	390	4.5
	Turkey	47 03	84 25	360	37.0
British Columbia[1]	Lost	50 08	112 56	690	11.0
	Deer	49 22	121 40	60	5.0
	Hicks	49 20	121 42	60	17.0
New York, U.S.[6]	Big Moose	43 49	74 51	556	18.0
	Deep	43 37	74 40	789	22.5
	Brooktrout	43 36	74 40	722	20.0
	U. Wallface	44 09	74 03	948	8.0
	Windfall	43 48	74 50	597	5.5
New Brunswick[7]	Black	46 06	64 22	34	6.0
	Wood's	45 53	64 24	25	2.0
	Portey	45 51	64 25	45	1.5
Grassland & Parkland[8]	Pasqua	50 47	104 00	475	13.3
	Tyrrell	49 23	112 36	970	4.4
	Rhinehart	49 01	113 15	1265	1.6
	Chappice	50 10	110 22	730	2.7
	Bow Island Sl.	49 55	111 36	780	<0.3

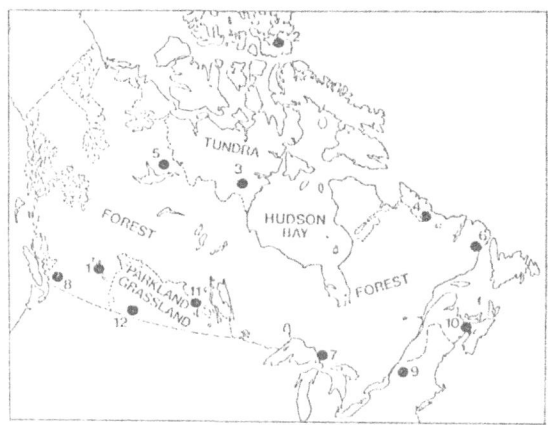

Fig. 1. Map of Canada illustrating distribution of surface sample sites with respect to major climate-vegetation zones (1. B.C. & Alberta (alpine), 2. Devon Island (high arctic), 3. Western low arctic, 4. Eastern low arctic, 5. Western subarctic, 6. Labrador, 7. Ontario, 8. British Columbia, 9. Northern New York, 10. New Brunswick, 11. Pasqua Lake, Saskatchewan, 12. Alberta).

range farther north. Hudson Bay may also have prevented the postglacial dispersal of *Corynocera* to eastern arctic sites.

Many common arctic/alpine taxa also occur in the cold profundal of very large, deep, oligotrophic lakes of temperate latitudes, including *Parakiefferiella* sp.A, *Paracladius*, and *Heterotrissocladius subpilosus*. These temperate populations may be considered relicts (e.g. Sæther, 1975). *Pseudodiamesa* has not been collected from lakes outside of arctic, alpine, and subalpine regions, but relict populations exist in cold springs in temperate areas (e.g. Siciński, 1988). *Abiskomyia* and *Oliveridia* are also characteristic taxa of arctic lakes (Wiederholm, 1983).

Heterotrissocladius, *Protanypus*, and *Sergentia* are common in the cold profundal of stratified lakes from temperate forest regions as well as arctic sites. Although *Heterotrissocladius* was not recorded from the Devon Island samples that I have examined (Fig. 2), evidence from other studies (e.g. de March *et al.*, 1978; Oliver, 1964; Welch, 1976) indicates that this genus is a dominant component of the lacustrine fauna throughout most of the arctic. At temperate latitudes, *Heterotrissocladius* and *Protanypus* tend to

be most abundant in deep, oligotrophic lakes, where temperatures remain cool throughout the year, and significant oxygen depletion does not occur. *Sergentia coracina* also requires cool waters, but thrives in lakes with moderate oxygen depletion.

The most distinctive aspect of the arctic/alpine fauna is the absence of many taxa common in temperate littoral habitats. Some genera, including *Chironomus*, *Procladius*, and *Psectrocladius* are common in shallow, relatively warm arctic ponds, but are excluded from the colder, deeper habitats. Since most arctic lakes do not stratify, warm littoral habitats are not available in arctic lakes. Most genera of Chironomini, many Orthocladiinae, Pentaneurini, *Pseudochironomus*, *Stempellina* and *Stempellinella* are common in temperate lakes, but have distributions which do not extend beyond arctic treeline, or have distributions which barely extend onto the southernmost margin of the tundra. It is probable that arctic/alpine summer temperatures are not sufficiently warm for these taxa to either pupate or emerge (Walker & Mathewes, 1989).

The fauna of grassland and aspen parkland lakes in southwestern Canada is also distinct from that of arctic/alpine sites, and most forested regions. Most of these lakes are eutrophic and/or saline. *Chironomus* and *Procladius* are common in the more dilute lakes. At high salinities, the fauna is restricted to only three chironomid taxa, *Chironomus* nr. *annularis*, *Cricotopus ornatus*, and *Tanypus nubifer* (Timms *et al.*, 1986). In addition to the marked distinctions between arctic and temperate faunas, these salinity-related differences should prove useful in future paleoclimatic studies.

The surficial sediment fauna of an ephemeral grassland pond, Bow Island Slough, was dominated by soil-inhabiting Chironomidae (Fig. 2). It is probable that the ratio of this group to lacustrine taxa will, in the future, provide a useful index of pond permanence. Thus records of precipitation balance should be recorded by chironomids in the sediments of grassland lakes.

Studies of chironomid remains in the surficial sediments of lakes provide the best means for assessing potential analogues. For many late-

Fig. 2. Percentage chironomid diagram illustrating the distribution of chironomid taxa among lakes and among major climate-vegetation regions in Canada and adjacent northeastern United States. Open bars indicate expanded horizontal scale used for four taxa. Asterisks indicate samples where data regarding Ceratopogonidae and *Chaoborus* are not available.

glacial sites, arctic/alpine lakes provide the best analogues. For example, in Atlantic Canada, arctic/subarctic vegetation and climates are inferred, from palynological evidence, to have existed during late-glacial time. The depauperate Chironomini fauna in early late-glacial deposits (Walker & Paterson, 1983) also implies a low arctic or high subarctic environment.

References

Danks, H. V., 1981. Arctic Arthropods: A Review of Systematics and Ecology with Particular Reference to the North American Fauna. Ent. Soc. Can. Ottawa. 608 pp.

de March, L., B. de March & W. Eddy, 1978. Limnological, fisheries, and stream zoobenthic studies at Stanwell-Fletcher Lake, a large high arctic lake. Arctic Islands Pipeline Program, Preliminary Report 1977. Dept. of

Indian and Northern Affairs Publ. QS-8160-004-EE-A1, Ottawa.

Johnson, M. G. & O. C. McNeil, 1988. Fossil midge associations in relation to trophic and acidic state of the Turkey lakes. Can. J. Fish. aquat. Sci. 45 (Suppl. 1): 136–144.

Oliver, D. R., 1964. A limnological investigation of a large arctic lake, Nettilling Lake, Baffin Island. Arctic 17: 69–83.

Sæther, O. A., 1975. Nearctic chironomids as indicators of lake typology. Verh. int. Ver. Limnol. 19: 3127–3133.

Siciński, J., 1988. New data on the rare species *Pseudodiamesa nivosa* (Goetghebuer) (Diptera, Chironomidae). Aquatic Insects 10: 73–76.

Timms, B. V., U. T. Hammer & J. W. Sheard, 1986. A study of benthic communities in some saline lakes in Saskatchewan and Alberta, Canada. Int. Revue ges. Hydrobiol. 71: 759–777.

Uutala, A. J., 1986. Paleolimnological assessment of the effects of lake acidification on Chironomidae (Diptera) assemblages in the Adirondack region of New York. Ph.D. thesis, State Univ. of N.Y., Coll. Environ. Sci. For., Syracuse, U.S.A.

Walker, I. R., 1982. The Chironomidae (Diptera) of shallow, humic lakes and bog pools, and their value as palaeoenvironmental indicators. M.Sc. thesis, Univ. Waterloo, Waterloo, Ont.

Walker, I. R. & R. W. Mathewes, 1989. Chironomidae (Diptera) remains in surficial lake sediments from the Canadian Cordillera: analysis of the fauna across an altitudinal gradient. J. Paleolimnol 2: 61–80.

Walker, I. R. & C. G. Paterson, 1983. Post-glacial chironomid succession in two small humic lakes in the New Brunswick – Nova Scotia (Canada) border area. Freshwat. Invertebr. Biol. 2: 61–73.

Warwick, W. F., 1982. The palaeolimnology of Pasqua Lake, southeastern Saskatchewan. Can. Dep. Envir., Inland Wat. Dir., Nat. Wat. Res. Inst. Tech. Rep. W.N.R.-82-1, 70 pp.

Welch, H. E., 1976. Ecology of Chironomidae (Diptera) in a polar lake. J. Fish. Res. Bd Can. 33: 227–247.

Wiederholm, T. (ed.), 1983. Chironomidae of the Holarctic Region: Keys and Diagnoses, Part I, Larvae. Ent. scand. Suppl. 19: 1–457.

Hydrobiologia **214**: 229–238, 1990.
J. P. Smith, P. G. Appleby, R. W. Battarbee, J. A. Dearing, R. Flower, E. Y. Haworth, F. Oldfield & P. E. O'Sullivan (eds), 229
Environmental History and Palaeolimnology.
© *1990 Kluwer Academic Publishers.*

Paleolimnology of Neusiedlersee, Austria

I. The Succession of Ostracods

Heinz Löffler
Biocenter, Althanstrasse 14, A-1090 Vienna, Austria

Key words: Paleoecology, Neusiedlersee, Ostracoda

Abstract

Neusiedlersee is a shallow, alkaline lake (Table 1), which came into existance by tectonic subsidence about 12 000–14 000 yr ago. At present, half of it is covered by *Phragmites australis* which developed after the lake fell dry for the last time in 1868.

Due to its astatic character, with about 100–200 dry periods since the lake came into existence, most of the sediment of the open lake overlying a thick Tertiary layer presents a mixture of terrestrial and lacustrine components. Because of disturbance by wind, this lacks stratification. Preliminary investigations, however, demonstrated that within the fringing *Phragmites* belt – especially landward – such a stratification does still exist. In order to learn more about the history of the lake 18 deep freeze cores of between 135 and 190 cms in depth, were collected from the *Phragmites* belt and investigated for chemical data (see Gunatilaka, this volume) and the succession of ostracods. Most of the cores contain Tertiary material (foraminifera and ostracods) at their base, followed by a cold water fauna as indicated by the presence of *Cytherissa lacustris*. Indicators of high alkalinity (i.e. the remains of male *Limnocythere inopinata* and high salinity (*Heterocypris salina*) were found in only a few cores.

Introduction

Neusiedlersee is a shallow lake (Löffler, 1982) covering 321 km^2 (Table 1), of which 178 km^2 are covered by a *Phragmites* belt and isolated *Phragmites* stands. It is situated in the easternmost part of Austria with a small part extending into Hungary (Fig. 1, Table 1). It occupies a tectonic depression in deep Pannonian sediments deposited during the last stage of the Paratethys Sea, which came into existence during the end of the Pleistocene, some 12 000–14 000 yrs ago. Originally without outflow and including a large area in the southeast (the so called 'Hanság'), most likely representing the precursor of the present lake, later part of it and finally drained by man and claimed for agriculture, the lake has been extremely astatic throughout its history, with 100–200 desiccations and as many high-water episodes covering an area almost twice as large as at present. During high water levels the lake discharged into the lower Raab-Rabnitz river system and finally the Danube. Due to the construction of dams and an artifical outlet, the Hanság became dry and, as a result of a bilateral agreement between Hungary and Austria in 1965, the level of the present lake is kept permanently at 115.4 m altitute by a sluice and only slight fluctuations may occur. Since the last dessication in 1968, the area occupied by *Phragmites australis* has expanded, especially along the western and southern shores, and resulted in an accumulation of approximately 100–150 million cubic meters of sediment within its vegetation zone. The limn-

Fig. 1. The area of Neusiedlersee and Seewinkel with the core sites and the indication of the depth of the onset of lacustrine remnants (The Zitzmannsdorf core is not presented since it did not contain any lacustrine material). (Mö: Mörbisch; R1,2: Rust; Og: Oggau; Do: Donnerskirchen; Po: Site south of Purbach; Pur 1,3,5: Purbach; Br 1,2: Breitenbrunn; W 2,1: Winden; J 1,2: Jois; Ns: Neusiedl; Wei: Weiden).

Table 1. Neusiedlersee, background data

Lat.:	47°38′–47°57′
Long.:	16°41′–16°52′
Altitude (m):	115.4
Length (km):	33.5
Width (km):	12.0
Max. depth (z) (m):	2.2
Mean depth ($\bar{z}$) (m):	0.8
Ratio z/$\bar{z}$:	0.36
Volume (10^6 m^3):	250
Lake Area (A) (km^2):	321 (Phragmites: 178)
Catchment (A′) (km^2):	1400
Ratio A′/A:	4.8
Theoretical retention time:	3–4 years
Secchi depth (m, average):	0.2–0.3
Ice cover (days):	0–60
Conductivity (μS/20 °C):	1000–2800
Alkalinity (meq l^{-1}):	9–13

ological features of the open lake and the *Phragmites* belt differ greatly. If not frozen over, the open lake is rather turbid with transparencies often less than 20 cm whereas the water within the *Phragmites* belt is clear but strongly coloured by organic matter. Apart from other parameters (oxygen conditions, nutrients etc.), the plant (including algae) and animal communities of the open lake and the *Phragmites* belt are completely different (Table 2).

The sediments of the open lake, which rest on top of 400–600 m of Pannonian deposits almost lacking in any remains of organisms, represent a mixture of allochthonous (washed-in terrestrial), autochthonous (lacustrine) and eroded Pan-

Table 2. Open lake and Phragmites marsh species (list of important remnants).

Ostracoda:	with the exception of the
Limnocythere inopinata	*Cytherissa lacustris* all the other
Ilyocypris bradyi	species in Table 3
Cladocera:	
Moina ephippia	*Simocephalus* ephippia
Gastropoda:	
	all 10 species confined to the
	Phragmites belt

nonian material. The thickness ranges from zero to sixty centimeters and, due to ongoing transfer of turbid material into the *Phragmites* belt area, is obviously continuously eroded to an unknown extent. Frequent disturbance of this sediment inhibits any orderly stratification and therefore attempts to analyze it have been a failure. So far the only reliable core data from the whole Neusiedlersee-Seewinkel area has been obtained from the Hanság area, limited to a small section representing approximately 2000 years of the latest Pleistocene (Bobek, Löffler & Schultze, 1978). Therefore the *Phragmites* belt was thought to be a promising site for further information about the details of the lake's history such as:

– the assessment of the ecological conditions of the early lake stage during the late Pleistocene
– the consequence of the spreading of *Phragmites* since the last desiccation of Neusiedlersee in 1868
– the amount of sediment accumulated within the vegetation belt
– the influence of changing salinity (alkalinity)
– recent impacts of man.

Methodology

In order to cope with the coring difficulties caused by the rhizome layer of *Phragmites* an *in situ* deep-freeze technique was applied (see Gunatilaka, this volume) to 18 cores taken between the villages of Weiden and Mörbisch (Fig. 1). Since previous tests indicated that the Upper Tertiary boundary is at a depth of 150 to 200 cm, the length of the cores was selected accordingly. For technical reasons, however, the appropriate depth could not be attained at nine of the sites. From the cores obtained, 2 cm^3 samples were collected at intervals of 5 cm and rinsed through a filter with a mesh-size of 50 μm for the inspection of plant (mainly *Chara* (Fig. 5) and large diatoms) and animal remnants. The ostracods and foraminifera were preserved in 'Franke-slides'.

Results and discussion

Four main combinations can be distinguished from the ostracod remains together with some remnants of other organisms.

1. The Tertiary component comprises a wide range of Miocene stages from the Badenian to the Pannonian. The early remnants (forams and pieces of corals) are mainly contributed by washed-in material.

2. An early lacustrine stage of the late Pleistocene is represented by the holarctic species *Cytherissa lacustris*, a long-lived ostracod (at least two years) which does not tolerate long periods of temperatures above 18 degrees C. At present it occurs only in deep lakes at low altitude, whereas in shallow lakes it is restricted to higher altitude or latitude. It is therefore now extinct in the Neusiedlersee area but the shells occur in 11 of the 18 cores (e.g. Mörbisch, Purbach 1 and 3, Breitenbrunn 1, Jois 1 and 2, Winden 2, see Fig. 2 and 3). These *Cytherissa* sections of the cores possibly comprise the last two millenia of the Pleistocene. In contrast to the Hanság (Bobek, Löffler & Schultze, 1978) the cold water species, *Limnocythere sanctipatricii*, could not be found at any of the sites and has never been found in the present Neusiedlersee area. This supports the hypothesis that this former part of Neusiedlersee – the Hanság – is much older (1000–2000 yr) than the present lake.

3. The present open lake fauna is poor in ostracod species. Only *Limnocythere inopinata*, an eurythermic species, and *Ilyocypris bradyi* occur. *L. inopinata* abounds in the open lake but is absent from the *Phragmites* belt (Table 2). It is, however, present in 17 of the 18 cores, indicating a phase when *Phragmites* was possibly absent early in the period since the last dessication in 1868. The theory is strongly supported by the presence of other organisms such as *Moina* sp. (ephippia) and the diatom *Campylodiscus clypeus*, that are now restricted to open water. There exists, however, no information about the extent of the *Phragmites* stands during and before the last desiccation of the lakes. The only known painting of Neusiedlersee from the early 19th century, by Schnorr-Carolsfeld, does not give any indication of the presence of *Phragmites*.

In 10 cores, the upper *Limnocythere inopinata* section (Fig. 3, 4) coincides with remains of organisms such as gastropods (mainly *Planorbidae*), which are today confined to the *Phragmites* stands, and the ephippia of *Simocephalus* sp., *Notodromas* sp. and various *Candona* spp. Obviously this co-occurence is caused by seiche movements transferring remnants of open lake species into the spreading *Phragmites* vegetation. The rare occurence of the males of *Limnocythere inopinata* (Fig. 3) is a special feature. They have been observed only in strongly alkaline bodies of water, so their presence in the cores of Purbach 5 (95–85 cm) and Oggau (70–65 cm) may reflect an increased alkalinity about sixty years ago.

4. The *Phragmites* marshes are the site of a great diversity of both plants and animals. In contrast to the open lake with only two ostracod species, a community of more than 20 species is distributed in the *Phragmites* belt (Table 3). Among these the Candoninae clearly dominate (14 species), followed by Cyclocyprinae (4 species), Cyprinae (5 species) and others. As very little is known about the ecology of most of the species, only a few points may be stressed. The presence of *Notodromas* (probably two species: *N. monacha* and *N. persica*) clearly indicates protection from wind – induced wave movements. *Potamocypris unicaudata* and *Heterocypris salina* are halophilic species; the former, described also from saline ponds east of Neusiedlersee, is confined to the upper layer of 4 cores at sites (Morbisch, Purbach and Jois 1 and 2) where the impact of tourism and C1-containing detergents may be responsible. Similarly, the occurence of *H. salina* at the same sites suggests the recent influence of man-induced salinity. *H. salina*,

CYTHERISSA LACUSTRIS (■), FORAMINIFERA (☐), OSTRACODA, Pannonian (▨)

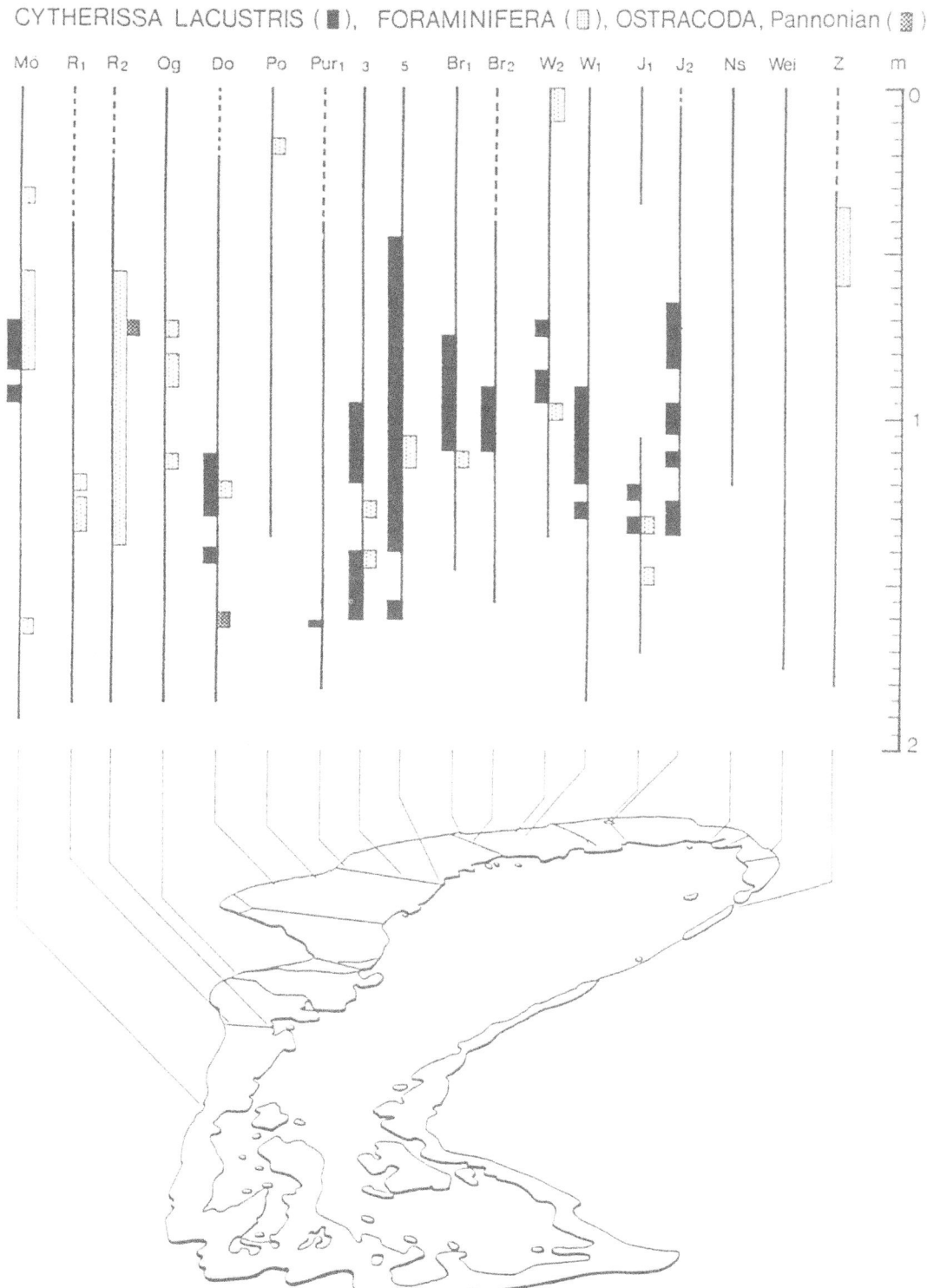

Fig. 2. The distribution of *Cytherissa lacustris*, Foraminifera and Pannonian ostracods.

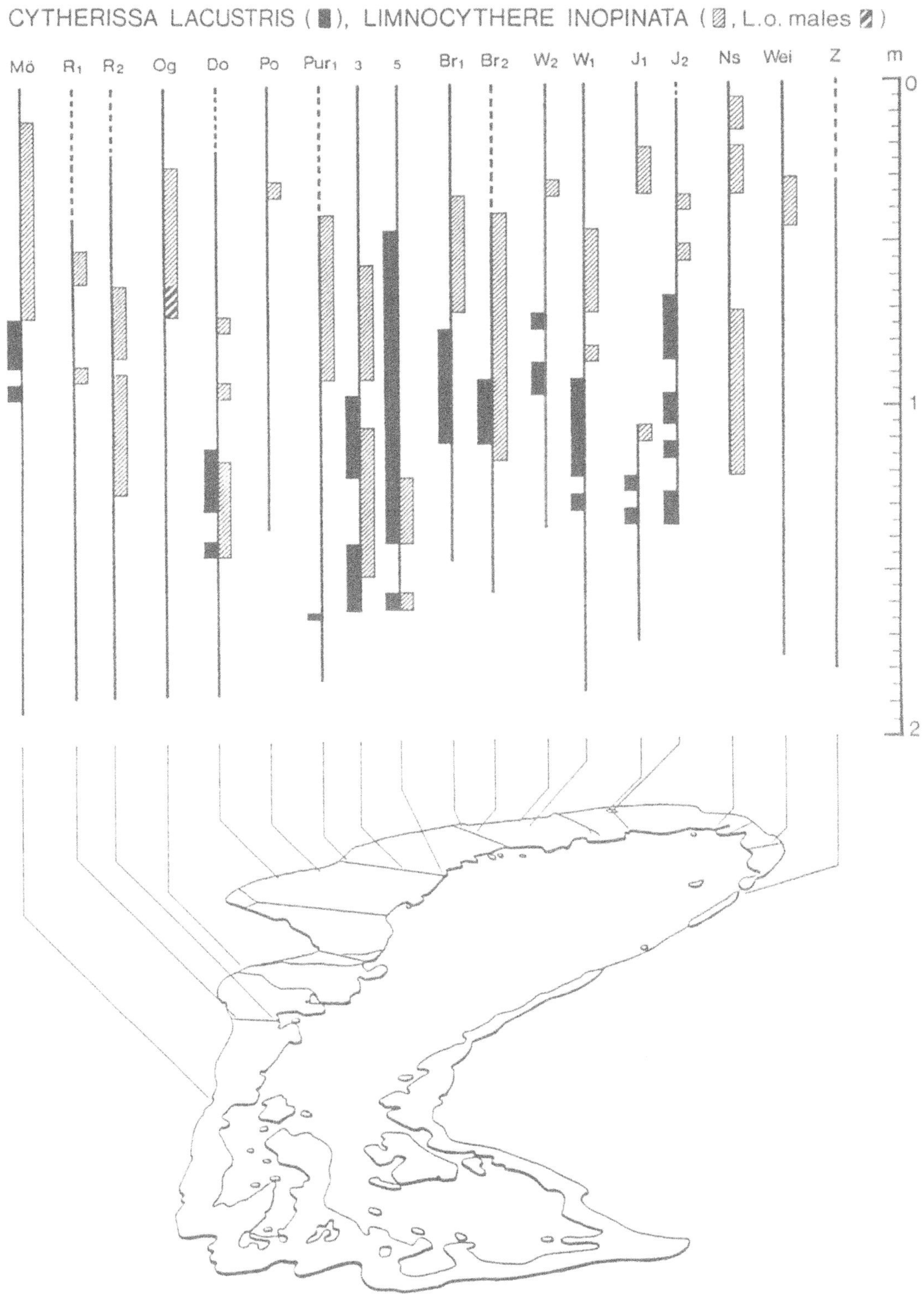

234

Fig. 3. The distribution of *Cytherissa lacustris* and *Limnocythere inopinata*.

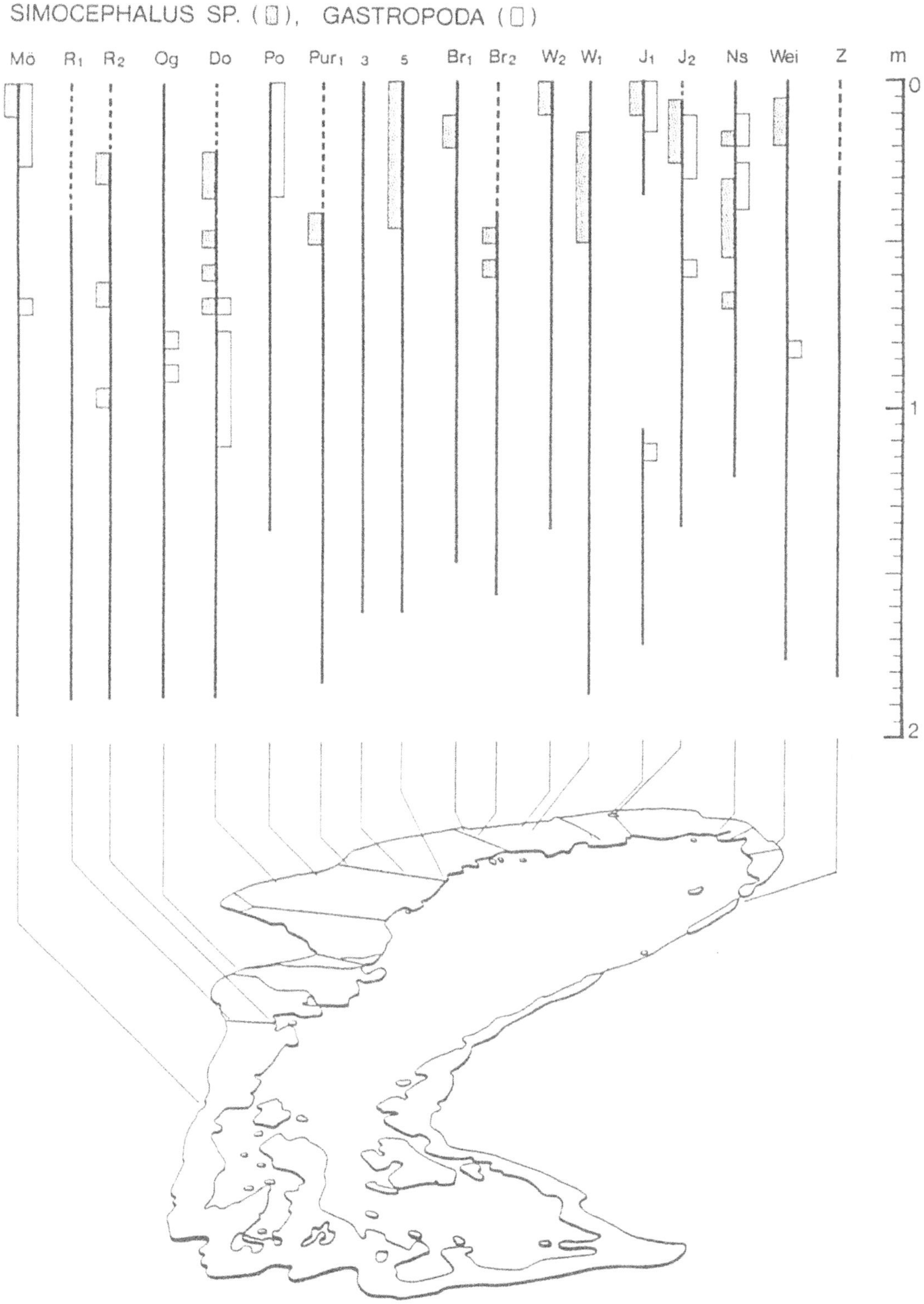

235

SIMOCEPHALUS SP. (▨), GASTROPODA (▢)

Fig. 4. The distribution of *Simocephalus sp.* and Gastropoda.

CHARA SP.

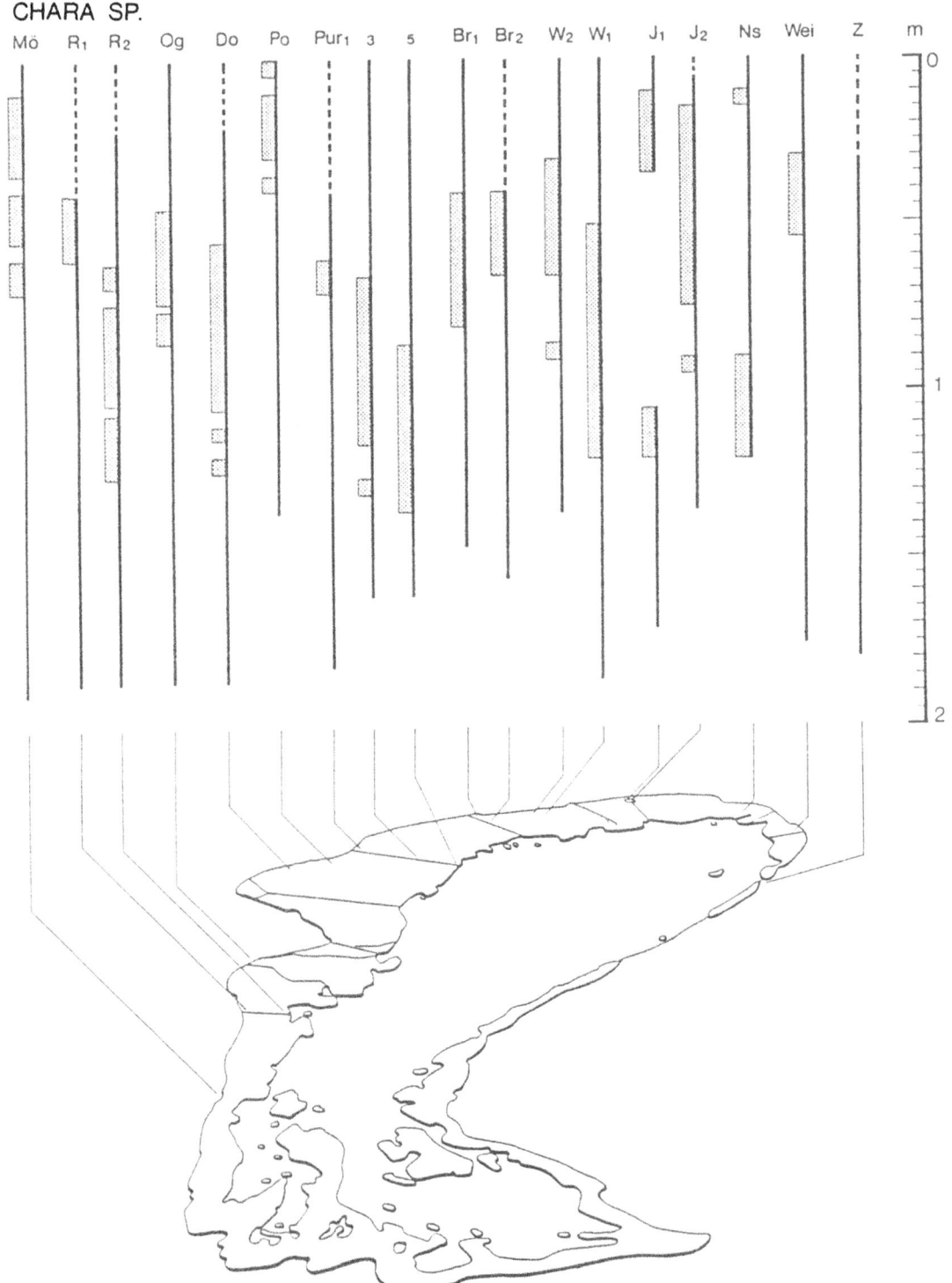

Fig. 5. The distribution of *Chara* sp.

HETEROCYPRIS SALINA (▯), POTAMOCYPRIS UNICAUDATA (▮)

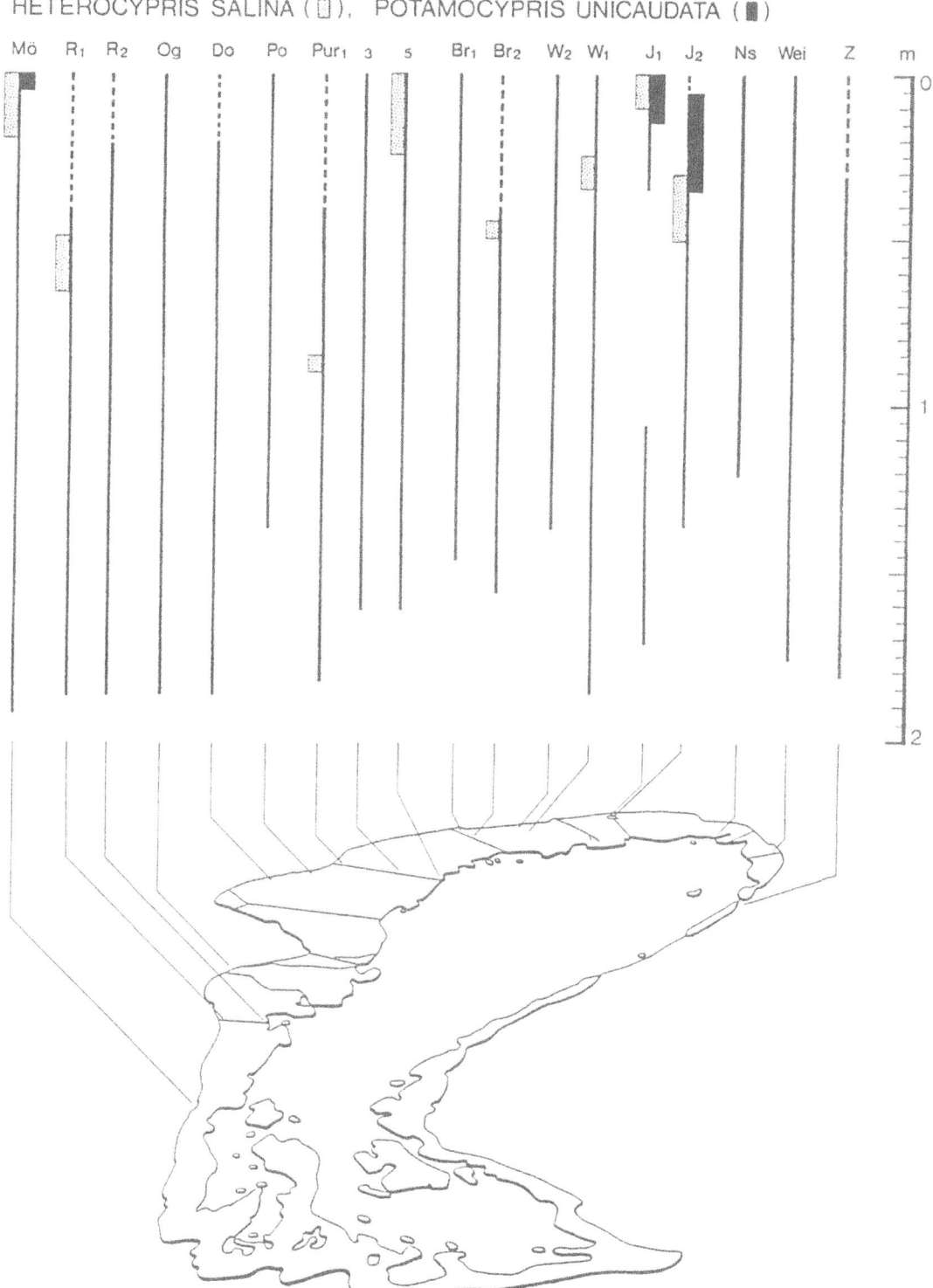

Fig. 6. The distribution of *Heterocypris salina* and *Potamocypris unicaudata*.

238

Table 3. List of ostracod species found in the Phragmites belt cores.

---

* *Candona candida* (O. F. Müller)
  Candona neglecta G. O. Sars
* *Candona costrata* Brady & Norman
  Candona marchica Hartwig
* *Candona hartwigi* G. W. Müller
* *Candona compressa* (Koch)
  Candona pratensis Hartwig
  Candona sucki Hartwig
  Candona fabaeformis Fischer
* *Candona fragilis* Hartwig
* *Candona protzi* Hartwig
  Candonopsis kingsleii (Brady & Robertson)
  Cyclocypris laevis (O. F. Müller)
  Cyclocypris ovum (Jurine)
* *Cypria ophthalmica* (Jurine)
* *Physocypria fadeewi* Dubovskij
* *Ilyocypris gibba* (Ramdohr)
  Notodromas monacha (O. F. Müller)
  Cypris pubera O. F. Müller
* *Heterocypris salina* (Brady)
  Erpetocypris sp.
* *Cypridopsis vidua* (O. F. Müller)
* *Potamocypris unicaudata* H. W. Schäfer
* *Darwinula stevensoni* (Brady & Robertson)
  Limnocythere inopinata (Baird)
  Cytherissa lacustris G. O. Sars
* *Metacypris cordata* Brady & Robertson

---

already mentioned, the corer did not penetrate into the solid tertiary sediment at 9 of the sites but attained only a depth close to the upper mixed layers found in the other cores. Fig. 1 presents the approximate depth of the occurence of the first remnants of organisms of Neusiedlersee. It appears that the sediment within the *Phragmites* belt, influenced by the dynamics of the open lake, comprises a stratum between 1 and 2 m thick. Its extent is influenced by the inflows, by terrestiral erosion and by seiche movements.

The investigation so far carried out has resulted only in a tentative description of the lake's history; to learn more, dating is of course essential. However, the disturbance of the sediment by wind and water action, and even more by bioturbation of the rhizomes of *Phragmites*, make such attempts rather problematic. It is hoped then that dating of ostracod and mollusc shells will provide some satisfactory data in the near future.

Acknowledgements

The author expresses his thanks to Dr. A. Gunatilaka for collecting the core-material and to Mr. Klejch for the elaboration of the figures.

however, occurs also at deeper strata at Rust (60–50 cm), Breitenbrunn (50–45 cm) and Weiden (35–30 cm) and is therefore an additional indication of increased salinity some 60 yr ago. Finally, the presence of *Physocypria fadeewi*, a Ponto-Caspian species, demonstrates the hydrographic connection with the Danube.

If the disturbance of the uppermost Pannonian layers and their mixing with the early and later lacustrine stages of Neusiedlersee sediments are taken into consideration, the upper boundary of the Tertiary sediment can be roughly estimated from the animal stratigraphy results obtained. As

References

Bobek, M., H. Löffler & E. Schultze, 1978. Neue Daten zur Geschichte des Neusiedlersees. Biol. Forsch. Inst. Burgenld. Ber. 29: 5–10.

Gunatilaka, A., 1991. Palaeolimnology of Neusiedlersee, Austria. II: The distribution of nutrients and trace metals. In J. P. Smith *et al.* (eds.), Environmental History and Palaeolimnology. Kluwer Academic Publishers, Dordrecht: 239–244. Reprinted from Hydrobiologia 214.

Löffler, H. (ed.), 1979. Neusiedlersee. Limnology of a shallow lake in Central Europe. Dr W. Junk, The Hague, 543 pp.

Löffler, H., 1982. Limnological aspects of shallow lakes. In: D. O. Logofet and N. K. Luckyanov (eds), Proc. of the International Scientific Workshop on Freshwater Ecosystem Dynamics in Wetlands and Shallow Water Bodies, 1: 37–62. Center of International Projects GKNT, Moscow.

Hydrobiologia **214**: 239–244, 1990.
J. P. Smith, P. G. Appleby, R. W. Battarbee, J. A. Dearing, R. Flower, E. Y. Haworth, F. Oldfield & P. E. O'Sullivan (eds), 239
Environmental History and Palaeolimnology.

Palaeolimnology of Neusiedlersee, Austria

II. *The distribution of nutrients and trace metals*

A. Gunatilaka

Dept. Limnology, Inst. Zool., Univ. Vienna, Althanstr. 14, A-1090 Vienna, Austria

Key words: organic phosphorus, mineralisation, trace metals, ancient sediments, Neusiedlersee

Abstract

Eighteen sediment cores from the Neusiedlersee reed belt were analysed for organic and inorganic carbon, nitrogen, organic and inorganic phosphorus and trace metal content. The depth variation of the distribution patterns of *C*, *P* and *N* in the cores are influenced by: 1. temporal variability of organic matter deposition (e.g. burial of *Phragmites* litter) 2. the impact of a number of astatic phases in the history of the lake, and 3. early palaeo-ecological changes associated with the alteration of marine to lacustrine environment. At some sites, due to mineralisation, the organic phosphorus compartment in the Pannonian sediments is completely exhausted; accordingly there is a parallel decrease in carbon and nitrogen content in the profiles. These changes in the pattern of the distribution of the nutrients (especially organic phosphorus) can be used to differentiate between the recent and the upper boundary layer of the tertiary sediments. There is no clear trace metal distribution pattern observed in the cores but the Pannonian sediments recorded the lowest levels. In contrast, at some locations surface sediments register a two to three fold increase. The highest *C*, *P*, *N*, values and trace metal concentrations are recorded in the surface sediments deposited after the drying up of the lake a hundred and twenty-two years ago.

Introduction

The large Phragmites belt in the Neusiedlersee represents the ultimate stage of hydroseral succession in the lake. This shallow lake in Austria ($47°82'$ N, $16°77'$ E, $z_{max} = 2.2$ m) came into existence during the end of the Pleistocene as a tectonic depression in Pannonian sediments. Since then, dry periods has reduced the area of the lake at about half its original size. After the last draw down in 1868, *Phragmites australis*, a highly productive wetland species, has colonized its shores forming monospecific, dense reed beds (see Fig. 1 in Löffler, 1991). The broad reed stands can extend up to 7–11 km, especially along its western and southern shores. Recent aerial mapping show that *P. australis* covers now 178 km² of the lake littoral, which is nearly half of the lake area.

The reed extension in the littoral has been promoted by a number of natural and anthropogenic factors. *P. australis* is propagated mostly vegetatively and the presence of a suitable anaerobic substrate favours rhizome expansion. The major processes in substrate building are: 1. natural erosion of sediments from the open lake (through strong wind action) and subsequent redeposition in the reed belt; 2. erosion of soils from the catchment due to intensive agricultural practices; 3. high autochthonous production of plant biomass [11–16 t.ha^{-1} (d.w) below ground biomass is nearly twice that above ground] that undergoes annual burial. Due to prevailing anaerobic conditions, the decomposition of this material is relatively slow (Gunatilaka, 1985). These processes have resulted in rapid sediment accumulation in the reed belt and the building up of a thick sediment-rootmat zone above the Pannonian sediments. In this study we focused our attention on the *Phragmites* belt to get more insight into the eutrophication history of the lake and to get more information on: 1. sediment accumulations since the last desiccation; 2. clues to the original lake boundaries; 3. differences between the recent and Pannonian sediments.

However, because of the sediment dynamics between the open lake and the reed belt, and due to the internal nutrient and metal cycles in the reed plants (transfer of nutrients to the rhizome and back is common in most of the perennial marsh plants) the use of ^{137}Cs and ^{210}Pb dating techniques are questionable; as the sediments are carbonate rich they are unsuitable for ^{14}C dating. Therefore the dating of sediments here is based on the associated Ostracod fauna which served as discrete time markers (Löffler, 1991). This paper presents a comparison of the nutrient and metal levels in recent and ancient sediments.

Materials and methods

Eighteen cores have been taken from the reed belt of the Neusiedlersee, along four landward to lakeward transect at Winden, Breitenbrunn, Purbach, Rust and at the outer reed fringe [from Weiden (NE) to Mörbisch (SW), which covers the areas of dense reed growth; place names and abbreviations are same as in Figs 1 and 2 in Löffler, 1991]. To overcome sampling difficulties and to obtain undisturbed sediment cores, a specially designed freeze-corer was used (Gunatilaka & Niedereiter, in prep.). The freeze-corer comprised of a long aluminum shaft ($1 \times 8 \times 250$ cm) and a head with a gas inlet and outlet. In the shaft there are four parallel interconnected canals to facilitate the rapid distribution of liquid nitrogen in the metal column (for a better freezing effect). To obtain a core, the shaft was pressed down carefully to the hard sediment (tertiary) and liquid nitrogen was pumped in (using a portable compressor or an oxygen cylinder) for 4–6 min, and then the frozen shaft was lifted with a help of a crank and a pulley fixed on to a tripod mounted on a wooden platform. The frozen sediment core was cleaned by removing the outer layers, and subsequently the sediment was detached from the aluminum shaft in 50 cm long 2 cm thick frozen pieces which were stored in dry ice boxes to be transported to the laboratory.

In the laboratory sediment cores were sectioned at 1 cm intervals and samples at every five cm were analysed in detail for nutrient and metal content. Total phosphorus (TP), organic phosphorus (OP) and inorganic phosphorus (IP) in the sediments were analysed according to Aspila *et al.* (1976) and Gunatilaka (1988); organic carbon and total nitrogen were determined using a Carlo Erba (NA-1500) CHN analyser. For metal analysis, sediments were digested in a mixture of nitric (Suprapur, Merck) and hydrofluoric (spectroscopic grade, Baker) acids (3:1) in Teflon lined Tölg type pressure vessels at 140 °C for 6 hrs. The contents were made up to 10 ml and analysed for Cd, Pb, Zn and Cu content using standard AAS methods. IAEA quality control service reference materials (lake sediments) IAEA/SL3 and IAEA/SL1 were used as controls.

Results and discussion

Sediment cores collected from the reed belt show the following zonation: 1. accumulation horizon (AH), which is comprised of buried *Phragmites* litter (several years old stems + leaves), 2. rhizome-root mat zone (RZ), 3. a blackish grey transition zone (TZ); (RZ and TZ together forms a sapropel) and 4. a zone of yellowish grey Pannonian sediments (PS). The distribution of redoxpotential in the sediment profiles (Fig. 1) show that the plant litter undergoes decomposition in a highly reduced environment. The anaerobic conditions prevail throughout the year and hence the decomposition is slow (see Gunatilaka, 1985).

The sedimentary organic material in the cores

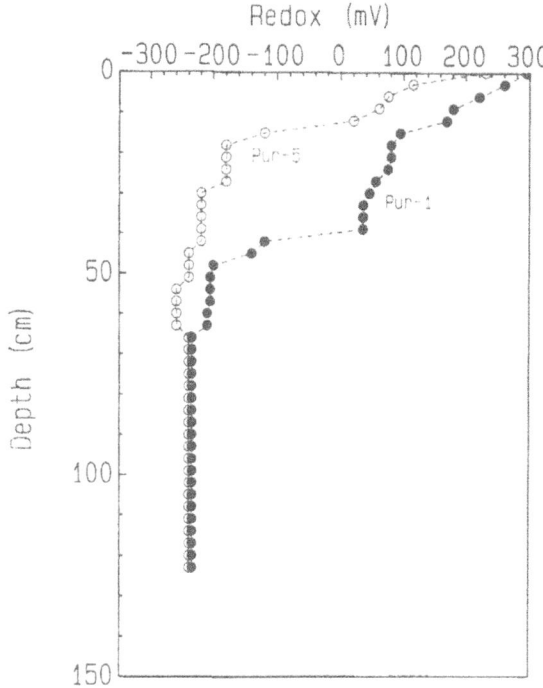

Fig. 1. The redoxpotential in lakeward and landward sites in the reed stands which are situated 100 m (Pur-5) and 3500 m (Pur-1) away from the open lake.

The main input of carbon and nitrogen to the sediment is from the *Phragmites* litter; nevertheless the contribution from other plants (eg. *Chara*) and organisms (mollusks, zooplankton and algae) cannot be neglected. The deeper, tertiary sediments also include refractory organic remains from lake organisms (e.g. plankton, ostracods and forams) and plant remains (e.g. *Chara*) and materials from sedimentary rocks and soils washed in from the surrounding basin.

There are marked differences in the sediment profiles from landward and lakeward areas. In lakeward areas the AH is compact and the litter is often buried in eroded sediments from the open lake which are comprised of fine clays (illite, chlorite, montmorillonite). They are deposited mainly in the near shore areas and have given rise to large sediment accretions (100–125 cm above RZ) in western and southern parts of the lake. It was observed that during strong winds sediments can be transported sometimes up to 1.5 km in the western part of the reed belt. In landward areas the AH is comprised mainly of plant debris mixed with small percentage of sediments (30–40 cm above RZ). This coarse particulate organic matter (CPOM) in the AH, which is ultimately converted to fine particulate organic matter (FPOM) through physical (wave action, animal movements), chemical (chemical withering) and biological processes (microbial colonization; break-

constituted both autochthonous and allochthonus material. However, our observations indicate that the contribution of the autochthonous component is relatively high in the AH, RZ and TZ.

Table 1. The nutrient concentration in sediment profiles from landward and lakeward areas (transect at Purbach). Concentrations are expressed as $\mu g \cdot g^{-1}$ dry weight sediments.

	% Org. C	% N	TP $\mu g/g$	IP $\mu g/g$	OP $\mu g/g$	C : N
Landward (3500 m from the lake)						
Accumulation horizon (40–50 cm)	31.5 ± 6.87	0.43 ± 0.07	759 ± 101	277 ± 48	482 ± 87	62
Rhizone-rootmat region (30–40 cm)	13.8 ± 2.28	0.40 ± 0.04	514 ± 74	297 ± 39	218 ± 40	35
Transition zone (ca. 30 cm)	6.2 ± 1.89	0.26 ± 0.07	354 ± 74	213 ± 58	140 ± 41	32
Tertiary sediment	5.2 ± 1.01	0.11 ± 0.02	283 ± 50	151 ± 61	132 ± 47	43
Lakeward (100 m from the lake)						
Accumulation horizon (100–130 cm)	14.6 ± 9.10	0.31 ± 0.12	650 ± 150	285 ± 26	390 ± 112	59
Rhizone-rootmat region (15–25 cm)	8.7 ± 1.15	0.20 ± 0.04	152 ± 52	270 ± 28	152 ± 52	45
Transition zone (ca. 20 cm)	5.8 ± 0.60	0.10 ± 0.01	312 ± 57	266 ± 57	46 ± 26	59
Tertiary sediment	4.2 ± 0.78	0.09 ± 0.01	296 ± 64	263 ± 41	30 ± 14	43

242

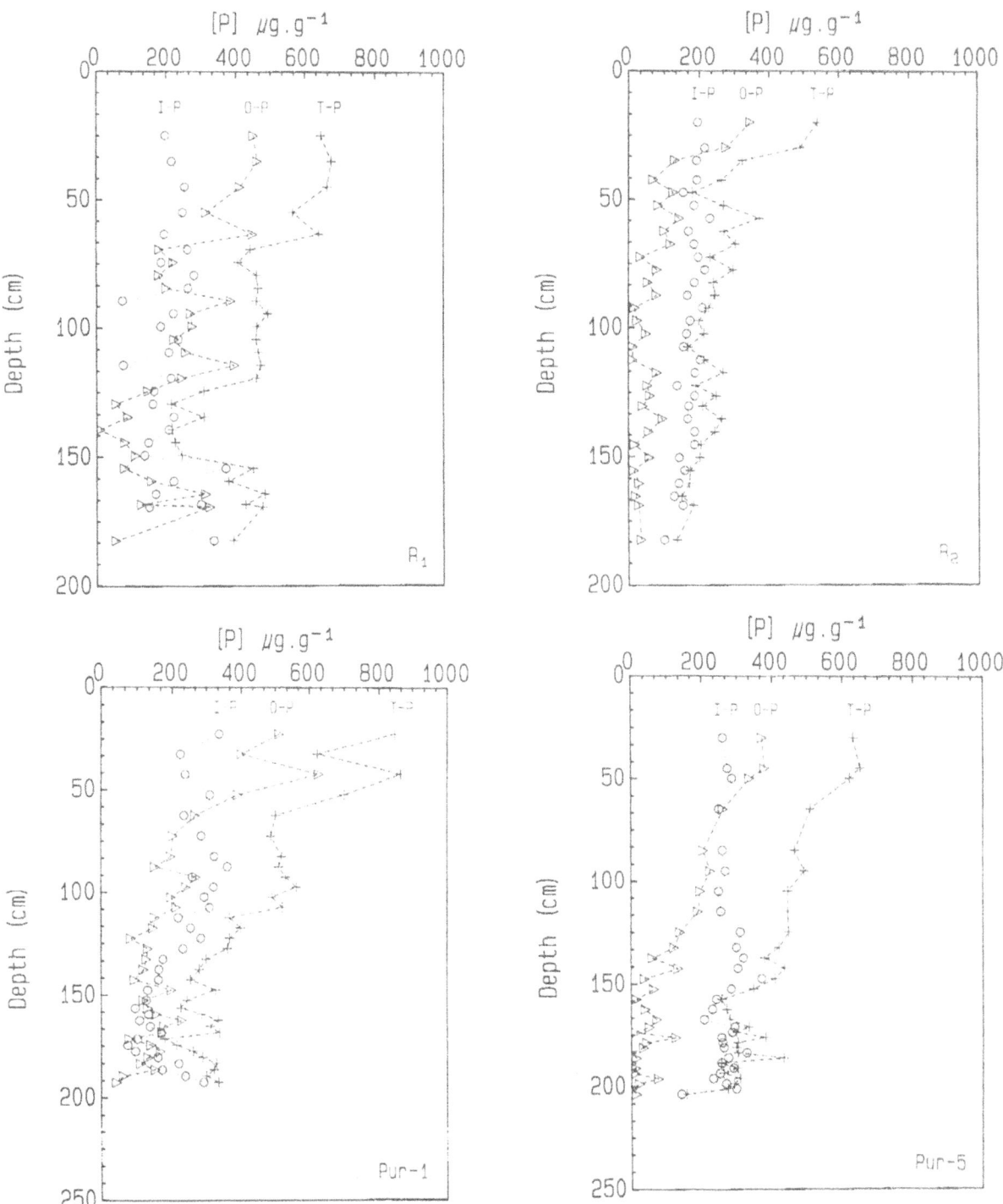

Fig. 2. The vertical distribution of total, inorganic and organic phosphorus (TP, IP and OP resp.) in lakeward (Pur-5 and R2) and landward (Pur-1 and R1) cores. Concentrations are expressed as μg/g dry weight sediments.

down through shredders, scrapers and grazers) support a number of food webs in the lake littoral (see Mendez, 1987). The resulting refractory organic material that passes through the RZ, is accumulated in the TZ. The transition zone also acts as a repository for organic matter from the RZ and eroded sediments from the open lake.

Organic carbon, nitrogen and total phosphorus content in the sediments ranged widely from 2–65%, 0.1–0.4% and 0.2–1.2 mg/g respectively. They showed clear differences between landward and lakeward sediments, and also between AH, RH, TZ and TS in respective cores (Table 1). The highest concentrations were recorded in the upper layers and decreased with increasing depth, with lowest levels in the Pannonian sediments. This pattern was characteristic for all the littoral cores but one. The cores taken at Zitzmannsdorf (see Figs 2 and 3, Löffler, 1991) showed a completely a reversed pattern with fully mineralised sediments at the top with very low C, P, N, content. This indicates that there was no top cover of

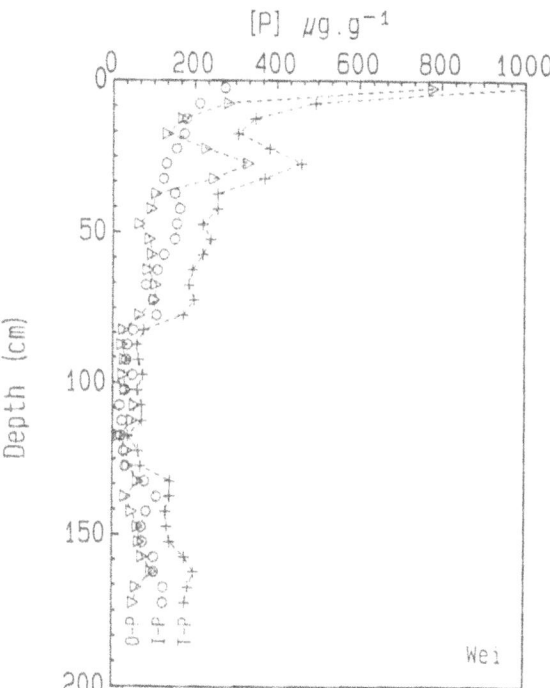

Fig. 3. The vertical distribution of total, inorganic and organic phosphorous (TP, IP and OP resp.) in landward (1500 m from open lake) core from Weiden.

Table 2. Comparison of Cd, Pb, Zn and Cu content expressed as $\mu g \cdot g^{-1}$ dry weight) in deep sediments from the reed belt (Pannonian sediments) with surface sediments from the open lake.

Stn.	Depth cm	Cd ppm	Pb ppm	Zn ppm	Cu ppm
Jois (J1)	170	0.87	10	39	28
Breitenbrunn (Br 1)	155	0.66	20	81	41
Breitenbrunn (Br 2)	145	0.78	17	71	45
Purbach 0 (Pur 0)	135	0.48	11	58	34
Purbach 1 (Pur 1)	185	0.47	7	67	31
Purbach 5 (Pur 5)	185	0.81	9	88	46
Oggau (Og)	185	0.74	12	85	47
Donnerskirchen (Do)	185	0.32	14	93	53
Winden 1 (W 1)	175	0.58	14	60	41
Winden 2 (W 2)	180	0.54	14	60	35
Open lake surface sediments					
Wulka mouth	0–2	0.74	30	382	80
Neusiedl	0–2	0.47	32	93	67
Illmitz	0–2	0.45	30	69	58

Phragmites and also that old lake boundaries may be did not extend up to this point. In cores where an open lake fauna was found (Mö, Do, Pur-1, W2, W1, J2, J1), the sediment was totally mineralized with low OP, organic C and N levels. However, lakeward cores from Purbach (Pur-5), and Br-1 cores are exceptions (see Löffler, 1991).

The distribution of inorganic and organic phosphorus (IP, OP resp.) fractions in the cores show a definite pattern with OP diminishing with depth and IP remaining more or less evenly distributed along the core. The OP fraction seems to influence the distribution pattern of total phosphorus in all cores and the quantity depends on the degree of mineralization. The typical distribution patterns of TP, IP, and OP are shown in Fig. 2 (Purbach & Rust) and Fig. 3 (Weiden). In some parts of the lake by taking into consideration the OP, organic C and N distribution patterns along with that of Ostracods (characteristic for reed belt and cold adupted open lake species; see Löffler, 1991) it may be possible to trace the layers that separte recent and tertiary sediments. However, these findings have to be confirmed with isotope analysis (age of Ostracod fauna).

There was no clear trace metal distribution

244

pattern in the sediment profiles but the levels of Cd, Pb, Zn and Cu in Pannonian sediment were comparatively low and can be considered as the background level (Table 2). However, in the open lake, sediments collected close to the reed stands (at the Wulka mouth, Neusiedel and Illmitz; see Fig. 1, Löffler, 1991) showed in general a 2–3 fold higher metal concentration. The highest levels were recorded near the Wulka, mouth (Zn up to 6 times higher) but the concentrations decreased with increasing distance from the river mouth (Table 2, Neusiedl and Illmitz). It appears that the trace metal flux to the marsh is related to the discharges from the Wulka and other inflows. The highest C, P, N, values and trace metal concentrations are recorded in the surface sediments deposited after the drying up of the lake a hundred and twenty-two years ago.

Acknowledgments

Thanks are due to R. Niederreiter, E. Niederreiter, H. Keckeis and S. Dudzinski for field assistance; H. Kraill and J. Luitz for analytical help in the laboratory. This work was supported by Fonds zu Förderung Wissenschaftlichen Forschung grant Nr. P6028.

References

Aspila, K. I., H. Agemian & A. S. Y. Chau, 1976. A semi-automated method for the determination of inorganic, organic and total phosphate in sediments. Analyst 101: 187–197.

Gunatilaka, 1985. Nährstoffkreisläufe im Schilfgürtel des Neusiedlersees – Auswirkungen des Grünschnittes. Wiss. Arb. Burg. 72: 225–310.

Gunatilaka, 1988. Estimation of the available P-pool in a large freshwater marsh. Arch. Hydrobiol. Beih. 30: 15–24.

Löffler, H., 1991. Palaeolimnology of Neusiedlersee, Austria. I: The Succession of Ostracods. In J. P. Smith et al. (eds), Environmental History and Palaeolimnology. Kluwer Academic Publishers, Dordrecht: 229–238. Reprinted from Hydrobiologia 214.

Mendez, M., 1987. The role of macrofauna in the decomposition processes of Phragmites australis in the Neusiedlersee reed belt. Ph.D. thesis, Univ. Vienna. 143 p.

Hydrobiologia **214**: 245–252, 1990.
J. P. Smith, P. G. Appleby, R. W. Battarbee, J. A. Dearing, R. Flower, E. Y. Haworth, F. Oldfield & P. E. O'Sullivan (eds), 245
Environmental History and Palaeolimnology.
© 1990 *Kluwer Academic Publishers.*

Paleolimnological investigation of three manipulated lakes from Sudbury, Canada

Sushil S. Dixit, Aruna S. Dixit & John P. Smol
Department of Biology, Queen's University Kingston, Ontario K7L 3N6, Canada

Key words: diatoms, chrysophytes, inferred pH, neutralization, fertilization, acidification, Sudbury

Abstract

Stratigraphic changes in diatoms and chrysophytes from three manipulated Sudbury lakes were explored in an attempt to examine the influence of fertilization and/or neutralization on algal microfossil assemblages. Both diatom- and chrysophyte-inferred pH profiles indicate that the pH of Labelle Lake has remained fairly stable in the past. The study of Labelle and Middle lakes indicates that the addition of nutrients to acidic and non-acidic oligotrophic lakes did not directly influence diatom and chrysophyte species composition, perhaps because pH remained stable. The diatom and chrysophyte assemblages of Middle Lake only changed when the pH was raised. In Mountaintop Lake the recent shift in chrysophyte species composition and the resulting inferred pH decline is most likely related to a decline in mid-summer epilimnetic pH. Reliable paleolimnological inferences are difficult in lakes such as these because it is difficult to track limnological conditions in the absence of modern analogues.

Introduction

Over the last decade, a large proportion of paleolimnological research in North America and Western Europe has focused on studies relating to past lakewater pH change (e.g. papers in Smol *et al.*, 1986 and see Charles *et al.*, 1989 for a recent review). Collectively, these studies have shown that quantitative analyses of diatom valves and chrysophyte scales can provide important and reliable records of past pH change.

Predating the early acidification work, diatoms and chrysophytes have also been used to infer past changes in lake trophic status (reviewed in Battarbee, 1986; Smol, 1987); the quantification of these relationships, however, has not yet reached the same level of sophistication as pH inferences. Recently, attempts have been made to quantify trophic relationships. Results from re-

cent and ongoing studies are encouraging (e.g. Brugam & Vallarino, 1989; Christie, 1988; Whitmore, 1989), and suggest that reliable inferences of trophic variables (e.g. total phosphorus and Secchi depth) will be available soon.

Lakes in the Sudbury region of Ontario (Canada) represent some of the most anthropogenically altered aquatic systems in the world. Because of local smelting activity (especially for Cu and Ni), the area is severely affected by high levels of sulphate deposition, as well as by the deposition of many metals (Conroy *et al.*, 1974; Nriagu *et al.*, 1982; Palmer *et al.*, 1989) Calibration equations for both diatoms and chrysophytes are now available for the Sudbury region (Dixit *et al.*, 1989a, and unpublished data). Not surprisingly, the optima of many algal taxa are somewhat different in this acid and high metal impacted region than in other less-stressed regions. Nonetheless, these differences can be quantified, and paleolimnological studies have been successfully implemented in Sudbury, where aspects of lake acidification (Dixit *et al.*, 1987; 1989a), recovery (Dixit *et al.*, 1989b), and the response of lakes to liming events (Dixit *et al.*, 1987) have been documented. Sudbury lakes offer an excellent research area for these studies because: the region contains a plethora of different types of lakes (often reflecting the orientation of wind trajectories to the known point sources of emissions); records are available for past changes in emissions; and, in some lakes, long-term limnological records are also available.

In this paper, we use paleolimnological techniques to compare diatom and chrysophyte species successions over the last century in three Sudbury lakes with differing water chemistries and manipulation histories. We attempt to extend our inference techniques to decipher past successional changes in lakes that experienced changes in both pH (as a result of acidification and liming) and trophic status (as a result of fertilization). Specifically, we assess how diatom and chrysophyte assemblages have responded to nutrient inputs in acidic and non-acidic lakes.

Study area and study lakes

The Sudbury region contains several manipulated lakes with long-term limnological data, three of which (Labelle, Mountaintop, and Middle) were selected for this study. [210]Pb analysis of the cores indicated that at least 2 cm of sediment has accumulated since the manipulations were initiated by the Ontario Ministry of the Environment in the 1970's.

Labelle Lake is located 27 km northwest of Sudbury (46° 42′ N, 81° 07′ W). It is a small (6.2 ha) single basin lake with a maximum depth of 10.2 m (Ontario Ministry of Environment, 1982). The lake is presently circumneutral, and limnological records do not indicate any recent acidification. In 1977, the lake was fertilized by the addition of nitrogen (N) as ammonium nitrate and phosphorus (P) as phosphoric acid (Yan &

Lafrance, 1984). The average lakewater pH, and N and P concentrations for the period 1976 to 1987 are presented in Table 1.

Mountaintop Lake is located 52 km north of Sudbury (46° 55′ N, 80° 53′ W). This single basin lake is 4.9 ha in size, with a maximum depth of 9.5 m. Lakewater pH was about 4.6 prior to fertilization in 1976. Between July 8 and September 8, 1976, P and N were added as 20:40:0 fertilizer, and N was also added as ammonium

Table 1. Past average pH, total phosphorus, and total nitrogen concentrations in Labelle, Mountaintop, and Middle lakes (1973–1979 data from Yan & Lafrance, 1984; 1980–1986 unpublished data from the Ontario Ministry of the Environment; 1987 data from Dixit *et al.*, 1989a).

		pH	Total phosphorus μg/L	Total nitrogen μg/L
Labelle	Pre-fertilization			
	1977 spring	5.95	11	319
	Post-fertilization			
	1977*	6.26	64	1226
	1978*	6.21	68	1335
	1979	6.20	18	144
	1987	6.78	14	280
Mountaintop	Pre-fertilization			
	1976 spring	4.62	7	290
	Post-fertilization			
	1976*	4.40	58	1837
	1977*	4.49	66	1211
	1978*	4.96	82	1130
	1979	5.44		
	1987	4.75	9	230
Middle	Pre-treatment			
	1973 spring	4.40	7	622
	Post-treatment			
	1973[+]	6.87	4	417
	1974	7.01	7	600
	1975*	6.56	7	560
	1976*	6.41	15	451
	1977*	6.37	13	405
	1978*	6.65	11	349
	1979	6.54	6	281
	1980	6.51	5	
	1981	6.44	4	
	1982	6.48	4	
	1983	6.51	5	
	1984	6.55	5	
	1985	6.69	5	
	1986	6.78		
	1987	7.11	14	210

* nutrients added
[+] base added

nitrate (Yan & Lafrance, 1984). After September 1976, P was added as 85% technical grade phosphoric acid. The measured lakewater pH remained low after the fertilization events (Table 1). Over recent years (e.g. 1987) both P and N have declined to pre-manipulation levels.

Middle Lake is located 6 km southwest of Sudbury (46° 26′ N, 81° 02′ W). The lake has a surface area of 28.2 ha, and a maximum depth of 15 m. Lakewater pH was measured at about 4.4 in 1973. Thereafter, the lake was limed using calcium carbonate and hydroxide, and its pH was raised to its present value of about 7 (Yan & Dillon, 1984). The lake was also fertilized in 1975 and 1976 by only adding P as phosphoric acid (Yan & Lafrance, 1984). The average pH, and N and P concentrations for Middle Lake (1973–1987) are in Table 1.

Methods

In June 1987 a short sediment core (ca. 25 cm long) was taken from each of the three study lakes using a modified K.B. gravity corer. Because we were mainly interested in detecting stratigraphic changes that have occurred within the last two decades, the upper portion of the core (0–5 cm) was sectioned in 0.25 cm intervals, using a vertical extrusion system similar to the one described by Glew (1988). The core was sectioned at 1 cm intervals between the 5–10 cm depth, and the remainder of the core was sliced into 2 cm segments.

The sediment samples were acid cleaned (Dixit, 1986) and non-quantitative slides were plated on coverglasses and mounted in Hyrax. A minimum of 500 diatom valves and/or chrysophyte scales was counted and identified from each sample in random fields at 1250 X magnification. The cores were dated using ^{210}Pb chronology, applying the CRS model (Appleby & Oldfield, 1978). From the percent species distribution data, the diatom- and chrysophyte-inferred pH values were computed using weighted-averaging-calibration models developed for Sudbury lakes (Dixit et al., 1989a, 1990a). A number of recent studies have shown

that weighted-averaging is ecologically more realistic and statistically stronger than other calibration methods (Stevenson et al., 1989; ter Braak & van Dam, 1989; Birks et al., 1990).

Results and discussion

Labelle Lake

The stratigraphic distribution of common diatom taxa indicates that little compositional change has occurred during this century (Fig. 1). However, since about the mid 1970's, chrysophyte species composition has changed, with increases in Mallomonas caudata, M. punctifera, Synura echinulata, and S. sphagnicola. These species changes are not likely associated with the inputs of nutrients because the available autecological data suggest that, with the possible exception of M. caudata (Smol, 1980, as M. fastigata), these taxa are usually most abundant in oligotrophic waters (Smol et al., 1984). Similarly, the influence of pH change can also be ruled out because: a) no marked shift was measured in lakewater pH between 1977 and 1978 when the lake was fertilized (Table 1); and b) the pH indicator status of these taxa are different (e.g. M. caudata and M. punctifera are indicators of circumneutral to alkaline waters, whereas S. echinulata and S. sphagnicola are more common in low pH waters). One plausible reason for these species changes may relate to decreased transparency after the lake was fertilized (Yan & Lafrance, 1984), as at least some of these taxa can maintain large populations in coloured waters.

Profiles of both diatom-inferred pH (hereafter referred to as DI-pH) and chrysophyte-inferred pH (hereafter referred to as CI-pH) indicate that the pH of Labelle Lake has remained fairly stable over the last century, and there is no evidence of recent acidification (Fig. 2). Although DI- and CI-pH profiles infer similar trends, at selected depths the difference between DI- and CI-pH was as much as one pH unit. The pH inferences for post-1976 sediments (pH 5.7 to 6.5) resemble the measured lakewater pH data for that period

248

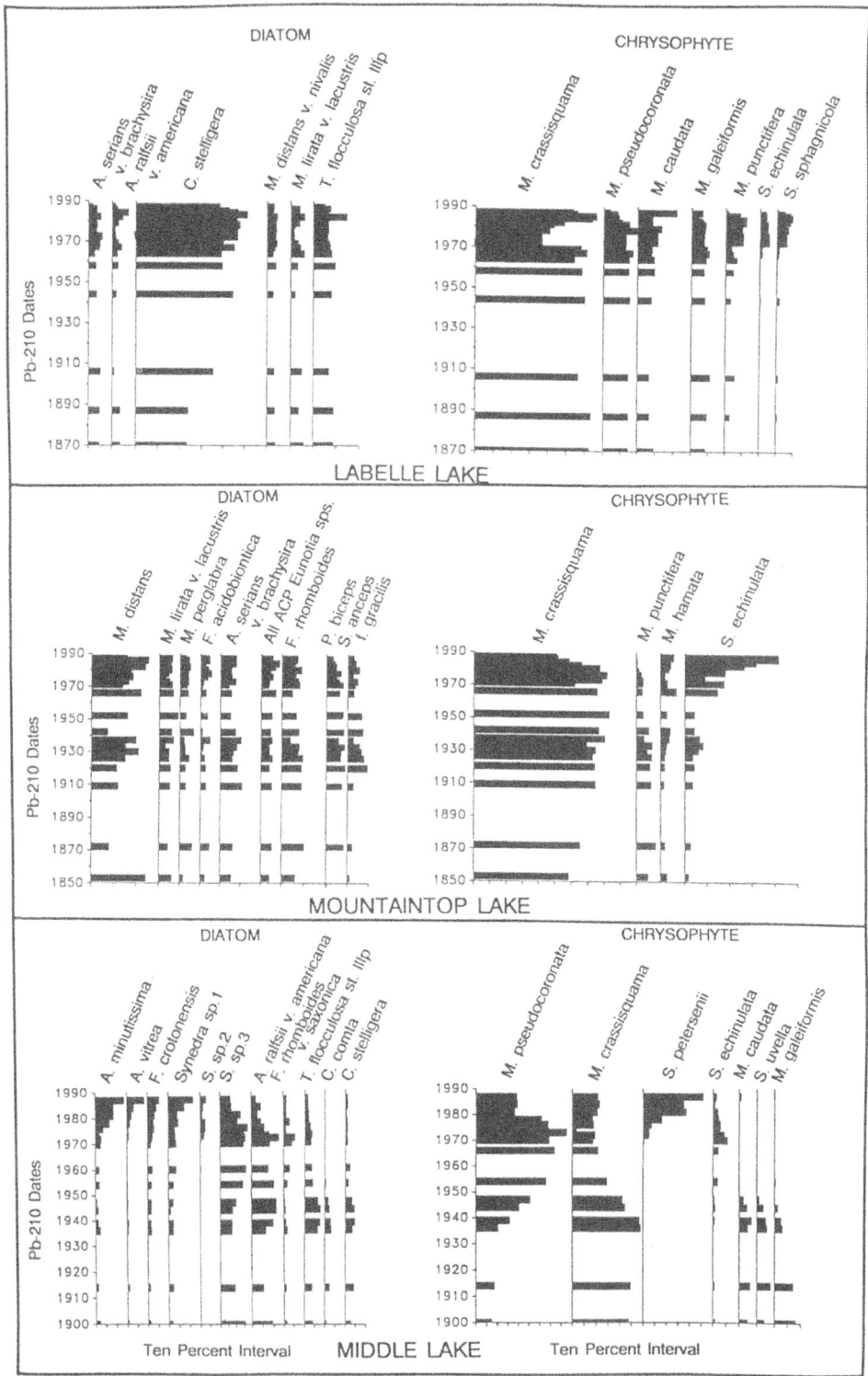

Fig. 1. Diatom and chrysophyte stratigraphies for Labelle, Mountaintop, and Middle lakes.

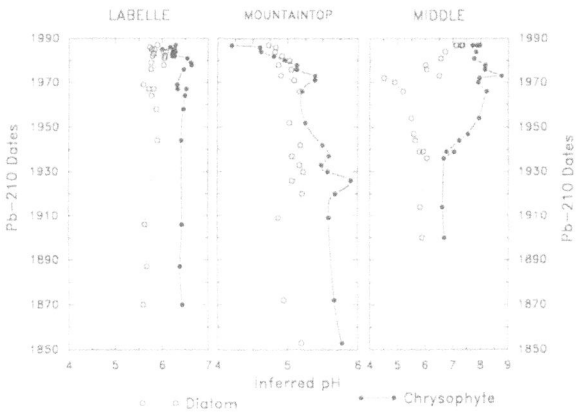

Fig. 2. Diatom- and chrysophyte-inferred pH profiles for Labelle, Mountaintop, and Middle lakes.

(Table 1). Our study of Labelle Lake suggests that the short-term increase in nutrients in 1977 and 1978 (Table 1) was not recorded by any marked diatom or chrysophyte species change in the sedimentary stratigraphy.

Mountaintop Lake

Although there were small species shifts in the downcore distribution of diatoms in Mountaintop Lake, no definite trends can be identified (Fig. 1). All the common diatom taxa are indicators of acidic waters, which is consistent with the low pH readings recorded for the lake since 1976 (Table 1). The DI-pH profile suggests that over the last 100 yr the lake has remained acidic, with a background pH of about 5 (Fig. 2). The diatom stratigraphy, and consequently DI-pH, did not show any marked deviation during the fertilization period.

In contrast to the diatom data, chrysophyte species composition has changed in the lake's recent history. Until about the mid 1970's, *Mallomonas crassisquama* was the dominant chrysophyte, but since then it exhibited a marked decline. The latter was mirrored by a dramatic increase in *S. echinulata*. This species change coincides with the fertilization of the lake in 1976 (Table 1, Fig. 1). Although the pre-1975 CI-pH trend is similar to the one inferred from diatoms,

CI-pH values are consistently higher (Fig. 2). Since about 1975 the CI-pH has indicated a marked pH decline.

Initially, the CI-pH results may seem unusual because the average annual pH of Mountaintop Lake did not change much since 1976 (Table 1). Moreover, the CI-pH decline does not appear to be directly related to the fertilization of the lake, but this may relate to the decline in mid-summer epilimnetic pH after the nutrients were added.

In spring, the pH of poorly buffered lakes generally declines due to input of acidic meltwater, but then the pH often rises again with the increases in algal productivity. However, following fertilization, the epilimnetic pH of Mountaintop Lake continued to decline in the mid-summers of 1977, 1978, and 1979; from 4.5 to 4.0, 4.6 to 4.2, and 4.7 to 4.3, respectively (Yan & Lafrance, 1984). We tentatively hypothesize that, because chrysophytes are euplanktonic and generally bloom in late spring and early summer, they responded to these pH depressions in Mountaintop Lake. Diatoms perhaps did not track these short term fluctuations in pH because most of the taxa are benthic, and not vernal-blooming planktonic forms.

Middle Lake

Diatom assemblages have undergone two major changes in Middle Lake over the last 50 years (Fig. 1). The first major change occurred between 1935 and the mid 1970's, when taxa characteristic of more acidic waters increased in percent abundance (e.g. *Asterionella ralfsii* v. *americana*, *Frustulia rhomboides* v. *saxonica*, and *Synedra* species 3). The second species shift occurred soon after the lake was neutralized in 1973, when circumneutral and alkaline taxa (e.g. *Achnanthes minutissima*, *Anomoeoneis vitrea*, *Fragilaria crotonensis*, and *Synedra* sp. 1) became more common, with compensatory declines in acidic and pH indifferent taxa (Fig. 1).

Chrysophyte species composition in Middle Lake also displayed two well-defined shifts (Fig. 1). Until about 1935 *Mallomonas crassis-*

quama dominated the assemblage, but thereafter *M. pseudocoronata* became very abundant. The second major shift started about 1970, when *S. petersenii* showed a distinct increase.

The DI-pH profile indicates that the lake's background pH was about 6 (Fig. 2), and the onset of acidification started about 1940. Between ~1940 and 1973 the lakewater pH declined from a pH of ~6.0 to 4.5. The lowest recorded DI-pH (4.5) corresponds closely with the measured lakewater pH (4.4) for that period (1973, see Table 1). The acidification of Middle Lake was primarily caused by the atmospheric deposition of sulphates from the smelters in Sudbury, which occurred at this time.

Since about 1973 the DI-pH indicated a marked recovery (from pH 4.5 to 7.4). The close relationship between the inferred pH recovery and the post-neutralization pH rise (Table 1) confirms that diatoms provide a reliable pH profile when compared to the known rapid change in lakewater pH change. Our results are similar to the DI-pH recovery reported after the neutralization of Hannah Lake in Sudbury (Dixit *et al.*, 1987). The rapid response in diatom communities after the neutralization supports our previous observation (Dixit *et al.*, 1990) that in order to use our diatom models to assess recent lakewater pH shifts in Sudbury, the changes may have to be greater than half a pH unit.

In comparison to Labelle and Mountaintop lakes, the increase in lakewater phosphorus concentration was small in Middle Lake (Table 1). Because lakewater pH had already risen significantly and the diatom species composition was already shifting when fertilization occurred (Table 1), the effect of nutrient addition on diatom species composition seems to have been masked by the pH changes. Moreover, if one compares these results to those obtained for Labelle and Mountaintop lakes, where diatom assemblages and inferred pH remained stable after fertilization, one could argue that diatom changes in Middle Lake were primarily pH related. However, it is possible that fertilization may have helped in maintaining the lake's higher pH.

Although diatoms provided a useful inferred pH history for Middle Lake, the post-1930 CI-pH values are erroneous (Fig. 2). Between 1930 and 1973 diatoms indicated a marked acidification, whereas chrysophytes suggest a reverse trend, and infer an increase of about two pH units. The latter is almost entirely due to the striking increase of *M. pseudocoronata*, an alkaliphilic taxon in the Sudbury region (Dixit *et al.*, 1989a). However, other more modest stratigraphic changes, such as declines in other circumneutral/alkaline taxa (*M. caudata*, *M. galeiformis*, and *S. uvella*) and an increase in the acidophilic taxon *S. echinulata* would suggest an acidifying trend. Nonetheless, the striking increase in *M. pseudocoronata* overpowers our CI-pH profile. Never before have we recorded such a discrepancy between these two indicator groups.

In many lake regions, including Sudbury (Dixit *et al.*, 1989a), *M. pseudocoronata* has been found to be an indicator of circumneutral to alkaline waters. To our knowledge, this is the first study where this taxon has been identified in such high abundance in acidic waters. Because metal inputs to this lake have increased dramatically since the beginning of this century due to its proximity to Copper Cliff Smelter (Dillon & Smith, 1984), we are tempted to hypothesize that this atypical increase of *M. pseudocoronata* might be associated with metals. Although it is known that in high pH waters this taxon is tolerant of high metal concentrations (Dixit *et al.*, 1989a), we further suggest that *M. pseudocoronata* can also prosper in acidic waters if metal concentrations are also very high.

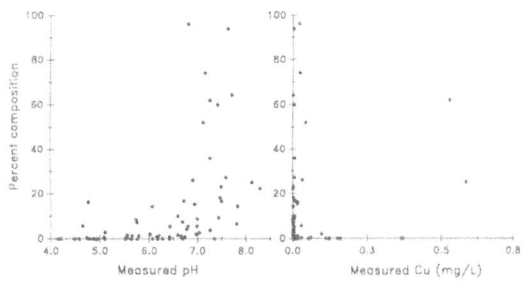

Fig. 3. The percent composition of *M. pseudocorononata* in surface sediment samples of 72 calibration lakes plotted against measured lakewater pH and Cu concentrations.

For example, prior to neutralization in 1973, the lakewater Cu, Ni, Zn, and Fe concentrations were 496, 1060, 91 and 143 μg/L, respectively (Yan & Dillon, 1984). A re-examination of our earlier chrysophyte calibration work in Sudbury (Dixit et al., 1989a) provides some interesting supportive data. Figure 3 illustrates the percent composition of *M. pseudocoronata* against measured lakewater pH and Cu concentration for our 72 calibration lakes (Dixit et al., 1989a). Although *M. pseudocoronata* is clearly most common in alkaline waters, it did in fact maintain sizeable populations in 2 very acid Sudbury lakes. These lakes also have high lakewater metal concentrations.

Although more data are needed to substantiate our hypothesis, we tentatively suggest that high metal concentrations may have ameliorated the impact of high acidity by lowering the availability of hydrogen ions to *M. pseudocoronata*. This taxon would be an ideal candidate for laboratory culture experiments. Following neutralization of the lake in 1973, the metals precipitated out of the water column (Yan & Dillon, 1984), and the CI-pH approached the known lakewater pH (Fig. 2).

Conclusions

Surprisingly, both diatom and chrysophyte species composition seem to have been little influenced by the direct effects of fertilization. This is surprising because in other nearby lake regions, both diatoms (Smol & Dickman, 1981) and chrysophytes (Smol, 1980) have been shown to respond (or at least were believed to have responded) to even modest eutrophication events, such as road construction in the drainage. We recognize that in Sudbury we are dealing with a highly perturbed lake region. We further recognize that we are attempting to decipher complicated lake trajectories, with few (if any) modern analogues in our calibration lakes which correspond to conditions that existed during periods of increased nutrients and low pH waters. Metals could have also altered the nutrient uptake and availability to algae. Some of our interpretations and inferences may be speculative, but the recorded species changes are real and will hopefully encourage further research in these areas.

The study supports various surface sediment calibration studies that have shown pH to be the most important variable influencing diatom and chrysophyte species distribution. As observed for Middle Lake, the role of trace metals on algal distributions should be carefully examined when attempting to quantify an alga's tolerance or optimum along environmental gradients.

Acknowledgements

This study was supported by a Strategic Grant (G1996) from the Natural Sciences and Engineering Research Council of Canada. Many thanks to W. Keller and N. Yan (Ontario Ministry of the Environment) for providing water chemistry data.

References

Appleby, P. G. & F. Oldfield, 1978. The calculation of ^{210}Pb dates assuming a constant rate of supply of unsupported ^{210}Pb to the sediment. Catena 5: 1–8.

Battarbee, R. W., 1986. Diatom analysis. In B. E. Berglund (ed.), Handbook of Holocene Palaeoecology and Palaeohydrology. John Wiley & Sons: 527–570.

Birks, H. J. B., J. M. Line, S. Juggins, A. C. Stevenson & C. J. F. Ter Braak, 1990. Diatoms and pH reconstruction. Phil. Trans. R. Soc. Lond., B, 327: 263–278.

Brugam, R. B. & J. Vallarino, 1989. Paleolimnological investigations of human disturbance in western Washington lakes. Arch. Hydrobiol. 116: 129–159.

Charles, D. F., R. W. Battarbee, I. Renberg, H. van Dam & J. P. Smol, 1989. Paleoecological analysis of diatoms and chrysophytes for reconstructing lake acidification trends in North America and Europe. In S. A. Norton, S. E. Linberg & A. L. Page (eds), Acid Precipitation, Vol. 4. Soils, Aquatic Prousses, and Lake Acidification. Springer Verlag, N.Y. (N.Y.); 207–276.

Christie, C. E., 1988. Surficial diatom assemblages as indicators of lake trophic status. M.Sc. Thesis, Queen's Univ., Kingston, Ontario.

Conroy, N., D. S. Jeffries & J. R. Kramer, 1974. Acid shield lakes in Sudbury, Ontario region. Proc. 9th Can. Sym. Wat. Pollut. Res.: 45–60.

Dillon, P. J. & P. J. Smith, 1984. Trace metal and nutrient accumulation in the sediment of a lake near Sudbury, Ontario. In K. Nriagu (ed.), Environmental Impacts of Smelters. John Wiley: 375–416.

Dixit, S. S., 1986. Algal microfossils and geochemical reconstructions of Sudbury lakes: a test of the paleo-indicator potential of diatoms and chrysophytes. Ph.D. Thesis, Queen's University, Kingston, Ontario.

Dixit, S. S., A. S. Dixit & R. D. Evans, 1987. Paleolimnological evidence of recent acidification in two Sudbury (Canada) lakes. The Science of the Total Environment 67: 53–67.

Dixit, S. S., A. S. Dixit & J. P. Smol, 1989a. Relationship between chrysophyte assemblages and environmental variables as examined by canonical correspondence analysis (CCA). Can. J. Fish. aquat. Sci. 46: 1667–1676.

Dixit, S. S., A. S. Dixit & J. P. Smol, 1989b. Lake acidification recovery can be monitored using chrysophycean microfossils. Can. J. Fish. aquat. Sci. 46: 1309–1312.

Dixit, A. S., S. S. Dixit & J. P. Smol, 1990. Algal microfossils provide high temporal resolution of environmental change. Wat. Air Soil Pollut. (submitted).

Glew, J. R., 1988. A portable extruding device for close interval sectioning of unconsolidated core samples. J. Paleolimnology 1: 235–239.

Nriagu, J. O., H. K. T. Wong & R. D. Coker, 1982. Deposition and chemistry of pollutant metals in lakes around metal smelters at Sudbury, Ontario. Envir. Sci. Tech. 16: 551–560.

Ontario Ministry of Environment, 1982. Studies of lakes and watersheds near Sudbury, Ontario: final limnological report, SES 009/82. Ontario Ministry of the Environment, Ontario.

Palmer, G. R., S. S. Dixit, J. D. MacArthur & J. P. Smol, 1989. Elemental analysis of lake sediment from Sudbury, Canada, using PIXE. The Science of the Total Environment 87/88: 141–156.

Smol, J. P., 1980. Fossil Synuracean (Chrysophyceae) scales in lake sediments: A new group of paleoindicators. Can. J. Bot. 58: 458–465.

Smol, J. P., 1987. Methods in Quaternary ecology 1. Freshwater algae. Geoscience Canada 14: 208–217.

Smol, J. P. & M. D. Dickman, 1981. The recent history of three Canadian Shield lakes: a paleolimnological experiment. Arch. Hydrobiol. 93: 83–108.

Smol, J. P., R. W. Battarbee, R. B. Davis & J. Meriläinen (eds.), 1986. Diatoms and Lake Acidity. Dr W. Junk Publ., Dordrecht. 307 p.

Smol, J. P., D. F. Charles & D. R. Whitehead, 1984. Mallomonadacean (Chrysophyceae) assemblages and their relationships with limnological characteristics in 38 Adirondack (N.Y.) lakes. Can. J. Bot. 62: 611–630.

Stevenson, A. C., H. J. B. Birks, R. J. Flower & R. W. Battarbee, 1989. Diatom based pH reconstruction of lake acidification using canonical correspondence analysis. Ambio 18: 228–233.

ter Braak, C. J. F. & H. van Dam, 1989. Inferring pH from diatoms: a comparison of new and old calibration methods. Hydrobiologia 178: 209–223.

Whitmore, T. J., 1989. Diatom assemblages as indicators of trophic state and pH. Limnol. Oceanogr. 34: 882–895.

Yan, N. D. & P. J. Dillon, 1984. Experimental neutralization of lakes near Sudbury, Ontario. In J. Nriagu (ed.), Environmental Impacts of Smelters. John Wiley: 417–456.

Yan, N. D. & C. Lafrance, 1984. Responses of acidic and neutralized lakes near Sudbury, Ontario, to nutrient enrichment. In J. Nriagu (ed.), Environmental Impacts of Smelters. John Wiley: 457–521.

Hydrobiologia **214**: 253–258, 1990.
J. P. Smith, P. G. Appleby, R. W. Battarbee, J. A. Dearing, R. Flower, E. Y. Haworth, F. Oldfield & P. E. O'Sullivan (eds), 253
Environmental History and Palaeolimnology.

Dominant diatoms in the interglacial lake sediments of the Middle Pleistocene in Central and Eastern Poland

Barbara Marciniak
Institute of Geological Sciences, Polish Academy of Sciences, Al. Żwirki i Wigury 93, 02–089 Warszawa, Poland

Key words: diatoms, interglacial lakes, Middle Pleistocene, Poland

Abstract

Diatom studies carried out in Central and Eastern Poland have shown a diversified composition and considerable changes of relative frequency of dominant diatoms in profiles of lake sediments of the Mazovian and Ferdynandovian interglacials. *Cyclotella* spp. are dominant in a large part of the sediment profiles at Adamówka, Ławki-7 and Wola Grzymalina-59 and, at Krępiec and Podlodów, *Stephanodiscus* is also common while at Biała Podlaska and Falęcice, *Aulacoseira* is most abundant. *Cyclotella comta* var. *lichvinensis* and *C. vorticosa* are the typical diatoms in the Mazovian Interglacial. In the sediments representing the Ferdynandovian Interglacial, the characteristic diatoms are *Cyclotella* cf. *reczickiae* and the *Stephanodiscus rotula/niagarae* complex.

Introduction

The aim of this contribution is to compare the distribution of the most common diatoms occurring in seven profiles of lake sediments of the Middle Pleistocene (sensu Lindner, 1984) in Central and Eastern Poland (Fig. 1). Three of these profiles, Adamówka, Biała Podlaska and Krępiec, represent the Mazovian Interglacial (= Holstein, Likhvin) and the other four, Falęcice, Ławki-7, Podlodów, Wola Grzymalina-59, represent the older Ferdynandovian Interglacial (= Voigtstedt, Byelovezhian, Cromerian 3 + 4 in the Netherlands). The geology, paleogeomorphology, and lithology, as well as the results of pollen analyses of the majority of the above profiles, have been given in papers by Harasimiuk and Henkiel (1981), Janczyk-Kopikowa (1981), Krupiński *et al.* (1986, 1988), Krzyszkowski and Kuszell (1987), Bińka *et al.* (1987), Laskowska-Wysoczańska (1987) and Lindner *et al.* (in press).

In the present paper examples are given of the occurrence of characteristic diatoms that are of importance for the stratigraphy of lake sediments in the Middle Pleistocene of Poland. Certain diatoms have been selected as biochronological indicators to allow between site correlation of sediments of the interglacial lakes.

Mazovian interglacial

Studies of diatoms in the Mazovian Interglacial sediments were started in Poland in 1980. They included sediments from fossil lakes of the Mazovian Intreglacial at Krępiec, Adamówka and Biała Podlaska, in which a very diverse

254

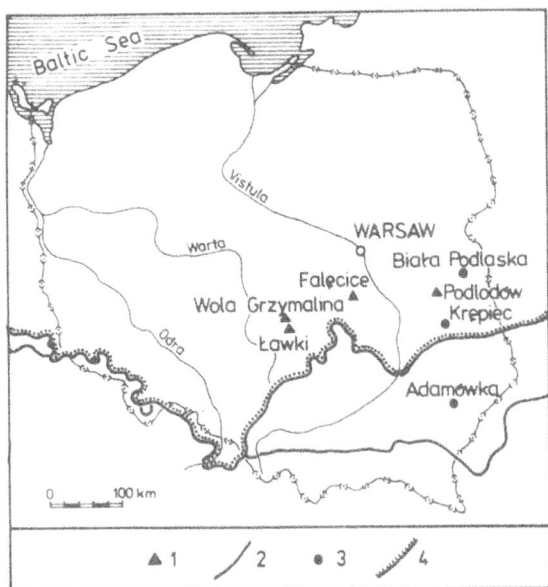

Fig. 1. Location sketch-map of the localities studied, 1 – Localities of the Ferdynandovian Interglacial, 2 – extent of the South-Polish Glaciation, 3 – localities of the Mazovian Interglacial, 4 – exent of the Middle-Polish Glaciation.

diatom flora has been found but with varying degrees of frustule preservation (Marciniak, 1980, 1983, 1986a; Bińka *et al.*, 1987). The best preserved diatom frustules have been found in sediments c. 20 m deep, at Krępiec. These sediments are so rich in diatoms that they are defined as diatomites (Harasimiuk & Henkiel, 1981). The poor preservation observed at Biała Podlaska, however, may be connected with strong compaction of the sapropel sediments which make bituminous shales that have been strongly glaciotectonically disturbed in places (see Krupiński *et al.*, 1986, 1988).

Cyclotella is the most common diatom in the Adamówka profile and both *Cyclotella* and *Stephanodiscus* are equally common in the Krępiec sediments. *Cyclotella comta* var. *lichvinensis* and *C. vorticosa* are the most typical elements of the diatom flora during the Mazovian Interglacial at Adamówka and Krępiec.

The lacustrine sediments of the interglacial at Biała Podlaska have revealed a different spectrum of dominant diatoms in which *Aulacoseira* sp. cf. *ambigua?*, *A. ambigua* and *A. granulata* show the

greatest frequencies. *Fragilaria* spp. and *Navicula scutelloides* prevail in the early stages of the lake development (Marciniak, 1986a).

In the sediments of the Likhvin Interglacial, in the USSR, *Cyclotella comta* var. *lichvinensis* and *C. vorticosa* are regarded as Pliocene relics and occur together with *C. comta* var. *pliocaenica*, *C. iris* and *C. temperei* and some intermediate forms (Loginova, 1979, 1982). The *Cyclotella* taxa mentioned above have not, so far, been found in the fresh water basins of the younger Pleistocene and Holocene in Poland, despite the frequent occurrence of *Cyclotella* flora (see Marciniak, 1973, 1987, 1988; Kaczmarska, 1976; Marciniak & Kowalski, 1978; Dąbrowski *et al.*, 1987, Przybyłowska-Lange *et al.*, 1989).

Specimens very similar to *Cyclotella comta* var. *lichvinensis*, which were preliminarily classified to *Cyclotella* cf. *quadriiuncta?*, occur only in the Holocene and late-glacial sediments of a highland, oligotrophic lake the Przedni Staw in the Tatra Mts (Marciniak, 1982, 1986b; Marciniak & Cieśla, 1983). Their internal valve structure is similar to that of *Cyclotella comta* var. *lichvinensis* but the marginal zone of the exterior differs in having more delicate ornamentation of the striae. In the light of new SEM investigations of *Cyclotella* species of the *Cyclotella bodanica/comta* and *C. bodanica/radiosa/ comensis* complex (Kling & Håkansson, 1986; Håkansson, 1988), the preliminary identification of *Cyclotella* cf. *quadriiuncta* appears incorrect.

Ferdynandovian interglacial

The lacustrine sediments of the Ferdynandovian Interglacial in the Bełchatów outcrop (Central Poland) occur there as several isolated lenses (profiles Wola Grzymalina-59, Ławki-7) that can be correlated by means of diatomite beds several metres thick. These index beds contain remains of fish, molluscs, insects and macrofossils of plants which according to pollen analyses, represent the lower optimum of the Ferdynandovian Interglacial (Krzyszkowski & Kuszell, 1987).

Diatom analysis of profiles Wola Grzymalina-

59 and Ławki-7 has furnished examples of a very diverse and abundant *Cyclotella* flora. Such a *Cyclotella* flora has not yet been found in sediments of the Mazovian Interglacial, the younger Pleistocene, or the Holocene. Although *Cyclotella distinguenda* has been noted in the Mazovian Interglacial and in the youngest Quaternary (Marciniak, 1979, 1980, 1986a, Bińka *et al.*, 1987, Przybyłowska-Lange *et al.*, 1989).

The assemblage of small forms of *Cyclotella* is difficult to recognize via LM. There is considerable variability in valve diameter and in the numbers of the elements of the valve structure (striae). It is particularly difficult to identify the valves in sediments which are 40–80% calcium carbonate, especially in the upper part of the Ławki-7 profile, as these are very poorly preserved (Krzyszkowski & Kuszell, 1987; Marciniak, in press).

Preliminary SEM study of the *Cyclotella* flora dominant in sediment profiles from Wola Grzymalina-59 and Ławki-7, shows four types of *Cyclotella* valves. Three types of valves are similar to *Cyclotella reczickiae* (Figs 2–9), which is a characteristic or indicator species abundant in sediments of the Byelovezhian Interglacial (in the profiles from Rassvet, Krasnaya Dubrova 13B and Chkalovo) and which, in Byelorussia, defines these sediments as belonging to the lower Pleistocene (sensu Khursevich & Loginova, 1984, 1986).

Samples of sediments from Podlodów, supplied by Dr. J. Rzechowski from the State Geological Institute in Warszawa, represent the lower optimum of the Ferdynandovian Interglacial (Janczyk-Kopikowa, personal communication), which is regarded as equivalent to the Shklov or Roslavl and Byelovezhian Interglacial in Byelorussia (Janczyk-Kopikowa, 1975; Makhnach *et al.*, 1982; Khursevich & Loginova, 1986). These sediments contain abundant and well preserved *Stephanodiscus* spp. similar to *Stephanodiscus rotula* and *S. niagarae* as described by Round (1981) and by Theriot *et al.* (1988). The same forms also occur in sediments of the Byelovezhian Interglacial in Byelorussia (Makhnach *et al.*, 1982, Khursevich & Loginova, 1986).

Many intermediate or transitional forms have

been observed between *Stephanodiscus rotula* and *S. niagarae* and these are difficult to separate with LM. These transitional forms need defining more clearly, using SEM and identifying additional diagnostic characters. Statistical studies are also required to define taxa as species rather than complex aggregations, e.g. *Stephanodiscus 'niagarae* complex' (Theriot & Stoermer, 1984, 1986; Theriot *et al.*, 1988). The presence of numerous unidentified intermediate taxa closely related to *Stephanodiscus niagarae* complex strongly reduces their values as biochronological indicators for correlation of Middle Pleistocene sediments. New taxa named by Khursevich and Loginova (1986), such as *Stephanodiscus determinatus, S. peculiaris, S. styliferum, S. niagarae* var. *insuetus, S. rotula* var. *distinctus, S. rotula* var. *paucus*, are typical of material from lakes of the Byelovezhian Interglacial and these could be valuable indicators if they were recognized more widely.

The profile of the Ferdynandovian Interglacial at Fałęcice contains the poorest diatom assemblages, both quantitatively and qualitatively, and is quite distinct from those found at Bełchatów and Podlodów. Above the lower optimum of this interglacial, the sediments are characterized by *Aulacoseira ambigua* and *A. granulata*, with *Fragilaria* spp. becoming more prominent in the younger, overlying, minerogenic clay section. The succession of diatoms at Fałęcice reflects a peculiar stage of lake development in which diatoms typical of the Ferdynandovian Interglacial found near Bełchatów and Podlodów were absent. This may be a result of differential destruction of the assemblage but the taxa present do indicate the existence of a small initially eutrophic lake associated with the warm and humid period of the Ferdynandovian Interglacial and during the final phase of the lake decline (Lindner *et al.*, in press). It should be pointed out here that a similar prevalence of *Aulacoseira* spp. has been noted in sediments representing an analogous part of the decline of the Ferdynandovian Interglacial in the ławki-7 profile in which there are no indicator diatoms, such as *Cyclotella* cf. *reczickiae* and the *Stephanodiscus rotula/niagarae* complex, despite the fact that they are abundant in older

256

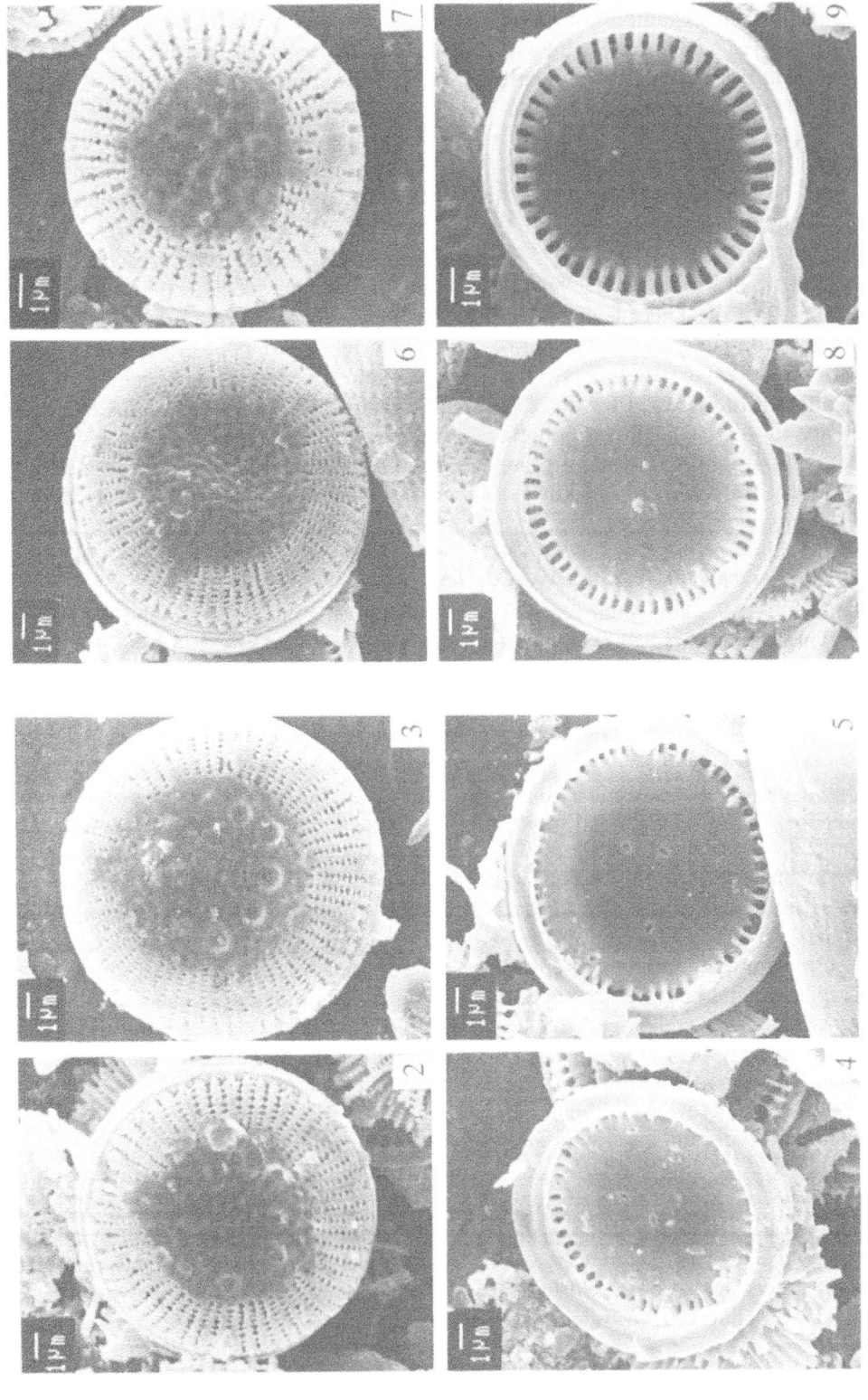

Figs 6–9. Cyclotella sp. similar or allied to *C. reczickiae,* external and internal views of valves found in the Ferdynan-dovian interglacial.

Figs 2–5. Cyclotella cf. *reczickiae* Khursevich et Loginova, external and internal views of valves found in the Ferdynandovian interglacial.

257

sediments of the same profile (Marciniak, in press).

The qualitative and quantitative differences in the diatom assemblages from the sediments of these interglacial lakes may result from different geographic locations and differences in the local geology. Other influential factors may include the size and depth of the lakes in question, as well as the chemistry of the waters. The various climatic conditions ascribed to the Ferdynandovian and the Mazovian interglacials were probably also important.

Acknowledgements

I am very grateful to Dr. Elizabeth Haworth and Dr. Roger Flower, for constructive comments about this manuscript.

References

Bińka, K., B. Marciniak & M. Ziembińska-Tworzydło, 1987. Palynologic and diatomologic analysis of the Masovian Interglacial deposits in Adamówka (Sandomierz Lowland). Kwart. Geol. 31: 453–474.

Dąbrowski, S., J. Dzierżek, K. M. Krupiński, L. Lindner & B. Marciniak, 1987. On the occurrence of two series of interglacial sediments in the Piła section (Northern Poland). Bull. Acad. Pol. Earth Sci. 35: 379–390.

Håkansson, H., 1988. A88. A study of diatom species belonging to the *Cyclotella bodanica/comta* complex (*Bacillariophyceae*). Proc. 9th Int. Diatom Symp. 1986 (ed.) F. E. Round Biopress, Bristol & Koeltz, Koenigstein, 329–354.

Harasimiuk, M. & A. Henkiel, 1981. Fossil valley forms in the vicinities of Łęczna and their importance for paleogeography of the Wieprz river system. Kwart. Geolog. 25: 147–161.

Janczyk-Kopikowa, Z., 1975. Flora of the Mazovian Interglacial at Ferdynandów. Biul. Inst. Geol. 290: 5–94.

Janczyk-Kopikowa, Z., 1981. Pollen analysis of the Pleistocene sediments at Kaznów and Krępiec. Biul. Inst. Geol. 321: 249–258.

Kaczmarska, I. 1976. Diatom analysis of Eemian profile in freshwater deposits at Imbramowice near Wrocław. Acta Paleobot. 17: 3–33.

Khursevich, G. K. & L. P. Loginova, 1984. Novyy vid roda *Cyclotella* Kütz. iz nizhneantropogenovykh otlozheniy drevneozernykh Byelorussii i Latvii. Doklady Akademii Nauk BSSR. 28: 52–55.

Khursevich, G. K. & L. P. Loginova, 1986. Vozrast i paleo-geographicheskiye uslovya phormirovaniya drevnoozernykh otlozheniy Rechitskogo Pridneprovya. Pleistotsen Rechitskogo Pridneprovya Byelorussii. Mińsk, Nauka i Tekhnika, 76–142.

Kling, H. & H. Håkansson, 1988. A light and electron microscope study of *Cyclotella* species (*Bacillariophyceae*) from Central and Northern Canadian lakes. Diatom Research 3: 55–82.

Krupiński, K. M., L. Lindner & W. Turowski, 1986. Sediments of the Mazovian Interglacial at Biała Podlaska (Eastern Poland). Bull. Acad. Pol. Earth Sci. 34: 365–373.

Krupiński, K. M., L. Lindner & W. Turowski, 1988. Geologicfloristic setting of the Mazovian Interglacial sediments at Białka Podlaska (E. Poland). Acta Palaeobot. 28: 29–47.

Krzyszkowski, D. & T. Kuszell, 1987. Nowe stanowisko interglaciału ferdynandowskiego w odkrywce Bełchatów. In Przewodnik II Sympozjum Czwartorzęd Rejonu Bełchatowa. Wrocław-Warszawa: 125–134.

Laskowska-Wysoczańska, W., 1987. Stratigraphic position of the interglacial deposits in Adamówka near Tarnogród (Sandomierz Lowland). Kwart. Geol. 31: 441–452.

Lindner, L., 1984. An outline of Pleistocene chronostratigraphy in Poland. Acta Geol. Pol. 34: 27–49.

Lindner, L., B. Marciniak & M. Ziembińska-Tworzydło, 1989. Osady interglacjalne w Falęcicach oraz ich znaczenie dla stratygrafii plejstocenu w dorzeczu dolnej Pilicy. Annal. Soc. Geol. Polon. In press.

Loginova, L. P., 1979. Paleogeographiya likhvinskogo mezhlednikovya srednej polosy vostochno-evropeyskoy ravniny. Nauka y tekhnika, Minsk. 138 pp.

Loginova, L. P., 1982. The Likhvin diatom flora from the Central Part of the East-European Plain, its paleogeographical and stratigraphic significance. Acta Geol. Hung. 25: 149–160.

Makhnach, N. A., G. K. Khursevich, L. P. Loginova & L. N. Bogomolova, 1982. Novye paleobotanicheskiye issledovaniya drevnozernykh pleistotsenovykh otlozheniy razreza Krasnaya Dubrova. In Noegenovye otlozheniya Byelorussii. Nauka y Tekhnika, Minsk: 37–53.

Marciniak, B., 1973. The application of the diatomological analysis in the stratigraphy of the Late-glacial deposits of the Mikołajskie Lake. Studia Geolog. Pol. 39: 1–159.

Marciniak, B., 1979. Dominant diatoms from late-glacial and Holocene lacustrine sediments in Northern Poland. Nov. Hedwig. Beih. 64: 411–416.

Marciniak, B., 1980. Middle Pleistocene diatoms from lacustrine deposits from Krępiec (Lublin Upland). Kwart. Geol. 24: 349–360.

Marciniak, B., 1982. Late-glacial and Holocene new diatoms from a glacial Lake Przedni Staw in the Pięć Stawów Polskich Valley, Polish Tatra Mts. Acta Geol. Acad. Sci. Hungar. 25: 161–171.

Marciniak, B., 1983. Diatoms in the Mazovian Interglacial of the Lublin Upland. Bull. Acad. Pol. Sci., Sciences de la terre 30: 77–85.

Marciniak, B., 1986a. Late Quaternary diatoms in the sediments of Przedni Staw Lake (Polish Tatra Mts.). Hydrobiologia 143: 255–265.

Marciniak, B., 1986b. Diatoms in the Mazovian (Holstein, Likhvin) Integracial sediments of South-eastern Poland. Proc. 8th Int. Diatom-Symp. 1984. (ed.), M. Ricard, Koeltz, Koeningstein: 483–494.

Marciniak, B., 1987. Diatoms from the Holocene sediments of Lake Steklin (Dobrzyń Lake District). Acta Palaeobot. 27: 319–334.

Marciniak, B., 1988. Diatoms from the Late Quaternary sediments of the Błędowo Lake,Central Poland. Preliminary report. Proc. Nordic Diatomist Meeting 1987. (ed.) U. Miller and A.-M. Robertsson. University of Stockholm, Report 12: 57–65.

Marciniak, B., 1990. Diatoms of the Ferdynandovian Interglacial in the Bełchatów region Central Poland (preliminary report). Folia Quat. 61: in press.

Marciniak, B. & A. Cieśla, 1983. Diatomological and geochemical studies on late-glacial and Holocene sediments from the Przedni Staw Lake in the Dolina Pięciu Stawów Polskich Valley (Tatra Mts.). Kwart. Geol. 27: 123–150.

Marciniak, B. & W. W. Kowalski, 1978. Dominant diatoms, pollen, chemistry and mineralogy of the Eemian lacustrine sediments from Nidzica (Northern Poland): A Preliminary Report. Pol. Arch. Hydrobiol. 25: 269–281.

Przybyłowska-Lange, W., I. Kaczmarska, B. Marciniak & J. Siemińska, 1989. Gromada Chrysophyta. In Rühle & E. Rühle (eds), Budowa geologiczna Polski t. III, atlas skamieniałości przewodnich i charakterystycznych cz. 3b Kenozoik, Czwartorzęd. Wydawnictwa Geologiczne, Warszawa: 128–214.

Round, F. E., 1981. The diatom genus *Stephanodiscus*: an electron-microscopic view of the classical species. Arch. Protistenk. 124: 455–470.

Theriot, E. & E. F. Stoermer, 1984. Principal component analysis of *Stephanodiscus*: Observations on two new species from the *Stephanodiscus niagarae* complex. Bacillaria 7: 37–58.

Theriot, E. & E. F. Stoermer, 1986. Principal component analysis of *Stephanodiscus*: Field evidence for two varieties of *S. niagarae*. Proc. 8th Int. Diatom-Symp. 1984. (ed.) M. Ricard, Koeltz, Koenigstein: 385–394.

Theriot, E., Y. Qi, J. Yang & L. Ling, 1988. Taxonomy of the diatom *Stephanodiscus niagarae* from a fossil deposit in Jingyu County, Jilin Province, China. Diat. Res. 3: 159–167.

Hydrobiologia **214**: 259–266, 1990.
J. P. Smith, P. G. Appleby, R. W. Battarbee, J. A. Dearing, R. Flower, E. Y. Haworth, F. Oldfield & P. E. O'Sullivan (eds), 259
Environmental History and Palaeolimnology.
© 1990 *Kluwer Academic Publishers.*

Fossil diatom inferred reconstruction of the pH history of two acidic, clear water lakes from insular Newfoundland, Canada

David A. Scruton[1], Janet K. Elner[2] & Sankar N. Ray[1]
[1] *Fisheries and Oceans, Science Branch, P.O. Box 5667, St. John's, Newfoundland, Canada A1C 5X1;*
[2] *Fisheries and Oceans, Fisheries and Habitat Management Branch, 867 Lakeshore Drive, P.O. Box 867, Burlington, Ontario, Canada L7R 4A6*

Key words: Palaeolimnology, acidification, diatoms, clear water, lake, pH, regional calibration, Newfoundland

Abstract

Palaeoecological research has been used to evaluate the impact of acidic deposition on lakes in insular Newfoundland. Terrestrial organic deposits in the region have considerable influence on freshwaters and have placed constrains on interpretation of the degree of anthropogenic acidification. In this paper, a region-specific calibration equation unique to clear water lakes (colour values ≤ 15) is developed from diatom assemblages in surface sediments from 22 lakes. The inferred pH history of two acidic, clear water lakes is then developed with a view to eliminating the influence of organic acidity from the interpretation of historical acid-base chemistry. The pH histories of the two lakes suggest modest declines (0.3 to 0.4 unit) in the most recent strata (i.e. since the 1930's). Both lakes demonstrate an increase in inferred pH in the surface horizon, which is consistent with declines in acidic deposition in the region since the mid 1970's. The magnitude and timing of the pH trends in these two lakes is common to those previously developed from more highly coloured lakes. The similar magnitude and onset of pH declines in lakes with varying amounts of organic influence provides no palaeolimnological evidence to suggest a contribution to, or modification of, lake acidification by organic acids.

Introduction

Synoptic water quality surveys and long term aquatic monitoring have confirmed the extreme susceptibility of much insular Newfoundland's freshwaters to potential anthropogenic acidification (Scruton, 1983; Scruton & Taylor, 1989). Evidence of acidification is primarily apparent in an erosion of buffering capacity as indicated from deficits in alkalinity (Howell & Brooksbank, 1987; Scruton & Taylor, 1989). Acid sensitive rivers have also demonstrated episodic acidification during peaks in the hydrological regime

(Scruton, 1986). Evidence of chronic acidification of lakes and rivers and perturbation of biological communities is not apparent.

A major problem in interpretation of regional water chemistry and evaluation of effects of anthropogenic acidification has been the considerable influence of organic deposits on the acid-base chemistry of many lakes. In these lakes it has been difficult to quantitatively ascribe the contribution of acids of atmospheric origin, or naturally occurring weak organic acids, to low contemporary pH values. Organic influences can be difficult to isolate because organic acids can contribute significantly to the acidity of dilute waters independent of other sources of hydrogen ions (Oliver *et al.*, 1983) or may play a role in the buffering regime of bicarbonate poor lakes receiving strong acidic deposition (Jones *et al.*, 1986). Organic contributions to lake acidity has been demonstrated to vary both geographically and seasonally, making it difficult to systematically evaluate the influence (Jones *et al.*, 1986; Scruton & Taylor, 1989).

In the absence of long-term historical water chemistry data for insular Newfoundland lakes, palaeoecological research has been used to interpret lake pH histories over the period of acidic deposition in Eastern Canada. Region-specific relationships between fossil diatoms in surface sediments and contemporary pH for insular Newfoundland lakes have been quantified, refined, and the inferred pH histories of eight lakes developed (Scruton *et al.*, 1987a and b, Rybak *et al.*, 1989). This research has suggested modest pH declines (0.2 to 0.5 unit) in the recent horizons (i.e. since the 1940's) of 5 lakes. The inferred pH declines have largely been within the standard error of the transfer functions, however the similarity in the magnitude and onset of trends does suggest progressing acidification. Watershed humification is considered a possible alternative (or contributing) cause of the inferred pH declines and this consideration is more important in highly coloured lakes (Scruton *et al.*, 1987a and b). A further complication in evaluating inferred trends in Newfoundland lakes is the absence of detailed information on watershed history.

This paper recreates the pH history of two acidic, clear water lakes with a view to removing the potential organic contribution from the interpretation to isolate regional pH change attributable to anthropogenic causes. A calibration equation specific to clear water lakes is developed, from surficial diatoms and contemporary pH in 22 lakes, and used to trace the pH history of the two lakes from the distribution of fossil diatoms in 0.5 cm strata from 25-cm long, 10-cm diameter sediment cores.

Methods

Short sediment cores (25 to 60 cm) were collected from the deepest portion of 44 lakes in insular Newfoundland in 1984 and 1986 using a light weight, 10 cm diameter Kajak corer. Lakes selected for surface samples only (calibration series) had the top 1.0 cm (1984 samples) or 0.5 cm (1986 samples) retained, while cores from downcore lakes (203 and 630) were sectioned at 0.5 cm intervals for the length of the consolidated core.

Sediment sub-samples for diatom analyses were cleaned by acid digestion, rinsed with distilled water, and dehydrated in ethyl alcohol. The diatom suspension was mounted on slides and ringed after the method of Battarbee (1973). Diatom identifications was based on the floras of Cleve-Euler (1951–1955), Hustedt (1930, 1937–1939), Mölder & Tynni (1967–1973), Patrick & Reimer (1966, 1975), Renberg (1977) and Tynni (1975–1981).

Water samples were collected at the time of core collection and during lake inventory/monitoring programs. Parameters measured, methods of analysis, and limits of detection have been previously described (Scruton, 1983; Scruton & Taylor, 1989). Measurements of pH (mean values, values collected at the time of core collection, minima/maxima) were evaluated for suitability for calibration. It was determined that pH measurements taken during core collections produced most statistically significant equations (Scruton *et al.*, 1987a) and these values were subsequently employed in development of the clear water transfer function.

Diatom data and contemporary water chemistry for the 22 clear water lakes were used to develop a transfer function specific to clear water lakes (colour values $\leq$ 15). Taxa were assigned to pH tolerance categories of Hustedt (1930) in relation to frequency of occurrence in the surface samples of the 44 lakes. Multiple regression (MR, forward selection, using maximum r^2 method) of pH tolerance groups was selected as the method for use in calculating the colour-specific calibration equation (Scruton *et al.*, 1987a). The new transfer function was evaluated in relation to pre-

Fig. 1. Location of the 22 lakes used in developing the clear water calibration equation and the downcore lakes (#203 and 630).

semblages in the 0.5 cm sediment samples from lakes 203 and 630 using the MR transfer function developed in this paper. Cores were dated from alpha-ray spectral analysis of [210]polonium activity in the uppermost five horizons to put a chronology on the inferred pH trends. Dates were determined from unsupported [210]lead activity (as determined from the daughter decay product [210]Po by isotope dilution) using the Constant Rate of Supply (CRS) model of Appleby & Oldfield (1978). Unsupported ^{210}Pb activity indicated typical log-linear relationship with depth suggesting the cores were undisturbed (absence of mixing zones).

Study lakes

Lakes selected for use in developing the regional clear water calibration equation were located in remote areas removed from present and historical human disturbance (Fig. 1). Lakes were predominantly small (all < 200 ha, 12 < 100 ha), located high in river drainage basins (1° and 2° order lakes), and had low relative drainage basin size (ratio of basin to lake areas were less than 10:1 in 20 lakes). The lakes represented a wide range of contemporary pH (4.78 to 7.72) while maintaining a water colour of ≤ 15. The calibra-

viously used equations (Scruton et al., 1987a) by comparing difference plots of inferred and measured pH.

Inferred pH was calculated from diatom as-

Fig. 2. The proportional distribution of fossil diatoms grouped as (Hustedt) pH tolerance groups in the 22 clear water lakes.

Table 1. Multiple regression equations and statistics for pH *versus* proportional representation of pH tolerance groups. Equations developed are for (1) all 34 lakes sampled in 1984, (2) 13 clearwater lakes sampled in 1984, and for (3) 22 clearwater lakes (13 sampled in 1984 and 9 sampled in 1986).

Equation	MR regression model
(1)	pH = 7.07 − 0.028ACIDB − 0.015ACIDP + 0.006ALKP + 0.010ALKB (r^2 = 0.95, F-ratio = 45.20, Se = + − 0.18, $P < 0.001$)
(2)	pH = 6.83 − 0.026ACIDB − 0.011ACIDP + 0.010ALKP + 0.013ALKB (r^2 = 0.96, F-ratio = 51.04, Se = + − 0.23, $P < 0.001$)
(3)	pH = 7.39 − 0.037ACIDB − 0.019ACIDP + 0.003ALKP + 0.015ALKB (r^2 = 0.91, F-ratio = 44.43, Se = + − 0.29, $P < 0.001$)

ACIDB = acidobiontic
ACIDP = acidophilic
ALKP = alkaliphilic
ALKB = alkalibiontic

tion series lakes also represented a range of buffering regimes from acidic, extremely acid-sensitive lakes (alkalinity $\leqslant 40\ \mu$eqL^{-1}, $n = 14$), through less acidic, highly acid-sensitive lakes (alkalinity from 40 to 200 μeqL^{-1}, $n = 4$) to circumneutral and basic lakes that are not acid-sensitive (alkalinity $> 500\ \mu$eqL^{-1}, $n = 4$).

Results

Clear water transfer function

A total array of 115 diatom taxa from 22 lakes were used in developing the clear water calibration equation. The proportional distribution of taxa,

after assignment to pH tolerance groups, is presented in Fig. 2.

The coefficients for multiple regression (MR) of proportional representation of pH tolerance groups in the 22 clear water lakes (independent variables) with contemporary pH (dependent variable) produced a regional transfer function specific to clear water lakes. This model is compared with the (MR) transfer function developed from all 34 lakes sampled in 1984 and with the clear water transfer function from the same data set (Scruton *et al.*, 1987) in Table 1. Difference plots compare inferred pH values from the three equations with contemporary pH values (Fig. 3). The three difference plots produced similar statistics, however the clear water equation provides

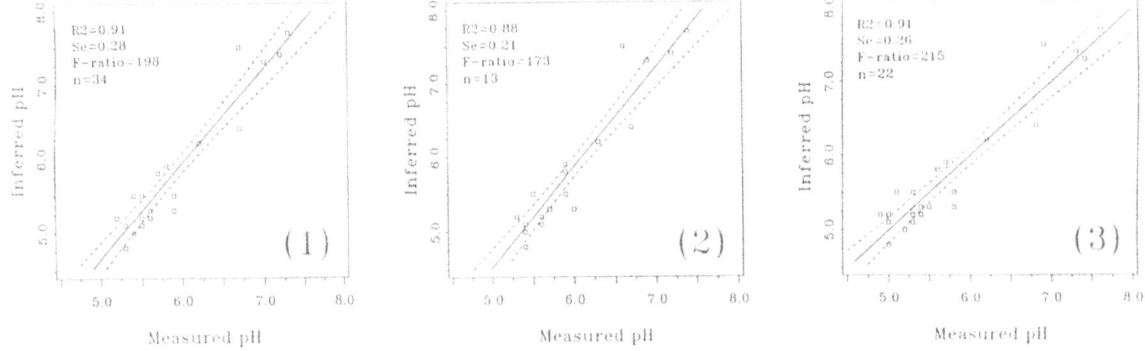

Fig. 3. Difference plots of inferred versus measured pH. Inferred pH is calculated from (1) a transfer function developed from 34 lakes sampled in 1984, (2) a transfer function developed from 13 clear water lakes sampled in 1984 and (3) the transfer function developed in this paper from 22 clear water lakes.

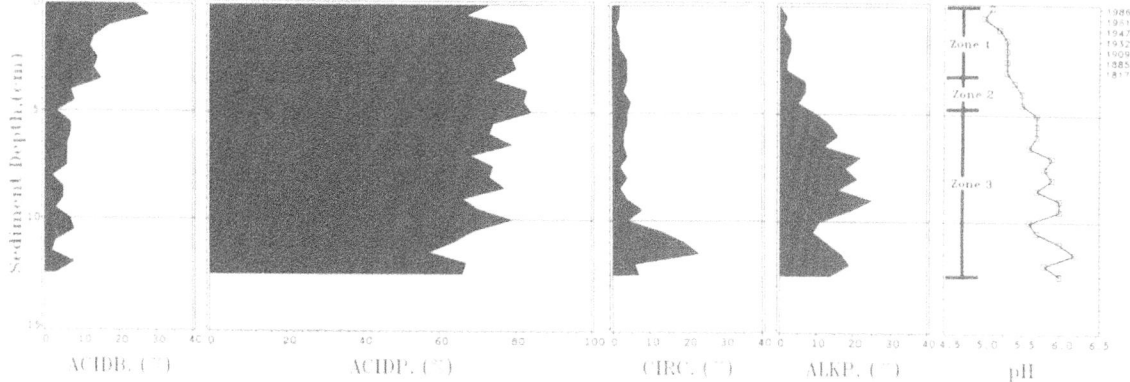

Fig. 4. Stratigraphic distribution of fossil diatoms grouped as (Hustedt) pH tolerance groups and the inferred pH history of Lake 203.

the best correspondence between inferred and measured pH values (closest to the 1:1 line). The other two equations demonstrate some under-estimation of inferred pH values in the lower ranges and overestimation in the higher pH ranges.

Inferred pH history of lake 203

The stratigraphic distribution of diatoms (by pH tolerance group) and the historical pH profile for lake 203 is presented in Fig. 4. The top of hori-zon 2 (0.5 to 1.0 cm) is dated at 1961 A.D. and the most recent 160 years are represented in the top 3.5 cm.

The horizons from the lake 203 core can be divided into 3 zones. The third zone comprised horizons from 13 cm to 5.0 cm. During this period the inferred pH of lake 203 fluctuated from 5.6 to 6.2. The flora of this period is characterized by low proportions of acidobiontic taxa and sub-stantial proportions of alkaliphilic diatoms (10 to 25%). During this period, centric diatoms were dominant, with at least 50% of the diatom as-semblages in these strata belonging to 6 species. *Melosira italica* (Ehr.) Kütz. declined throughout this period whereas species of the genus *Coscino-discus*, *Cyclotella meneghiniana* Kütz., *C. stelligera* (Grun.) Cl. and *Melosira distans* (Ehr.) Kütz. peaked in zone 3 with notable declines in the latter years of this period.

The second zone comprises the next 3 horizons (4.5 cm to 3.5 cm, c. 1817 at the top horizon at

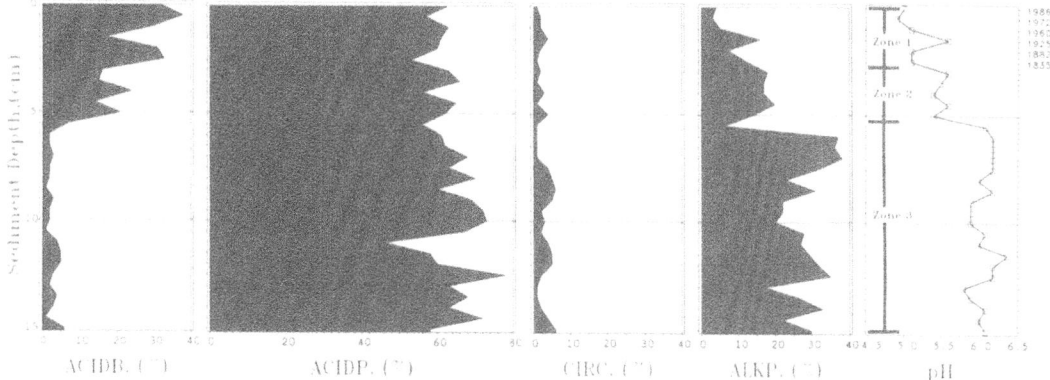

Fig. 5. Stratigraphic distribution of fossil diatoms grouped as (Hustedt) pH tolerance groups and the inferred pH history of Lake 630.

Table 2. Morphometric, limnological and drainage basin characteristics of lakes #203 and 630.

Parameter	Lake 203	Lake 630
Lake Area (LA)(ha)	24.0	32.0
Drainage Area (DA)(ha)	127.0	246.0
Ratio (DA:LA)	5.3	7.7
Elevation (m above SL)	419.0	434.3
Distance from Sea (km)	8.5	18.5
Drainage Order	1	1 (true pond)
Secchi depth (m)	3.5	N/A
Apparent Max. Depth (m)	13.0	6.0
Bedrock Type*	4	6

* Type 4 -gneiss, schist, and foliated granite.
  Type 6 -granite.

3.5 cm) and demonstrated an inferred pH decline of 0.2 units to pH 5.4. This zone is characterized by an increase in acidobiontic taxa and concomitant declines in alkaliphilic taxa. The dominant centric taxa of Zone 3 are largely replaced by acidobiontic and acidophilic species of the genera *Eunotia* and *Tabellaria*.

The first zone represents the top 7 strata (3.0 to 0 cm, c. 1817 to 1986). The top 3 horizons (c. 1947 to 1986) demonstrate a decline to pH 5.1 with the minimum inferred pH value (5.0) at horizon 2 (0.5 cm, c. 1961 at the top of the horizon). A dramatic increase in acidobiontic taxa (from 12 to 29%) with a concurrent decline in alkaliphils is evident. The increase in acidobionts is accounted for primarily by the increased abundance of *Asterionella ralfsii* W. Sm.. Centrics, excepting *Coscinodiscus*, have completely declined by this stage in the history of the lake with recent horizons being dominated by pennates.

Inferred pH history of lake 630

The pH tolerance groups and the inferred pH profile for lake 630 is presented in Fig. 5. The top of horizon 2 (0.5 to 1.0 cm) is dated c. 1972 A.D. and the top 6 horizons represent a period of 150 years.

Principal component analysis of the flora in the 31 horizons divides the strata into 3 zones. The third zone includes the 20 horizons from the base of the core to a depth of 5.5 cm, with inferred pH fluctuating about 6.1 during this period, and alkaliphilic taxa being as important (17–38%) as acidophilic taxa. Species of the genus *Melosira* were predominant in this zone comprising up to 80% of each assemblage.

The second zone (5.0 to 3.0 cm) demonstrated an initial decline in pH to 5.4 (0.7 units) and fluctuation between pH 5.4 to 5.6. Alkaliphilic taxa decline by approximately 50% over this period with concurrent increases in acidobiontic taxa. The most notable floristic change over this time period is the decline of *Melosira lirata* and *M. italica*.

In the most recent six horizons (zone 1, c. 1885 to 1986), the pH declines further to 5.0 at the surface with the minimum pH at the top of horizon 2 (pH of 4.9, c. 1972). Floristic changes causing this inferred pH decline include further increases in acidobiontic taxa (to 37% at horizon 2) and declines in alkaliphilic diatoms (less than 5% in the top 2 horizons). The *Melosira* spp., dominant in zone 3, have declined to about 10% in the uppermost horizons. The acidobionts *Anomoeoneis serians*, *Tabellaria quadriseptata* and *T. binalis* have increased dramatically in recent strata.

Discussion

Palaeoecological research in Newfoundland has suggested modest pH declines from 0.2 to 0.5 of a unit in the recent strata (since 1930's 1940's) of several lakes. Inferred pH declines have largely been within the standard error of transfer functions, however similarity in magnitude and onset of trends does suggest progressing acidification. Lakes with inferred pH histories studied to date have included clear water, brown water and highly coloured systems (Scruton *et al.*, 1987a).

Previous research (Scruton *et al.*, 1987a) had determined that frequency of occurrence data for taxa, in Hustedt (1930) pH tolerance groupings, from calibration lakes is preferable to literature assignment of pH tolerance. This assessment has resulted in several taxa being assigned to a different tolerance category than literature reference had suggested. Region-specific assignment of taxa to pH tolerance categories has been adopted in developing the clear water transfer function and in inferring pH in the strata from lakes 203 and 630.

Multiple regression (MR) of pH tolerance groups was selected as the preferred method for use in calculating the clear water calibration equation. Index B (Renberg & Helberg, 1982), employed previously in historical pH recreation of Newfoundland lakes (Scruton *et al.*, 1987a and b, Rybak *et al.*, 1989), was not used in this study as weighting factors employed in the Index B calculations have no ecological basis (Charles, 1985). Additionally, alkaliphilic taxa were either absent or in low proportions in surface strata of calibration lakes making it difficult to derive an Index B model. In previous reconstructions of lake pH histories, MR and Index B transfer functions gave a similar pattern of results and MR equations were statistically superior (higher regression coefficient and lower standard error).

Research has also demonstrated the profound influence that lake water colour has on the slope and y-intercept of calibration equations and demonstrated the value of calculating and applying colour-specific transfer functions to inferring the pH histories of lakes within each colour classification (Scruton *et al.*, 1987a). This is apparent in Fig. 3 where functions previously developed (equations 1 and 2) demonstrated departure from the 1:1 line with underestimation of inferred pH values in the lower ranges and overestimation in the higher pH ranges. This bias in the lower ranges (< 6.0) is particularly important in evaluation of anthropogenic lake acidification and could lead to overestimation.

The inferred pH histories and proportional representation of diatoms as pH tolerance groups in the recent strata from lakes 203 and 630 show very similar trends. Both lakes demonstrate substantial increases in acidobiontic taxa and declines in alkaliphilic taxa in the uppermost strata, with concurrent declines in pH (in the order of 0.3 to 0.4 pH unit). The watersheds of both lakes are afforested and, owing to the low organic influence

on the chemistry of these two lakes, watershed processes such as humification and forest fires are unlikely to have caused or contributed to the inferred declines. The most plausible explanation for pH decline is as a response the anthropogenic deposition of acids. The timing of the onset of pH change and the recovery in inferred pH in the surface stratum, consistent with regional trends in acidic deposition and declines since the mid-1970's (Martin & Brydges, 1986), provides further evidence to suggest the impact of acid precipitation. Sulphate deposition in the region in recent years (since 1985) has declined to below 10 kg ha^{-1} yr^{-1} (Ryan *et al.*, 1990).

The similarity in the inferred histories of the clear water lakes studied in this paper and those previously studied with more organic influence (Scruton *et al.*, 1987a, Rybak *et al.*, 1989) suggests that organic processes in lake watersheds have not contributed to, or modified, the impact of anthropogenic acidification. Efforts to regionally refine calibration equations and develop transfer functions in relation to degree of organic influence have improved the confidence of the inferred lake histories.

Conclusions

Inferred pH histories of lakes 203 and 630 indicate modest pH declines (0.3 to 0.4 unit) in recent strata (since circa 1932 and 1925, respectively). Floristic changes contributing to the inferred pH declines are predominantly a dramatic increase in acidobiontic taxa with concomitant declines in alkaliphilic taxa. Both lakes demonstrate an increase in pH from horizon 2 to the surface horizon. This is consistent with results from other regional historical pH reconstructions in insular Newfoundland and is likely caused by declining levels of acidic deposition since the mid-1970s. Similarity in the magnitude and timing of inferred pH changes between the two clear water lakes in this study and other more highly coloured lakes, provides no evidence to suggest that organic acidity has contributed to, or modified, the degree of lake acidification attributable to acidic deposition.

References

Appleby, P. G. & F. Oldfield, 1978. The calculation of lead-210 dates assuming a constant rate of supply of unsupported lead-210 in sediment. Catena 5: 1–8.

Battarbee, R. W., 1973. A new method for the estimation of absolute microfossils with a special reference to diatoms. Limnol. Oceanogr. 18: 647–653.

Charles, D. F., 1985. Relationships between surface sediment diatom assemblages and lakewater characteristics in Adirondack lakes. Ecology 66: 994–1011.

Cleve-Euler, A., 1951–1955. Die Diatomeen von Schweden und Finnland. Kunglia Svenska Velenskapakademien Avhandlinger. Stockholm. 5: 2(1); 3(30); 4(1); 4(5); 5(4).

Howell, G. D. & P. Brooksbank, 1987. An assessment of LRTAP acidification of surface waters of Atlantic Canada. Environment Canada, Conservation and Protection Service, Inland Waters Directorate, Water Quality Branch, Moncton, N.B. Report IW/L-AR-WQB-87-121: 292 p.

Hustedt, F., 1930. Bacillariophyta (Diatomae). In: A. Pascher (ed.), Die Süsswasserflora Mitteleuropas 10. Fischer, Jena. pp 466.

Hustedt, F., 1937–1939. Systematische und ökologische Untersuchungen über die Diatomeen-Flora von Java, Bali und Sumatra. Arch. Hydrobiol. Suppl. 15–16: 274–394.

Jones, M. L., D. R. Marmorek, B. S. Reuber, P. J. McNamee & L. P. Rattie, 1986. Brown waters: Relative importance of external and internal sources of acidification on catchment biota. Review of existing knowledge. ESSA (Environmental and Social Systems Analysis) Ltd., Toronto, Ontario. 85 p.

Martin, H. C. & T. G. Brydges, 1986. Stress Factors: Trends and Recovery. P. 125–139. In: T. Schneider (ed.), The International Conference on Acidification and its Policy Implications. 125–139. Elsevier Science Publ. B.V. Amsterdam.

Mölder, K. & R. Tynni, 1967–1973. Über Finlands rezente und subfossile Diatomeen I–VII. Geol. Soc. Finland. Bull. 39: 199–217, 40: 151–170, 41: 235–251, 42: 129–144, 43: 203–220, 44: 141–149, 45: 159–179.

Oliver, B. G., E. M. Thurman & R. Malcolm, 1983. The contribution of humic substances to the acidity of coloured natural waters. Geochim. Cosmochim. Acta 47: 2031–2035.

Patrick, R. & C. W. Reimer, 1966. The diatoms of the United States. V. 1, Acad. Nat. Sci. Philadelphia Monogr. 13: 388 p.

Patrick, R. & C. W. Reimer, 1975. The diatoms of the United States. V. 2, Acad. Nat. Sci. Philadelphia Monogr. 13: 213 p.

Renberg, I., 1977. Flagilaria lata, a new diatom species. Bot. Notiser 130: 315–318.

Renberg, I. & T. Helberg, 1982. The pH history of lakes in southwestern Sweden, as calculated from the sub-fossil diatom flora of the sediments. Ambio 11: 30–33.

Ryan, P. M., D. A. Scruton & D. E. Stansbury, 1990. Acidification monitoring by the Department of Fisheries and Oceans in Insular Newfoundland, Canada. Verh. int. Ver. Limnol. 24 (in press).

Rybak, M., I. Rybak & D. A. Scruton, 1989. The impact of atmospheric deposition on the aquatic ecosystem with special emphasis on lake productivity. Hydrobiologia 179: 1–16.

Scruton, D. A., 1983. A survey of headwater lakes in insular Newfoundland, with special reference to acid precipitation. Can. Tech. Rep. Fish. Aquat. Sci. 1195: 110 p.

Scruton, D. A., 1986. Spatial and temporal variability in the water chemistry of Atlantic salmon rivers in insular Newfoundland. An assessment of sensitivity to and effects from acidification and implications for resident fish. Can. Tech. Rep. Fish. Aquat. Sci. 1451: 143 p.

Scruton, D. A., J. K. Elner & M. Rybak, 1987a. Regional calibration of fossil diatom – contemporary pH relationships for insular Newfoundland, Canada, including historical pH reconstruction for five lakes with an assessment of palaeo-inferred productivity changes in one lake. P. 457–464. In: R. Perry, R. M. Harrison, J. W. B. Bell and J. N Lester (eds.). Acid Rain: Scientific and Technical Advances, Selper Ltd. (London, U.K.). 821 p.

Scruton, D. A., J. K. Elner & G. Howell, 1978b. Palaeolimnological investigation of freshwater lake sediments in insular Newfoundland. Part 2. Downcore diatom stratigraphies and historical pH profiles for seven lakes. Can. Tech. Rep. Fish. Aquat. Sci. 1521: (Part 2), 67 p.

Scruton, D. A. & D. Taylor, 1989. A survey of dilute lakes on the south coast of insular Newfoundland in relation to natural and anthropogenic acidification. Can. Tech. Rep. Fish. Aquat. Sci. 1711: 58 p.

Tynni, R., 1975–1981. Über Finlands rezente and subfossil Diatomeen VII–XI. Geol. Surv. Finland Bull. 274 (1975); 1–555, 284 (1976): 1–37, 296 (1977): 1–55, 312 (1978): 1–93.

Hydrobiologia **214**: 267–272, 1990.
J. P. Smith, P. G. Appleby, R. W. Battarbee, J. A. Dearing, R. Flower, E. Y. Haworth, F. Oldfield & P. E. O'Sullivan (eds), 267
Environmental History and Palaeolimnology.
© 1990 *Kluwer Academic Publishers.*

Taphonomy and diagenesis in diatom assemblages; a Late Pleistocene palaeoecological study from Lake Magadi, Kenya

Philip Barker[1], Françoise Gasse[2], Neil Roberts[1] & Maurice Taieb[3]
Department of Geography, University of Technology, Loughborough, U.K.[1]; Laboratoire d'Hydrologie et de Géochemie Isotopique, Université de Paris-Sud, Orsay, France[2]; CNRS, Laboratoire de Géologie du Quaternaire, Marseille, France[3]

Key words: diatoms, taphonomy, dissolution, diagenesis, saline lakes

Abstract

Many fossil diatom assemblages do not possess a direct modern analogue as a result of taphonomic processes and diagenesis within the assemblage. Some of these problems are illustrated with reference to core material collected from hypersaline Lake Magadi, Kenya, which during the Late Pleistocene experienced major fluctuations in water chemistry and depth. Competing multiple hypotheses are proposed for no analogue assemblages, with selection between these hypotheses being based on the results of interdisciplinary research.

Introduction

The study of fossil diatom assemblages provides a powerful means of reconstructing the palaeo-limnology of closed basin, saline lakes (e.g. Holdship, 1976; Gasse, 1977; Bradbury, 1989). The interpretation of these depends on modern aut- and syn-ecological studies of diatoms from the investigated region (e.g. Gasse *et al.*, 1983; Gasse, 1986a; Fritz & Battarbee, 1986). But, despite the increasing amount of modern ecological data by no means all fossil assemblages have modern analogues at the regional scale.

The search for suitable analogues is further hindered by taphonomic processes which can mix diatoms from different habitats or from different periods in time. These processes are particularly important in tropical lakes where seasonal differences in the hydrological regime are often very pronounced and recharge by the river system occurs ephemerally. Fluvial influx of dilute water

may produce a periodically stratified lake thus increasing the heterogeneity of lake habitats and/or introduce assemblages that were living within the river system. Death assemblages will be subject to mixing by limnological processes, involving not only the incorporation of planktonic and periphytic species but also the dispersal of diatoms from micro habitats such as hot springs. Once the sediment is deposited it becomes liable to physical mixing processes and bioturbation often resulting in breakage and/or dissolution of diatom frustules. Thus the fossil assemblage represents a combination of environmental conditions occuring spatially in the lake catchment as well as those operating temporally during distinct hydrological seasons (Gasse, 1988).

In highly saline environments diatom frustules are liable to be dissolved (e.g. Badaut & Risacher, 1983) and can even be diagenetically transformed into zeolites (e.g. Stoffers & Holdship, 1976; Gasse & Seyve in Tiercelin *et al.*, 1987). Dissolu-

268

Fig. 1. The Magadi-Natron catchment (after Hillaire-Marcel and Casanova, 1987).

tion of frustules can occur either during deposition or much later as a result of the circulation of concentrated groundwater circulation. Such chemical dissolution could preferentially dissolve weakly silicified forms (Shemesh *et al.*, 1989) and can therefore lead to biased palaeoecological data. These problems are present in all lake-based diatom studies but take a different and more complex form in the context of saline tropical lakes.

Some of these problems will be illustrated here with reference to a new diatom sequence from Lake Magadi, Kenya. Permanent water in Magadi only occupies 40% of the lake area, the remainder being covered by a thick deposit of trona (sodium carbonate). Water levels are maintained from a groundwater reservoir that discharges through a series of hot springs. Magadi has no perennial inflowing streams at present, the Ewaso Ngiro drainage system entering the Magadi-Natron catchment but passing to the west of Magadi (Fig. 1).

Materials and methods

A 9 m core was taken from a mudflat on the north western edge of Lake Magadi named Flamingo Nursery. This was collected in 1987 as part of the EQUARIFT programme coordinated by the CNRS, Laboratoire de Géologie du Quaternaire, Marseille. A wide range of analyses have been undertaken on this core including sedimentology, mineralogy, palynology, and palaeomagnetics (Taieb *et al.*, 1989). Five radiocarbon dates have been obtained, placing the sequence in the Late Pleistocene period (Fig. 2). These have been made on organic matter from the upper and lower sections of the core using both conventional and accelerator mass-spectrometer dating methods.

Fifty five samples were taken from the core for diatom analysis and 500 valves were counted where possible. Standard treatments with 10% HCl and 30% H_2O_2 to remove the carbonates and the organic matter respectively were used to prepare the samples for observation. Routine counting was made with a light microscope ($\times 1250$) whilst evidence for dissolution was studied using scanning electron microscopy.

A tripartite zonation of the core is indicated by many of the analyses which have been undertaken and the diatom stratigraphy will be treated accordingly. The fossil data have been compared to modern assemblages included in a database incorporating diatom and environmental information from 210 sites in East AFrica (Gasse *et al.*, 1983). Transfer functions for pH (Gasse & Tekaia, 1983; Gasse, 1986b) and conductivity (Gasse, in prep.) have been used to reconstruct palaeoenvironmental conditions.

Zone 3; 18000–17000 BP (870–760 cm)

This lowest section of the core is characterized by the abundance of *Anomoeoneis sphaerophora*, with *Rhopalodia gibberula*, *Hantzschia amphioxys*, *Nitzschia 'groups latens'*, *N. frustulum* and *Navicula* spp. (including *N. mutica*, *N. tenella* and *N.* sp. af. *jakhalsensis*). The sample analysed at the very base of the core (869 cm) differs from the rest of

269

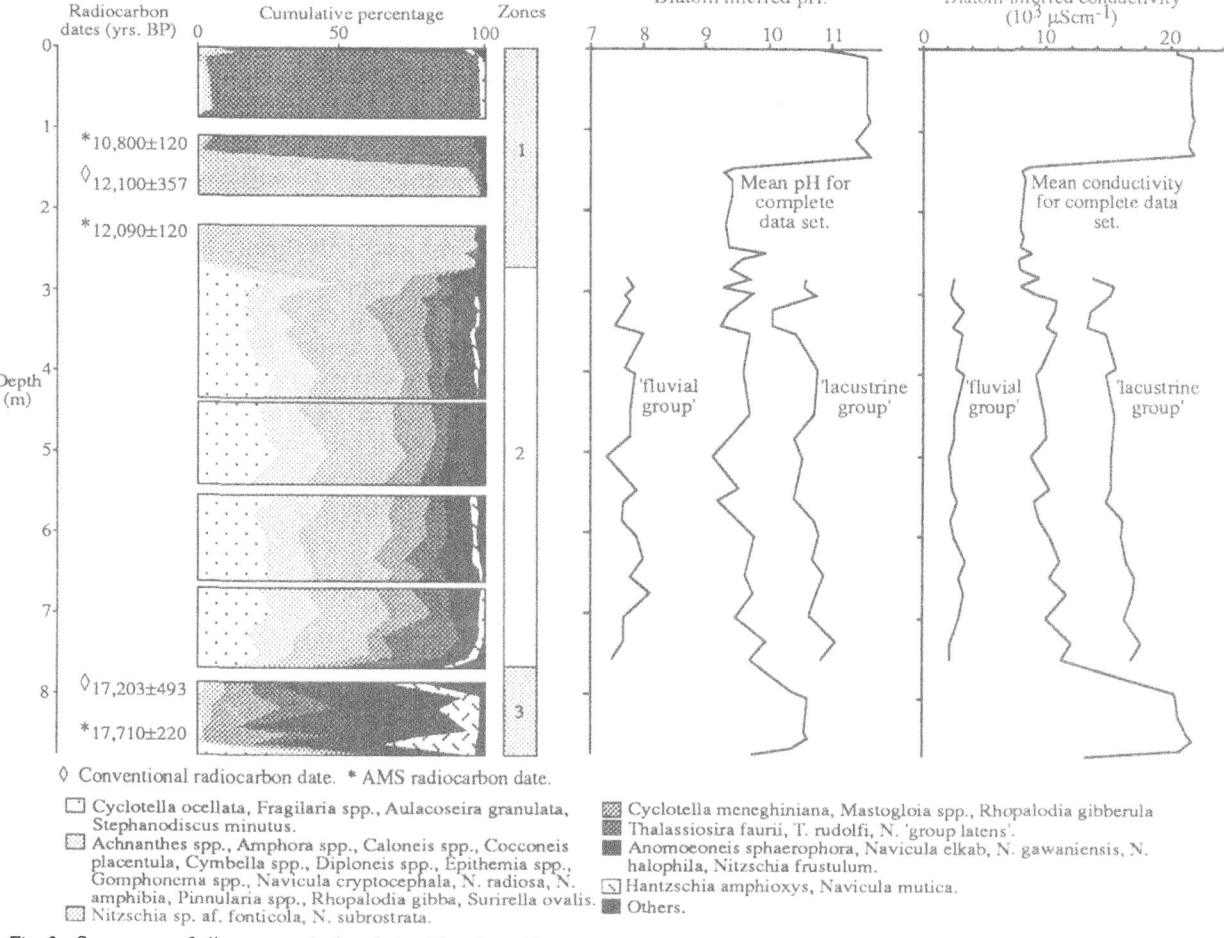

◊ Conventional radiocarbon date. * AMS radiocarbon date.

☐ Cyclotella ocellata, Fragilaria spp., Aulacoseira granulata, Stephanodiscus minutus.
▨ Achnanthes spp., Amphora spp., Caloneis spp., Cocconeis placentula, Cymbella spp., Diploneis spp., Epithemia spp., Gomphonema spp., Navicula cryptocephala, N. radiosa, N. amphibia, Pinnularia spp., Rhopalodia gibba, Surirella ovalis.
▨ Nitzschia sp. af. fonticola, N. subrostrata.
▨ Cyclotella meneghiniana, Mastogloia spp., Rhopalodia gibberula
▨ Thalassiosira faurii, T. rudolfi, N. 'group latens'.
▨ Anomoeoneis sphaerophora, Navicula elkab, N. gawaniensis, N. halophila, Nitzschia frustulum.
▨ Hantzschia amphioxys, Navicula mutica.
▨ Others.

Fig. 2. Summary of diatom analysis of the Flamingo Nursery core, Lake Magadi, Kenya (left) with estimates for pH and conductivity (right).

the zone as it contains a more diverse diatom flora including several oligosaline species, it is therefore considered as a sub-zone (B) of zone 3. The dominance of the assemblage by *A. sphaerophora* makes it comparable to mud samples collected from Embagai Crater Lake (pH 10.1, conductivity 13 500 μS cm^{-1}, Gasse et al., 1983), and in surface mud samples collected from the Magadi hot springs (pH > 9, conductivity > 20 000 μS cm^{-1}). These analogues are all from shallow lakes or hot springs where the living assemblage is periphytic. Sub-zone 3B has a more diverse flora and includes many species with greater affinity to water of moderate or low salinity alongside the more saline elements characteristic of sub-zone 3A.

The autecological attributes of species in this zone suggest an environmental interpretation similar to that indicated by the synecology. The mean conductivity of zone 3A is 20 800 μS cm^{-1} whilst mean pH is estimated at 9.6 (Fig. 2). Zone 3A is largely comprised of species classified as littoral epipelic-epilithic (> 72% excluding aerophilous species). Very few epiphytic or planktonic species were found, which is taken to indicate shallow water conditions with little fringing vegetation close to the coring site. A significant number of species encountered are often, although not exclusively, found in aerophilous habitats. This could suggest that the lake virtually dried up or was merely moist periodically during this interval.

Zone 2.: 17 000–?12 500 (750 cm–280 cm)

Zone 2 incorporates diatoms typical of saline environments (e.g. *Navicula elkab*, *Thalassiosira faurii*, *Rhopalodia gibberula*, *Cyclotella meneghiniana*) alongside those usually found in more dilute conditions (*Aulacoseira granulata*, *Epithemia sorex*, *E. zebra*, *Fragilaria* spp. and *Stephanodiscus minutus*). This situation is unusual although similar mixed assemblages have been found elsewhere, for example in the Guidimouni salt pond, Niger (Gasse, 1987). For lake Magadi three alternative hypotheses will be discussed below for this conjunction of species, although several combinations of these hypotheses may also be envisaged.

The combination of high diversity within, but close similarity between samples, is difficult to explain unless either the environment was very stable or the sediment accumulated very quickly. The sediment appears to include volcanic glass at certain parts of this zone and it could be that there was an increase in volcanism within the area which contributed greatly to the rate of accumulation via an aeolian deposition of ash. Organic matter and diatom abundance are both quite low throughout this zone (although the number of diatoms/g does increase above 4 m). This could be a result of the 'diluting' effect of an allochthonous input into the lake. Zone 2 finishes as suddenly as it begins with a striking change in the fossil diatom assemblage taking place, only *Nitzschia subrostrata* and *N.* sp. af. *fonticola* continuing to be present in zone 1C. The sediment stratigraphy again mirrors this change with the fine homogenous clay changing to laminated sediments via an intervening sandy horizon. Other analyses including mineralogy, sedimentology and magnetics also show marked changes close to this point, perhaps indicating an hiatus in deposition.

Zone 1: 12 500–10 000 BP (270 cm–0 cm)

This zone is divided into three sub-zones, 1C from 270 cm–139 cm, 1B from 130 cm–6 cm, 1A a single sample at the surface. Sub-zone 1C comprises almost entirely *N.* sp. af. *fonticola* which accounts for over 69% of the diatom assemblage, although *N. subrostrata* is significant in the lower part of this zone. Assemblage 1B begins at 130 cm, where the diatom flora becomes almost entirely comprised of *Nitzschia* 'group latens' (>90%). Problems of the taxonomy of *N.* sp. af. *fonticola* and *Nitzschia* 'group latens' have been discussed by Gasse (1986a) and environmental interpretation using these species is difficult. The core top sample shows another change in the diatom stratigraphy with a return of *A. sphaerophora* and *Navicula gawaniensis* which correspond well to other modern samples collected from around the Magadi area today.

The water chemistry of the lake during the deposition of sub-zone 1C is difficult to estimate because of the variance of the morphology of *N.* sp. af. *fonticola* which may include taxa of different autecology. The calculation here is based on a weighting derived from an average of all the morphological types within this species. The estimated mean pH and conductivity for this zone is 9.3 and 8 150 μS cm^{-1}. A further problem occurs in trying to categorise the life-form of this species as it can occur in both the periphyton and the plankton of lakes. The dominance of *N.* sp. af. *fonticola* suggests it was living planktonically as fewer niches are available to competitors in the plankton than the periphyton. The interpretation of sub-zone 1B is also difficult because of the taxonomy of *N.* 'group latens'. By making the assumption that the autecology of *N. latens sensu stricto* is representative of the various morphological types within this group it is possible to estimate mean pH at 11.2 whilst average conductivity is 21 500 μS cm^{-1}. These values should be treated with caution given that they are based largely on the autecology of a single taxon. *N.* 'group latens' can live in planktonic or periphytic life-forms in lakes, swamps and hot springs. The morphological type which dominates here is very similar to that living today in the chemically concentrated, yet moderate to very deep lakes Elmenteita and Shala (Gasse, 1986a). This sub-zone therefore would seem to represent an increase in chemical concentration following sub-zone 1C.

Discussion

The importance of taphonomic processes in palaeoenvironmental interpretation is well illustrated by zone 2, where the conjunction of diatoms is indicative of different habitats. One hypothesis to explain this juxtaposition of species is that the lake was stratified seasonally and that a freshwater lens developed during the wet season over a more saline brine; indeed Eugster (1986) has reported periodic stratification in Magadi today. If this were the case it would be expected that the freshwater species were largely planktonic, but as 30–40% are in fact periphytic, this casts doubt on this hypothesis. What seems to be a more likely explanation is that the (relatively saline) lake received periodic inputs of freshwater from a river entering the lake close to the coring site. This would explain both the different life-forms and an increase in sediment accumulation rate as is suggested by the various analyses. A third hypothesis is that the mixture of species is due to the erosion of older sediments from elsewhere in the catchment. However, the preservation of diatoms within this zone is good which would not seem to indicate reworked material.

Adopting as a working hypothesis the second explanation (ie. a river entered close to the site) the data have been tentatively split into freshwater 'fluvial' and saline-alkaline 'lacustrine' components and an environmental interpretation is attempted accordingly. The autecology of the complete data set produces estimates of pH and conductivity with a large degree of variance, the calculation was therefore repeated using the split data (Fig. 2). Statistically these represent respectively minimum and maximum values since the division into the two groups was based largely on chemical preferences. The life-forms of the species in the lacustrine group are divided quite evenly between the three categories of littoral epipelic- epilithic, facultative planktonic and euplanktonic. In contrast the species thought to have been brought to the site by a river include many epiphytic forms as would be expected if this hypothesis were correct.

The problem of diatom frustule dissolution is crucial to the reliability of the environmental interpretation given for Zone 1. An indication that some dissolution has taken place comes from the poor preservation of valves and the low abundance of diatoms within this zone. Indirect evidence of silica dissolution and subsequent diagenesis is provided by the mineralogy of this zone. Magadiite (Eugster, 1967) was noted in varying amounts on the diatom slides and was found to entirely comprise the samples where diatoms were absent, a similar situation was found north of Lake Chad by Maglione (1970) where saline diatoms were dissolved and their silica diagenetically transformed into magadiite. It is possible that the diatom frustule silica has been dissolved and has contributed along with non-biogenic silica to the formation of magadiite and other zeolites (eg. erionite, analcime, Nicolas Quash, pers. comm.). Similar observations of frustule diagenesis in hyperalkaline Lake Bogoria have been made by Gasse & Seyve (1987). If dissolution and diagenesis occurs differentially with respect to different species then the percentage calculations upon which the transfer functions are based become unreliable.

Conclusions

The precise taxonomy of diatoms in living communities and in fossil assemblages is essential to any palaeoenvironmental interpretations based on fossil diatom assemblages. Problems are often exacerbated in saline lakes where the taxonomy may be poorly known and the species diversity is often rather low. Taphonomic processes can contribute to the problems of interpretation and may bring together assemblages from different environments. However, within the context of a multi-disciplinary study it may be possible to distinguish these assemblages if support is available from other analyses. Dissolution of diatom frustules is also a common problem in saline lakes but this can be detected from the preservation of the remaining frustules and supporting mineralogical analysis, therefore avoiding erroneous palaeoenvironmental interpretations being made.

272

The processes discussed here must be taken into account when reconstructing palaeoenvironments generally but especially when using transfer functions to give quantitative estimates for environmental variables. The apparent precision derived from these may mask underlying inaccuracies. It is only when these issues are fully addressed that the potential of diatoms in reconstructing palaeoenvironments of salt lakes will be fully realised.

Acknowledgements

This paper has benefitted greatly from discussion with other members of the EQUARIFT project. This work has been funded by INSU-CNRS programme PIRAT and NERC.

References

Badaut, D. & F. Risacher, 1983. Authigenic smectite in Bolivian saline lakes. Geochim. Cosmochim. Acta 47: 363–375.

Bradbury, J. P., 1989. Late Quaternary lacustrine palaeoenvironments in the Cuenca de Mexico. Quat. Sci. Rev. 8: 75–100.

Casanova, J. & C. Hillaire-Marcel, 1987. Isotopic hydrology and palaeohydrology of the Magadi (Kenya) – Natron (Tanzania) basin during the late Quaternary. Palaeogeog., Palaeoclim., Palaeoecol. 58: 155–181.

Eugster, H. P., 1967. Hydrous sodium silicates from Lake Magadi, Kenya: Precursors of bedded chert. Science 157: 1177–80.

Eugster, H. P., 1986. Lake Magadi, Kenya: a model for rift valley hydrochemistry and sedimentation. In Frostick, L. E. et al. (eds) Sedimentation in the African Rifts. Geol. Soc. Lond. Sp. Publ. 25: 177–189.

Fritz, S. & R. W. Battarbee, 1986. Sedimentary diatom assemblages in freshwater and saline lakes of the northern Great Plains, North America: Preliminary results. Proc. IX Int. Symp. on living and Fossil Diatoms.

Gasse, F., 1977. Evolution of Lake Abhe (Ethiopia and TFAI) from 70000 BP. Nature 265: 42–45.

Gasse, F., 1986a. East African diatoms: taxonomy and ecological distribution. Bibliotheca diatomologica Vol 2. Cramer, Berlin.

Gasse, F., 1986b. East African diatoms and water pH. In Smol, J. P., Battarbee, R. W., Davis, R. B. and Merilainen, J. (eds) Diatoms and lake acidity. Junk.: 149–168.

Gasse, F., 1987. Diatoms for reconstructing palaeoenvironments and palaeohydrology in tropical semi-arid zones. Hydrobiologia 154: 127–163.

Gasse, F., 1988. Diatoms, palaeoenvironments and palaeohydrology in the western Sahara and the Sahel. Wurzb. Geogr. Arb. 69: 233–254.

Gasse, F. & F. Tekaia, 1983. Transfer functions for estimating palaeoecological conditions (pH) from East African diatoms. Hydrobiologia 103: 85–90.

Gasse, F., J. F. Talling & P. Kilham, 1983. Diatom assemblages in East Africa: classification distribution and ecology. Revue d'Hydrobiologie Tropicale. 16: 3–34.

Holdship, S. A., 1976. The palaeolimnology of Lake Manyara, Tanzania: a diatom analysis of a 56 meter sediment core. Ph.D. Dissertation, Duke University.

Maglione, G., 1970. La magadiiite, silicate sodique de neoformation des facies evaporitiques du Kanem (Littoral Nord-Est du Lac Tchad). Bull. Serv. Carte. Geol. Als. Lorr. 23 3–4: 177–189.

Shemesh, A., L. H. Burckle & P. N. Froelich, 1989. Dissolution and preservation of Antarctic diatoms and the effect of sediment thanatocoenoses. Quat. Res. 31: 288–308.

Stoffers, P. & S. A. Holdship, 1975. Diagenesis of sediment in an alkaline lake: Lake Manyara, Tanzania. IXth Int. Cong. of Sediment. Nice.

Taieb, M., C. Hillaire-Marcel, P. Barker, F. Gasse, C. Goetz, M. Icole, G. Krempp, R. Lafont, N. Roberts & D. Williamson, 1989. New results on paleohydrological and palaeoecological changes between 18000 and 10000 years in the intertropical region (Lake Magadi, Kenya). Terra Abstracts 1,1, 225.

Tiercelin, J.-J. & A. Vincens, 1987. Le demi graben de Baringo-Bogoria, Rift Gregory, Kenya. Bull. Cent. Rech. Explor. -Prod. Elf-Aquitaine 11: 249–540.

Hydrobiologia **214**: 273–278, 1990.
J. P. Smith, P. G. Appleby, R. W. Battarbee, J. A. Dearing, R. Flower, E. Y. Haworth, F. Oldfield & P. E. O'Sullivan (eds), 273
Environmental History and Palaeolimnology.
© 1990 *Kluwer Academic Publishers.*

Palaeolimnological aspects of a Late-Glacial shallow lake in Sandy Flanders, Belgium

Luc Denys [1], Cyriel Verbruggen [1] & Patrick Kiden [2]
[1] *Laboratorium voor Regionale Geografie & Landschapskunde, Rijksuniversiteit Gent, Krijgslaan 281, B-9000 Gent, Belgium;* [2] *Laboratorium voor Fysische Aardrijkskunde, Rijksuniversiteit Gent, Krijgslaan 281, B-9000 Gent, Belgium*

Key words: Late-Glacial, lake ontogeny, palaeoclimate, palaeohydrology, Sandy Flanders, diatoms

Abstract

A summary account is given of the development of a small Late-Glacial lake at Snellegem-St. Andries, Belgium. Sedimentation, hydrology, water quality and biotic succession clearly depended on climatic conditions and catchment processes (soil stability and leaching, vegetation). Special attention is drawn to a period of low water level near the end of the Allerød and the abundance of *Fragilaria* in certain periods.

Introduction

After melting of the permafrost, shallow lakes and mires were formed during the Late-Glacial at suitable sites in Sandy Flanders (NW-Belgium). Although many of them were only short-lived, an almost complete Late-Glacial sequence accumulated in some of them. In the frame of palaeoclimate research several such sites are being investigated in detail. As the inference of palaeoclimatic data from limnic environments requires knowledge about basin and catchment processes, lake ontogeny is studied. Here we give a preliminary summary account of the development of a small palaeolake at Snellegem-St. Andries. A first palynological investigation at this site, indicating its interest, was made by Verbruggen (1979).

Site and core description

The former lake (lat. = 51° 10′ 06″ N, long. = 3° 08′ 48″ E; altitude 12 m above m.s.l.; surface approx. 7 ha; catchment approx. 300 ha) is situated in a shallow, closed depression, developed in Tertiary and Pleistocene sandy deposits. The lacustrine sediments are largely Late-Glacial. Holocene deposits are almost totally lacking.

At the coring site, close to the deepest part of the depression, the lowermost Pleistocene sediment consists of fine- to medium-grained sand grading into silty sand at about 188 cm depth. Around 170 cm a gradual transition to lake marl is observed, which gives way to a rather pure peat at 102.5 cm. Above 95 cm the peat becomes more and more sandy and silty. Some sand laminae of presumed aeolian origin are present between 92 and 86 cm. From 70 cm to the top of the core, the sediment consists of a rather homogeneous silty organic clay with a few thin fine-sandy lenses.

The Allerød/Dryas III boundary was dated at $10\,940 \pm 60$ y. B.P. (GrN 6033) by Verbruggen (1979). Further radiocarbon dating is in progress.

Methods

A 250 cm long, 10 cm diameter undisturbed and continuous core was retrieved by driving a 10 cm diameter tube into the sediment. Half of the core was cut into 0.5 cm slices and used for microfossil, sediment and isotope analyses. The remaining half was used for extraction of macrofossils, micromorphology and determination of bulk density.

Carbonate content was measured manometrically and loss-on-ignition was determined at 500 °C. Oospores of Characeae were recovered by sieving 3 g of dry or 4 cm³ of wet sediment through a mesh of 210 μm. Calcareous samples were treated with HCl before sieving. Concentrations of diatom remains, chrysophyte statospores, pollen and palynomorphs were estimated with Stockmarr's (1971) method. At least 500 diatom valves were counted when possible from peroxide-cleaned material mounted in pleurax. The counts were increased considerably at levels where *Fragilaria* dominated.

Diatom zones were established by evaluating the clusters generated by a stratigraphically constrained classification (minimum variance, Euclidian distance; Grimm, 1987) of the samples. Herein only the taxa reaching 5% or more in at least one sample were used (N = 46). Inferred pH was calculated as pH = 6.4 − 0.85logB (Renberg & Hellberg, 1982). Considering the context, this estimation probably gives only a rough picture of the former pH-trends rather than yielding accurate values.

Results (Figs. 1–2)

The basal fine sand from diatom zone 1 contained few diatoms, many of which originate from reworked marine-littoral Pleistocene deposits. Some subaerial taxa ar present as well. The presence of small-ripple laminations suggest deposition by gentle currents.

Zone 2 shows an assemblage of mainly *Rhopalodia operculata* (Ag). Håkansson, *Mastogloia grevillei* W. Sm. and *Navicula cryptotenella* Lange-Bertalot, reflecting mineral-rich, wet subaerial to aquatic conditions.

This pioneer assemblage is soon replaced by one dominated by *Fragilaria* (mainly *F. pinnata*

Ehr. in zone 3a and *F. brevistriata* Grun. in 3b). Concentrations of diatoms, *Chara* oospores and chrysophyte cysts increase steadily, as do the number of diatom taxa, indicating more limnic conditions. Oospores reach their highest concentrations in 3b and, at the same time, diatoms tolerant of moist subaerial conditions disappear almost completely.

The transition to zone 4, where there is a diverse assemblage of epiphytic and epipelic taxa and *Cyclotella kuetzingiana* Thwaites reaches modest frequencies, coincides well with the onset of the Bølling. Although alkaliphilous diatoms (*Mastogloia smithii* Thwaites var. *lacustris* Grun., *Denticula inflata* W. Sm.) remain most abundant, circumneutral ones reach considerable levels and some halophobous, acidophilous taxa appear. Oligotraphentous diatoms attain almost 20% and cyst concentrations are fairly high, perhaps indicating conditions slightly less rich in nutrients (Smol, 1985). These changes could be due to increased soil stabilisation. The decline in oospore concentrations may result from a replacement of the *Chara* beds by other aquatics, as the concentrations of their pollen suggest. However, the return of diatoms from moist habitats and the later decline of aquatic pollen indicate that a lowering of the water-table could also be responsible.

Diatom, cyst and pollen concentrations fall at the beginning of zone 5, as the older Dryas II period begins. In zones 5 and 7 only very few diatom valves could be counted, mainly of *Fragilaria*. The intermediate zone 6 at the end of the Dryas II yielded a somewhat higher number with mainly *Mastogloia smithii* var. *lacustris* and some *Cymbella microcephala* Grun. and *Fragilaria elliptica* Schumann. Considerably more cysts than frustules are found while oospores are very rare.

The first part of the Allerød shows an alternation of layers with hardly any diatom remains (zones 7 and 9) and fairly well preserved assemblages (zones 8 and 10) suggesting shallow, circumneutral to slightly acid, meso-oligotrophic conditions. Dominants are *Navicula jaagii* Meister, *N. bryophila* Petersen var. *lapponica* Hust. (zone 8), *Eunotia arcus* Ehr., *E. praerupta*

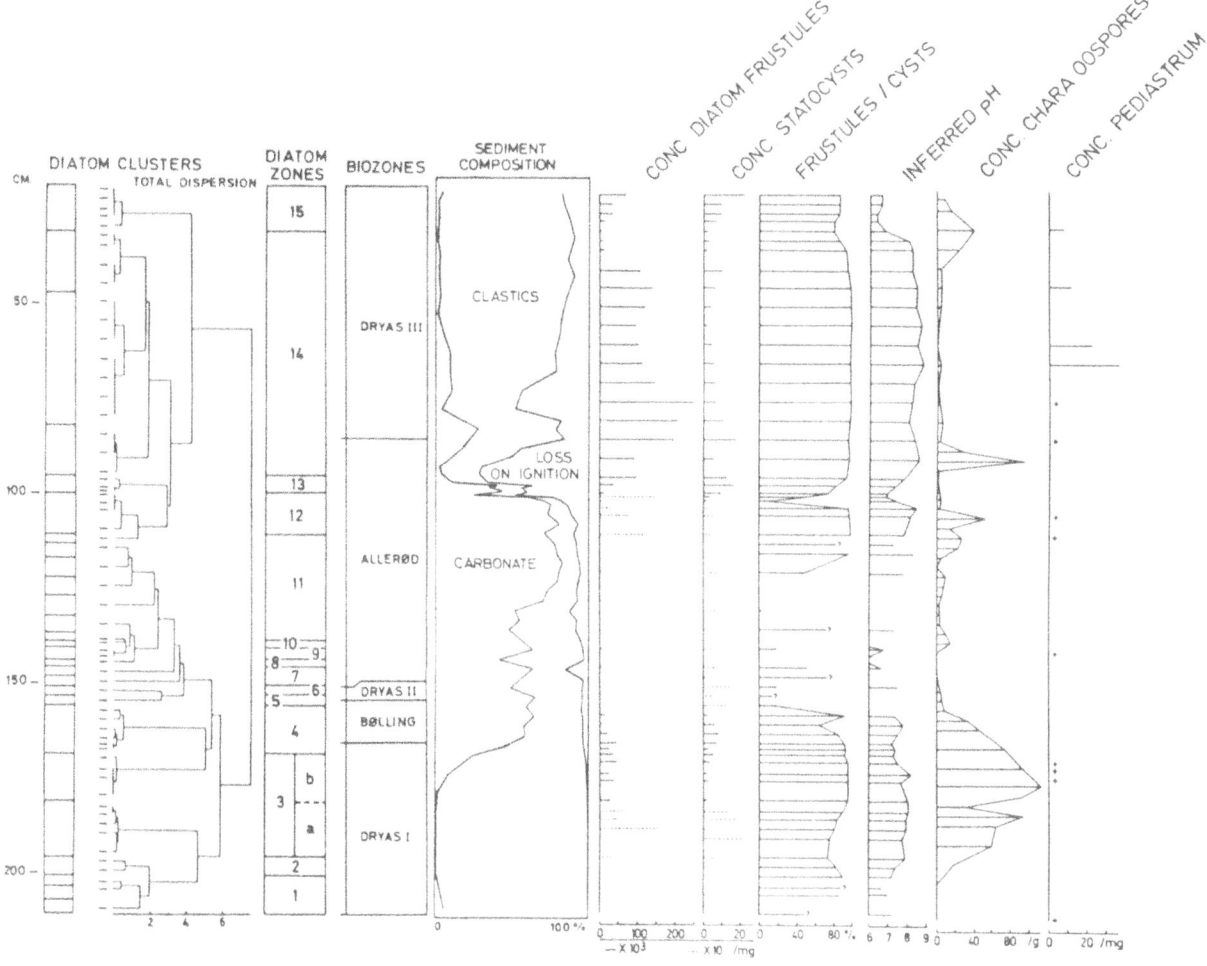

Fig. 1. Zonation, sediment composition, algal remains and diatom inferred pH of the Snellegem-St. Andries core.

Ehr., *Tabellaria fenestrata* (Lyngb.) Kütz. and *Fragilaria brevistriata* (zone 10). Aquatic pollen are badly represented and oospores largely absent, demonstrating also that the water-level has dropped. Stable isotope data (unpublished) point to increased evaporation at the time these deposits formed. If may be that the intermittent diatom preservation results from periodic exposure of the sediment to the air.

Zone 11 is again very poor in diatoms. There is evidence for leaching of biogenic silica from this part of the core (unpublished) which again could imply a fluctuating water-table. Oospore concentrations remain low throughout, but aquatic

pollen is more frequent in the lower part of the zone.

In zone 12 diatom preservation is quite well again. The assemblages, dominated by *Fragilaria elliptica* and *F. pinnata*, are indicative of an alkaline and rather nutrient-rich limnic environment. The diatom/chrysophyte cyst ratio is very high. Oospores become more abundant and a peak of aquatic pollen is reached. The permanent inundation inferred from all this agrees well with the wetter climate derived from the stable isotope ratios.

As the sediment changes from marl to a rather pure and compact fen peat (zone 13), a number of

276

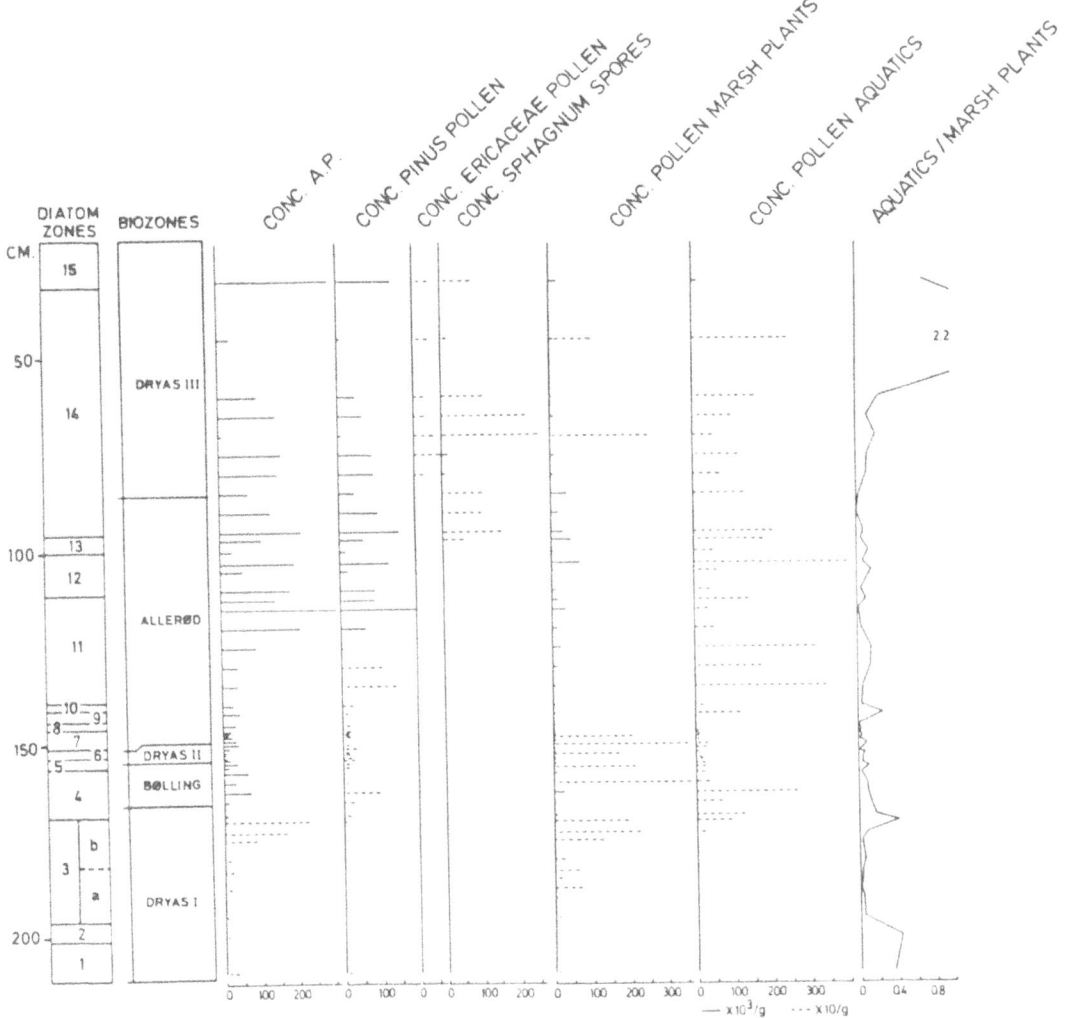

Fig. 2. General pollen and spore date from the Snellegem-St. Andries core.

other diatoms also become frequent (*Fragilaria construens* (Ehr.) Grun. var. *venter* (Ehr.) Grun., *Opephora martyi* Hérib., *Gomphonema angustatum* (Kütz.) Rabenh., *Achnanthes minutissima* Kütz.). The pH declines (cfr. also *Sphagnum*) and a more mesotrophic marshy situation arises. Statocysts predominate over diatoms and oospores decline markedly. The peat occurs throughout the whole lake basin and most probably reflects a lowering of the water-level. A break in the presence of aquatic pollen at the beginning of the next diatom zone also points in this direction. The well-sorted fine sand interspersed in the peat was most likely

brought in by wind and indicates decreased soil stability on higher grounds.

At the start of the Dryas III the water rises anew and more minerogenic material is washed in. An organic silty clay with low carbonate content is deposited from now on. *Fragilaria* has regained predominance (zone 14) and attains very high concentrations. *Pediastrum* flourishes as well. Ericaceae (mainly *Empetrum*) profit from the climatic deterioration. Their expansion here is much more marked than in the adjacent alluvial plain of Scheldt and Lys (see Verbruggen, 1979), which can be related to a more intense local soil decalcification. Initially the oospores show peak concentrations but their numbers soon fall back which may be due to unfavourable light conditions caused by increased turbidity and to shading by other aquatics, especially *Ranunculus* sect. *Batrachium*, whose pollen are present in quantity. At the top of zone 14 a minimum in diatom and statospore concentrations accompanies a transient increase of oospores.

In zone 15 diatom and chrysophyte cyst concentrations gradually recover. The *Fragilaria* are replaced in part by epiphytic and epipelic taxa resistant to periodically moist conditions. Besides the predominant diatoms, among which are *Fragilaria elliptica*, *F. brevistriata*, *Achnanthes minutissima* and *Gomphonema angustatum*, a number of acidophilous subaerial moss dwellers are rather frequent (*Pinnularia subcapitata* Greg., *Eunotia paludosa*. Grun., *E. exigua* (Bréb.) Rabenh.). Together with an increase of circumneutral forms they suggest microbiotope differentiation and a less alkaline situation connected to a reduction of the open water, immediately prior to the onset of the Holocene.

Discussion

Cores from very shallow lakes offer a far more local, and often more complicated and less well preserved record, than those taken in deep and large basins. On the other hand the rapid and marked response of small waterbodies to environmental changes justifies their palaeolimnological study. As could be expected from such a hydrologically and ecologically sensitive system, the ontogeny of this small lake strongly depended on regional climatic changes and basin processes such as vegetation and soil development.

Of special interest is the lowering of the watertable and the consequent formation of peat near the end of the Allerød. Usinger (1981) has drawn attention to a widespread sedimentary hiatus occurring near the Allerød/Dryas III transition in northern German and Danish lakes which he explains as the result of an important lowering of contemporary lake levels. Low lake levels in north-western Europe near the end of the Allerød were reported by several other authors as well (see Bohncke & Wijmstra, 1988). Although a climatic cause seems obvious we feel that a second, vegetational possibility requires examination as well. Indeed the 'dry' conditions shortly follow the local expansion of *Pinus* which by then had accumulated its greatest biomass before its cover was degraded by the intense cold of the Dryas III. Lockwood (1983) suggested that *Pinus* stands have even greater evapotranspirative capacities than deciduous woods. It is therefore not unreasonable to suspect that the dense cover of mature *Pinus* trees attributed to, or even caused, a lowering of the groundwater table of regional importance, affecting susceptible sites such as wetlands and lakes.

In the diatom record, the abundance of *Fragilaria* in certain Late-Glacial environments stands out once again. Here this genus was especially abundant during the Dryas periods. Several hypotheses can be formulated in this respect, including competitive development in a short vegetation season (extensive ice cover) (Smol, 1988), outstanding minerotrophy (Haworth, 1976; Round, 1957; Stabell, 1985) and adaptation to ecologically unstable 'extreme' environments (cfr. Denys, 1990).

References

Bohncke, S. & L. Wijmstra, 1988. Reconstruction of Late-Glacial lake-level fluctuations in the Netherlands based on palaeobotanical analyses, geochemical results and pollen-density data. Boreas 17: 403–425.

Denys, L., 1990. *Fragilaria* blooms in the Holocene of the western Belgian coastal plain. In H. Simola (ed.), Proc. 10th Int. Diatom Symp. 1988. Koeltz, Koenigstein & Biopress, Bristol (in press).

Grimm, E. C., 1987. CONISS: a fortan 77 program for stratigraphically constrained cluster analysis by the method of incremental sum of squares. Computers & Geosc. 13: 13–35.

Haworth, E. Y., 1976. Two Late-Glacial (Late-Devensian) diatom assemblage profiles from northern Scotland. New Phytol. 77: 227–256.

Lockwood, J. G., 1983. Modelling climatic changes. In K. J. Gregory (ed.), Background to palaeohydrology: a perspective. Wiley, Chichester: 33–50.

Renberg, I. & T. Hellberg, 1982. The pH history of lakes in southwestern Sweden, as calculated from the subfossil diatom flora of the sediments. Ambio 11: 30–33.

Round, F. E., 1957. The Late-glacial and Post-glacial diatom succession in the Kentmere Valley deposit. New Phytol. 56: 98–126.

Smol, J. P., 1985. The ratio of diatom frustules to chrysophycean statospores: a useful paleolimnological index. Hydrobiologia 123: 199–208.

Smol, J. P., 1988. Paleoclimate proxy data from freshwater arctic diatoms. Verh. int. Ver. Limnol. 23: 837–844.

Stabell, B., 1985. The development and succession of taxa within the diatom genus *Fragilaria* Lyngbye as a response to basin isolation from the sea. Boreas 14: 273–286.

Stockmarr, J., 1971. Tablets with spores used in absolute pollen analysis. Pollen Spores 13: 615–621.

Usinger, H., 1981. Ein weit verbreiteter Hiatus in spätglazialen Seesedimenten: mögliche Ursache für Fehlinterpretation von Pollendiagrammen und Hinweis auf klimatisch verursachte Seespiegelbewegungen. Eiszeitalter Ggw. 31: 91–107.

Verbruggen, C., 1979. Vegetational and palaeoecological history of the Late-Glacial period in Sandy Flanders (Belgium). Acta Univ. Oul. A 82 Geol. 3: 133–142.

Hydrobiologia **214**: 279–292, 1990.
J. P. Smith, P. G. Appleby, R. W. Battarbee, J. A. Dearing, R. Flower, E. Y. Haworth, F. Oldfield & P. E. O'Sullivan (eds), 279
Environmental History and Palaeolimnology.
© 1990 *Kluwer Academic Publishers.*

Palaeolimnological studies of laminated sediments from the Shropshire–Cheshire meres

K.M. Farr[1], D.M. Jones[2], P.E. O'Sullivan[3], G. Eglinton[4], D.H. Tarling[5] & R.E.M. Hedges[6]
Current addresses: [1]*School of Applied Sciences, Wolverhampton Polytechnic, Wolverhampton WV1 1SB, UK*; [2]*Newcastle Research Group, Fossil Fuels & Environmental Chemistry, Drummond Building, The University, Newcastle NE1 7RU, UK*; [3]*Dept. of Environmental Sciences, Polytechnic Southwest, Plymouth PL4 8AA, UK*; [4]*Organic Geochemistry Unit, The University, Bristol, BS8 1TS UK*; [5]*Dept. of Geological Sciences, Polytechnic Southwest, Plymouth PL4 8AA, UK*; [6]*Research Laboratory for Archaeology, The University, 6, Keble Road, Oxford OX1 3JQ, UK*

Key words: Holocene, palaeolimnology, diatom, lipid, palaeomagnetism, varves

Abstract

Studies of frozen and soft mud cores from Ellesmere Mere, Rostherne Mere and Berrington Pool, the three deepest of the Shropshire–Cheshire meres of the English Midlands, reveal the presence of laminations which may be varves. This hypothesis is being tested by means of fine resolution diatom and other microfossil analysis after the method of Simola (1977). Even where the laminations are faint and disrupted, it appears that seasonal signals from algal blooms are preserved.

Organic geochemical analyses of sediments from Ellesmere show that the uppermost layers contain abundant organic matter (over 17% total dry matter) and that the extractable lipid fractions from different horizons exhibit clear compositional differences. These are produced by temporal changes in the organic inputs to the sediments, and also by diagnetic effects.

The clastic and organic content of the cores also provides evidence for lake level variations in this area over the last 250 years.

Introduction

The Shropshire–Cheshire meres are a group of ca. fifty small, mostly eutrophic, lowland lakes, situated in the north west part of the English Midlands (Fig. 1). The lake basins were formed during the deglaciation of the area between 10000 and 12000 years ago. The limnology and palaeolimnology of some of the principal meres has been extensively investigated and is well understood (Beales, 1976, 1980; Gaskell & Eglinton, 1975; Livingstone & Cambray, 1978; Reynolds, 1979; Cranwell, 1984a; Livingstone, 1984; Smith, 1986).

One of the original aims of this research was to combine palaeoecological, magnetic and organic geochemical approaches to develop a method of calibrating both the Holocene radiocarbon time-scale, and the West European calendar of secular palaeomagnetic variations, using a varve chronology developed from the laminations in the sediments. The meres at Ellesmere and Rostherne were known to be floored by more than 6 m of sediment (Nelms, 1984), which at Rostherne contains annual laminations (Haworth, 1980). Consequently, these meres were selected for study. Also included is Berrington Pool, a small but deep kettle hole at the drift limit in south

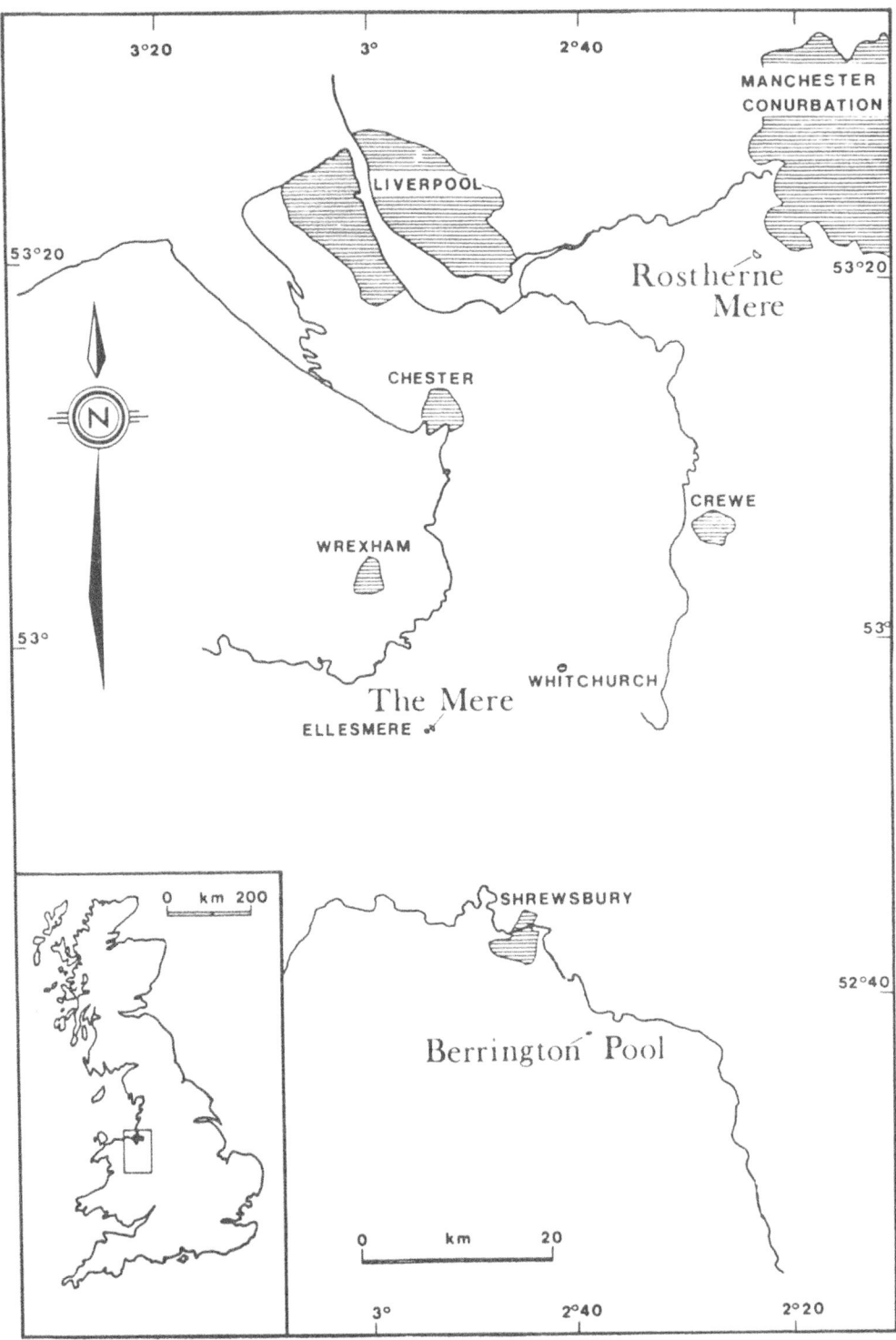

Fig. 1. The Shropshire–Cheshire meres.

Shropshire. The mere is situated in private farmland and has not been greatly affected by human activity other than its use as the village reservoir until the 1930's, and recent eutrophication caused by fertiliser run-off. Palaeolimnological data collected to date are described and reviewed below.

Palaeolimnology of the lake sediments

A number of 1 m and 6 m soft-sediment cores and 1 m frozen cores have been taken from Ellesmere Mere, Rostherne Mere and Berrington Pool. The sediments comprise mainly detrital, organic,

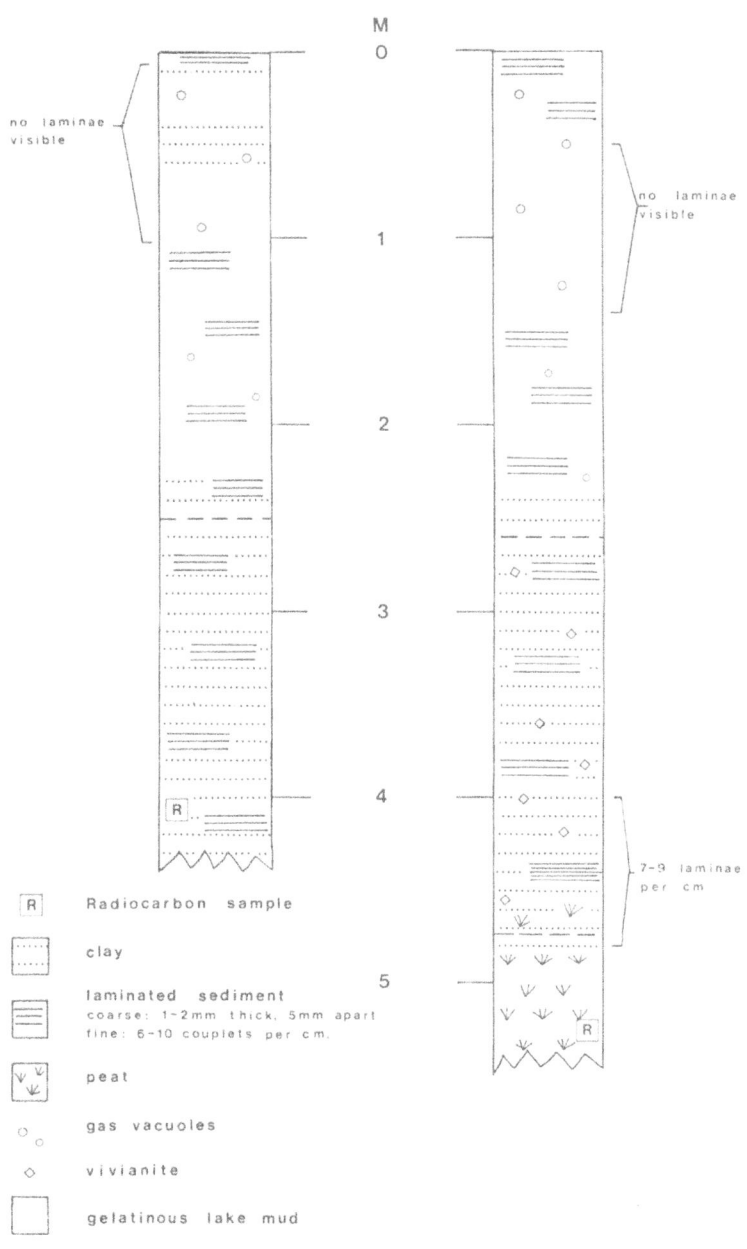

Fig. 2. Stratigraphic logs of Livingstone cores from Ellesmere Mere and Berrington Pool.

ELLESMERE CORE F5

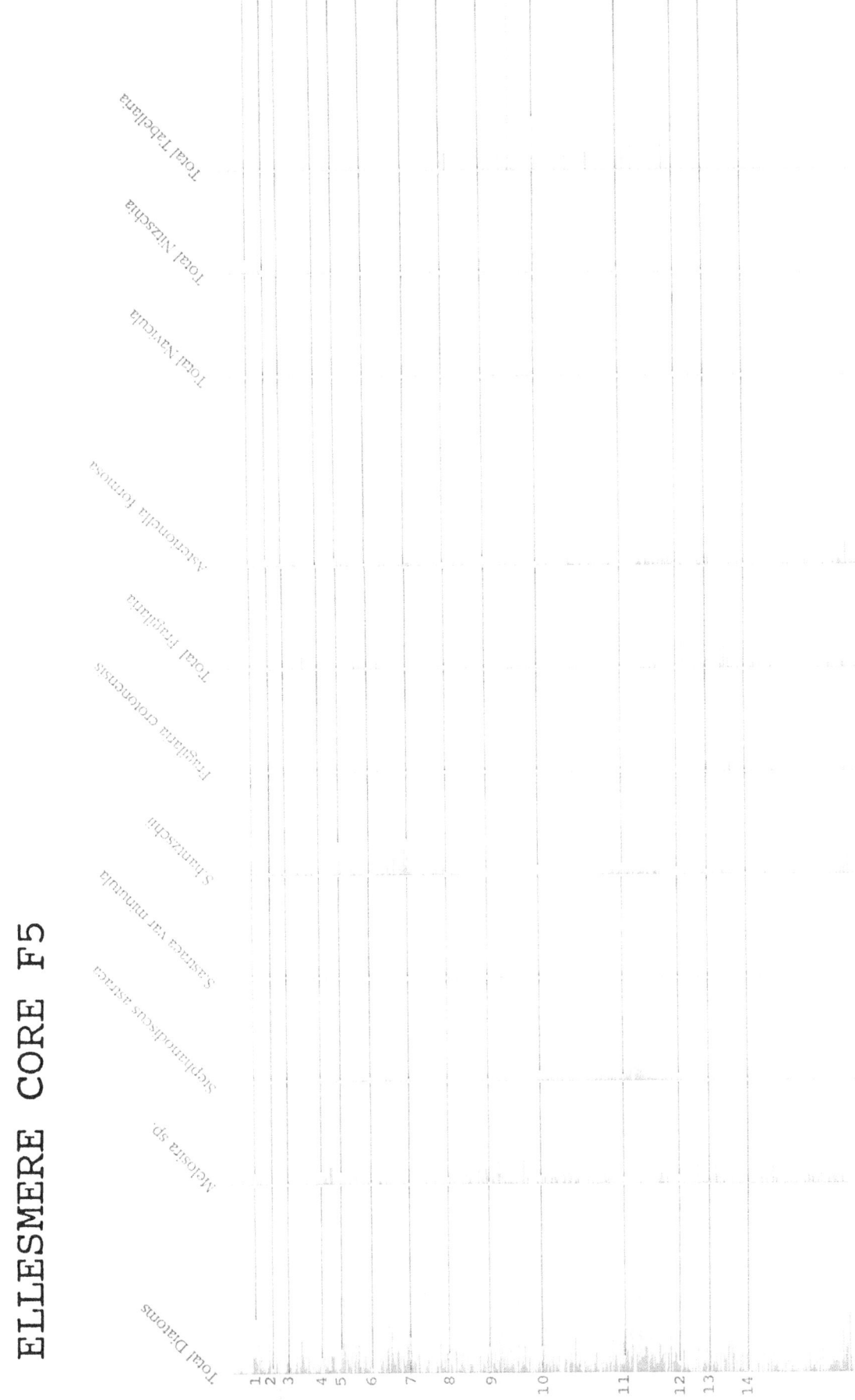

Fig. 3. Diatom profile (adhesive tape peel) from the uppermost sediments of Ellesmere Mere. Results only from top 8 cm displayed.

283

brown-black lake muds and, at Ellesmere and Berrington, pink-brown silty clay. Work on the sediments has consisted of

a) identification, photography and description of sediment stratigraphy
b) mineral magnetic and palaeomagnetic analyses of the cores (not discussed here)
c) microfossil analyses of 1 m and 6 m soft-sediment cores including fine resolution microfossil analyses of adhesive tape peels from slow freeze-dried material
d) organic geochemical analyses of core material

Sediment stratigraphy

Mini-Mackereth, 6 m Mackereth, 4–5.5 m Livingstone, and 1 m freezer coring techniques have been used to sample the sediments from the deepest parts of the meres at Ellesmere, Rostherne and Berrington. The sedimentary succession in each lake consists of gelatinous, dark brown, organic, detrital lake mud (actually fine sand and silt), rich in gas vacuoles, the top metre or so of which is very unconsolidated (Fig. 2). At Ellesmere and Berrington the mud overlies plastic, pink-brown, fine silty clay at about 2.6 m depth. The Ellesmere sediments also contain occasional bands of fine silt and clay-grade material near the top of the succession. At Berrington, a peat was encountered at the base of a Livingstone core (4.67 m) through which the corer would not penetrate. All deposits become more consolidated with depth.

Real colour and black and white infra-red photography of frozen cores from Ellesmere, and observation of frozen cores from the other two meres, has confirmed the presence at the top of the cores of clear laminae whose thickness, texture and colour are consistent with the hypothesis that they are annual or seasonal laminations. Further down the cores these laminae become much less readily visible although they are still present; the darkness of the sediment contrasts poorly with the dark colour of the laminations. They are clearer in frozen cores from Rostherne Mere and Berrington Pool. No laminae are apparent in the clayey portions of the Ellesmere frozen core.

The top 40 cm of the Berrington and Rostherne frozen cores, and the top 2 cm of the Ellesmere frozen cores, are laminated at ca. 1–4 mm intervals. Faint laminae with a thickness of ca. 1–2 mm and a spacing of ca. 5 mm are just discernible in Livingstone cores from Berrington and Ellesmere from ca. 1.5 m and 1 m respectively down to about 2.0 m, where an additional set of finer laminations become visible (5–7 light/dark couplets per cm). However, many of these are lenticular dark flecks which are difficult to record. It is possible that they have been disrupted by bioturbation (Haynes et al., 1989). Occasional dark fibrous layers, ca. 1 mm thick, also occur which may cut across laminae. Sub-horizontally bedded, partly decomposed twigs, leaf and avian bone fragments are present. Detailed stratigraphic studies have yet to be made of the Rostherne cores.

At about 2.5 m the muds grade into pink clay which also contains the coarser and finer sets of laminae, with the same spacing as in the overlying mud, except that below 4.0 m there are 7–9 fine couplets per cm. At 4.67 m in the Berrington succession there is a change to more compact, more friable, dark brown silty peat. The proportion of clay-grade material is much lower here than in the overlying clay. No further laminae are visible.

All stratigraphic boundaries are gradational. The fine grain size and pink colour of the sediments indicate the presence of iron oxides, possibly washed in from the surrounding red-coloured drift or country rock. X-ray diffraction analysis shows the dominant minerogenic component of all the sediments to be quartz. No clay minerals are present except for a little illite/muscovite and chlorite. A little biogenic pyrite is present in the Ellesmere deposits and bright blue crystals of vivianite are common in the lower, clayey portion of the Berrington Livingstone core.

Microfossil counts

The base of a 1 m Mackereth core from Ellesmere contains frustules of *Cyclotella* spp. and *Stephanodiscus rotula* var. *rotula*, whereas the top is dominated by *S. hantzschii* and *Asterionella formosa*, indicating an increase in lake productivity in recent years. This corresponds well with diatom assemblages noted from 1 m Ellesmere Mackereth cores by Nelms (1984).

Fine resolution diatom counts have been carried out in continuous 200 μm fields along the length of adhesive tape peels taken from frozen cores from each of the meres after the method of Simola (1977). The top 30 cm of the Ellesmere frozen core (Fig. 3) contain a diverse microflora, including *Stephanodiscus* spp. (mainly

Fig. 4. Diatom profile (adhesive tape peel) from a frozen core from Berrington Pool.

S. hantzschii and *S. rotula* var. *rotula*), *Aulacoseira granulata* (var. *granulata* and var. *angustissima*), *Fragilaria* spp. and *Asterionella formosa*. Diatom assemblages from the top 18 cm of Berrington Pool are dominated by *Aulacoseira granulata* var. *granulata*, *Cyclostephanus dubius*, *S. hantzschii* and *S. parvus* (Fig. 4). Very few diatom frustules are present on tape peels made from Rostherne cores, perhaps because of the high sedimentation rate in this mere (Nelms, 1984), and no further micropalaeontological studies have been made of Rostherne sediment. No calcite crystals or chrysophyte cysts are present in any of the cores. Green algae and dinoflagellate cysts are present but rare. Pollen grains, spores, invertebrate eggs and integument occur in low numbers throughout the cores.

Signals of changing diatom abundance in these cores from Ellesmere and Berrington wax and wane regularly. Numbers of *Aulacoseira* spp. reach a peak just after *Stephanodiscus* spp., which may reflect the earlier seasonal blooming of the latter. Preliminary observations indicate that the signals of changing diatom abundance obtained are, at times, closely related to the microstratigraphy identified from infra-red photographs and from the cores themselves. In particular, peaks in total diatoms, and in taxa such as *S. hantzschii*, *S. rotula* var. *rotula*, *Cyclostephanus dubius* and *Aulacoseira* spp., coincide with dark laminae. It has been concluded that these may represent the spring and summer layers.

Distinct laminae present in the top 2 cm of a frozen core from Ellesmere (Plate 5, Fig. 5) have been dissected out and analysed for their microfossil and lipid content. The top green layer of *Microcystis aeruginosa*, known to bloom in the autumn, also contains abundant *Nitzschia palaea* (up to 50% of total microfossils, Fig. 6). The relative proportion of this frustule decreases rapidly down the core: 1 cm below the sediment water interface it forms just 1% of the total microfossils. It does not reappear in the lower green layer, possibly because this is not another autumnal bloom of *M. aeruginosa* or because the delicate frustules do not survive gentle compaction; it was not found in preparations from further down the core.

The dominant diatoms in the immediate subsurface layers are *Aulacoseira* spp., which is most abundant just below the *N. palaea* layer and is known to bloom in late summer, and *Stephanodiscus hantzschii*, most abundant in the lower layers and known to bloom in spring. Between these layers there is a small rise in the relative frequency of other diatoms (e.g. *S. rotula* var. *rotula*, *Fragilaria* spp.) and green algae (e.g. *Pediastrum* sp., *Staurastrum* sp.). The latter are known to reach a peak in abundance during the summer.

It appears that the six laminae present in the top 2 cm of the core may represent seasonal algal succession during one year. Further down the core, these seasonal signals appear to be com-

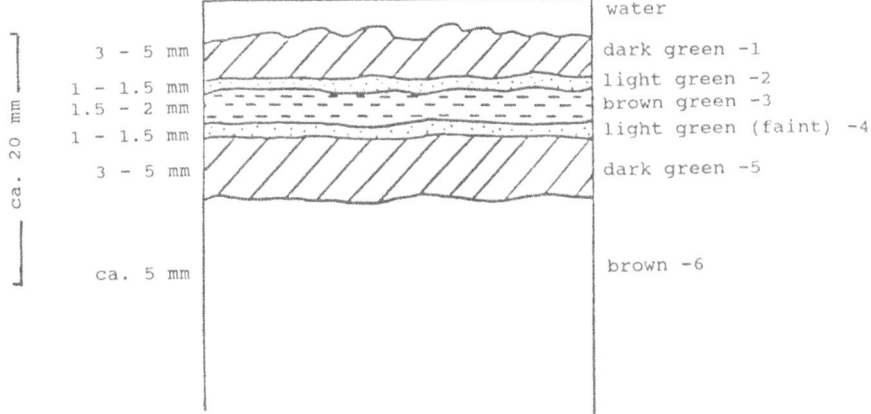

Fig. 5. Stratigraphy of top 2 cm of frozen core F5 from Ellesmere Mere.

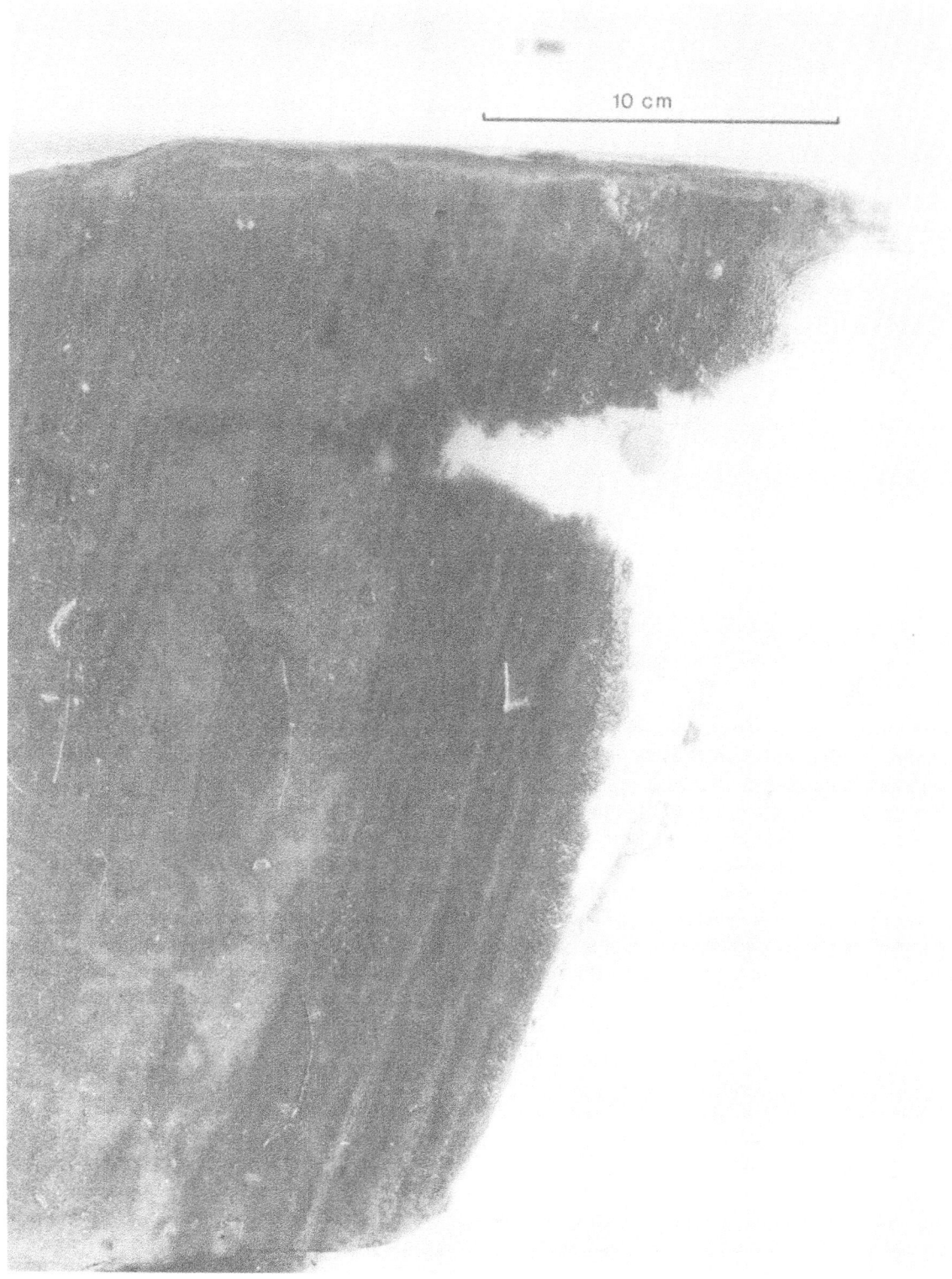

10 cm

Plate 1. Infra-red black & white photograph of laminae at the mud–water interface of a frozen core from Ellesmere Mere.

288

Fig. 6. Diatom profile of top 2 cm of frozen core F5 from Ellesmere Mere.

pressed into annual ones. Different organic geochemical compounds have been detected in the separate laminae but this work is still at an early stage and data are not presented here. Further samples have been dissected from the core and are currently being examined.

Microfossil analyses were also made of a 5.5 m Livingstone core from Berrington Pool. The basal peat is rich in pollen and spores, mainly *Alnus glutinosa*, *Betula* sp. and *Quercus* sp., with minor *Corylus avellana*, *Salix* sp., *Rumex* sp., Graminae, *Tilia* sp. and *Sphagnum* (spores and tissue). The majority of the grains both here and higher up the core in the clayey sediment, are highly oxidised. Common microfossils include sponge spicules, invertebrate eggs, *incertae sedis* cysts, phytoclasts, fungal hyphae and diatoms (e.g. *Navicula* spp., *Pinnularia* sp., *Stephanodiscus hantzschii*, *Amphora ovalis*), although diatoms are not as abundant as higher up the core. No charcoal was observed.

Initial observations of the Berrington core show the upper gelatinous dark-brown muds (ca. 1.0–2.5 m) to be exceptionally rich in centric diatoms, mainly *Cyclostephanus dubius*, with *S. rotula* var. *rotula*, *S. hantzschii*, *S. parvus*, and also *Aulacoseira granulata* var. *granulata*. Siliceous sponge spicules are abundant. Although laminations in the cores are difficult to record visually, it appears that seasonal signals from diatoms are present. If continuous diatom counts could be made from the length of a long sediment core it might be possible to gain a good indication of the age of the sediment.

Organic geochemistry

Organic geochemical analyses of wet and frozen sedimentary core material show that the upper sediments of Ellesmere Mere contain abundant organic matter (over 17% total dry matter) and that the extractable lipid fractions from different

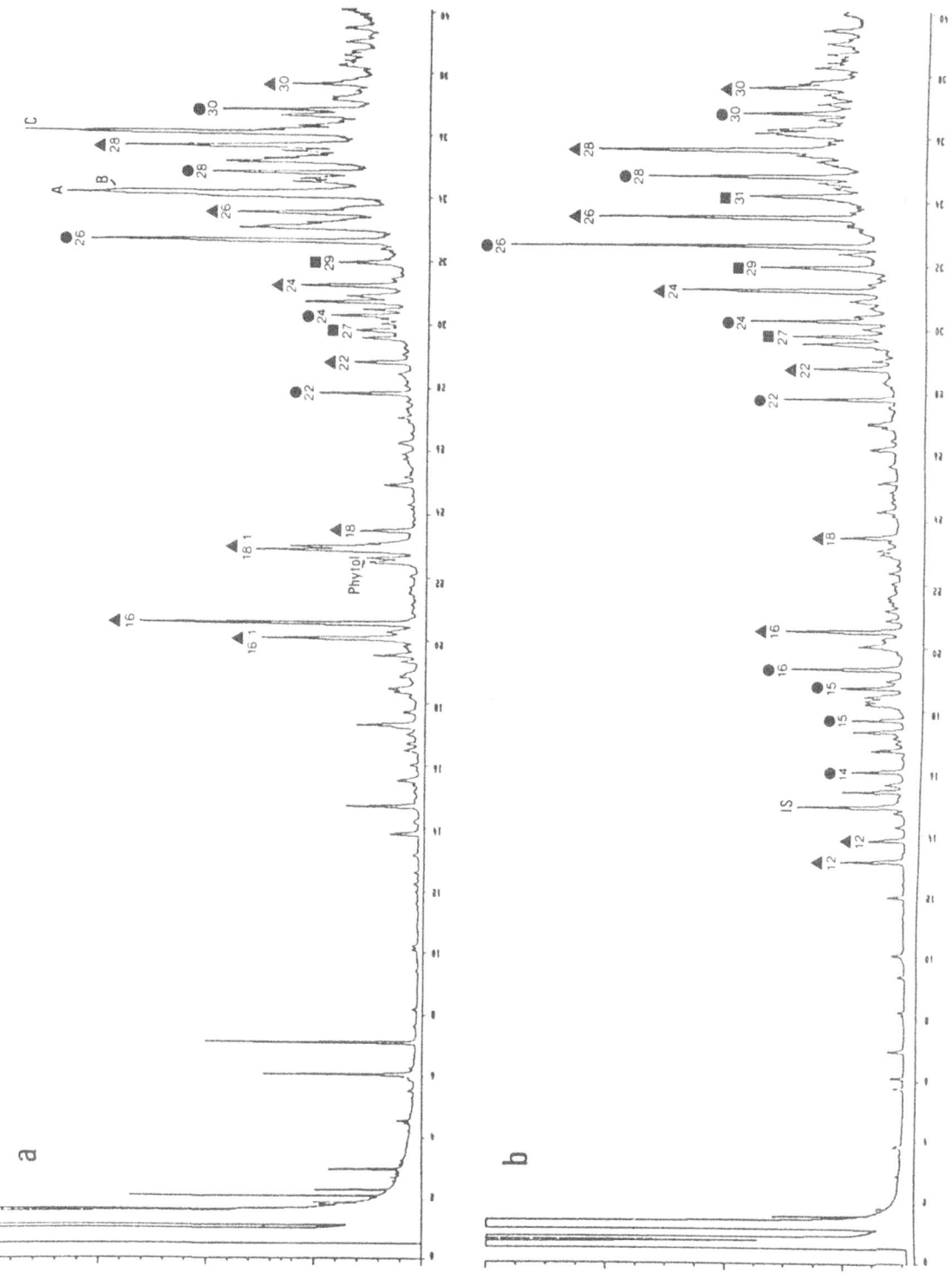

Fig. 7. Gas chromatograms of silyated (BSTFA) total solvent-extractable organic matter from Ellesmere sediment 0–5 cm (a) and 10–40 cm (b). Circles, triangles and squares represent alcohols and fatty acids as their TMS derivatives, and *n*-alkanes, respectively, with their carbon numbers also indicated. The peak marked IS is an internal standard. Compounds containing a double bond are shown by : 1. Identities of peaks A, B and C are given in Fig. 10.

290

horizons exhibit clear compositional differences. These relate to temporal changes in the organic inputs to the sediments and also to post-depositional diagenesis.

The lipid extracts of the uppermost layers are dominated by components (e.g. Fig. 7a, peak A, cholest-5-en-3β-ol; peak C, 24-ethylcholest-5-en-3β-ol) derived mainly from algae and higher plants (Huang & Meinschein, 1979; Volkmann, 1986 and references therein). The presence of phytol in these mixtures is probably derived from the phytyl side chain of chlorophyll, and is therefore a marker for an algal (or possibly higher plant) input to the sediments (e.g. Prahl *et al.*, 1984). Other components present include C_{14}–C_{18} fatty acids and alcohols, which represent probable inputs from aquatic microorganisms and bacteria as well as algae (Robinson *et al.*,

1984, 1987; Cranwell *et al.*, 1984b, 1987). Analyses of the lipid extracts of individual layers near to the surface show that compositional changes occur within a very small (<2 cm) depth interval. These changes include a gradual decrease in the relative abundance of phytol and of the C_{27} to C_{29} sterols, and sterol to stanol ratios.

The lipid composition of the deeper (10–40 cm) sediments (Fig. 7b) is clearly different from that of the surface layers (Fig. 7a), in that it is dominated by terrestrial higher plant derived long chain *n*-alcohols, carboxylic acids and *n*-alkanes, with a secondary maximum of C_{12}–C_{16} *n*-alcohols and carboxylic acids (probably bacterially derived) also present. These differences result, in part, from the greater resistance to degradation of the higher plant-derived components but they may also be evidence of the changing trophic status of

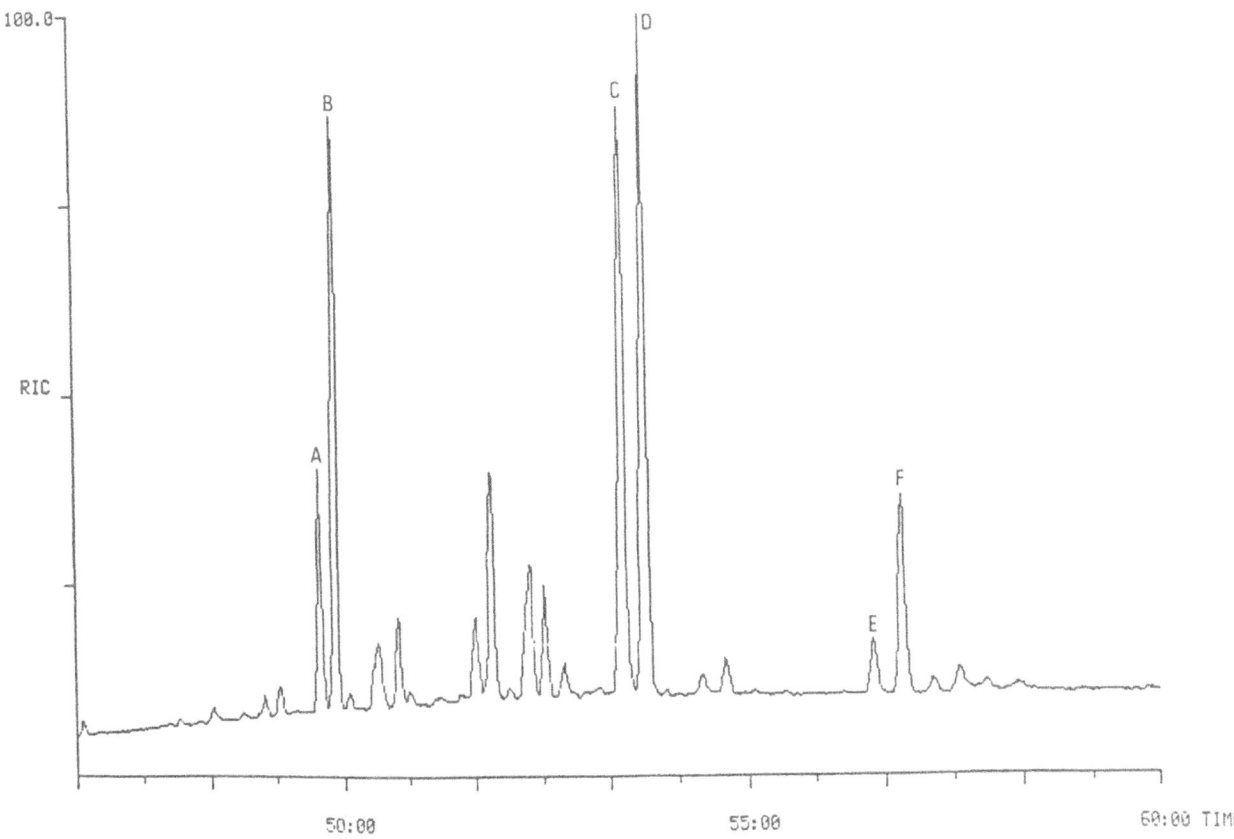

Fig. 8. Partial reconstructed ion chromatogram (RIC) of Ellesmere 10–40 cm sediments silyated (BSTFA) sterol fraction. Peak assignments for the TMS ethers are as follows: A, cholest-5-en-3βol; B, cholestan-3β-ol; C, 24-ethylcholest-5-en-3β-ol; D, ethylcholestan-3β-ol; E, gorgosterol; F, gorgostanol.

the lake (as evidenced by changes in diatom assemblages).

Comparison of the distribution of sterols in the surface sediment layer with those from the deeper layer (Fig. 8), shows that the ratio of the C_{27} Δ^5 cholesterol (peak A) to the C_{29} Δ^6 24-ethylcholesterol (peak C) is much lower in the deeper sediment. This may indicate an increased abundance of higher plant relative to algal components which is consistent with the higher relative abundance of higher plant derived straight chain lipids in the deeper sediments. Also noticeable are the lower sterol to stanol (e.g. peak A to peak B) ratios in the deeper sediment which are consistent with the lower stability of the sterols and with the well-known microbially-mediated sterol to stanol reduction which occurs with depth in recent sediments (Gaskell & Eglinton, 1975; Nishimura, 1977).

Other sterols present in this sediment extract which give information on their sources include dinosterol (indicating an input from dinoflagellate algae) and gorgosterol. The first report of the latter compound (and its related stanol) in freshwater lacustrine sediments appears to have been by Cranwell (1984b). It has been discovered in coelenterates (gorgonians) containing marine dinoflagellate symbionts (Wunsche et al., 1987 and references therein).

In summary the Ellesmere sediments contain abundant organic matter, the composition of which varies with depth. These variations are caused by temporal changes in the inputs and also by diagenetic effects. Detailed analysis of the changes in the compositions of the lipid fractions in the sediment layers allow some estimation of these variations in source input and post-depositional processes.

Interpretation of the palaeolimnological data

The silty peat at the base of the Berrington core appears to represent a subaqueous lake-marginal deposit, where there was sufficient standing water to support diatom communities (including planktonic *S. hantzschii*). The depositional environment was probably close to a terrestrial source of *Sphagnum* tissue and other plant macerals, spores from mosses and ferns, and pollen from shrubs including willow and alder, and aquatic plants. An emergent lakeside depositional environment appears to be less likely because of the lack both of plant macrofossils (whole leaves, twigs etc.) and coarser grains of sand, although it is possible that the subsoil drift deposits may locally comprise only fine-grained clays and marls. Deposition of palynomorphs in shallow water would be conducive to their partial oxidation and fungal infection prior to their incorporation into the sediment column.

In the upper parts of the cores diatom frustules and phytoclasts are well preserved, indicating a calm, anoxic depositional environment. Further down the core (ca. 2.5–4.0 m) the proportion of fine sediment ($<5\,\mu$m) is higher and organic matter is not so well preserved. The same diatom species are present but are less abundant. It therefore appears that the meres at Berrington and Ellesmere were once much shallower and less productive than today.

If the coarser laminae in the cores are annual, the pink clay at Berrington Pool should represent about 350 years of deposition and the overlying lake mud about 400 years. If the finer couplets are varves then the pink clay may have been laid down in about 1350 years and the brown muds in about 1200. The first hypothesis indicates a sedimentation rate of ca. 0.6 cm yr^{-1}. This is fast for lakes without a major inflowing stream. The second hypothesis gives a sedimentation rate of ca. 0.2 cm yr^{-1}. It would also indicate an age of ca. 2500 years for the top of the peat (4.67 m), which would date this part of the core as early Iron Age.

Since *Alnus glutinosa* is the dominant pollen grain at the base of Livingstone cores from both Ellesmere and Berrington, they must be younger than 7000 BP and, based on lamination counts, probably do not pre-date the British Iron Age (2500 BP). A radiocarbon date of 2670 ± 70 BP (OXA-2374) recently obtained from the basal core material supports this hypothesis.

There is mounting evidence for substantial

Holocene lake-level changes in this area. Pollen analysis of material from Newton Mere indicates a depositional hiatus between ca. 7000 and 2000 BP (Smith, 1986) when the lake may have dried out. The time-depth curve plotted by Nelms (1984) for Ellesmere suggests that there was a sharp change in sedimentation rate around 2200 BP and studies by Twigger (1988) of a number of Shropshire meres apparently indicate low lake water levels prior to the Iron Age. Some ice-formed depressions in the area may never have held much water until this time. However, archaeological evidence shows that during the Iron Age the water level in Ellesmere was several metres higher than today (Hamlin, 1988 and pers. comms). As it appears that we were unable to recover sediments dating from before ca. 2500 BP, we cannot speculate as to the early Holocene history of these meres.

Acknowledgements

We are grateful to Ms J.M. Wilmshurst for her assistance both in the laboratory and the field. G. Logan helped to analyse lipid fractions from the upper Ellesmere sediments. Field assistance from T.G. Acott, T. Allott, Miss F. Fitzpatrick, Dr Roger Flower and L. Madureira is also appreciated.

References

Beales, P. W., 1976. Palaeolimnological studies of a Shropshire mere. Unpublished Ph.D. thesis, University of Cambridge.

Beales, P. W., 1980. The late-Devensian and Flandrian vegetational history of Crose Mere, Shropshire. New Phytologist 85: 133–161.

Brassell, S. C., G. Eglinton, I. T., Marlowe, U. Pflaumann & M. Sarnheim, 1986. Molecular stratigraphy: a new tool for climate assessment. Nature 320: 129–133.

Cranwell, P. A., 1984a. Organic geochemistry of lacustrine sediments: triterpenoids of higher plant origin reflecting postglacial vegetational succession. In E. Y. Haworth & J. W. G. Lund (eds), Lake sediments and environmental history, Leicester University Press.

Cranwell, P. A., 1984b. Lipid geochemistry of sediments from Upton Broad, a small productive lake. Org. Geochem. 7: 25–37.

Cranwell, P. A., G. Eglinton & N. Robinson (1987). Lipids of aquatic organisms as potential contributors to lacustrine sediments II. Org. Geochem. 11: 513–527.

Gaskell, S. J. & G. Eglinton, 1975. Sterols of a contemporary lacustrine sediment. Geochim. Cosmochim. Acta 40: 1221–1228.

Hamlin, A. G. (ed.), 1988. Ellesmere historical town walk. The Ellesmere Society, Ellesmere, 16 pp.

Haworth, E. Y., 1980. Comparisons of continuous phytoplankton records with diatom stratigraphy in the recent sediments of Blelham Tarn. Limnol. Oceanogr. 25: 103–1103.

Haynes, C. V., C. H. Eyles, L. A. Parlish, J. C. Ritchie & M. Rybnak, 1989. Holocene palaeoecology of the Eastern Sahara: Selima Oasis. Quat. Sci. Rev. 8: 109–136.

Huang, W-Y & W. G. Meinschein, 1979. Sterols as ecological indicators. Geochim. Cosmochim. Acta 43: 739–745.

Kawamura, K., R. Ishiwatari & M. Yamazaki, 1980. Identification of polyunsaturated fatty acids in surface lacustrine sediments. Chem. Geol. 28: 31–39.

Livingstone, D., 1984. The preservation of algal remains in recent lake sediments. In E. Y. Haworth & J. W. G. Lund, (eds), Lake sediments and environmental history. Leicester University Press.

Livingstone, D. & R. S. Cambray, 1978. Confirmation of ^{137}Cs dating by algal stratigraphy in Rostherne Mere. Nature 276: 259–261.

Nelms, R. J., 1984. Palaeolimnological studies of Rostherne Mere and Ellesmere Mere. Unpublished Ph.D. thesis, Liverpool Polytechnic: 154 pp.

Nishimura, M., 1977. The geochemical significance in early sedimentation of geolipids obtained by saponification of lacustrine sediments. Geochim. Cosmochim. Acta 41: 1817–1823.

O'Sullivan, P. E., K. M. Farr, A. L. Heathwaite & J. P. Smith, 1989. South west England and the Shropshire-Cheshire meres. Guide to Excursion A, Vth International Symposium on Palaeolimnology, Ambleside, U.K., September, 1989.

Prahl, F., G. Eglinton, E. D. S. Corner & S. C. M. O'Hara, 1984. Copepod faecal pellets as a source of dihydrophytol in marine sediments. Science 224: 1235–1237.

Reynolds, C. S., 1979. The limnology of the eutrophic meres of the Shropshire-Cheshire plain. Field Studies 5: 93–173.

Robinson, N., P. A. Cranwell, B. J. Finlay & G. Eglinton, 1984. Lipids of aquatic organisms as potential contributors to lacustrine sediments. Org. Geochem. 6: 143–152.

Simola, H., 1977. Diatom succession in the formation of annually-laminated sediments in Lovojärvi, a small eutrophicated lake. Ann. Bot Fennici, 14: 143–148.

Smith, J. P., 1986. Mineral magnetic studies on two Shropshire-Cheshire meres. Unpublished Ph.D. thesis, University of Liverpool.

Twigger, S., 1988. Late holocene palaeocology and environmental archaeology of six lowland lakes and bogs in North Shropshire. Unpublished PhD thesis, University of Southampton, 402 pp.

Wunsche, L., F. O. Galacar & A. Buchs, 1987. Several unexpected marine sterols in a freshwater sediment. Org. Geochem. 11: 215–219.

Hydrobiologia **214**: 293–303, 1990.
J. P. Smith, P. G. Appleby, R. W. Battarbee, J. A. Dearing, R. Flower, E. Y. Haworth, F. Oldfield & P. E. O'Sullivan (eds), 293
Environmental History and Palaeolimnology.

Paleolimnological studies using sequential lipid extraction from recent lacustrine sediment: recognition of source organisms from biomarkers

P.A. Cranwell
Institute of Freshwater Ecology, The Ferry House, Ambleside, Cumbria LA22 OLP, UK

Key words: Sediments, lipids, biological markers, paleolimnology

Abstract

Free lipids were isolated from recent sediment of Loch Affric by solvent extraction; hydrolysis of residual sediment, initially with dilute alkali and then with mineral acid, gives two additional bound lipid components.

The distribution patterns of fatty acids, hydroxyacids and total neutral constituents in these chemically-distinct lipids show that the mode of occurrence contains much information. The molecular compositions of the neutral and acidic fractions obtained from the three lipid extracts were determined by gas chromatography – mass spectrometry. Acidic and neutral free lipids show a dominance of long-chain ($> C_{20}$) compounds characteristic of the wax constituents of higher plants; n-alkan-2-ones and α-hydroxyacids may be microbial metabolites of wax constituents. Base hydrolysis liberates C_{16} and C_{18} ω-hydroxyacids occurring widely in the cutins and suberins of higher plants together with higher homologues occurring in plant suberins. β-Hydroxyacids liberated by acidic hydrolysis show a molecular size range (C_{10}–C_{18}) and abundance of branched chain compounds typically occurring in lipopolysaccharides of gram-negative bacteria. The biological sources of c. 50% of the sediment lipids were identified using this organic geochemical approach.

Introduction

The organic matter of recent sediments includes lipid components that serve as marker molecules of input from specific organisms (Johns, 1986). The information content of biological markers resides in their structure and stereochemistry, the molecular size range and predominance of odd or even carbon chain length within a homologous series containing the same functionality, and the mode of occurrence (i.e. in a free or chemically-bound state). The stability of a compound in sedimentary deposits and its resulting value as a biological marker may also be determined by its chemical form. Free sterols, for example, undergo

microbially-mediated hydrogenation while esterified sterols do not (Eyssen *et al.*, 1973); consequently the composition of esterified sterols may resemble that of terrestrial plants more closely than the free sterols (Cranwell, 1986). Bound forms of other lipids are also better preserved than the free compounds (Kawamura & Ishiwatari, 1984; Albaigés *et al.*, 1984).

Until recently, single-stage hydrolysis with either acid or base was used to release bound carboxylic acids from sediments (Mendoza *et al.*, 1987a), however sequential hydrolysis with base under increasingly rigorous conditions revealed different modes of lipid incorporation into geopolymers (Kawamura & Ishiwatari, 1984).

294

Bacteria contain both ester and amide-linked fatty acids (Goldfine, 1982), labile to base hydrolysis or acid hydrolysis, respectively, thus sequential hydrolysis by base and then acid has been used to recognize bacterial inputs to sedimentary organic matter (De Leeuw, 1986; Ten Haven *et al.*, 1987; Mendoza *et al.*, 1987b; Wünsche *et al.*, 1988; Goossens *et al.*, 1989). This paper reports the application of this isolation procedure to a lacustrine sediment having a long-term homogeneous terrestrial input. Separation of the neutral free lipids enabled long-chain esters to be studied directly, whereas saponification had been used previously to show the presence of esters, with some loss of molecular information. The analytical data is used to determine the percentage source composition of the sediment lipids.

Experimental methods

Sampling site

Loch Affric (Grid Reference: NH 160225) is an oligotrophic lake (Area, 2.12 km², Max. Depth, 67 m), unmodified for water storage, situated in the Beauly basin, N. Scotland. The loch drains a catchment (Area, 121 km²) in which the Caledonian pine forest has formed the predominant vegetation for the last 8000 years, based on pollen analysis of a long core (Pennington, private communication). A sediment core 1 m in length was sectioned and stored at $-20°$ until required; the 2–27 cm section was homogenized and an aliquot gave, after drying, the elemental analysis: C, 17.2; H,2.6; N,1.0% (mean of duplicate analyses).

Extraction and isolation

Free lipids were obtained from wet sediment (155 g, 60% w/w water) by successive extraction with chloroform-methanol (1:2, 1:1 and 2:1 v/v respectively) and chloroform by stirring at room temperature, centrifuging and decanting the extract each time. The extracts were combined, diluted with water and the organic layer was dried

and evaporated. Residual lipids were extracted from the sediment with chloroform under reflux (Soxhlet) and combined with the former extract. Acidic compounds were separated by chromatography on KOH-impregnated silica gel (McCarthy & Duthie, 1962).

Base-liberated lipids were obtained by hydrolysis of solvent-extracted sediment with 5% KOH in methanol (300 ml) for 12 h under reflux. The mixture was filtered, the filtrate was concentrated, acidified with dil. HCl and extracted with ether. The material insoluble in methanol was extracted with chloroform (Soxhlet) and the product was combined with the ether extract and separated into acidic and neutral fractions by partitioning between ether and aqueous KOH (1 N).

Acid-liberated lipids were released from the residual sediment by boiling under reflux with 4 N HCl for 6 h while stirring. The mixture was cooled, insoluble material was removed by centrifuging and washed with water by centrifuging and decanting. The combined aqueous solutions were extracted with chloroform; the sediment was extracted as described for the free lipids. The combined organic product was saponified with KOH (5%) in aqueous methanol. Neutral and acidic fractions were isolated from the hydrolysate as described for the base-liberated lipids.

Carboxylic acids from each procedure were converted into methyl esters and separated into monocarboxylic esters (R_f 0.70–0.90) and hydroxy esters (R_f 0.10–0.40), as described previously (Cranwell, 1984).

Neutral free lipids were separated by column chromatography followed by TLC, (Cranwell, 1988). *n*-Alkanes, wax esters and alkan-2-ones were separated from branched/cyclic components of similar functionality by urea adduction. Neutral lipids released by hydrolytic methods were analysed without prior separation.

Gas chromatography (GC) and mass spectrometry (MS)

GC analyses were performed as described by Cranwell (1988). Compounds containing hydroxyl groups were derivatized as trimethylsilyl (TMS) ethers prior to analysis.

Sedimentary constituents, except *n*-alkanes, were identified by computerized gas chromatography-mass spectrometry (C-GC-MS) (Cranwell, 1988). Mass fragmentography using structurally-diagnostic ions facilitated recognition of the following classes: hopanes (m/z 191), alkan-2-ones (m/z 58,59) and carboxylic acids containing a hydroxyl group at the α (m/z 129, M-59), β (m/z 175), ω(M-15, M-47) and ω-1 (m/z 117) positions.

Results and discussion

The yields of neutral and acidic lipids obtained at each stage of the sequential extraction process are shown in Table 1. Neutral lipids dominate the free lipids, however the relative abundance of acidic lipids increases as hydrolysis proceeds being particularly marked for the hydroxyacids released by acidic hydrolysis.

Carboxylic acids

The carbon number distribution of monocarboxylic acids occurring in extractable, base- and acid-hydrolysable forms is shown in Fig. 1. Extractable acids were dominated by $C_{22}-C_{32}$ *n*-alkanoic acids showing a strong even-carbon preference. The carbon number distribution of

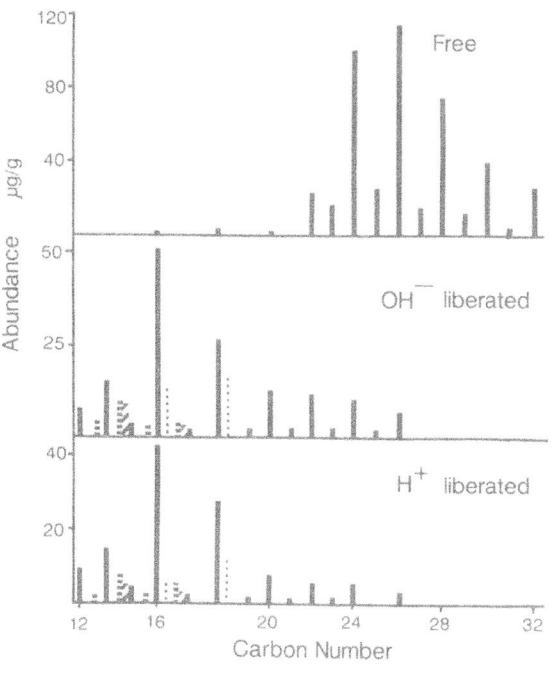

Fig. 1. Composition of monocarboxylic acids in the three modes present in the sediment.

free fatty acids in surficial lacustrine sediments is correlated with trophic status, C_{16} being dominant in productive lakes, the sediments of which are enriched in detritus from aquatic biota, and C_{24} or C_{26} in unproductive lakes, in which terrestrial plant detritus is the main source of sediment organic matter (see review by Cranwell, 1982).

Table 1. Yield of total neutral lipids, monocarboxylic acids and isomeric hydroxyacids obtained by sequential extraction.

Sediment treatment	Lipids[1]		Carboxylic acids[1]						
	Neut	Acid	Mono-[2]	Hydroxyacids[3]					
				α	β	ω	$\omega - 1$	phen	total[2]
Solvent extraction	6360	570	460	20	–	50	10	–	90
Base hydrolysis	970	680	220	–	30	265	10	75	380
Acid hydrolysis	400	1180	195	40	510	190	–	55	795

[1] All yields expressed as μg/g dry wt initial sediment. Neut = Neutral.
[2] Difference between abundance of identified and total acids reflects intermediate TLC band, not analysed.
[3] Abundances estimated from GC/MS data (Fig. 2), assuming equal response for all structures. Phen = Phenolic acids.

296

The base- and acid-liberated fatty acids show similar distributions and abundance levels of n-alkanoic acids maximizing at C_{16} (Fig. 1). Previously reported two stage hydrolysis procedures (Mendoza et al., 1987a, b; Wünsche et al., 1988) gave uni- or bimodal distributions of bound fatty acids in which homologues centered on C_{16} were dominant, with enhanced levels of iso- and anteiso-branched C_{14}–C_{17} acids reflecting a microbial contribution and also of unsaturated C_{16} and C_{18} fatty acids; similar distributions of branched-chain and unsaturated constituents occur in Loch Affric (Fig. 1). The dominance of short-chain fatty acids may represent a greater autochthonous input to bound acids or may reflect the increased stability of the bound form (Cranwell, 1981a).

Hydroxyacids

Extractable hydroxyacids comprise mainly α- and ω-substituted C_{20}–C_{28} constituents (Table 1), maximizing in abundance at C_{24} and C_{22}, respectively (Fig. 2). The lack of a diagnostic mass spectral fragment ion at m/z 175 (Eglinton & Hunneman, 1968) showed the absence of β-hydroxyacids. These α- and ω-substituted constituents are probably formed by microbial attack under aerobic conditions on the corresponding fatty acids (Table 2). In one sediment profile, the stereochemistry of α-hydroxyacids was consistent with this origin (Cranwell, 1981b). ω-Hydroxylation of fatty acids has also been reported (Fulco, 1983), although plant waxes may provide an additional source of $>C_{20}$ ω-hydroxyacids.

Extractable ω-hydroxyacids isolated from Rostherne Mere (Cardoso & Eglinton, 1983) and Coniston Water (Robinson et al., 1987) showed a similar distribution to Loch Affric, but a sub-surface sediment from Lake Leman showed a greater dominance of C_{22} (Mendoza et al., 1987b). Distributions of free α-hydroxyacids similar to that reported here have been noted in other lakes (Cranwell, 1981b, 1984; Wünsche et al., 1988).

Base-liberated hydroxyacids are also dominated by ω-substituted compounds, here showing a

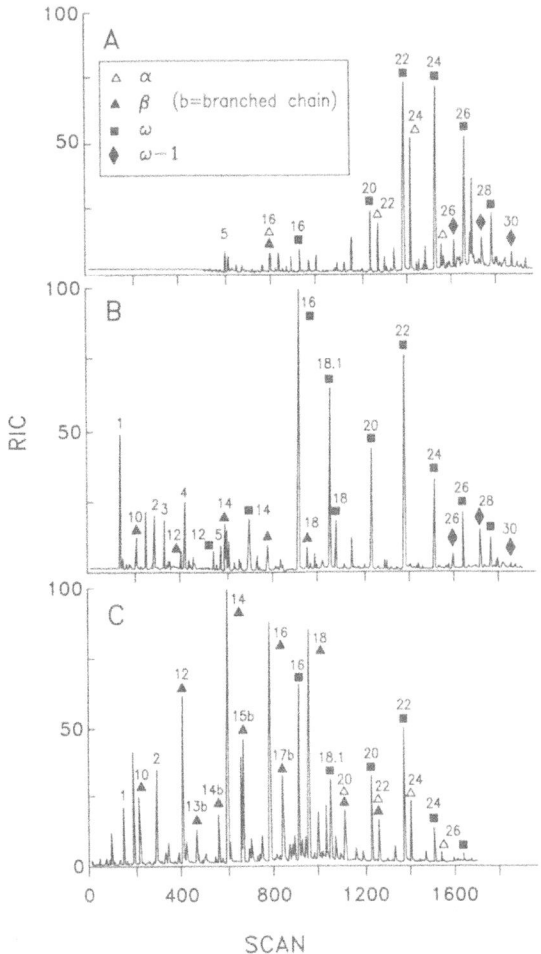

Fig. 2. Hydroxyacid composition of free (A), base-labile (B) and acid-labile (C) lipids, as shown by reconstructed ion current (RIC) trace obtained during GC/MS analysis.

bimodal distribution maximizing at C_{16} and C_{22}. Sources of homologues below C_{20} (Table 2) include the estolide waxes of pine needles, probably a major source in Loch Affric. Higher homologues ($>C_{18}$) may originate from suberin, a protective polymer of plants.

Bimodal distributions of ω-hydroxyacids liberated by alkaline hydrolysis were reported for sediment of Lake Biwa (Kawamura & Ishiwatari, 1984) and Lake Leman (Mendoza et al., 1987b). Variations in relative abundance of C_{16} and C_{22} with sediment depth were noted in Rostherne Mere (Cardoso & Eglinton, 1983); a suberin input to this sediment was confirmed by the recognition

Table 2. Abundance and sources of straight-chain lipids.

Compound class	Carbon number[§]		CPI[+]	Amount ($\mu g/g$)[*]	Source	References
	Range	Maximum				
n-Alkanes	17–22	22	0.8	30	Microorganisms	Grimalt & Albaigés, 1987
n-Alkanes	23–35	31	10.5	340	Vascular plants	Eglinton & Hamilton, 1967
n-Alkyl esters	30–36	36		20	Algae, bacteria	Weete, 1976; Russell & Volkman, 1980
	38–50	42 = 44	10.5	310	Vascular plants	Kolattukudy, 1980b
n-Aldehydes	22–34	28	4.4	265	Vascular plants	Kolattukudy, 1980b
n-Alkan-2-ones	21–35	27	6.7	220	Microbial metabolites	Volkman *et al.*, 1983
n-Alkan-1-ols	20–32	26 = 28	9.7	200	Vascular plants	Kolattukudy, 1980b
	14–28	22 (OH^-)	10.5	300		
n-Alkanoic acids	22–32	26	5.7	460	Vascular plants	
		22 (OH^-)	5.1	50		
		22 (H^+)	3.7	30		
	12–20	16 (OH^-)	10.8	170	Microorganisms	
	12–20	16 (H^+)	9.3	165		
α-Hydroxyacids	22–26	24		20	Microbial metabolites	Harwood & Russell, 1984
	20–26	24 (H^+)		40	Vascular plants	Harwood & Russell, 1984
β-Hydroxyacids[‡]	10–18	14 (OH^-)		30	Prokaryotes + eukaryotes	Goldfine, 1982
	10–22	14 (H^+)		510	Bacterial cell wall lipids	Lüderitz *et al.*, 1982
ω-Hydroxyacids	16–28	22		50	Microbial metabolites	Fulco, 1983
	12–28	16 + 22 (OH^-)		265	Cutin, suberin Plant wax	Holloway, 1982; Kolattukudy, 1980a Herbin & Sharma, 1969
	14–26	16 + 22 (H^+)		190	Unknown	
(ω-1)-Hydroxy-acids	26–30	28 (Free + OH^-)		20	Microbial metabolites	Miura & Fulco, 1975
Phenolic acids		(OH^-, H^+)		130	Vascular plants	Kolattukudy, 1980a

§ Free lipid, unless stated otherwise: (OH^-) = base-liberated, (H^+) = acid-liberated lipids.

[+] Expressed as odd/even predominance for alkanes and alkan-2-ones and as even/odd predominance for other compounds. Among hydroxyacids, only α-isomers show odd-carbon homologues.

* Sediment dry weight.

‡ Class includes *iso-* and *anteiso*-branched constituents.

of cork cells by microscopic examination (Cardoso *et al.*, 1977).

Phenolic acids liberated with alkali included two isomeric coumaric acids and ferulic acid, that occur esterified mainly to suberin (Kolattukudy, 1980a), *p*-hydroxybenzoic and vanillic acids; the last two acids are commonly released on hydrolysis of soil humic matter. The presence of vanillyl phenols and absence of syringyl phenols is consistent with a major gymnosperm (conifer) input (Hedges & Parker, 1976). These phenolic compounds have been previously identified in saponified extracts of sediments (Matsumoto & Hanya, 1980).

Trace levels of C_{26}, C_{28} and C_{30} (ω-1)-hy-

droxyacids occurring in both extractable and base-hydrolysable lipids (Table 2) were previously reported in freshwater (Mendoza *et al.*, 1987b; Cranwell *et al.*, 1987; Wünsche *et al.*, 1988) and marine (Boon *et al.*, 1977) sediments.

β-Hydroxyacids are minor products of base hydrolysis, but dominate the hydroxy-acids released by acidic hydrolysis (Table 1), among which branched-chain C_{15} and C_{17} compounds are abundant (Fig. 2). Among the known sources (Table 2), amide-bound β-hydroxyacids (liberated by acidic hydrolysis) are regarded as specific markers of bacterial input to sediments (Boon *et al.*, 1977), being only found in bacterial LPS and ornithine lipids (Harwood & Russell, 1984).

298

The stereochemistry of sedimentary bound β-hydroxyacids was that reported for amide-bound constituents (Cranwell, 1981b). β-Hydroxyacids released by acid hydrolysis of lacustrine (Mendoza et al., 1987b, c) and marine (De Leeuw, 1986) sediments showed a similar composition to Loch Affric (Fig. 2).

Acid-liberated ω-hydroxyacids show a similar distribution pattern (Fig. 2) to the products of alkaline hydrolysis, a feature also noted in Lake Leman sediment (Mendoza et al., 1987b). As cutin and suberin resist acid hydrolysis but are decomposed by base (see above), an alternative unidentified source of acid-labile ω-hydroxyacids is postulated.

Acidic hydrolysis release α-hydroxyacids showing a distribution range similar to the free α-hydroxyacids. A stereochemical study (Cranwell, 1981b) suggested that sedimentary bound α-hydroxyacids originate from cerebrosides, complex lipids of plants and fungi, in which the α-hydroxyacid is attached through an amide bond.

Acidic hydrolysis released additional amounts of phenolic acids, mainly comprising p-hydroxybenzoic and vanillic acids.

Neutral lipids

GC analysis of neutral lipids from each extraction stage showed the broadest molecular size range occurring in the extractable lipids, from which compound classes differing in functionality were isolated prior to GC-MS.

The extractable lipids identified during GC/MS analysis consist mainly of homologous straight-chain compounds differing in functional group (Table 2). Each series shows a marked alternation in abundance of consecutive members, expressed as a carbon preference index, and a molecular size range typical of the surface waxes of vascular plants (Table 2). The n-alkane distribution, dominated by $n\text{-}C_{31}$ (Fig. 3) is typical of sediments derived from acidic soils and their acid-tolerant vegetation (Cranwell, 1973) and closely resembled that found in sediment from Loch Clair (Cranwell, 1981a), the drainage basin of which

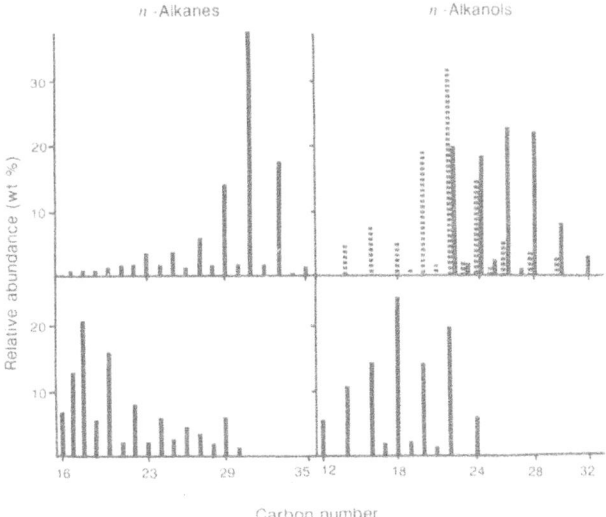

Fig. 3. Histograms showing homologue distributions of n-alkanes (left) and n-alkanols (right) in the extractable (upper) and acid-released (lower) neutral lipids. Broken bars in upper right histogram show the distribution of n-alkanols liberated by base hydrolysis; no corresponding alkanes were detected.

also contains remnants of the Caledonian pine forest. n-Alkanes in the C_{17}–C_{22} range show a slight even carbon number preference typical of hydrocarbons biosynthesized by microorganisms (Grimalt & Albaigés, 1987 and references therein), but only constitute 8% of the total. The major C_{42}, C_{44} and C_{46} alkyl esters show the same dominant alkyl-acyl pairings, calculated from mass spectral data (Aasen et al., 1971), as the corresponding constituents in a peat deposit, the arboreal pollen of which was dominated by pine (Cranwell, 1988). Sources of shorter-chain esters are given in Table 2.

n-Aldehydes and alkanols derived from vascular plants occur widely in sediments (Prahl & Pinto, 1987; Wünsche et al., 1988). n-Alkan-2-ones occurring in soils and sediments are considered to be microbial metabolites (Volkman et al., 1983).

Extractable lipids also included branched/cyclic (b/c) hydrocarbons, esters, ketones and hydroxyl derivatives, identified by GC/MS, that contain many source-diagnostic constituents (Table 3). The major cyclic compounds, other

Table 3. Branched/cyclic neutral lipids: abundance and sources.

Compound class	Structural type[1]	Abundance (μg/g dry wt sediment)	Major constituent[2]	Source
Hydrocarbons	Acyclic	65	2,6,10-Trimethyl-7-(3-methylbutyl)-dodecane	Algae
	Hopanoid	200	$17\alpha H,21\beta H,22R$-C_{31}	Bacterial precursor
Esters	Steroid	55	$C_{29}\Delta 5$-16 : 0	Vascular plants
	Triterpenoid	80	Urs-12-en-3-yl-16 : 0	Vascular plants
Ketones	Steroid	65	C_{29} 5αH-stanone	Microbial metabolites
	Triterpenoid	170	Friedelin	Vascular plants
	Hopanoid	160	C_{29} unidentified	Breakdown of bacterial precursor
Alcohols	Steroid	40	C_{29} 5αH-stanol	Vascular plants
	Steroid (OH$^-$)	60	C_{29} Δ^5sterol	Vascular plants
	Triterpenoid	220	Urs-12-en-3-ol, Friedelinol	Vascular plants
	Hopanoid	70	Diplopterol[3]	Bacteria

[1] Triterpenoid refers to pentacyclic carbon skeleton of oleanane, ursane, lupane, taraxerane or friedelane type. Free lipids except for base-liberated sterols (OH$^-$).

[2] C_{29} steroids have 24-ethylcholestane carbon skeleton. Esters are identified as alkyl-acyl moieties.

[3] Tentative identification.

than hydrocarbons, are derivatives either of the sterols or of pentacyclic triterpenoids widely distributed in vascular plants (Pant & Rastogi, 1979).

Steroidal derivatives identified by GC/MS (Cranwell, 1986) include, among the esters, cholest-5-en-3β-yl-16 : 0, 24-ethylcholest-5-en-3β-yl-16 : 0 and -18 : 0 and 24-ethylcholesta-5,22-dien-3β-yl-18 : 0, together totalling c. 40% of b/c esters, the C_{29} steryl derivatives predominating. Steroidal ketones consisted mainly of 5β(H)- and 5α(H)-24-ethylcholestan-3-ones (α/β ratio = 4.5); the 24-methyl-5α(H) analogue was also detected. 3-Keto steroids are intermediates in the microbial conversion of Δ^5-3β-hydroxy sterols into the corresponding stanol in sediments (Mermoud *et al.*, 1984). Free sterols include Δ^5-stenols and their related 5α-stanols, in which the stanol : stenol ratio > 1.5 for the C_{28} and dominant C_{29} components, but is 1.2 for the minor C_{27} analogues. 24-Ethylcholesta-5,22-dien-3β-ol and its related 5α(H)-Δ^{22}-stenol were also present at more than trace levels. Although 24-ethyl sterols occur in algae more widely than formerly recognized (Volkman, 1986), extractable

biolipid distributions noted above, typical of vascular plant sources, suggest the same dominant source of sterols.

Among pentacyclic compounds, urs-12-en-3-yl-16 : 0 was the major ester and was accompanied by the analogous stearate. The major constituent was identified by comparison of GC retention and mass spectral data, including appropriate M^+ and M-15 ions, with a synthetic standard, (Cranwell, 1986). Esters of the oleanane, ursane and lupane series occur in plant waxes (Madrigal *et al.*, 1975). The abundance of these esters in lacustrine sediments shows a positive correlation with that of other terrestrial biomarkers (Cranwell, 1986). Pentacyclic triterpenoid ketones included friedelin, taraxer-14-en-3-one and olean-13(18)-en-3-one as abundant constituents, olean-12-en-3-one and urs-12-en-3-one as minor peaks. These ketones are primary products in the more primitive plant divisions and also occur in tree bark (Sainsbury, 1970). Triterpenoid alcohols occurring in a free state were structurally analogous to corresponding ketones and characteristic of vascular plants.

In contrast to the above functionalized classes,

300

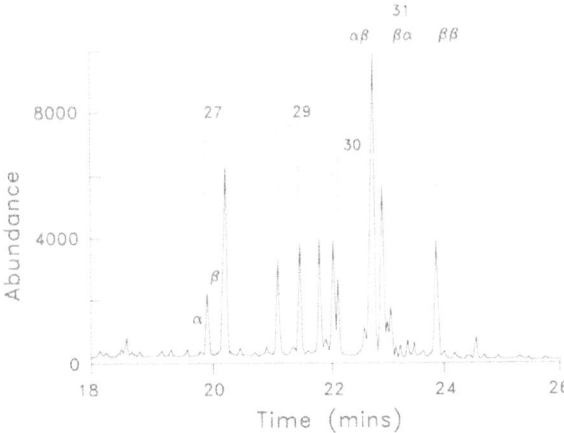

Fig. 4. Fragmentogram of m/z 191 obtained during GC/MS analysis of branched/cyclic alkanes. Elution order of C_{29} and C_{30} isomers is as shown for C_{31}.

the main branched/cyclic hydrocarbons (Fig. 4) are C_{27}–C_{31} hopanes showing the $17\beta H$, $21\beta H$ configuration, derived from prokaryotic precursors (Ourisson *et al.*, 1979). In peat environments, however, a rapid microbially-mediated formation of $17\alpha H$, $21\beta H$, 22R-homohopane, the major constituent in Loch Affric, has been reported (Quirk *et al.*, 1984). The observed hopane distribution (Fig. 4) resembles that of other sediments deposited at sites unpolluted by petroleum-derived hydrocarbons e.g. sub-surface deposits in Rostherne Mere (Cardoso *et al.*, 1983) or a pre-industrial age deposit in Windermere (Cranwell & Koul, 1989).

Minor hopanoid ketones included two C_{32} compounds, the earlier-eluting of which gave a mass spectrum indicative of a bishomohopan-31-one (Dastillung *et al.*, 1980), and a C_{33} ketone showing a mass spectrum identical with that of a $\beta\beta$-trishomohopan-32-one isolated from sediment of the Baltic Sea (Dastillung *et al.*, 1980). The major b/c ketone was unidentified but was thought to be a norhopenone, based on its GC retention and mass spectrum.

Hopanols were detected in the branched/cyclic alcohols, a C_{30} compound tentatively identified as diplopterol being dominant. Two isomeric C_{32} extended chain hopanoids showed mass spectral fragmentation consistent with $17\alpha H$, $21\beta H$- and

$17\beta H$, $21\beta H$-configurations, respectively, and a C_{33} analogue of the former was detected.

Neutral lipids released by alkaline hydrolysis showed a simpler composition than the extractable neutrals, consisting mainly of *n*-alkan-1-ols (C_{12}–C_{28}, Fig. 3), 24-ethylcholest-5-en-3β-ol and the related 5α-stanol. In the sediment the *n*-alkanols are esterified to insoluble polymeric material such as suberins (Kolattukudy, 1980a); the sterols may be released from polar sulphates, as suggested by Ten Haven *et al.* (1987).

Acidic hydrolysis released neutral lipids in low abundance (Table 1), including many minor constituents. *n*-Alkanols (C_{12}–C_{24}) were again major constituents; their molecular distribution is compared with those of the corresponding extractable and base-hydrolysable components in Fig. 3. *n*-Alkanes (C_{16}–C_{20}) were abundant and homologues extending to C_{29} were detected during GC/MS analysis. The alkane distribution showed an even-carbon predominance below C_{25} (Fig. 3). Low levels of C_{15}–C_{27} *n*-alkan-2-ones were recognised by GC/MS. Steroidal compounds included cholesta-3,5-dien-7-one, its 24-ethyl substituted analogue and 24-ethylcholest-5-en-3β-ol; the ketones, autoxidation products of sterols, may be natural or may be artifacts.

Little published data is available comparing the molecular compositions of neutral lipid classes occurring in sediments in extractable, base-hydrolysable and acid-hydrolysable forms. Total lipid profiles for these chemically-distinct forms were obtained from a Mediterranean sapropel containing organic matter of a mixed marine, terrigenous and bacterial origin (Ten Haven *et al.*, 1987), and, more recently, from a lacustrine sediment (Goossens *et al.*, 1989). In the latter, the extractable lipids showed, as expected, relatively more abundant biomarkers typical of vascular plants, eg *n*-alkanes and alkanols, whereas algal biomarkers were dominant in the marine sediment. ω-Hydroxyacids occurring in cutins and suberins were enriched in both base- and acid-liberated lipids isolated from the lake sediment, relative to the Mediterranean sediment.

Alkanols released by acidic hydrolysis of sediments (Cranwell, 1981a; Robinson *et al.*, 1987)

Fig. 5. Approximate assignment of biological sources of sediment lipids based on abundances of biological markers.

may be derived from decomposer organisms, based on a comparison with the molecular distributions in fresh and decayed algae (Cranwell, 1981a). Distributions of extractable *n*-alkanes showing even carbon number preference up to C_{22} or C_{24}, obtained from several depositional environments (Nishimura & Baker, 1986; Grimalt & Albaigés, 1987), were attributed to microorganisms (bacteria, fungi and yeasts). The binding mechanism for alkanes is uncertain, but may result from physical trapping within clay minerals.

Approximately 50% of the total lipids have been chemically identified and their biological sources recognized (Tables 2 and 3), thus the relative sediment composition shown in Fig. 5 can be inferred. Owing to differences in stability of biological markers and the continuation of bacterial activity after deposition, the input balance may differ from that shown in Fig. 5. The organic geochemical approach does, however, extend the scope of paleolimnological information available from the study of morphological microfossils alone.

Conclusions

Sequential extraction of sedimentary lipids facilitates recognition of discrete input sources from the mode of occurrence of lipid constituents together with their structural identity. The biological marker approach is a useful technique for paleolimnological research.

Acknowledgements

GC-MS facilities were provided by grants GR3/2951, GR3/3758 and GR3/6619 from the Natural Environment Research Council to the Organic Geochemistry Unit, Bristol University. The work was supported by the NERC. Miss K. Ross typed the manuscript and Mr T. Furnass prepared Figs. 2, 4 and 5.

References

Aasen, A. J., H. H. Hofstetter, B. T. R. Iyengar & R. T. Holman, 1971. Identification and analysis of wax esters by mass spectrometry. Lipids 6: 502–507.

Albaigés, J., J. Algaba & J. Grimalt, 1984. Extractable and bound neutral lipids in some lacustrine sediments. Org. Geochem. 6: 223–236.

Boon, J. J., F. De Lange, P. J. W. Schuyl, J. W. De Leeuw & P. A. Schenck, 1977. Organic geochemistry of Walvis Bay diatomaceous ooze. II Occurrence and significance of the hydroxy fatty acids. In Advances in Organic Geochemistry 1975, Campos R. & Goni J., (eds) ENADIMSA, Madrid: 255–271.

Cardoso, J. N. & G. Eglinton, 1983. The use of hydroxyacids as geochemical indicators. Geochim. Cosmochim. Acta 47: 723–730.

Cardoso, J. N., G. Eglinton & P. J. Holloway, 1977. The use of cutin acids in the recognition of higher plant contribution to recent sediments. In Advances in Organic Geochemistry 1975, Campos R. & Goni J., (eds) ENADIMSA, Madrid: 273–287.

Cardoso, J. N., S. J. Gaskell, M. M. Quirk & G. Eglinton, 1983. Hydrocarbon and fatty acid distributions in Rostherne lake sediment (England). Chem. Geol., 38: 107–128.

Cranwell, P. A., 1973. Chain length distribution of *n*-alkanes from lake sediments in relation to post-glacial environmental change. Freshwat. Biol., 3: 259–265.

Cranwell, P. A., 1981a. Diagenesis of free and bound lipids in terrestrial detritus deposited in a lacustrine sediment. Org. Geochem. 3: 79–89.

Cranwell, P. A., 1981b. The stereochemistry of 2- and 3-hydroxy fatty acids in a recent lacustrine sediment. Geochim. Cosmochim. Acta 45: 547–552.

Cranwell, P. A., 1982. Lipids of aquatic sediments and sedimenting particulates. Prog. Lipid Res. 21: 271–308.

Cranwell, P. A., 1984. Lipid geochemistry of sediments from Upton Broad, a small productive lake. Org. Geochem. 7: 25–37.

Cranwell, P. A., 1986. Esters of acyclic and polycyclic isoprenoid alcohols: biochemical markers in lacustrine sediments. In D. Leythaeuser & J. Rullkötter (eds), Advances in Organic Geochemistry 1985. Pergamon Press, Oxford: 891–896.

Cranwell, P. A., 1988. Lipid geochemistry of late-Pleistocene lacustrine sediments from Burland, Cheshire, U.K. Chem. Geol. 68: 181–197.

Cranwell, P. A., G. Eglinton & N. Robinson, 1987. Lipids of aquatic organisms as potential contributors to lacustrine sediments – II. Org. Geochem. 11: 513–527.

Cranwell, P. A. & V. K. Koul, 1989. Sedimentary record of polycyclic aromatic and aliphatic hydrocarbons in the Windermere catchment. Wat. Res. 23: 275–283.

Dastillung, M., P. Albrecht & G. Ourisson, 1980. Cetones aliphatiques et polycycliques dans les sediments: aldehydes et cetones en C_{27}–C_{35} de la serie hopanique. J. Chem. Res. (M): 2319–2327.

De Leeuw, J. W., 1986. Sedimentary lipids and polysaccharides as indicators for sources of input, microbial activity, and short-term diagenesis. In Organic Marine Geochemistry, Sohn M. L., (ed.) Am. Chem. Soc., Washington DC: 33–61.

Eglinton, G. & R. J. Hamilton, 1967. Leaf epicuticular waxes. Science 156: 1322–1334.

Eglinton, G. & D. H. Hunneman, 1968. Gas chromatographic-mass spectrometric studies of long chain hydroxy acids. III The mass spectra of the methyl esters trimethylsilylethers of aliphatic hydroxy acids. A facile method of double bond location. Org. Mass Spectrom. 1: 593–611.

Eyssen, H. J., G. G. Parmentier, F. C. Compernolle, G. De Pauw & M. Piessens-Denef, 1973. Biohydrogenation of sterols by Eubacterium ATCC 21,408- Nova Species. Eur. J. Biochem. 36: 411–421.

Fulco, A. J., 1983. Fatty acid metabolism in bacteria. Prog. Lipid Res. 22: 133–160.

Goldfine, H., 1982. Lipids of prokaryotes – structure and distribution. In Current topics in Membranes and Transport Vol. 17 Razin S. & Rottem S. (eds), Academic Press, New York: 1–43.

Goossens, H., R. R. Düren, J. W. De Leeuw & P. A. Schenck, 1989. Lipids and their mode of occurrence in bacteria and sediments-II. Lipids in the sediment of a stratified, freshwater lake. Org. Geochem. 14: 27–41.

Grimalt, J. & J. Albaigés, 1987. Sources and occurrence of C_{12}–C_{22} n-alkane distributions with even carbon-number preference in sedimentary environments. Geochim. Cosmochim. Acta 51: 1379–1384.

Harwood, J. L. & N. J. Russell, 1984. Lipids in Plants and Microbes. Allen and Unwin, London: 162 pp.

Hedges, J. I. & P. L. Parker, 1976. Land-derived organic matter in surface sediments from the Gulf of Mexico. Geochim. Cosmochim. Acta 40: 1019–1029.

Herbin, G. A. & K. Sharma, 1969. Studies of plant cuticular waxes-V. The wax coatings of pine needles: a survey. Phytochemistry 8: 151–159.

Holloway, P. J., 1982. The chemical constitution of plant cutins. In: The Plant Cuticle, Cutler D. F., Alvin K. L. & Price C. E. (eds), Academic Press, London, 45–85.

Johns, R. B. (ed.), 1986. Biological markers in the sedimentary record. Elsevier, Amsterdam, 364 pp.

Kawamura, K. & R. Ishiwatari, 1984. Tightly bound aliphatic acids in Lake Biwa sediments: their origin and stability. Org. Geochem. 7: 121–126.

Kolattukudy, P. E., 1980a. Biopolyester membranes of plants: cutin and suberin. Science 208: 990–1000.

Kolattukudy, P. E., 1980b. Cutin, suberin and waxes. In The Biochemistry of Plants, Vol. 4, Lipids: Structure and Function, P. K. Stumpf, (ed) Academic Press, New York: 571–645.

Lüderitz, O., M. A. Freudenberg, C. Galanos, V. Lehmann, E. T. Rietschel & D. H. Shaw, 1982. Lipopolysaccharides of gram-negative bacteria. In Current Topics in Membranes and Transport, Vol. 17, Razin S. & Rottem S., (eds) Academic Press, New York: 79–151.

Madrigal, R. V., R. D. Plattner & C. R. Smith, 1975. Carduus nigrescens seed oil – a rich source of pentacyclic triterpenoids. Lipids 10: 208–213.

Matsumoto, G. & T. Hanya, 1980. Gas chromatographic-mass spectrometric identification of phenolic acids in recent sediments. J. Chromatogr. 193: 89–94.

McCarthy, R. D. & A. H. Duthie. 1962. A rapid quantitative method for the separation of free fatty acids from other lipids. J. Lipid Res. 3: 117–119.

Mendoza, Y. A., F. O. Gülacar & A. Buchs, 1987a. Comparison of extraction techniques for bound carboxylic acids in recent sediments 1. Unsubstituted monocarboxylic acids. Chem. Geol., 62: 307–319.

Mendoza, Y. A., F. O. Gülacar, Z-L. Hu & A. Buchs, 1987b. Unsubstituted and hydroxyl substituted fatty acids in a recent lacustrine sediment. Int. J. Envir. Anal. Chem. 31: 107–127.

Mendoza, Y. A., F. O. Gülacar & A. Buchs, 1987c. Comparison of extraction techniques for bound carboxylic acids in recent sediments 2. β-Hydroxyacids. Chem. Geol. 62: 321–330.

Mermoud, F., L. Wünsche, O. Clerc, F. O. Gülacar & A. Buchs, 1984. Steroidal ketones in the early diagenetic transformations of Δ^5-sterols in different types of sediments. Org. Geochem. 6: 25–29.

Miura, Y. & A. J. Fulco, 1975. ω-1, ω-2, and ω-3 Hydroxylation of long-chain fatty acids, amides and alcohols by a soluble enzyme system from Bacillus megaterium. Biochem. Biophys. Acta 388: 305–317.

Nishimura, M. & E. W. Baker, 1986. Possible origin of n-alkanes with a remarkable even-to-odd predominance in recent marine sediments. Geochim. Cosmochim. Acta 50: 299–305.

Ourisson, G., P. Albrecht & M. Rohmer, 1979. The hopanoids. Palaeochemistry and biochemistry of a group of natural products. Pure Appl. Chem., 51: 709–729.

Pant, P. & R. P. Rastogi, 1979. The triterpenoids. Phytochemistry 18: 1095–1108.

Prahl, F. G. & L. A. Pinto, 1987. A geochemical study of long-chain n-aldehydes in Washington coastal sediments. Geochim. Cosmochim. Acta 51: 1573–1582.

Quirk, M. M., A. M. K. Wardroper, R. E. Wheatley & J. R. Maxwell, 1984. Extended hopanoids in peat environments. Chem. Geol. 42: 25–43.

Robinson, N., P. A. Cranwell & G. Eglinton, 1987. Sources of the lipids in the bottom sediments of an English oligo-mesotrophic lake. Freshwat. Biol. 17: 15–33.

Russell, N. J. & J. K. Volkman, 1980. The effect of growth temperature on wax ester composition in the psychrophilic bacterium *Micrococcus cryophilus* ATCC 15174. J. Gen. Microbiol. 118: 131–141.

Sainsbury, M., 1970. Friedelin and epifriedelinol from the bark of *Prunus turfosa* and a review of their natural distribution. Phytochemistry 9: 2209–2215.

Ten Haven, H. L., M. Bass, J. W. De Leeuw & P. A. Schenck, 1987. Late quaternary Mediterranean sapropels I – On the origin of organic matter in sapropel S_7. Mar. Geol. 75: 137–156.

Volkman, J. K., 1986. A review of sterol markers for marine and terrigenous organic matter. Org. Geochem. 9: 83–9

Volkman, J. K., J. W. Farrington, R. B. Gagosian & S. G. Wakeham, 1983. Lipid composition of coastal marine sediments from the Peru upwelling region. In Advances in Organic Geochemistry 1981, M. Bjoroy (ed.), Wiley, Chichester: 228–240.

Weete, J. D., 1986. Algal and fungal waxes. In Chemistry and Biochemistry of Natural Waxes, Kolattukudy P. E. (ed). Elsevier, Amsterdam: 349–418.

Wünsche, L., Y. A. Mendoza & F. O. Gülacar, 1988. Lipid geochemistry of a post-glacial lacustrine sediment. Org. Geochem. 13: 1131–1143.

Hydrobiologia **214**: 305–310, 1990.
J. P. Smith, P. G. Appleby, R. W. Battarbee, J. A. Dearing, R. Flower, E. Y. Haworth, F. Oldfield & P. E. O'Sullivan (eds), 305
Environmental History and Palaeolimnology.

Vegetation change and pollen recruitment in a lowland lake catchment: Groby Pool, Leics (England)

Carol David & Neil Roberts
Department of Geography, Loughborough University of Technology, UK

Key words: pollen, palaeoecology, lake sediments, landscape history, lake-catchment ecosystem

Abstract

This investigation adopts an historical approach to pollen recruitment in a lowland English lake. The history of woodland and land-use has been reconstructed for the past 200 years from documentary sources and from the lake sediment pollen record. Comparison of these data sets are used to establish the 'effective' pollen catchment of the lake and to investigate how well these support predictive models of pollen input to lakes.

Introduction

Pollen preserved within lake deposits is amongst the most widely used biological indicator of landscape history, enabling the reconstruction of former plant communities at both the regional and local scale. To understand the palaeoecological implications of fossil pollen assemblages contemporary process studies have been undertaken to provide information on the production, dispersal and preservation of pollen grains in different depositional environments. The source of the pollen preserved within fossil assemblages is one of the most important factors to be considered in any interpretation and a number of studies have focused on this particular aspect of pollen analysis (Tauber, 1965, 1967; Berglund, 1973; Bradshaw & Webb, 1985). Methods adopted have included trapping of contemporary pollen rain within specially designed containers (Tauber, 1974), the use of moss polsters and lake surface sediments samples. The timescales of investigation have generally been relatively short (≤5 years) and involve periods of stable vegetation conditions. From these data models of pollen dis-

persal have been proposed relating pollen source area to basin size (Tauber, 1965; Jacobson & Bradshaw, 1981); Prentice has also attempted to incorporate the different dispersal characteristics of individual pollen types (Prentice, 1985). Whilst studies of pollen input to small basins (<20 m) within closed forest seem to confirm the validity of predicted inputs (Andersen, 1970: Bradshaw, 1981) validation of models for lake basins of intermediate size has proved more problematical. Reasons for this include

1) the composite nature of pollen rain.
2) the influence of inflow streams on final pollen composition.
3) the relatively short observation period (<5 years) of contemporary pollen rain studies.

Many of the sites used in pollen analytical studies are open basin lakes (i.e. water derived from inflow streams as well as precipitation). Studies of pollen recruitment to lakes with surface inflows indicate the dominant role played by catchment hydrology in influencing pollen influx at these sites (Peck, 1973; Bonny, 1978; Pennington,

1979). In contrast to mires, pollen deposition in lakes is further complicated by limnological processes which may cause re-suspension and re-deposition of the sediments (Davis, 1973; Davis & Brubacker, 1973). It is not clear to what extent current models of pollen dispersal may be applicable to sites with inflow streams given the more complex recruitment and depositional processes experienced by these types of lakes.

The preliminary results presented here form part of a programme of investigation of contemporary and historical pollen recruitment to a small lowland lake in Leicestershire, U.K. Adopting an historical perspective has allowed the timescale of the contemporary investigation to be extended to include periods of documented vegetation change within the landscape. It is hoped that comparison of palaeoecological and historical sources will allow pollen/vegetation relationships over longer periods to be evaluated and the sensitivity of the pollen record at this site to vegetation changes at different spatial scales to be identified.

Study site

Groby Pool is a shallow (12 ha) lake located on the south-western edge of Charnwood Forest, Leicestershire. The lake has a large catchment with a lake/catchment ratio of 1 : 71. It possesses one main and one subsidiary inflow stream and one outflow. At present the lake is surrounded by a dense *Alnus/Salix* carr to the north and west, with lake marginal beds of *Phragmites*, Cyperaceae and *Typha*. As an aid to interpreting the pollen record in terms of pollen sources, three hypothetical catchments have been identified from which pollen may be recruited to the lake from the 'local' vegetation (Fig. 1a). Although not mutually exclusive, these catchments have been separated for the purpose of discussion and it can be seen that the potential aerial and lake/stream/watershed catchments only partially overlap.

Archival record

The landscape history of the area lying within 2 km radius of the lake has been reconstructed in some detail for the past 200 years from a range of documentary sources (Government and Estate papers and maps, agricultural statistics, aerial photographs, descriptive accounts and secondary sources). Most of this area lies within the Manor of Groby, and all of it was owned by the same family until 1925. In addition, the landscape history of the area within c. 20 km radius of the lake, along with broader regional trends, has also been established but in less detail. To some extent changes affecting local woodland are mirrored by vegetation changes at the sub-regional scale (< 20 km radius) although not at the regional or county scale (> 20 km). It has therefore been difficult to establish precisely the source of some pollen types within the lake sediment record at Groby.

The main trends in local land-use and woodland history which have occurred since 1757 are illustrated in (Fig. 1, b–d). Since at least the 11th century AD arable cultivation was carried out within an 'open' landscape but in the 18th century this arable land was enclosed within small fields and much of it was converted to pasture (Woodward, 1984; Hoskins, 1949). These events occurred both at the local and regional (i.e. county) scale. Locally, Enclosure was also accompanied by a shift in the location of arable cultivation from land lying outside the lake's catchment (Catchment A, Fig. 1a & 1b) to fields within the lake/stream catchment (Catchment C, Fig. 1a & 1d). The Domesday Survey of 1086 records Groby as being well wooded in contrast to the rest of Leicestershire which it is estimated contained only 3% woodland in 1086 (Rackham, 1980). Until the mid-19th century the mixed oak woodlands located immediately around Groby Pool, in common with those throughout the rest of the county, were managed on a traditional basis which involved rotational felling of 'coppice with standards' (Rackham, *op. cit.*). Locally, there was very little loss of woodland during this period although at the sub-regional scale (i.e. within the common lands of Charnwood to the north of Groby) there appears to have been a gradual decline in woodland (Crocker, 1981). During the 18th and 19th century many small plantations

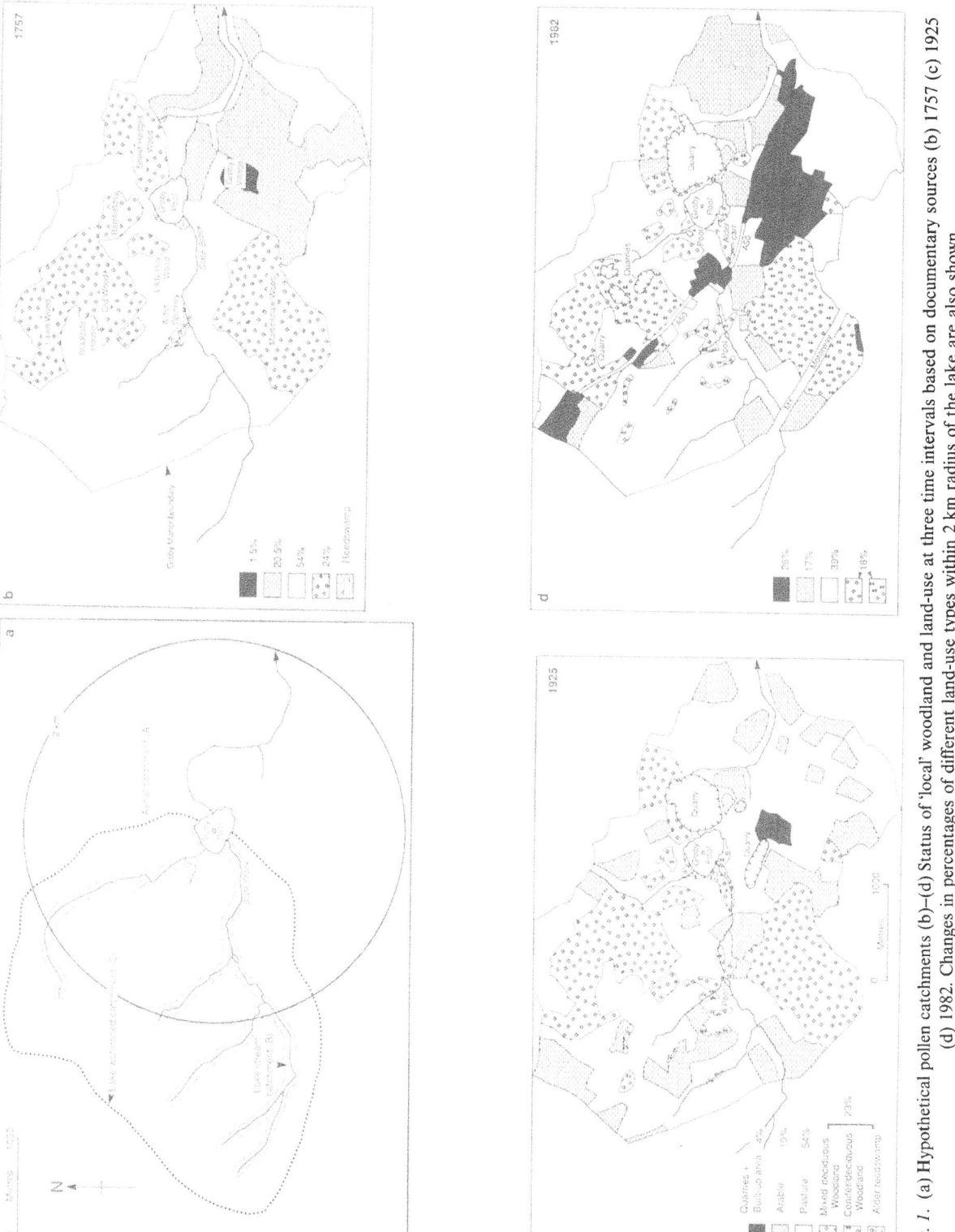

Fig. 1. (a) Hypothetical pollen catchments (b)–(d) Status of 'local' woodland and land-use at three time intervals based on documentary sources (b) 1757 (c) 1925 (d) 1982. Changes in percentages of different land-use types within 2 km radius of the lake are also shown.

308

were established on the larger estates within the sub-regional landscape. Management of local woods ceased in the mid-19th century. During the 20th century there has been large-scale deforestation of the native oak/ash woodlands. Around 60% of local woodlands were lost between 1925 and 1948. Much of the sub-regional woodland was also lost at this time to be replaced in the 1950's by pine plantations introduced by the Forestry Commission (Crocker, *op. cit.*). Locally, Martinshaw Wood was replanted in the 1950's and '60's, predominantly with *Pinus* and *Quercus* but also with *Thuja*, *Tsuga*, *Fagus* and *Betula*. An extensive area of woodland to the north of the lake (Old Wood), which had been cleared for quarrying after 1926, has been colonised by stands of *Betula*.

In 1895 the Board of Agriculture estimated the amount of woodland within Leicestershire to be 2.97% of land area, a figure which appears to have changed little since the Domesday Survey of 1086, and is very similar to present day estimates. The amount of sub-regional woodland, estimated to be between 5–7% in 1895, has also changed very little in the intervening period (Loughborough Naturalists' Trust survey Crocker 1981). Locally, woodland has declined from 23% in 1895 to 18% at the present day. Although there has been little change in the overall amount of woodland at the regional scale since 1086, its composition has changed from native mixed oak woodland to a landscape largely dominated by conifer woods.

Lake sediment-based pollen record
A number of 1 metre Mackereth cores have been retrieved from littoral and central areas of the lake and two of the central cores analysed for pollen using standard extraction techniques (Moore & Webb, 1978). A percentage pollen diagram has been prepared for Core 1, showing trends in the behaviour of selected taxa (Fig. 2). Pollen curves of all the major taxa recorded in the two diagrams are almost identical suggesting that the stratigraphic integrity of the sediments in this part of the lake has not been disturbed. In the following discussion the curves for arboreal pollen (AP) and

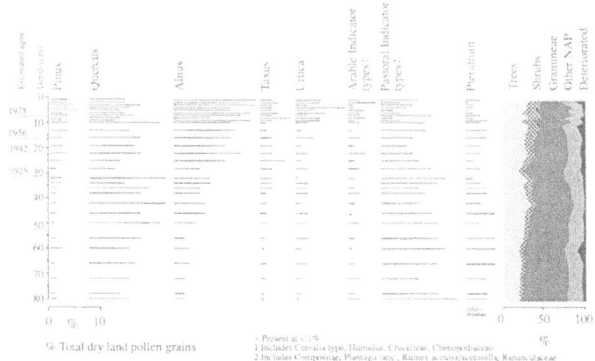

Fig. 2. Groby Pool, percentage pollen diagram, selected taxa.

non-arboreal pollen (NAP) will be treated separately as these two categories reflect the history of woodland over non-woodland areas respectively. Preliminary radiometric dates using Lead-210 are available for the core (N. Richardson, pers. comm.) and these have been used, in conjunction with a curve of carbonaceous particles, to provide a provisional chronology from which to undertake the comparative analysis.

Comparative analysis

Comparison of the pollen data with the historical record indicates a good overall correspondence between arboreal pollen curves and documented woodland history at the local and sub-regional scale. In the following discussion the behaviour of the pollen curves of *Alnus*, *Pinus* and *Quercus* is used to highlight recruitment of arboreal taxa. These trees were chosen because pollen production characteristics are similar (Andersen, 1973) and therefore it is assumed that relative differences between these taxa are primarily the result of factors other than these. *Alnus* and *Quercus* also constitute the dominant AP types recorded in the diagram.

Alnus: Prior to growth of the alder carr, the main source of *Alnus* pollen would appear to have been located in the upstream section of Slate Brook in the area known as Alder Spinney, c.

1.5 km from the lake (Fig. 1-b) and from *Alnus* growing along stream margins. The rise in *Alnus* pollen above 25 cm matches the development of the alder carr within 500 metres of the Pool since the 1930's. *Alnus* is not a common tree at the sub-regional or regional scale, being confined to streamside and riverside habitats. A high percentage of present day *Alnus* pollen input is likely to originate from the local alder carr and lake marginal trees (Fig. 1-d).

Pinus: *Pinus* is recorded at fairly low levels throughout the diagram ($<3\%$), increasing slightly from c. 36 cm upwards. There has only been a slight increase in the frequency of *Pinus* in the top of the profile to coincide with planting, since 1950, of *Pinus* within Martinshaw Wood and within the sub-regional woodland. *Pinus* pollen is known to disperse over long distances and has been found in surface samples >30 km distance from the nearest pollen source (Bradshaw & Webb, 1985). The behaviour of *Pinus* appears to indicate that the contribution of regional AP is not significant at this site. Further evidence for this may come from the behaviour of *Taxus* in the diagram. This tree constitutes only a very minor element of the sub-regional and regional flora although it is relatively abundant within 500 metres of the lake. It is recorded consistently throughout the diagram at around the same values as *Pinus*, which is locally and sub-regionally a much more abundant tree species. Further, if meteorological factors are biasing recruitment both taxa should suffer in the same way as they are not spatially distinct within the local landscape.

Quercus: The trend in *Quercus* values closely mirrors the history of mixed oak woodland both at the local and sub-regional scale. From relatively low values at the base of the diagram values rise temporarily between 32–50 cm, possibly in response to the cessation of woodland management in the mid to late 19th century. Extensive clearance after 1925 is reflected by a fall in *Quercus* values at c. 30 cm followed by a recovery to former levels. Given the similar trends in the history of native woodland at the local and sub-regional scale the origin of the *Quercus* pollen is

difficult to establish precisely. However, if *Quercus* pollen were to have originated from sub-regional woodland prior to its decline, one would expect *Pinus* pollen values to have increased as pine replaced oak within the landscape, particularly as *Pinus* produces abundant and well dispersed pollen. As this has not occurred, it suggests that most of the *Quercus* pollen originated from areas relatively close to the lake. The rise in *Quercus* values after 1940, which also coincides with regeneration of *Quercus* within the 'local' landscape, would seem to support such an interpretation.

Non-Arboreal Pollen

The 'pastoral' indicator types (Behre, 1981) reach relatively high values in the base and middle sections of the diagram (30–80 cm). *Pteridium* values follow a similar trend. Independent dates are needed but it is possible that the lower section of the diagram below c. 50 cm covers the period when extensive arable cultivation was carried out within the aerial catchment (Catchment A) but not within Catchments B & C (Fig. 1a). *Plantago* spp. and *Pteridium*, taxa which are thought to indicate pasture and 'open' conditions, are both anemophilous high pollen producers and therefore more likely to be represented than taxa included in the 'arable' category. However, the 'pastoral' group and *Pteridium* decline from c. 30 cm upwards when pasture increases in the former arable areas. Their decline coincides with the growth of streamside and lakeside trees and the increase in *Alnus* pollen. Both of Groby's streams flow through meadowland before reaching the lake. If it is assumed that the pastoral NAP types were derived from catchments 'B' and 'C' rather than catchment 'A' their decline could be explained by the growth in streamside and, later, by lakeside trees and shrubs. The role of vegetation in filtering airborne pollen influx to lakes and bogs is well documented (Tauber, 1967). Values of *Urtica dioica*, which grows abundantly along the banks of Slate Brook to-day, increase at the top of the diagram also suggesting a streamside origin for this taxon. Gramineae pollen is the single dominant taxa present throughout the diagram. Values decline from c. 50% TDLP in the lower section of the profile to c. 35% in the upper sediments although absolute influx figures are needed before this trend can be confirmed. This decline appears to coincide with the growth in the alder carr around the lake although it may also be attributable to a reduction in pollen production of this taxa at all spatial scales. However, some of the Gramineae pollen will be of 'local' origin as the reedbed areas must be contributing a proportion to the total input. Although not conclusive, trends in some of the NAP curves suggest an origin within catchments 'B' and 'C', possibly indicating spatial bias towards recruitment from local and stream marginal vegetation.

Conclusion

This study offers an historical perspective on the problem of pollen recruitment to lakes receiving inflowing streams. The evidence from historical and palaeoecological sources suggest total AP/NAP recruitment to Groby Pool to be predominantly of local origin (i.e. from within c. 2 km of the lake), with sub-regional and regional sources contributing a much smaller percentage of total pollen input. This conclusion contrasts with models of pollen recruitment to similar sized lakes without inflow streams, where a significant regional component is predicted (Jacobson & Bradshaw, 1981, Prentice, 1985). Some of the problems of recruitment and pollen sourcing may be clarified when the results of a monitoring programme are available, particularly in relation to the magnitude and composition of the stream-borne component.

Acknowledgements

This research was made possible by grants from the University of Technology, Loughborough. Thanks are due to Dr. Tony Brown of Leicester for advice and for the loan of equipment, to Vernon Poulter for assistance with field and laboratory work and to an anonymous reviewer for comments.

References

Andersen, S. T., 1970. The relative pollen productivity and pollen representation of north European, and correction factors for tree pollen spectra, Dan. Geol. Unders. 2,96: 1–99.

Andersen, S. T., 1973. The differential pollen productivity of trees and its significance for the interpretation of a pollen diagram from a forested region. In Quat. Plant Ecol. H. J. B. Birks & R. G. West (eds), Oxford: 109–115.

Behre, K. E., 1981. The interpretation of anthropogenic indicators in pollen diagrams, Pol. Spor. 23: 225–245.

Berglund, B. E., 1973. Pollen dispersal and deposition in an area of southeastern Sweden – some preliminary results. In Quat. Plant Ecol. H. J. B. Birks & R. G. West (eds), Oxford: 117–129.

Bonny, A. P., 1978. The effect of pollen recruitment processes on pollen distribution over the sediment surface of a small lake in Cumbria, J. Ecol. 66: 385–416.

Bradshaw, R. H. W., 1981. Modern pollen representation factors for woods in south-east England, J. Ecol. 69: 45–70.

Bradshaw, R. H. W. & Webb, Thompson, III., 1985. Relationships between contemporary pollen and vegetation data from Wisconsin and Michigan, USA, Ecology 66: 721–737.

Crocker, J., 1981. Charnwood Forest: A Changing Landscape. Loughborough Naturalist Trust Davis, M. B. 1973. Redeposition of pollen grains in lake sediments, Limnol. Oceanogr. 18: 44–52.

Davis, M. B. & L. B. Brubacker, 1973. Differential sedimentation of pollen grains in lakes, Limnol. Oceanogr. 18: 635–646.

Doharty, John, 1757. Pre-enclosure map of the Manor of Groby, Leicester Records Office PP443.

Hoskins, W. G., 1949. The Leicestershire Crop returns of 1801, Trans. Leics. Arch. Sc. 24: 127–153.

Jacobson, G. L. & R. H. W. Bradshaw, 1981. The selection of sites for palaeovegetation studies, Quat. Res. 16: 80–96.

Moore, P. D. & J. A. Webb, 1978. An illustrated guide to Pollen Analysis, London.

Peck, R. M., 1973. Pollen budget studies in a small Yorkshire catchment. In Quat. Plant Ecol. H. J. B. Birks & R. G. West (eds), Oxford: 43–60.

Pennington, W., 1979. The origin of pollen in lake sediments: an enclosed lake compared with one receiving inflow streams, New Phytol. 83: 189–213.

Prentice, I. C., 1985. Pollen Representation, Source Area and Basin Size: Toward a Unified Theory of Pollen Analysis, Quat. Res. 23: 76–86.

Rackham, J., 1980. Ancient Woodland: Its History, Vegetation and Uses in England. Edward Arnold.

Tauber, H., 1965. Differential pollen dispersion and the interpretation of pollen diagrams. Dans. Geol. Unders., II. 89: 1–69.

Tauber, H., 1967. Investigations of the mode of pollen transfer in forested areas, Rev. Palaeobot. Palynol. 3: 277–286.

Tauber, H., 1974. A static non-overload pollen collector, New Phytol. 73: 359–69.

Woodward, S., 1984. 'The landscape of a Leicestershire Parish', Leicester Museum Pub. 58.

Hydrobiologia **214**: 311–316, 1990.
J. P. Smith, P. G. Appleby, R. W. Battarbee, J. A. Dearing, R. Flower, E. Y. Haworth, F. Oldfield & P. E. O'Sullivan (eds). 311
Environmental History and Palaeolimnology.
© 1990 *Kluwer Academic Publishers.*

Seasonal changes in sedimenting material collected by high aspect ratio sediment traps operated in a holomictic eutrophic lake

R. J. Flower

Palaeoecology Research Unit, Dept. of Geography, University College, London WC1H OAP, UK

Key words: sediment traps, diatoms, phytoplankton

Abstract

Cylindrical sediment traps with an aspect ratio (height (60 cm): diameter (5.1 cm)) of 11.8 were located 1 m above the surface sediment by a rigid metal framework support. Traps were exposed in Lough Neagh for one year, from May 1978. Each trap collected between 11 and 12 cm of faintly laminated sediment. One 12 cm sediment column was examined using conventional palaeolimnological techniques of core extrusion and analysis. The algal record in trapped sediment is shown to correspond with successional changes in phytoplankton abundance in the lake during the year of study. The sediment accumulation rate measured by the traps is an order of magnitude greater than that measured in dated sediment cores and redeposited and inwashed sediment formed the bulk of trapped material. However, the value of these high aspect ratio traps is that they provide a continuous but qualitative account of compositional changes in sedimenting material through time. Their potential as long term biological monitoring devices is emphasized.

Introduction

It is a prerequisite in palaeolimnology that temporal changes in the water column are adequately recorded in accumulating contemporary sediment. Limnological changes are linked with the sedimentary record by changes in the quality and quantity of particulates settling from the water column. This sedimenting material (SM) can be readily intercepted using sediment traps and the collection technique offers a beguilingly easy way to study processes which control SM production, composition, and deposition. However, problems concerning trap efficiencies, redeposited sediment and decomposition processes usually confound simple interpretation of trapping results.

Studies of sediment trap dynamics (Gardner,

1980a & 1980b) have shown that traps with high aspect ratios (height : diameter) prevent loss of trapped SM and such traps with a simple tube geometry are recommended for most trapping studies (Bloesch & Burns, 1980; Blomqvist & Hakanson, 1981). The simple tube design was developed by providing a mechanism for adding marker beads periodically to the trapped SM (Anderson, 1977) so allowing resolution of trapped SM into discrete time intervals without repeated sampling. This paper examines the potential of simple tube traps, exposed for long periods (one year), to provide a continuous record of temporal changes in SM in a eutrophic wind stressed lake where sediment resuspension can be expected frequently. Furthermore, the relevance of SM composition to seasonal changes in water column variables is assessed qualitatively by comparing phytoplankton standing crops with algae, principally diatoms, collected in the traps.

Site details and methods

All sediment trapping experiments were carried out in Lough Neagh, a large (383 km^{-2}) but relatively shallow (mean depth 8.9 m) lake, in Northern Ireland. The lake is eutrophic and is wind stressed through much of the year with only occasional and transient thermocline development (Wood & Gibson, 1973). According to predictive equations of sediment transport in lakes (Håkanson, 1981) most of the benthic environment of Lough Neagh is a transporting zone.

Simple tube traps were constructed from plexiglas or black PVC tubing, height 60 cm and diameter 5.1 cm, giving an aspect ratio (h : d) of 11.8. This high aspect ratio should prevent disturbance of trapped contents in current velocities up to about 35 cm sec^{-1} (from Lau, 1979) and Bloesch & Burns (1980) suggest that aspect ratios above 10 should prevent trap contents from resuspension in most lake environments. Traps were attached to a rigid supporting frame work (Flower, 1980), to avoid problems of trap tilt (Gardner, 1985), and so that taps were located 1 m above the substratum. Four traps and framework supports were positioned on the sediment of Lough Neagh at the Battery site (see Battarbee, 1978a) some 3 km from the nearest shore in 14.7 m depth of water and marked by small buoys. Traps were exposed for one year, between May 1978 and May 1979.

On retrieval traps were detached from their framework support and extruded for sectioning as in conventional palaeolimnological analysis of short sediment cores. Trap contents were sampled at 1 cm intervals and analyzed for organic matter (loss on ignition at 550 °C) and diatoms. Diatom analysis followed Battarbee (1986). Subsamples of fresh SM were also examined microscopically at low magnification.

The phytoplankton of Lough Neagh was monitored during the period of trap deployment by Dr

C. E. Gibson. Sampling was approximately every two weeks, algae were fixed in Lugol's iodine and counted using an inverted microscope and sedimentation chamber.

Results

All four traps collected a similar amount of SM following exposure for one year and 11–12 cm of SM was extruded from each tube for which the mean dry weight was 18.4 g (S.D. 0.38 g). Detailed analysis of the contents of one tube trap was carried out (Fig. 1). The trap contained 12 cm of sediment which, in the fresh state, appeared faintly laminated. SM was a greenish colour between 0–2 and 10–12 cm depth but the middle section was mostly dark brown to black (3–10 cm) and one section (2–3 cm) contained grey flecks. Organic matter was highest in the 0–2 and 10–12 cm sections where values exceeded 30%. Examination of the fresh material showed that remains of blue-green algal filaments were very abundant in these sections. In the dark coloured middle section organic matter content of SM was around 25%. The diatom concentration profile in SM shows considerable variation, values decline by 50% between 12 and 11 cm depth and increase 3 fold between 4 and 3 cm depth. The point at which total valve concentration increases coincides with a sharp decline in abundance of periphytic diatoms and a major increase in the proportion of unbroken *Stephanodiscus* valves. The ratio of periphytic to planktonic valves is lowest in the upper and lower SM sections and exceeds 0.1 in the middle section.

The percentage frequency profiles of diatom taxa in the tube trap SM column (Fig. 2) show that *Melosira italica* ssp. *subarctica* (= *Aulacoseira subarctica* Howarth, 1988) dominates throughout. Frequencies of this taxon drop below 50% only in the 12–10 cm and 2–1 cm sections. *Stephanodiscus minutula* declines from > 20% in the bottom sample to < 10% at 9 cm depth and shows no consistent variation above this depth as does *S. astraea* (= *S. rotula*, Håkansson & Locker, 1981) throughout the profile. *S. minutula*

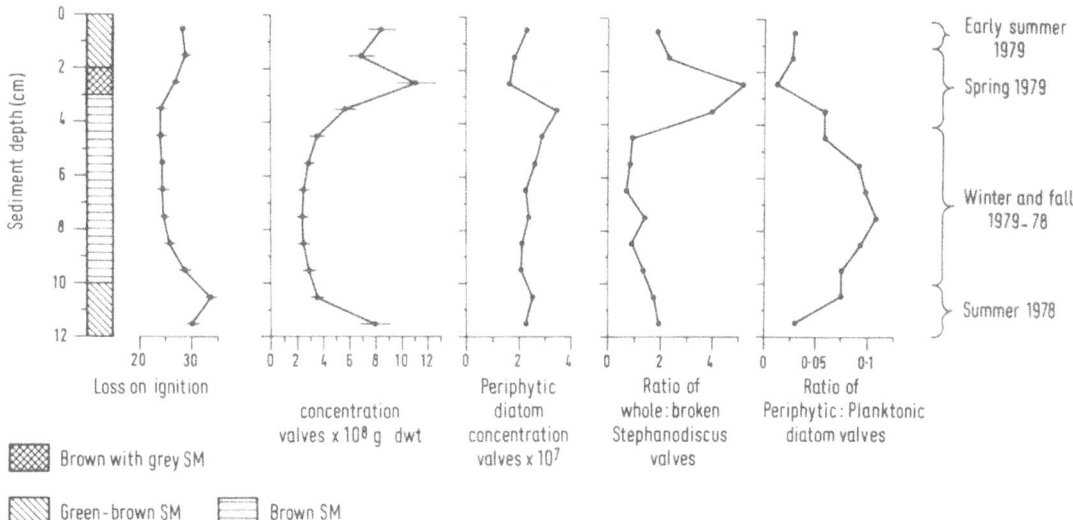

Fig. 1. Sequential analysis of a 12 cm deep sediment column collected by a high aspect ratio sediment trap deployed in Lough Neagh, May 1978 to May 1979. The visual stratigraphy is indicated left and profiles for percent loss on ignition, total and periphytic diatom concentrations, ratio of whole to broken *Stephanodiscus* valves, and ratio of periphytic to planktonic valves are shown right. The sediment column is divided into seasonal depositional periods according to biostratigraphy and phytoplankton succession (see text).

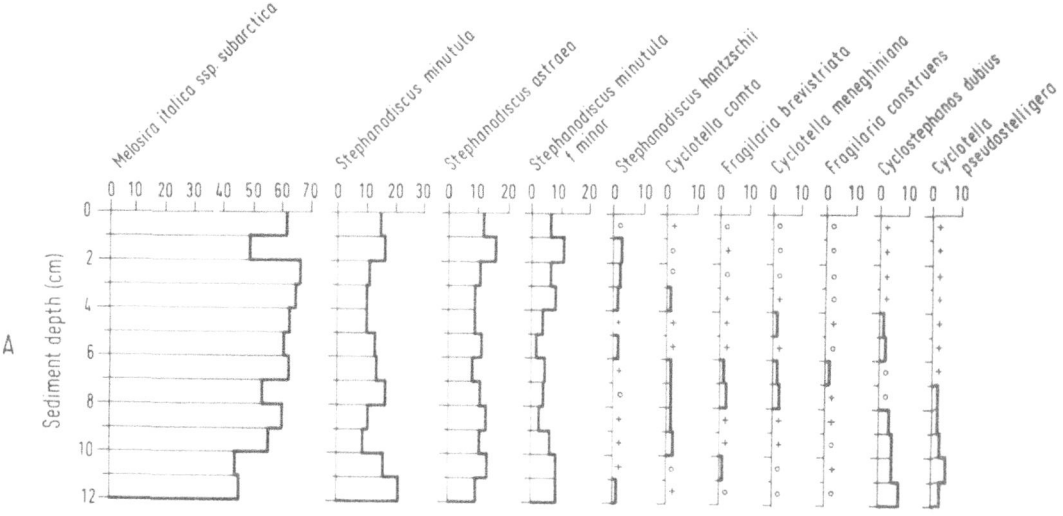

Fig. 2. Percentage frequency profiles for the common and mainly planktonic diatoms within the 12 cm deep sediment column described in Fig. 1. + indicates abundances <2% and 0 indicates absence.

fo. *minor* is more common in both the upper and lower 2 cm sections and *Cyclostephanos dubius* and *Cyclotella pseudostelligera* achieve highest frequencies in the lowest section. *C. comta* (= *C. radiosa*, Håkansson, 1988) and *C. mene-*

ghiniana are only present in significant frequencies in the middle section of the profile.

Changes in the phytoplankton of Lough Neagh during the trapping period are shown in Fig. 3. The period starts in May 1978 sometime after the

314

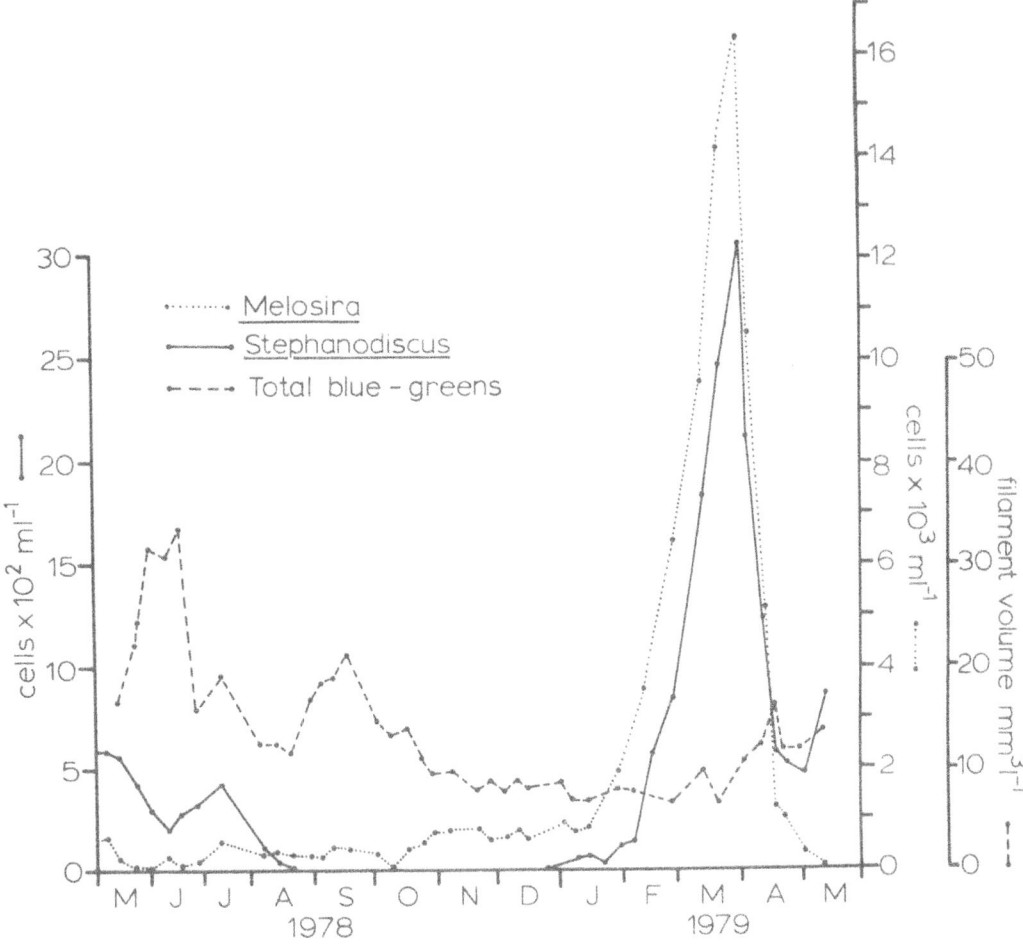

Fig. 3. Seasonal changes in phytoplankton abundance in Lough Neagh, May 1978 to May 1979. Note that diatom abundances are as numbers per ml and blue-green algae are as filament volume per L. Abundance data are from Dr C. E. Gibson.

main planktonic diatom crops of that year with a declining *Stephanodiscus astraea* population whilst *Melosira italica* ssp. *subarctica* was already at very low abundance. The diatom phytoplankton remained at low levels throughout the rest of 1978 but in late January 1979 the *Melosira* population began to bloom and was closely followed by *Stephanodiscus astraea*. Both populations peaked in late March before rapidly declining in the following month. The main blue-green crop developed in May-June, declining in late June and, after a second minor peak, declined further until growth accelerated in April the following year.

Discussion

Analysis of the sediment column accumulated over one year in a tube trap revealed a clear stratigraphy that was related to successional changes in phytoplankton of Lough Neagh during that year (May 1978 to May 1979). SM collected initially in the trap was high in organic matter and possessed abundant remains of blue-green algal filaments. These observations are consistent with the predominant source of this SM being from the decline of the June crop of blue-green algae. The fairly high number of diatoms in this material reflects an abundance of small centric forms which were not enumerated in routine phytoplankton counting but are often common in the early summer following crash of the main diatom crops (C. E. Gibson, pers. comm.). These forms however are well represented at the base of the diatom frequency profiles and the 10–12 cm section of SM are taken to represent material collected during the summer months (May–August) of 1978.

The bulk of SM collected (between about 10 and 4 cm depth) is fairly uniform in diatom composition but periphytic taxa usually exceed 10% of the total diatom assemblage. The organic content of this material is close to that of the natural surface sediment (Flower, 1980) and this section probably results mainly from redeposited sediment collected in the trap following resuspension of the natural surface sediment by frequent winter storms. This interpretation is supported by trapping experiments conducted at two-weekly intervals in the lake (Flower, 1991). Hence the middle 6 cm of SM was probably all collected between about September 1978 and mid-January 1979. The significant abundances of *C. comta* and *C. meneghiniana* in this section could indicate an influence of inflow river phytoplankton on SM quality during this period.

The sharp increases in total diatom valve concentration and in the proportion of unbroken *Stephanodiscus* cells combined with a decline in periphytic diatoms, all beginning c. 4 cm depth in the SM column, strongly suggest that material collected above this depth results mainly from settling of the 1979 spring diatom crop, beginning in late January 1979 as the crop began to increase. In the greyish layer (2–3 cm depth) total diatom concentration exceeds 10^{-9} valves g^{-1} SM and must reflect the diatom phytoplankton crop crash in late March 1979. The marked decline in diatom abundance and increasing quantity of blue-green algal remains and of organic content in the top 2 cm of SM clearly correspond with the post mid-April period when diatom phytoplankton populations were low but the blue-green crop was increasing. The approximate seasonal chronology established for the sequential changes in trap contents based principally on the timing of phytoplankton standing crops is indicated on Fig. 1.

Although close links exist between the composition of trapped SM and the contemporary phytoplankton crops, the gross quantities of SM constituents collected have little to do with annual rates of sediment accumulation or primary production within the lake. This is because sediment resuspension plays a particularly important role in controlling suspended solid concentrations in Lough Neagh (Flower, 1980) and hence the annual total downwards flux of particulates is much greater than the accumulation rate of 0.8–1.0 cm yr^{-1} measured in cores of natural sediment from the lake (Battarbee, 1978b). Periphytic diatoms derived from marginal habitats and degraded planktonic diatoms are present throughout the profiles indicating that SM is always a mixture of inwashed and resuspended sediment as well as of primary material. Interestingly, due to the similarity of diatom taxa frequencies in surface sediment and in spring crops in Lough Neagh there is little difference between the frequency composition of mainly redeposited winter SM and SM collected during diatom blooms periods (Fig. 2). The intimate mix of SM resulting from all particulate sources in Lough Neagh make use of traps for direct quantitative measurement of, for example, algal crop loss rate or net sediment accumulation rate impossible. Even when corrections are made for resuspended and inwashed components of trapped material (e.g. Gasith, 1975) resulting estimates of net fluxes will remain imprecise and unreliable.

Clearly, the main value of collecting SM in tube traps is to provide a qualitative and essentially continuous record of change in water column particulates, particularly of diatoms, through time. Long exposure periods provide enough material to be sampled using conventional palaeolimnological techniques and resulting intraannual time control provided by diatom analysis permits seasonality changes in other SM characterises such as ^{210}Pb and magnetic minerals to be readily studied. Furthermore, despite 'contamination' of primary material, these traps can provide a within year record of environmental change which is lost in the naturally lake accumulating sediment by resuspension and bioturbation processes. In other words, because SM collected in these traps is protected from resuspension effects, a relatively fine resolution stratigraphy is imposed on the trapped material. Although these traps have been evaluated in a holomictic eutrophic lake where particulate fluxes are high they are suitable for deployment in any lake type and even in flowing waters. Remote sites where regular sampling of algae is not possible these traps, possibly equipped with SM amplification devices (Anderson, 1977), offer an inexpensive and simple means of monitoring biological change without recourse to repeat coring strategies.

Acknowledgements

I am indebted to Dr C. E. Gibson for providing the 1978/79 phytoplankton data for Lough Neagh and for advice on sampling procedures. Mr K. Gilmore helped with field work and Prof. R. B. Wood freely made available laboratory facilities at the University of Ulster (Freshwater Laboratory). Dr. R. W. Battarbee initiated sediment trapping studies on Lough Neagh and kindly made constructive comments on the manuscript. The work was partly funded by the Natural Environmental Research Council and the Department of the Environment, UK.

References

Anderson, R. Y., 1977. Short term sedimentation response in lakes in Western United States as measured by automated sampling. Limnol. Oceanogr. 22: 423–433.

Battarbee, R. W., 1978a. Observations on the recent history of Lough Neagh and its drainage basin. Phil. Trans. R. Soc. Lond., B, 281: 303–345.

Battarbee, R. W., 1978b. Biostratigraphical evidence for variations in the recent pattern of sediment accumulation in Lough Neagh, Northern Ireland. Verh. Int. Ver. Limnol. 20: 624–629.

Battarbee, R. W., 1986. Diatom analysis. Handbook of Holocene Palaeoecology and Palaeohydrology B. E. Berglund (ed.). Wiley & Sons. Chichester: 527–570.

Bloesch, J. & N. M. Burns, 1980. A critical review of sedimentation trap technique. Schweiz. Z. Hydrol. 42: 15–55.

Blomqvist, S. & L. Hakanson, 1981. A review of sediment traps in aquatic environments. Arch. Hydrobiol. 91: 101–132.

Flower, R. J., 1980. A study of sediment formation, transport and deposition in Lough Neagh, Northern Ireland, with special reference to diatoms. Unpub. D. Phil. thesis. University of Ulster (Coleraine). 214 pp.

Flower, R. J., 1991. Field evaluation and calibration of sediment trap performance in an entrophic holomictic lake. J. Paleolimnol. 4: in press.

Gardner, W. D., 1980a. Sediment trap dynamics and calibration: a laboratory evaluation. J. Mar. Res. 38: 17–39.

Gardner, W. D., 1980b. Field assessment of sediment traps. J. Mar. Res. 38: 41–52.

Gardner, W. D., 1985. The effect of tilt on sediment trap efficiency. Deep-Sea Res. 32: 349–361.

Gasith, A., 1976. Seston dynamics and tripton sedimentation in the pelagic zone of a shallow eutrophic lake. Hydrobiologia 51: 225–231.

Håkanson, L., 1981. On lake bottom dynamics – the energytopography factor. Can. J. Earth Sci. 18: 899–909.

Håkanson, H., 1988. A study of species belonging to the *Cyclotella bodanica/comta* complex (Bacillariophyceae). In Proceedings of the 9th Diatom Symposium 1986 F. E. Round (ed.). Biopress Ltd., Bristol.

Håkansson, H. & S. Locker, 1981. *Stephanodiscus* Ehrenberg 1846, a revision of the species described by Ehrenberg. Nov. Hedw. 35: 117–150.

Howarth, E. Y., 1988. Distribution of diatom taxa of the old genus *Melosira* (now mainly *Aulacoseira*) in Cumbrian waters. In Algae and the aquatic environment F. E. Round (ed.). Biopress Ltd., Bristol.

Lau, Y. L., 1979. Laboratory study of cylindrical sedimentation traps. J. Fish. Res. Bd Can. 36: 1288–1291.

Wood, R. B. & C. E. Gibson, 1973. Eutrophication and Lough Neagh. Wat. Res. 17: 173–187.

Hydrobiologia **214**: 317–325, 1991.
J. P. Smith, P. G. Appleby, R. W. Battarbee, J. A. Dearing, R. Flower, E. Y. Haworth, F. Oldfield & P. E. O'Sullivan (eds), 317
Environmental History and Palaeolimnology.
© 1991 *Kluwer Academic Publishers.*

Paleolimnology of a polar oasis, Truelove Lowland, Devon Island, N.W.T., Canada

R.H. King
Department of Geography, The University of Western Ontario, London, Ontario, N6A 5C2, Canada

Key words: glacio-isostasy, coastal emergence, sediments, arctic lakes, environmental change

Abstract

Following a marine transgression which inundated the Truelove Lowland up to an elevation of 86 m by approximately 9700 years BP, lake development was initiated by glacio-isostatic rebound which isolated marine lagoons behind a series of raised beaches. The ages and elevations of whalebones and driftwood contained in these beaches permit the reconstruction of the progressive emergence of the Lowland from the sea and the timing of lake isolation. Further details of the environmental changes experienced by the lakes and their catchments during the Holocene are recorded in the chemistry of the lake sediments.

Introduction

The Truelove Lowland (75° 33′ N, 84° 40′ W) is one of a series of lowlands located on the northeastern coast of Devon Island, N.W.T. (Fig. 1). Possessing a distinctive local climate and underlain by continuous permafrost, the Lowland today constitutes an isolated area or oasis of relatively high biological diversity in the midst of the more typical Polar Desert of the Canadian High Arctic.

Because of its ecological importance Truelove Lowland was chosen as one of four major ecosystem studies within the Tundra Biome component of the International Biological Program (IBP) and the only one out of fourteen major arctic projects that was conducted within the High Arctic (Bliss, 1977). Based on field work undertaken over a four year period, 1970–74, the IBP project included among its main objectives the development of static and dynamic models of the ecosystem function. As a result of this research a considerable body of information was made available on the relatively short-term characteristics and performance of this ecosystem. However, the four years of environmental and ecological record obtained for the Truelove Lowland by the IBP project provides only a small picture of changes, both environmental and ecological, experienced by the Lowland since it has been in existence. What is presently lacking is information on the changes experienced by the Lowland over the relatively longer term of the postglacial period. The presence of numerous lakes in the Lowland offers the possibility of a continuous paleo-environmental record for much of the postglacial period preserved in the lake sediments and associated deposits. This paper examines the nature of this record.

Study area

Truelove Lowland is a strand-flat or emerged platform mantled with a complex of Quaternary beach and lacustrine deposits underlain by Precambrian high grade metamorphic rocks dominated by granitic gneiss. In places, the Pre-

318

Fig. 1. Location of the study area.

cambrian basement complex is unconformably overlain by carbonate and clastic rocks of Lower to Middle Cambrian age (King, 1969). These latter formations form a conspicuous 300 m high escarpment to the east of the Lowland.

In general, the Lowland represents the undulating and exhumed surface of the Precambrian/Cambrian unconformity with a relief ranging from approximately 10 to 20 m. Local relief is provided by a series of raised marine beaches rising in elevation inland from the present-day coastline and by bedrock highs in the Precambrian basement. Wolf Hill, the highest relief feature in the Lowland with an elevation of approximately 64 m a.s.l., is of uncertain provenance but may represent a complex terminal moraine reworked by marine processes and deposited by an outlet glacier that formerly occupied the Truelove Valley to the southeast of the Lowland.

The present vegetation and soils in the Lowland comprise a complex mosaic of cushion plant-lichen and sedge-moss and grass-moss communities (Bliss, 1977) underlain by Organo, Turbic and Static Cryosols (Canada Soil Survey Committee, 1978). Cushion plant-moss/lichen communities tend to dominate the more xeric sites such as the raised beach and the sedge moss/grass-moss communities dominate the more poorly drained sites surrounding the numerous lakes in the Lowland. The lakes, with a mean depth of 3 m, presently cover 22% of the Lowland.

In appearance and vegetational characteristics the Lowland closely resembles the Low Arctic of the arctic coastal plain of Alaska and areas of the Mackenzie Delta, and provides an adequate food base for muskox, lemming, waterfowl, and associated predators. Part of the reason for its relatively high biological diversity compared with surrounding areas of Polar Desert appears to be the thermal reservoir function of its numerous lakes (Courtin & Labine, 1977).

Methods

The heights of all the major raised beaches in the Lowland were determined using a Topcon level and a 4 m stadia rod. Levelling traverses were closed wherever possible and most beaches were included in more than one traverse. The survey datum point used was the extreme high water mark, as indicated by the upper line of flotsum and all levelled heights are related to this with an accuracy of ± 0.5 m. Samples of whalebone and driftwood partially buried in the raised beaches were dated by the radiocarbon laboratories at the University of Saskatchewan, the Geological Survey of Canada and Beta Analytic. In addition, basal sediment core samples from Phalarope Lake and Fish Lake were dated by accelerated mass spectrometry (AMS) at the Isotrace Laboratory, University of Toronto.

Sediment cores were collected from a number of the larger, deeper lakes in the Lowland using a modified Livingstone piston corer with an internal diameter of 5.75 cm. Cores were initially extruded in 20–24 cm sections in the field and then subsampled in 2 cm sections. Core samples were subsequently analyzed in the following ways. Chemical compositional data was obtained using Instrument Neutron Activation Analysis (INAA). In particular, trace elements including As, Cr, Mo, U, and selected macro elements such as Fe (Fe_t) were determined. In addition, total organic carbon was determined using the modified Walkley-Black technique (Nelson & Summers, 1982) and biogenic silica was determined colorimetrically (after DeMaster, 1981) using a Pye-Unicam model 5620 visible spectrophotometer. Fe and Mn (Fe_d and Mn_d) were extracted using a citrate-bicarbonate-dithionite solution (Mehra & Jackson, 1960) and determined by atomic absorption spectrophometry. Diatoms were mounted for microscopic identification in Hyrax (R.I. = 1.65) following sample preparation that included acidification with 10% HCl and peroxidation followed by extraction with concentrated H_2SO_4 and $K_2Cr_2O_7$ to remove organic matter. Diatoms were identified with reference to standard keys (Cleve-Euler, 1951–55; Germain,

1981; Peragallo & Peragallo, 1987–1908; Lichti-Federovich, 1983; Foged, 1972–74, 1981; and Patrick & Freese, 1961).

Results and discussion

The age of the emerged platform that forms the Lowland is uncertain, however it obviously predates the present mantle of deposits. Included in the raised beach deposits is a range of organic materials, including whalebones, marine shells, driftwood and seaweed. The oldest of this material, shells found at an elevation of 36 m on Wolf Hill, has been dated at approximately 30 000 years BP (sample Y-1733; Barr, 1971). Consequently, the possibility exists that the origin of the platform may date back to at least the last interstadial in the Canadian High Arctic. The platform most likely originated as a rock shelf produced by marine abrasion along a structural line of weakness provided by the Precambrian/Cambrian unconformity and associated with eustatic oscillations in sea level during the Quaternary.

Approximately 10 600 years BP the Lowland was inundated by a postglacial marine transgression which covered the Lowland up to an elevation of approximately 86 m a.s.l. The age of the marine transgression is indicated by two dates of 10 570 ± 200 years BP (TO-566) and 10 620 ± 160 years BP (TO-564) provided by basal sediments containing marine diatom floras in Fish Lake and Phalarope Lake respectively.

It is most likely that the marine transgression was a result of a general eustatic rise in sea level following deglaciation and that the altitudinal extent of the transgression was influenced by the close proximity of a diminishing Devon Ice Cap which continued to restrain glacio-isostatic crustal rebound for some time. The maximum elevation reached by the marine transgression, the so-called marine limit, can be identified by the presence of a well-developed wave-cut notch with associated deposits of well-rounded beach material at the mouth of the Truelove Valley. Previously identified as having a field elevation of 76 m a.s.l. (King, 1969; Barr, 1971), a subsequent re-survey in 1986 placed it 10 m higher at 86 m. This elevation has since been confirmed as being of regional significance with the identification of a marine limit at the same elevation at several other locations along the coast to the west of the Truelove Lowland. Although no dateable material has been found associated with the marine limit, the basal marine deposits in the larger Lowland lakes provide an indication of its age.

Progressive postglacial isostatic uplift of the Lowland subsequent to the marine transgression has resulted in the emergence of the Lowland from the marine environment and the formation of a series of raised beaches that today form much of the topography in the Lowland. Using the ages of organic materials included in the beaches, together with the elevation of the specific beaches above the present sea level it is possible to reconstruct the history of coastal emergence. King (1969) and Barr (1971) have previously attempted to reconstruct coastal emergence curves for the Lowland based on dates provided by an assortment of organic materials, but largely whalebone. Two additional whalebone samples found buried in beach materials at elevations of 21.80 m and 68.81 m have provided dates of 6790 ± 90 years BP (Beta-19201) and 8 920 ± 140 years BP (Beta-14819), respectively (Kelly, 1989). When combined with three whalebone dates and a date for a sample of driftwood previously reported by Barr (1971) the postglacial coastal emergence can be reconstructed (Fig. 2). Details of the dated samples used in the reconstruction of the postglacial emergence are included in Table 1. A linear regression of the ages and field elevations of these samples yielded the following equation:

$$y = 376.89 + 4\,810.96\,(\log_{10} x)$$
$$\text{where } x = \text{field elevation (m a.s.l.)}$$
$$y = \text{age } (^{14}\text{C years BP})$$
$$r = .96598,\ p = .002;\ r^2 = 93.31\%$$

The general shape of the postglacial emergence curve (Fig. 2) is very similar to that of other postglacial emergence curves for Arctic Canada (Andrews, 1968; Blake, 1975; England, 1976)

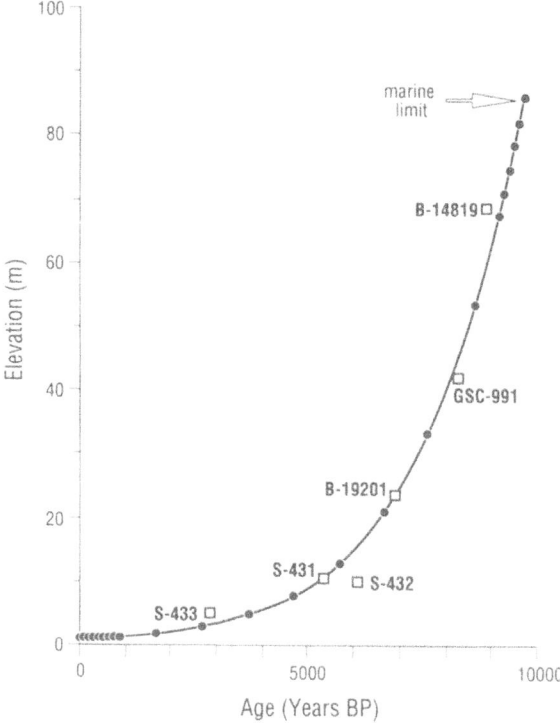

Fig. 2. Postglacial emergence curve for the Truelove Lowland.

and clearly attests to the similarity of the glacio-isostatic rebound process.

When extrapolated from the emergence curve, the predicted age of the marine limit at 86 m is approximately 9 700 years BP. When compared to the dates obtained for the basal marine deposits in the larger lakes in the Lowland this date is approximately 1 000 years younger. This suggests

Table 1. Radiocarbon dates of organic material related to the emergence of the Truelove Lowland.

Sample elevation (to nearest 0.5 m)	Material	Laboratory number	^{14}C (before 1950)
4.7	whalebone	S-433[1]	2900 ± 85
9.7	whalebone	S-433[1]	6100 ± 125
10.5	driftwood	S-431[1]	5280 ± 100
21.8	whalebone	B-1920[1]	6790 ± 90
42.4	whalebone	GSC-991[1]	8270 ± 150
68.6	whalebone	B-14819	8920 ± 140

[1] Data from Barr (1971)

that the marine limit is not synchronous with deglaciation, as suggested by Andrews (1968) for much of arctic Canada, and that some parts of the Lowland were covered by the postglacial sea much earlier than others.

Interpolated rates derived from the emergence curve reveal that a considerable proportion of the emergence (32.7 m or 38% of the total emergence) occurred within the first 1000 years. During the first 100 years the rate of emergence was approximately 4 meters per century, decreasing to approximately 3 meters per century within 1 000 years and to less than one meter per century after 4 000 years. The emergence has since continued at a smoothly decelerating rate, with approximately three quarters (76%) of the emergence taking place within the first 3 000 years. With the aid of the linear regression model it is possible to reconstruct the progressive emergence of the Lowland using isochrones or contours of equal age which provide a map of the shoreline displacement over time (Fig. 3). In this figure the shorelines are represented by the raised beaches and the heights of the beaches are used to predict their ages, assuming that the age derived from the elevation of the beach crest is equivalent to the time when that particular beach formed. The data are then contoured to provide a map of the relative age of the Lowland. Alternatively, the isochrones may be interpreted as representing the shoreline position at a specific time.

During the emergence of the Lowland from the sea a series of coastal marine lagoons were isolated by the emerging submarine bars. Although initially containing saltwater, these lakes have since been decanted and flushed with freshwater derived from snow and ice melt. Today, the larger lakes such as Fish Lake (21.00 m a.s.l.; 1.2 km²), Phalarope Lake (4.53 m a.s.l.; 1.9 km²) and Immerk Lake (13.71 m a.s.l.; 1.1 km²), have a maximum water depth of 9 m and are slightly alkaline and oligotrophic. Under contemporary conditions ice begins to form on these lakes in early September and lasts into late July or early August. In colder years, a central ice pan persists throughout the summer. Maximum lake ice thickness appears to be approximately

322

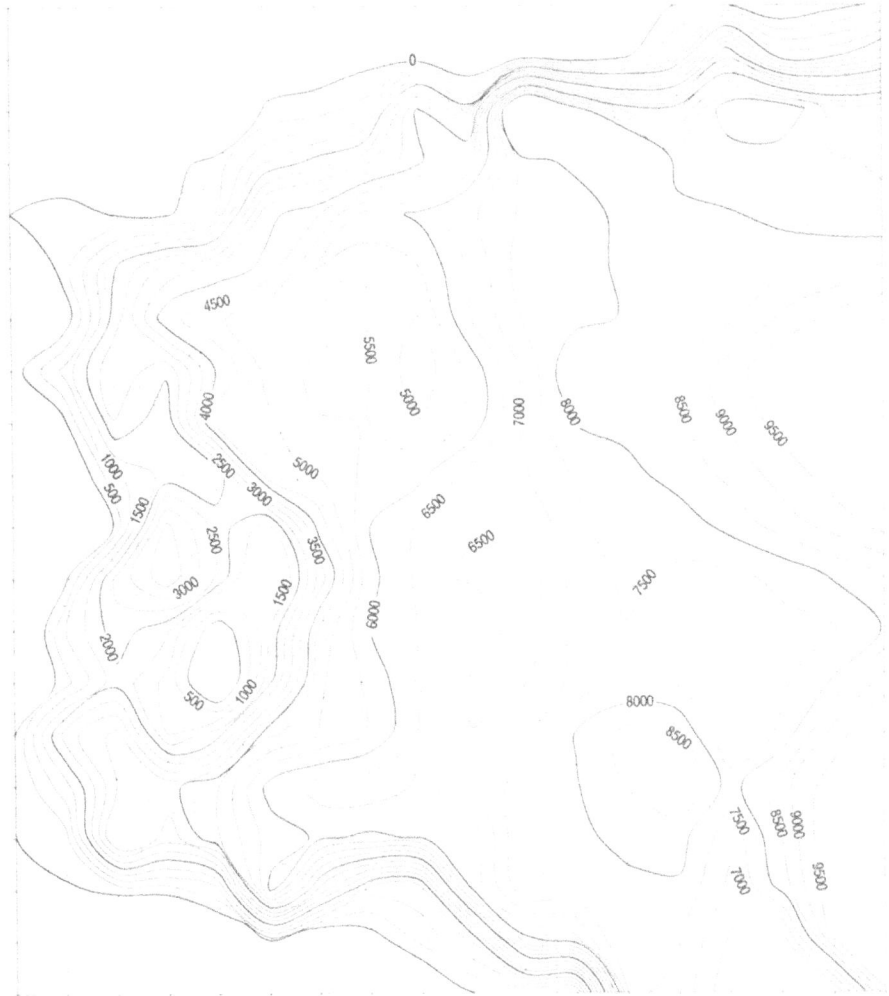

Fig. 3. Computer-generated isochrones (age contours, years BP) showing predicted changes in shoreline position for the Truelove Lowland over the last 9 500 years.

2.00 m. In winter, the deeper lakes are thermally stratified with colder, lighter water overlying warmer, denser water. During the spring melt period the lakes experience thermal overturning and become isothermal. This conditions persists throughout the summer.

To a large extent, the presence of an ice pan controls the degree to which lake water mixing occurs. Throughflow of lighter water immediately beneath the lake ice appears to be greatest when the ice pan is most extensive. As an ice-free moat develops and as the fetch increases, the potential for turbulent mixing of the lake water by the wind also increases. Consequently, the shallower lakes which warm up rapidly during the early summer, tend to be well mixed. Lake bathymetry, by controlling the thermal regime of the lakes, tends to be a major factor influencing the development and subsequent preservation of a stratigraphic record in the lake sediments. For this reason, studies of the environmental record preserved in the lake sediments of the Truelove Lowland have concentrated of the larger, deeper lakes (Young, 1988; Young & King, 1989).

A number of long (approximately 2 m) sediment cores have been obtained from the larger, deeper lakes and have been analyzed for diatoms and chemical composition. Diatoms together with allochthonous and autochthonous chemical components in the sediments have been used to reconstruct changes in paleoenvironmental conditions. Based on the presence of distinctive diatom assemblages, biostratigraphic zones in the sediments are identified as a basal marine zone representing the postglacial marine transgression and containing marine species such as *Cocconeis costata*, *Diploneis subsincta* and *Grammatophora angulosa* (Fig. 4), an intermediate and transitional brackish/marine zone representing the period when the lake became isolated from the sea and containing marine and brackish tolerant species such as *Diploneis interrupta*, *Navicula digitoradiata*, *N. protracta* var. *elliptica* and *N. salinarium*, and an upper freshwater zone dominated by *Fragilaria*, in particular, *F. pinnata*.

Deeper lakes have tended to retain a more complete record of the paleoenvironmental conditions than the shallower lakes. Changes in these conditions can be detected through the determination of specific chemical characteristics in the lake sediments. For example, Cr, Fe, U and Mo in the sediments are associated with the isolation phase when the marine lagoons were first isolated from the sea (Fig. 5). At this time lake sedimentation is strongly affected by the high energy environment associated with beach development and is sensitive to the presence of brackish water and erosion within the catchment.

The period of transition from marine to freshwater conditions also appears to be associated with the development of ephemeral hypolimnetic anoxia leading to the precipitation of Mo and MoS_2 and the presence of a distinctive black layer in the sediments. Calculations of the residence time of water in the larger, deeper lakes suggest that the persistence of such anoxic conditions would be very short lived; in the order of approximately 20 to 30 years at most. In the smaller, shallower lakes the residence time would be appreciably shorter.

During the early post-isolation phase in the Lowland the response of lake biota to an influx of nutrients is reflected in an increase in biogenic silica and organic carbon in the lake sediments (Fig. 5). However, throughout the Holocene the lakes have remained oligotrophic and lake sedimentation has been dominated by variations in non-biogenic factors and particularly by variations in the influx of allochthonous materials. Over time, the progressive stabilization of surface materials and pedogenesis within the lake catchments has been marked by decreasing Cr, As and U in the sediments and an increase in allochthonous Mn and Fe (Mn_d and Fe_d). Perhaps not surprisingly, the stratigraphic record provided by the sediments in these lakes provides more infor-

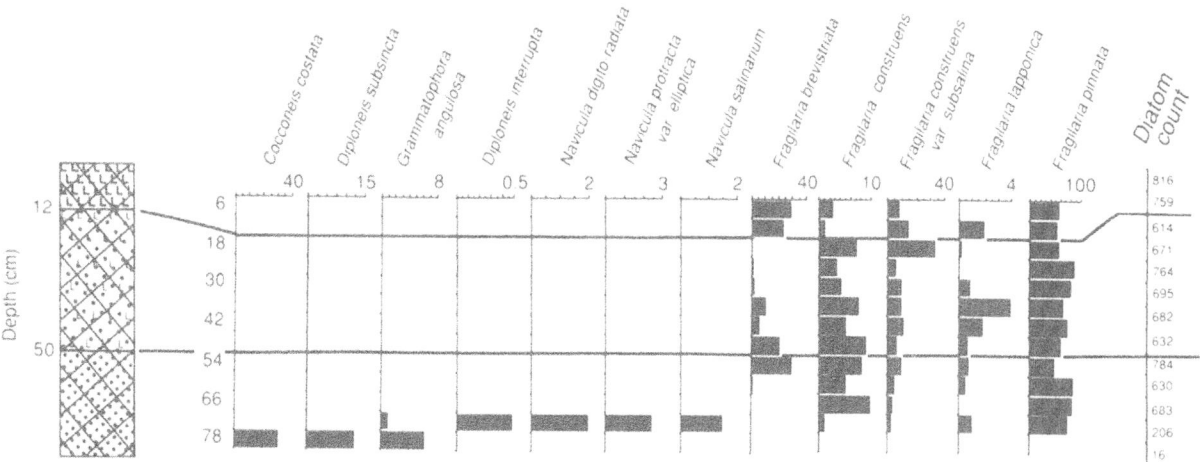

Fig. 4. Diatom stratigraphy of the FL3 core from Fish Lake reported in percentage of diatom sum.

324

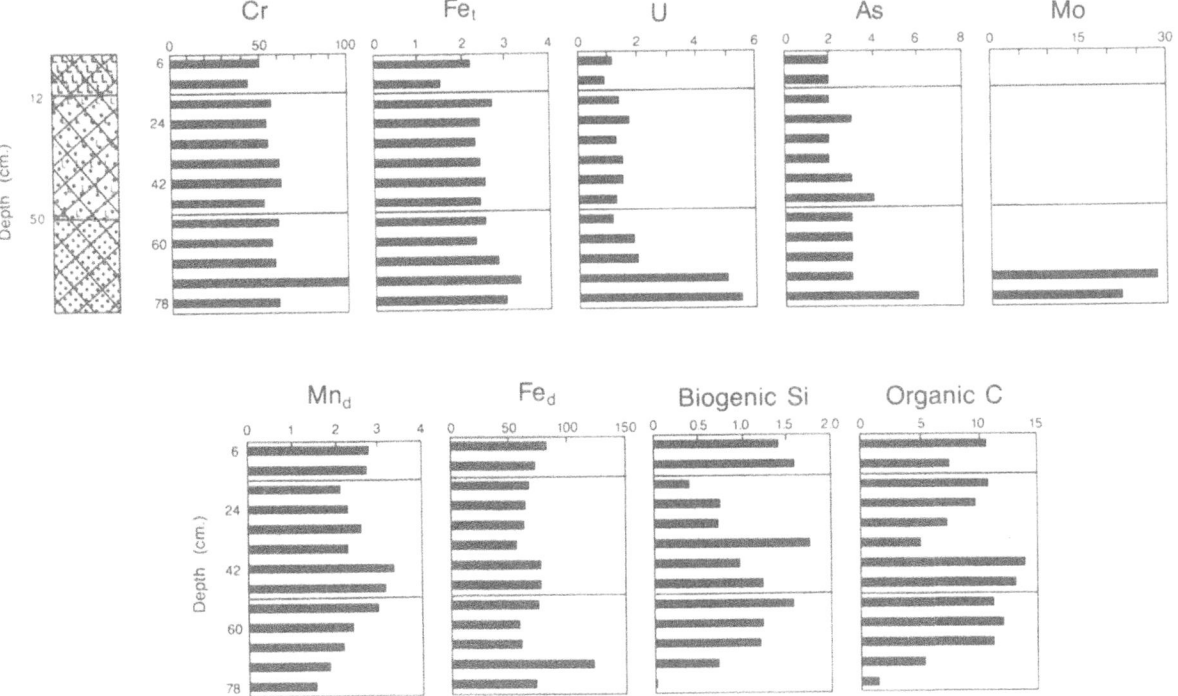

Fig. 5. Chemical stratigraphy of the FL3 core from Fish Lake reported in $\mu g\, g^{-1}$ (oven dried weight), with Fe_t, Organic C reported in %.

mation on the processes and changes occurring within the lake catchments than they do about events taking place within the lakes themselves.

Partly as a result of the progressive deceleration in the rate of coastal emergence there is a tendency for the younger Lowland lakes to be more shallow than the older lakes. Since water depth is a major factor influencing the thermal characteristics of these lakes and with the shallower lakes generally being warmer than the deeper lakes, the biological productivity of these shallow lakes also tends to be greater. As a consequence, algal gyttja tends to be a significant component in the sediments of these lakes. In places, unusually thick accumulations of algal material occur around the margins of some of the shallow lakes when free-floating algal mats are blown ashore during periods of high winds in summer.

As the more shallow lakes and tundra ponds have drained as a consequence of glacio-isostatic rebound and sediment infilling the permafrost has

aggraded into the saturated lake sediments. This, in turn, has led to the formation of epigenetic ice, in the form of both segregated ice and intrusive ice. Some of the ice lenses comprise relatively clear, clean ice, while others contain varying amounts of organic material and gas vesicles. Vertical heaving of the organic-rich deposits consequent to the development of the ground ice has been accompanied by the creation of fissures caused by thermal contraction and/or dilation and the subsequent formation of vertically foliated wedge ice to a maximum depth of 2.40 m. The net result has been the formation of polygonal peat plateaus (Permafrost Subcommittee, 1988) which appear to represent and end stage of lake development in the Lowland.

The stratigraphic record preserved in the sediments of the larger lakes of the Truelove Lowland thus permits a reconstruction of the major paleoenvironmental changes experienced both by the lakes and the lake catchments throughout the postglacial period. In time, given an improved resolution and greater chronologic control it may prove possible to obtain more precise information concerning the long term performance of this unique ecosystem.

Acknowledgements

This research has been supported by the Natural Sciences and Engineering Research Council of Canada through an operating grant to the author. Logistic support was provided by the Polar Continental Shelf Project of the Department of Energy, Mines and Resources, Canada and base camp facilities were provided by the Arctic Institute of North America.

References

Andrews, J. T., 1968. Postglacial rebound in arctic Canada: similarity and prediction of uplift curves. Can. J. Earth Sci. 5: 39–47.

Barr, W., 1971. Postglacial isostatic movement in northeastern Devon Island: A reappraisal. Arctic 24: 249–268.

Blake, Jr., W., 1975. Radiocarbon age determinations and postglacial emergence at Cape Storm, southern Ellesmere Island, Arctic Canada. Meddelanden från Naturgeografiska Institutionen vid Stockholms Universitet No. A62, Norstedts Tryckeri, Stockholm: 71 pp.

Bliss, L. C., 1977. Introduction. In L. C. Bliss (ed.) Truelove Lowland, Devon Island, Canada: A High Arctic Ecosystem. University of Alberta Press, Edmonton: 1–11.

Canada Soil Survey Committee, 1978. The Canadian System of Soil Classification. Canada Department of Agriculture Publication 1646, Supply and Services Canada, Ottawa: 164 pp.

Cleve-Euler, A., 1951–1955. Die Diatomeen Von Schweden Und Finnland. I–V. Kungl. Svenska Vetenskaps Acadamiens Handlingar, Fjärde Serien, Band 2 : 1, 4 : 1, 5, 5 : 4, 3 : 3. Almqvist & Wiksells Boktrykeri AB, Uppsala. Reprinted in 1968 as Bibliotheca Phycologica, Band 5, Verlag Van J. Cramer, Vaduz.

Courtin, G. M. & C. L. Labine, 1977. Microclimatological studies on Truelove Lowland. In L. C. Bliss (ed.) Truelove Lowland, Devon Island, Canada: A High Arctic Ecosystem. University of Alberta Press, Edmonton: 73–106.

DeMaster, D. J., 1981. The supply and accumulation of silica in the marine environment. Geochim. Cosmochim. Acta 45: 1715–1732.

England, J., 1976. Postglacial isobases and uplift curves from the Canadian and Greenland High Arctic. Arctic Alpine Res. 8: 61–78.

Foged, N., 1972. The diatoms in four postglacial deposits in Greenland. Meddelelser Om Grønland 194: 66 pp.

Foged, N., 1973. Diatoms from southwest Greenland. Meddelelser Om Grønland 194: 84 pp.

Foged, N., 1974. Freshwater Diatoms in Iceland. Verlag Van J. Cramer, Vaduz: 273 pp.

Foged, N., 1981. Diatoms in Alaska. Bibliothica Phycologica Band 53, Verlag Van J. Cramer, Vaduz: 317 pp.

Germain, H., 1981. Flora des Diatomées. Société Nouvelle Des Editions Boubee., Paris: 444 pp.

Kelly, P. E., 1989. Numerical analysis of a sequence of raised beach soils, Truelove Lowland, Devon Island, N.W.T. Unpublished M.Sc. thesis. Department of Geography, University of Western Ontario: 212 pp.

King, R. H., 1969. Periglaciation on Devon Island, N.W.T. Unpublished Ph.D. dissertation. Department of Geography, University of Saskatchewan: 470 pp.

Lichti-Federovich, S., 1983. A Pleistocene diatom assemblage from Ellesmere Island, Northwest Territories. Geological Survey of Canada Paper 83–9: 59 pp.

Mehra, D. P. & M. L. Jackson, 1960. Iron oxide removal from soils and clays by a dithionite-citrate system buffered with sodium bicarbonate. 7th National Conference on Clays and Clay Minerals: 317–327.

Nelson, D. W. & L. E. Summers, 1982. Total carbon, organic carbon and organic matter. In A. L. Page (ed.), Methods of Soil Analysis. Part 2: Chemical and Microbiological Properties. American Society of Agronomy, Madison: 539–579.

Patrick, R. & L. R. Freese, 1961. Diatoms (Bacillariophyceae) from northern Alaska. Proceedings of the Academy of Natural Sciences, Philadelphia 112: 129–293.

Peragallo, H. & M. Peragallo, 1897–1908. Diatomées Marines de France. Koeltz Scientific Books: 491 pp.

Permafrost Subcommittee, 1988. Glossary of Permafrost and Related-Ground Ice Terms. Associate Committee on Geotechnical Research, Technical Memorandum 142, National Research Council of Canada, Ottawa: 156 pp.

Young, R. B., 1987. Paleolimnology of two high arctic isolation basins, Truelove Lowland, Devon Island, N.W.T. Unpublished M.Sc. thesis, Department of Geography, University of Western Ontario: 148 pp.

Young, R. B. & R. H. King, 1989. Sediment chemistry and diatom stratigraphy of two high arctic isolation lakes, Truelove Lowland, Devon Island, N.W.T., Canada. J. Paleolimnol. 2: 207–225.

Hydrobiologia **214**: 327–331, 1991.
J. P. Smith, P. G. Appleby, R. W. Battarbee, J. A. Dearing, R. Flower, E. Y. Haworth, F. Oldfield & P. E. O'Sullivan (eds), 327
Environmental History and Palaeolimnology.

An environmental history of two freshwater lakes in the Larsemann Hills, Antarctica

D.S. Gillieson
Department of Geography and Oceanography, University of New South Wales, Australian Defence Force Academy, Canberra, Australia

Key words: Antarctica, coastal lakes, diatom, stratigraphy, holocene

Abstract

The Larsemann Hills are a series of rocky peninsulas and islands in Prydz Bay at 69° 24′ S and 76° 20′ E. There is about 2000 km² of ice free land with well over 150 freshwater lakes spread evenly over the granite and gneiss hills. The nearshore islands were ice free by 9500 BP, while the present coastline was exposed by 4500 BP. A relatively steady rate of ice retreat is indicated, around 0.3 ma^{-1}. The two freshwater lakes studied so far have evolved from oligotrophic, proglacial lagoons to fresh or brackish lakes affected by periodic influxes of salt water from sea spray and surges produced by glacial calving. The diatom assemblages increase in species diversity following marine incursion or influence. The major changes are therefore due to the postglacial recovery of sea level, rather than any intrinsic chemical evolution of the lake waters.

Introduction

In environments dominated by episodic, high magnitude processes, an historical approach may be the only viable route to understanding and predicting the dynamics of lacustrine ecosystems. Although a clear ecosystem boundary can be drawn at the shoreline for most lakes, inclusion of the lake's catchment is necessary to account for particulate (both biotic and inorganic) and solute flux. A total catchment approach allows definition of the chemical evolutionary series of lakes and may provide insights into their reaction to an imposed change. These series have been identified for freshwater lakes on sub-Antarctic islands (Priddle & Heywood, 1980) and for saline lakes on the Antarctic continent (Burton, 1981). The Larsemann Hills may provide an evolutionary series for freshwater lakes on the Antarctic continent. In this paper the preliminary environ-mental histories of two coastal lakes are presented.

Deglaciation of the Larsemann Hills

The ice sheet of the late Wisconsin glaciation reached to at least the nearshore islands of Prydz Bay. In this area the ice attained a thickness of between 200 and 500 m. Cyanobacterial and green algal mats have accumulated in all the lakes since ice retreat, with the thickest accumulations on the islands (about 60 cm over glacial debris or bedrock) and the thinnest in young lakes adjacent to the polar plateau (about 5 cm over bedrock). In all the lake sediment stratigraphies there are thick coarse sandy layers, the products of fluvioglacial sedimentation. In addition, thin fine sandy laminae are the products of annual aeolian deposition on snow and ice covered lakes.

328

Fig. 1. Location and topography of the Larsemann Hills, Antarctica. Shaded areas indicate ice-free land in 1986. Lakes larger than 0.1 ha. appear as solid black shapes.

Basal radiocarbon ages on the algal mats provide a minimum age for the retreat of glacial ice from the lake basin. The nearshore islands such as Kolloy were free of glacial ice by about 9500 BP, while the present coastline at Pup Lagoon was exposed by 4500 BP (Fig. 1). The large glacial troughs of the Progress and Nella Lakes were probably free of ice by 3500–4000 BP (Gillieson *et al.*, 1988).

A relatively slow, steady rate of ice retreat is indicated, about 0.3 ma^{-1}. This is lower than estimates for the nearby Vestfold Hills, where rates of ice margin retreat vary between 2.2 and 3.0 ma^{-1} (Adamson & Pickard, 1986). The lakes therefore provide a potential evolutionary sequence for freshwater ecosystems on the Antarctic mainland.

Environmental history of Kirisjes Pond, Kolloy Island

This small lake has an area of 12 ha and a maximum depth of 9 m. It is amictic and quite fresh, with low conductivity (261 mS/cm^{-1}) and sodium being the dominant cation (13 ppm). The water column is well mixed by katabatic winds, is well oxygenated and mainly fed by snowmelt (Burgess *et al.*, 1988). The lake is less than 5 m above sea level, and salt water may intrude following iceberg calving or as spray.

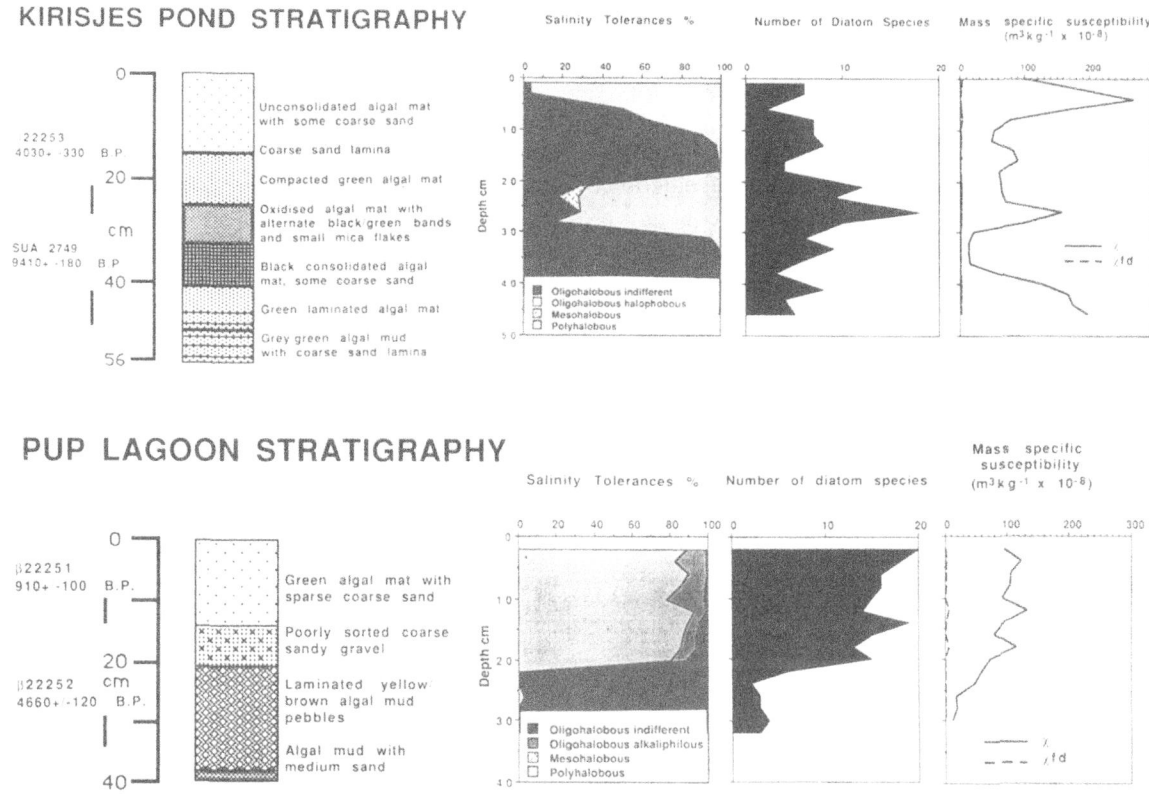

Fig. 2. Stratigraphy, reconstructed diatom salinity tolerances, species diversity and mineral magnetic parameters for Kirisjes Pond and Pup Lagoon.

The sediments of Kirisjes Pond preserve an environmental history of the last 9000 years. There is evidence of periodic disturbance to the catchment, with inwashing of coarse sand and mud (proglacial sediment) and finer sand (aeolian processes) (Fig. 2).

Salinity tolerances (based on Priddle & Fryxell, 1985; Johansen & Fryxell, 1985; Simonsen, 1987) indicate that prior to c. 6000 BP the diatom flora was dominated by oligohalobous species, consistent with a proglacial freshwater environment (Fig. 2). A major change to dominance by polyhalobous and mesohalobous species between 6000 and 4000 BP may reflect the postglacial rise in sea level. General ice shelf recession was caused by mid-Holocene warming, recorded in sub-antarctic deep sea cores. A consequent change in the grounding line of the ice sheet may have affected iceberg calving rates and position

(Denton & Hughes, 1981: 407). Marine incursion may have been continuous due to sea level rise or episodic due to surges from calving. Following stabilisation of sea level, the lake returned to a meltwater inflow regime reflected in dominantly oligohalobous diatoms. Recent inwashing of sea water and salt spray has prompted the resurgence of polyhalobous species.

The number of diatom species recorded increases markedly at the time of this marine incursion. This probably relates to increased nutrient availability in sea water (Priddle *et al.*, 1986). The oldest diatom assemblages are probably nutrient limited, with only a few species present notably *Pinnularia krasskei* var. *ventricosa*, *Nitzschia minuta* and *Amphora lineolata*. This is consistent with conditions at the contemporary ice margin. Since marine influence has been established, a wide range of marine species occur including *Thalassiosira antarctica*, *Chaetoceros hendeyi*, *Asteromphalus hookeri*, *Diploneis subcincta*, *Nitzschia curta* and *N. kerguelensis* (Table 1).

Mass specific magnetic susceptibility χ (Fig. 2) shows several peaks suggesting influx of catchment or nearshore material; the low frequency dependent susceptibility (χ^{fd}) values, all $<5\%$, support the notion of relatively unweathered regolith.

Table 1. Summary of diatom species identified from sediments of Pup Lagoon and Kirisjes Pond, Larsemann Hills.

Centrales:

Actinocyclus actinochilus (Ehrenb.) Simonsen
Actinocyclus curvulatus Janisch
Actinocyclus divisus (Grun.) Hustedt
Asteromphalus hookeri Ehrenb.
Biddulphia sinensis Hustedt
Chaetoceros hendeyi Manguin
Coscinodiscus oculoides Karsten
Coscinodiscus pseudodenticulatus Karsten
Cyclotella stelligera Cleve & Grun.
Eucampia antarctica (Castr.) Manguin
Thalassiosira antarctica Comber

Pennales:

Achnanthes brevipes var. *intermedia* (Kutz.) Cleve.
Amphora libyca Ehrenb.
Amphora lineolata Ehrenb.
Amphora veneta (Kutz.) var. *capitata* E.Y. Haworth
Cocconeis costata Greg.
Cocconeis pinnata Greg.
Cymbella tumida (Breb.) Van Heurck
Diploneis subcincta Cleve.
Frustulia rhomboides var. *saxonica* (Rabh.) De Toni
Navicula criophila (Castr.) De Toni
Navicula cryptocephala Kutz.
Navicula muticopsis Van Heurck
Nitzschia angulata Hasle
Nitzschia curta Hasle
Nitzschia kerguelensis (Castr.) Hustedt
Nitzschia minuta Bleisch.
Nitzschia obliquecostata (Van Heurck) Hasle
Nitzschia punctata W. Smith
Pinnularia krasskei var. *ventricosa* Hustedt
Pinnularia rectangulata Greg.
Stauroneis anceps Ehrenb.
Synedra inequalis Kobayashi
Tabellaria flocculosa (Rabh.) Kutz
Thalossionema bacillaris (Hustedt) Hasle & Mendiola
Trachyneis aspera (Ehrenb.) Cleve

Environmental history of Pup Lagoon, Stornes Peninsula

This small lagoon has an area of 1.0 ha and a maximum depth of 6 m. It is surrounded by fans of coarse sandy gravel, and is virtually at sea level. A deep snowpack separates the lagoon from the beach, where Weddell seals loiter and Adelie penguins have a rookery. The lagoon has a substantially higher conductivity ($856\ \mathrm{mS/cm^{-1}}$) and sodium ion content (66 ppm) than Kirisjes Pond. Iceberg calving from nearby glaciers sends periodic surges of saltwater onto the beach and presumably into the lagoon.

The environmental history of Pup Lagoon extends for the last 4500 years. About 40 cm of sediment has accumulated in the lake. The basal pebbly algal mud was probably a proglacial sediment. The salinity tolerances of the diatoms at that time were almost entirely oligohalobous (Fig. 2); the diatom assemblage was dominated by a very few species and was very similar to Kirisjes Pond (*Pinnularia krasskei* var. *ventricosa*, *Nitzschia minuta* and *Amphora lineolata*.).

A major change in the diatom assemblage at c. 20 cm depth indicates a shift to brackish conditions. Polyhalobous and alkaliphilous species dominated and have maintained this dominance to the present. Species present include *Achnanthes brevipes* var. *intermedia*, *Trachyneis aspera*, several species of *Nitzschia*, *Thalassiosira antarctica* and *Eucampia antarctica*. Species diversity increased, as did the level of environmental disturbance indexed by the magnetic susceptibility χ. Subsequently periodic influxes of saline water caused a change to relatively nutrient rich conditions. Nutrient influx from penguin guano and seal faeces may also have occurred, evidenced by the presence of *Amphora veneta* and the various species of *Nitzschia*.

Conclusions

The Larsemann Hills have been progressively exposed by glacial retreat since 9000 BP. The nearshore islands were ice free by that time, while the present coastline was exposed by 4500 BP. Lake basins next to the polar plateau have been exposed in the last few hundred years. The freshwater lakes studied so far have evolved from oligotrophic, proglacial lagoons to brackish lakes affected by periodic influxes of salt water from sea spray and surges produced by glacial calving. The early proglacial diatom floras are low in species diversity but high in numbers, suggesting good light availability. In both cases the diatom assemblages increase dramatically in species diversity following marine incursion or influence. The major changes are therefore due to the postglacial recovery of sea level and periodic inwashing of seawater, rather than any intrinsic chemical evolution of the lake waters from catchment weathering. Further studies will consider the environmental history of younger lakes closer to the present continental ice margin, where such marine influences are absent.

Acknowledgements

This research was supported by an Australian Research Council Grant No. A38615869 and by a grant from the Australian Antarctic Division. Mr. A.P. Spate and Dr. J. Burgess shared fieldwork and the lake sediment coring. I am grateful to Drs. D.R. Oppenheim and J.C. Ellis-Evans for discussions on diatom taxonomy and lake ecology; the author is responsible for any errors of identification and interpretation. Text and diagrams were produced by Anne Cochrane and Paul Ballard, while laboratory analysis was performed by Robyn Bartlett and Ross Robson.

References

Burgess, J., D. Gillieson & A. Spate, 1988. On the thermal stratification of freshwater lakes in the Snowy Mountains, Australia, and the Larsemann Hills, Antarctica. Search 19: 147–149.

Burton, H., 1981. Chemistry, physics and evolution of Antarctic saline lakes. A review. Hydrobiologia 82: 339–362.

Denton, G. H. & T. J. Hughes, 1981. The Last Great Ice Sheets. Wiley – Interscience, New York.

Gillieson, D., J. Burgess & A. Spate, 1988. Geomorphology and limnology of the Larsemann Hills, Antarctica. Paper presented to Glacial Geomorphology Section, International Geographical Union Congress, Sydney, August 1988.

Johansen, J. R. & G. A. Fryxell, 1985. The genus *Thalassiosira* (Bacillariophyceae): studies on species occurring south of the Antarctic Convergence Zone. Phycologia 24: 155–179.

Priddle, J. & R. B. Heywood, 1980. The evolution of Antarctic lake ecosystems. Biol. J. Linn. Soc. 14: 51–66.

Priddle, J. & G. Fryxell, 1985. Handbook of the Common Planktonic Diatoms of the Southern Ocean: Centrales except the Genus *Thalassiosira*. British Antarctic Survey, Natural Environment Research Council, Cambridge. 159 pp.

Priddle, J., I. Hawes & J. C. Ellis-Evans, 1986. Antarctic aquatic ecosystems as habitats for phytoplankton. Biol. Rev. 61: 199–238.

Simonsen, R., 1987. Atlas and Catalogue of the Diatom Types of Friedrich Hustedt, Vol. 1–3. J. Cramer, Berlin. 1741 pp.

Hydrobiologia **214**: 333–340, 1991.
J. P. Smith, P. G. Appleby, R. W. Battarbee, J. A. Dearing, R. Flower, E. Y. Haworth, F. Oldfield & P. E. O'Sullivan (eds), 333
Environmental History and Palaeolimnology.
© 1991 *Kluwer Academic Publishers.*

Paleolimnology of Qilu Hu, Yunnan Province, China

Mark Brenner,[1] Kathleen Dorsey,[1] Song Xueliang,[2] Wang Zuguan,[2] Long Ruihua,[2] Michael W. Binford,[3] Thomas J. Whitmore[1] & Allen M. Moore[4]

[1] *Florida Museum of Natural History, Gainesville, FL 32611, USA;* [2] *Yunnan Institute of Geological Sciences, Baita Road, Kunming 650011, Yunnan, China;* [3] *Graduate School of Design, Harvard University, 48 Quincy Street, Cambridge, MA 02138, USA;* [4] *Dept. of Biology, Western Carolina University, Cullowhee, NC 28733, USA*

Key words: China, limnology, paleolimnology, Pleistocene, sediment geochemistry

Abstract

Qilu Hu is a large (A = 36.9 km^2), shallow (z$_{max}$ = 6.8 m) lake that lies at an elevation of 1797 m above msl on the Yunnan Plateau, southern China. Lake waters are hard (Mg = 3.2 meq L^{-1}, Ca = 1.3 meq L^{-1}), fresh (conductivity = 380 μS cm^{-1}), and productive (Secchi < 40 cm). An 11-m sediment core has a basal ^{14}C age of 30 960 ± 860 B.P. Sediments between 11 m and 6 m are high in % dry weight, rich in clay components Al$_2$O$_3$, Fe$_2$O$_3$, K$_2$O, MgO, and low in organic C (≤6.1%), carbonate-C (<1.0%), total N (<3.2 mg g^{-1}), and total S (≤1.7 mg g^{-1}). Diatoms and pollen indicate open-water conditions between 9.0 m and 6.0 m (13 420–11 790 B.P.). Above 6.0 m, CaCO$_3$ and organic matter concentrations increase relative to clastics. The transition marks a change to shallow-water conditions as inferred from diatoms and pollen, and probably reflects a shift to drier climate. Uppermost (80–0 cm) red clays were deposited rapidly, probably as a consequence of recent (decades to centuries) riparian disturbances (e.g. agriculture, lake-bottom reclamation, urban development). Dates assigned to events in the Qilu Hu profile are tentative because of potential hard-water-lake error.

Introduction

The goal of our joint U.S.-China research program is to explore the sedimentary histories of selected lakes on the subtropical, karsted Yunnan Plateau, China. We are using a multidisciplinary approach to examine historical ecosystem alteration induced by climatic change and shifts in human land use. We hope to learn about the environmental impact of agriculture, industry, and urban development on watersheds in southern China. The work in Yunnan extends the geographic scope of paleolimnological research conducted by the U.S. investigators in several other subtropical and tropical karst regions (Binford *et al.*, 1987). This paper reports preliminary results from an 11-m core taken in Qilu Hu, a lake located about 100 km south of Kunming, Yunnan's capital (Fig. 1).

Kunming lies at about 2000 m above msl, 250 km north of Vietnam and Laos, and 400 km ENE of Burma. The area has a warm temperate climate and receives about 800 mm of rainfall annually. Like many karst regions, Yunnan's landscape is pocked with numerous lakes. The larger basins are of tectonic origin, whereas many of the smaller lakes were formed by solution processes. Near Kunming, the vegetation is charac-

334

Fig. 1. Map of the Yunnan Lake District showing the location of Qilu Hu. Upper inset map shows the location of Yunnan Province in southern China. Lower inset map is a bathymetric map of Qilu Hu, redrawn from the July 1980 map of the Tonghai County Hydrological Bureau. Water depth at the core site was 4.5 m in June 1987 when cores were collected.

terized as subtropical evergreen broadleaf forest, but varies with edaphic conditions (Li & Walker, 1986).

Previous paleoenvironmental studies in Yunnan have focused on the vegetational history of the region (Liu *et al.*, 1986; Lin *et al.*, 1986; Sun *et al.*; Walker, 1986). Pollen diagrams from cores taken in the northern basin of Lake Dianchi (Caohai), near Kunming (Fig. 1), yielded information on Pleistocene-Holocene climatic changes (Sun *et al.*, 1986). Not surprisingly, prior to 10 000 B.P., temperature on the Plateau was cooler than at present. Sun *et al.* (1986) suggest that Kunming was also wetter prior to 10 000 B.P., contrary to the arid late Pleistocene conditions inferred for lowland subtropical and tropical sites in the western hemisphere (Watts, 1975; Bradbury *et al.*, 1981; Deevey *et al.*, 1983; Leyden, 1984). The authors conclude that the Caohai pollen record lacks strong evidence for anthropogenic vegetation disruption over the past 10 000 years,

despite prehistoric settlement extending into the early Holocene. They speculate that Neolithic culture in southwest China may have developed and persisted in a forest context. Nevertheless, Li & Walker (1986) estimate that only about 10% of Yunnan Province is today covered by near-natural vegetation, most of it confined to alpine and subalpine areas.

Five lakes were cored by the U.S.-China team in June 1987 (Fig. 1). All five yielded short cores, 1–3 m long. Long cores were unretrievable in four of the lakes (Cao Dian Hai, Yue Hu, Chang Hu, Qing Shui Hai) because their deeper deposits were indurated and probably non-lacustrine. Shallow basins Cao Dian Hai and Yue Hu probably dried out in recent times, much as Dianchi was reported to have done nearly 200 years ago (Sun *et al.*, 1986). Qilu Hu yielded an 11 m profile that contains a 30 000-year record of sedimentation (Deevey *et al.*, 1988).

The study site

Qilu Hu lies at 24° 10′ N lat, 102° 45′ E long at an elevation of 1797 m (Fig. 1). The lake covers 36.9 km^2 and the watershed: lake ratio is nearly 10 : 1. Qilu Hu has a z_{max} of about 7 m and a $\bar{z}$ of 4 m (Fig. 1), but the lake level fluctuates and can vary by >2 m annually. The lake volume is 1.486 × 10^8 m^3 and water is supplied by rainfall (869 mm yr^{-1}), ten intermittent rivers, and 39 springs. Water residence time is about two years. A sluice gate built in 1939 at the edge of a sinkhole regulates the lake level, and discharged water ultimately flows into the Nanpanjiang River. Qilu Hu waters are hard and the Ca : Mg ratio indicates local dolomitization (Ca = 1.3 meq L^{-1}, Mg = 3.2 meq L^{-1}: Moore *et al.*, 1988). Lake waters are fresh (cond = 380 μS cm^{-1}) and dense algal blooms, dominated by *Microcystis* sp., restrict Secchi disk depth to <40 cm.

Recent excavations in Yunnan Province reveal a long history of hominid occupation. Middle Pleistocene *Homo erectus* teeth from Yuan-mou date to 0.5–0.6 × 10^6 years (Chang, 1986). At Kunming and other sites in the province,

H. sapiens remains are associated with paleolithic tools thought to be >20 000 years old (Li & Walker, 1986). Neolithic settlements were widespread by 4000 B.P. (Li & Walker, 1986), as was rice agriculture (Yen, 1987). Historical demographics in Yunnan were influenced by the invasion and settlement of exogenous cultural groups (e.g. Mongols), and villages populated by cultural minorities persist to the present.

Large numbers of people occupied Tonghai County, near Qilu Hu, at least since the Ming Dynasty (A.D. 1368–1644). During that period, population in the county varied from 1.5×10^4 to 3.5×10^4 persons. At the beginning of the Qing Dynasty (A.D. 1644–1911), only 5.2×10^3 people lived in the county, but the population grew to nearly 3.8×10^4 under the reign of Emperor Qian Long. By A.D. 1984, the population of Tonghai County exceeded 2.1×10^5 persons (Tonghai Office of Historical Records, 1985). Impacts of modern riparian settlement include reclamation of near-shore lake bottom for agriculture, discharge of industrial effluents, fishery development, and hydrologic control.

Field methods

A coring site was established at the east end of Qilu Hu (Fig. 1). Two mud-water interface cores were taken with a 4-cm diameter piston corer. The two profiles were designated cores 15-VI-87-1 and 15-VI-87-2. They were 100 cm and 99 cm long, respectively, and similar in gross stratigraphy. Cores were extruded and sectioned in the field at 2 cm intervals to 10 cm, at 1 cm intervals between 10 and 30 cm, and at 4 cm intervals from 30 cm to the base of the section. A long core (16-VI-87) was obtained with a steel-barrel square-rod corer (Wright *et al.*, 1984) beginning 0.5 m below the mud-water interface.

Laboratory methods

Percent dry weight was measured by weight loss on drying at 105 °C (Håkanson & Jansson,

1983). Total C was measured with a Coulometrics, Inc. Model 5011 coulometer (Huffman, 1977), and a System 120 preparation line set at 950 °C. Inorganic (carbonate) C was determined by coulometry using a System 140 prep line. CO_2 was evolved with 2N $HClO_4$. Organic C was figured as total C minus inorganic C. Total S was measured with a Coulometrics, Inc. Model 3200 coulometer and 3220 prep line using prep line temperatures of 1050 °C and 850 °C. Total N and P were measured by autoanalyzer following digestion (Parkinson & Allen, 1975). CaO, Al_2O_3, Fe_2O_3, K_2O, MgO, Na_2O, Mn, Pb, and Zn were determined by X-ray fluorescence with an American Research Laboratories, Inc. Model 8680 XRF.

Results

Chronology and sediment accumulation rates

Radiocarbon dates were provided by three laboratories: Guiyang, University of Pittsburgh, and the NSF-Arizona Accelerator Facility (Table 1). Low organic C content throughout the core required broad-interval stratigraphic sampling for ^{14}C analysis. The ages of the bottom two samples, PITT-0217 and GC-87045, are in reverse stratigraphic order. The great antiquity and very low organic C content of these deep deposits (Fig. 3) may place them beyond the reliable dating capability of the equipment.

Table 1. Uncalibrated radiocarbon dates from Qilu Hu Core 16-VI-87. The organic fraction of bulk sediments was dated following pretreatment to remove carbonates.

Depth (m)	Material	Lab-sample	Radiocarbon age (yr B.P.)
2.81–2.99	Bulk sediment	PITT-0214	5,450 ± 40
5.81–5.99	Bulk sediment	PITT-0215	11,790 ± 70
6.60	Mollusc shell	AA-3069	3,570 ± 65
7.36–7.39	Wood	AA-3070	10,740 ± 120
9.01–9.19	Bulk sediment	PITT-0216	13,420 ± 200
10.06–10.27	Bulk sediment	GC-87045	>40,000
10.81–10.99	Bulk sediment	PITT-0217	30,960 ± 860

The PITT dates on bulk sediment are all in stratigraphic order, but are subject to possible hard-water-lake error (Deevey & Stuiver, 1964). Additional dating anomalies may be a consequence of colluviation, if some lake sediment is redeposited soil of unknown isotopic content. The wood sample from 7.36–7.39 m is younger than the bulk sediment sample at 5.81–5.99 m (Table 1), suggesting that dates run on the organic fraction of bulk sediment are artificially old. Assuming that the magnitude of dating error is similar in PITT samples 0215 and 0216, and that net sediment accumulation was constant between 5.90 m and 9.10 m, a bulk sediment sample at 7.38 m would be 12 540 years old, or about 1 800 years older than wood from the same level. The shell date at 6.60 m (3570 ± 65 B.P.) is out of sequence and probably too young.

We estimated bulk sediment accumulation rates by assigning the uncorrected ^{14}C age of each PITT sample to the midpoint of the dated sample interval (Table 2). Net sediment accumulation rate near the bottom of the section (10.9–9.1 m) is slow, only 0.01 cm yr^{-1} (8.0 mg cm^{-2} yr^{-1}). Moving upward through the core, the sediment accumulation rate in the next interval (9.1–5.9 m) is more than ten times greater, 0.196 cm yr^{-1} (111.7 mg cm^{-2} yr^{-1}). Between 11 790 B.P. and 5 450 B.P. (5.9–2.9 m), the sedimentation rate declined relative to the interval below, dropping to 0.047 cm yr^{-1} (18.0 mg cm^{-2} yr^{-1}). Net sediment accumulation rate for the topmost section of the core (2.9–0.0 m: 0.053 cm yr^{-1}, 19.7 mg cm^{-2} yr^{-1}) is similar to the value computed for the interval below.

Bulk sediment accumulation rates were calculated by an alternative method, using the wood date at 7.36–7.39 m (10,740 B.P.), and the age at the base of the section (30 960 B.P.). Prior to 10 740 B.P., sediment accumulated slowly, at a rate of 0.17 cm yr^{-1} (12.4 mg cm^{-2} yr^{-1}), whereas after that date the rate increased to 0.69 cm yr^{-1} (28.3 mg cm^{-2} yr^{-1}).

Sediment geochemistry

Dry density (p) in the Qilu Hu core ranges from 0.17 to 0.87 g dry cm^{-3} wet sediment. High values were recorded near the base of the core (Table 2) where predominantly inorganic sediments are highly compacted. Percent dry weight parallels dry density and varies from a high of 57.3% (9.4 m) to a low of 12.6% at the mud surface (Figs. 2 & 3). Highest percent dry weight was measured in the most inorganic deposits, below 6.0 m, and in clayey sediments near the top of the section, between 80 and 15 cm.

Organic C concentrations are low ($\leq 6.1\%$) in the bottom 6 m of the core, but rise sharply, averaging $> 11\%$ between 5.0 m and 1.0 m depth (Fig. 2). Organic C content declines to $<5\%$ in the clay-rich deposits above 80 cm (Fig. 3). Inorganic C concentrations are also low ($<1\%$) below 6.0 m depth, but rise to a maximum value of 8.2% at 2.0 m (Fig. 2). If all inorganic C is bound in $CaCO_3$, carbonates account for 59–68% of the dry weight from 1.0 m to 2.0 m depth. Inorganic C content drops precipitously between 90 and 80 cm, and remains low to the top of the core (Fig. 3). Total N was measured on widely spaced samples, but shows similar trends to organic C. N values are low (0.15–0.32%) below 6 m in the core, but rise as high as 1.03%

Table 2. Sediment accumulation rates in Qilu Hu computed from PITT radiocarbon rates.

Depth interval (cm)	Interval thickness (cm)	Time interval (years)	Accumulation rate [depth] (cm yr^{-1})	Mean dry density (g dry cm^{-3} wet)	Accumulation rate [mass] [g cm^{-2} yr^{-1}]
0–290	290	5450	0.053	0.37	0.0197
290–590	300	6340	0.047	0.38	0.0180
590–910	320	1630	0.196	0.57	0.1119
910–1090	180	17540	0.010	0.78	0.0080

QILU HU

Long Core 16-VI-87

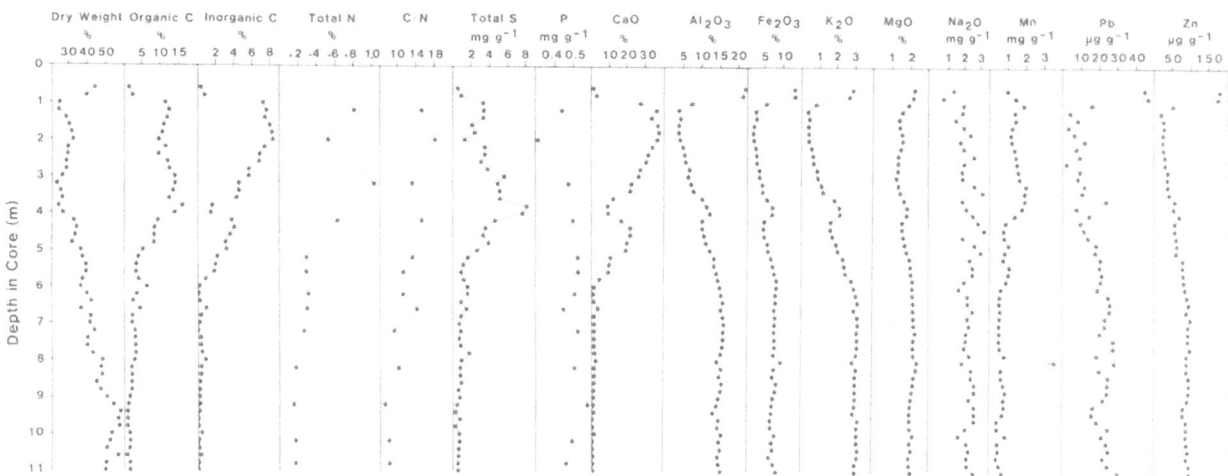

Fig. 2. Geochemistry of Qilu Hu long core 16-VI-87. Percent dry weight, organic C, inorganic C, total N, S, and P were measured at the Florida Museum of Natural History. CaO, Al_2O_3, Fe_2O_3, K_2O, MgO, Na_2O, Mn, Pb, and Zn were analyzed at Hubei Geological Laboratory. Concentrations are expressed on a dry weight basis.

in the organic section of the profile (5.0–1.0 m) before falling again in the clayey upper 80 cm (Figs 2 & 3). Organic C and total N concentrations are highly correlated ($r = 0.98$, $P < 0.01$). Organic C : total N weight ratios range from 7 to

18 in the core (Figs. 2 & 3). The mean ratio is 15.7 in the organic sediments (5.0–0.8 m), but only 10.4 in the basal 6 m of the section, and 10.3 in the uppermost 80 cm of the profile. Total S concentrations vary from 0.2 to 8.0 mg g^{-1} over the

QILU HU

Mud-Water Interface Cores

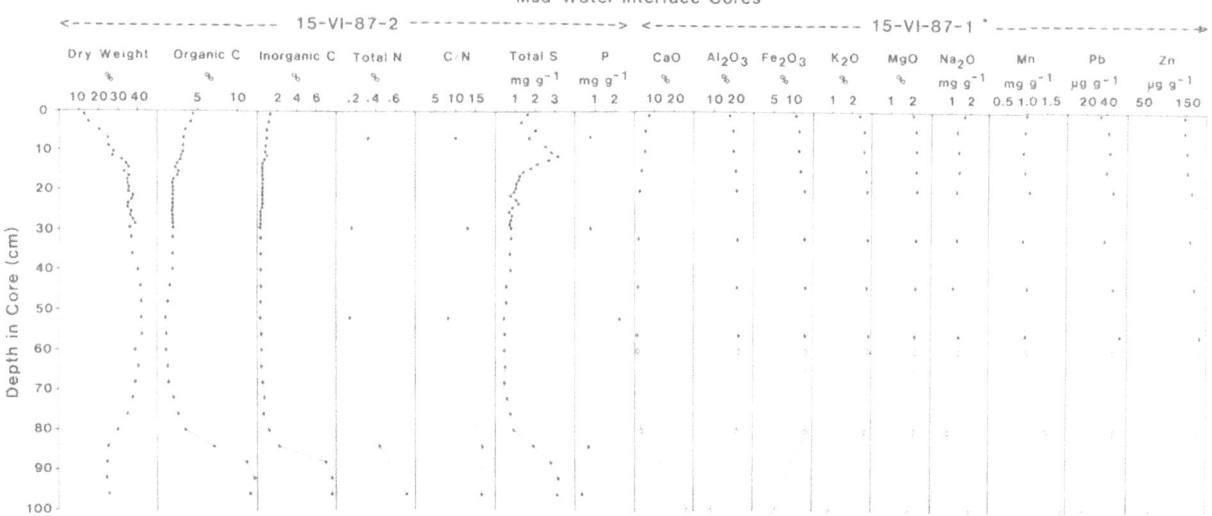

Fig. 3. Geochemistry of Qilu Hu mud-water interface cores 15-VI-87-1 and 15-VI-87-2. *Core 1 was analyzed at the Hubei Geological Laboratory. Core 2 was analyzed at the Florida Museum of Natural History. Open circles in Core 1 denote values obtained in the topmost section of long core 16-VI-87 (see Fig. 2). Concentrations are expressed on a dry weight basis.

length of the section (Figs. 2 & 3), and are highly correlated with organic C concentrations ($r = 0.88$, $P < 0.01$). Reduced inorganic S was undetectable. Total P content ranges from 0.31 to 2.23 mg g^{-1} (Figs. 2 & 3) and is uncorrelated ($P > 0.05$) with both organic C and inorganic C.

Plots of CaO (X-ray fluorescence) and inorganic C (coulometry) concentrations show nearly identical trends (Figs. 2 & 3). CaO : inorganic C weight ratios between 5 m and 1 m depth are close to 4.67, the expected value if all Ca and inorganic C is bound in $CaCO_3$. Above 1 m and below 5 m, Ca : inorganic C ratios are generally > 4.67, suggesting that some calcium is present in non-carbonate form. Al_2O_3, Fe_2O_3, and K_2O show similar concentration profiles (Figs. 2 & 3), and are probably associated with siliceous clastics. The clay components are inversely related to organic and inorganic C content. MgO is highly correlated with Al_2O_3 ($r = 0.88$, $P < 0.01$), suggesting that most Mg is associated with the clay, rather than the carbonate fraction of the sediment. Na_2O shows weak, but significant negative correlation with Al_2O_3 (-0.40, $P < 0.01$), and is uncorrelated with organic C ($P > 0.05$). Mn concentration is positively correlated with organic C ($r = 0.59$, $P < 0.01$), and even more so when the outlier Mn value at 8.0 m is removed ($r = 0.82$). Both Pb and Zn are associated with the inorganic, clay component of the sediment, as they are highly correlated with Al_2O_3 ($r = 0.91$, $P < 0.01$ and $r = 0.96$, $P < 0.01$, respectively).

Discussion

The Qilu Hu section is divisible into three principal stratigraphic zones based on lithology and geochemistry. The basal zone (11.0–6.0 m) is dominated by gray clay and has low organic C and carbonate content. In the middle zone of the core between 6.0 m and 1.0 m, sediments are richer in organic C and carbonates. Above 1.0 m, organic C and carbonates decline and are replaced by iron-rich, red clays. The general stratigraphy of the Qilu Hu section correlates well with core DZ18 from Caohai (Sun *et al.*, 1986),

although organic matter attains higher concentrations in Caohai than in Qilu Hu.

The lithologic shift at 6.0 m in the Qilu Hu core is probably the consequence of a drop in water level that was caused by a climate change. The timing of the event is difficult to establish because ^{14}C dates are subject to hard-water-lake error and some are out of stratigraphic order. Jumbled radiocarbon dates were also a problem at Caohai (Sun *et al.*, 1986). Preliminary diatom data show that the assemblage between 9.0 m and 6.0 m depth in the Qilu Hu core is dominated by planktonic centrics (e.g. *Cyclotella radiosa* [Grunow] Lem., *Stephanodiscus rotula* [Kütz.] Hendey, *Cyclostephanos dubius* [Fricke] Round, *Aulacoseira granulata* [Ehr.] Simonsen. These taxa indicate open-water conditions. Periphytic diatoms (*Fragilaria pinnata* Ehr. *pinnata, F. pinnata* var. *lancettula* [Schum.] Hust., *Opephora martyi* Hérib. var. *martyi, Pinnularia* sp., *Staurosira* sp. Q) are prevalent between 5 m and 1 m depth in the core and suggest shallow lake conditions. The inferred reduction in water level is supported by palynological evidence. Pollen of *Alisma* is very rare below 5.8 m, but becomes abundant (grains cm^{-3}) between 5.8 and 3.4 m. The emergent plant inhabits swamps and shallow waters. If we accept the ^{14}C age on terrestrial wood at 7.36–7.39 m, the lake level reduction detected at about 6.0 m depth occurred after 10 740 B.P.

Qilu Hu's hard waters and the low organic C content of deep deposits make ^{14}C dating of bulk sediments unreliable. Computed sediment accumulation rates based on bulk sediments are thus considered preliminary, and subject to error. Slow sediment accumulation between 30 960 and 13 420 B.P. (8.0 mg cm^{-2} yr^{-1}) may, in part, be a consequence of hiatuses in deposition. At some levels below 9.1 m in the core, gritty deposits contain low numbers of pollen grains and are devoid of diatoms. A high sedimentation rate was computed for the interval between 13 420 and 11 790 B.P. (9.1–5.9 m: 111.9 mg cm^{-2} yr^{-1}), and corresponds to the period of high water level in the lake. This wet event detected in Qilu Hu may coincide with the moist period recorded in Caohai that is said to have occurred before 10 000

B.P. (Sun *et al.*, 1986). Because ^{14}C dates on bulk sediments may be older than their true age, we cannot say with certainty that the wet event occurred in the late Glacial. Net accumulation rates computed for the time spans between 11790 B.P. and 5540 B.P. (18.0 mg cm^{-2} yr^{-1}, 0.47 mm yr^{-1}) and between 5540 B.P. and present (19.7 mg cm^{-1} yr^{-1}, 0.53 mm yr^{-1}) are nearly identical, and not unlike the Holocene (10000–1500 B.P.) rate of 0.38 mm yr^{-1} measured at Caohai (Sun *et al.*, 1986).

Sediments above 80 cm in Qilu Hu are clay-rich and differ from the underlying organic and carbonaceous deposits. The change may be associated with lake level fluctuation, but probably is a consequence of land clearance and lake bottom reclamation for agriculture. If we assume constant sediment accumulation since 5450 B.P. (2.9 m), the shift to clay deposition at 80 cm occurred about 1500 years ago. Nevertheless, several lines of evidence indicate that clay deposition began more recently, perhaps associated with the ten-fold increase in riparian human population since the turn of the century. The major change in sediment geochemistry suggests that the assumption of constant deposition is dubious. Furthermore, there is some evidence that the clays were deposted rapidly. Pollen concentrations are low at most levels above 80 cm, and diatoms are very scarce or absent in the clays. We attempted to determine whether Pb and Zn increases in the topmost 80 cm postdate industrial development. Throughout the Qilu Hu profile, the metals are correlated with the clay fraction (Al_2O_3) of the sediment. We looked at the Pb:Al_2O_3 and Zn:Al_2O_3 ratios in the uppermost clays (0–80 cm) and in the basal deposits (9.0–11.0 m) to see whether there might be a surplus of lead and zinc in the recent muds. Pb:Al_2O_3 ratios are 38% higher in recent muds than in deep deposits. Zn:Al_2O_3 ratios in the topmost clays are 30% higher than those computed for the basal sediments. The excess heavy metals in the recent sediments suggest an anthropogenic (industrial?) source, but the timing of their increase must await ^{210}Pb dating of the core.

Acknowledgements

This work was supported by NSF-INT grant 8802793 to E.S. Deevey, and supplemental funding from NSF and University of Florida to M. Brenner. Mr. Robert Dorion supported fieldwork in 1987. Financial assistance was provided by grants from NSF-China and the Yunnan Provincial Commission of Science and Technology to Yunnan University and the Yunnan Institute of Geological Sciences. Radiocarbon dates were kindly provided by Dr. Robert Stuckenrath and Ms. Qiao Yulou. We thank Bill DeBusk for assistance with geochemical analyses.

References

Binford, M. W., M. Brenner, T. J. Whitmore, A. Higuera-Gundy, E. S. Deevey & B. Leyden, 1987. Ecosystems, paleoecology and human disturbance in subtropical and tropical America. Quat. Sci. Rev. 6: 115–128.

Bradbury, J. P., B. Leyden, M. Salgado-Labouriau, W. M. Lewis, Jr., C. Schubert, M. W. Binford, D. G. Frey, D. R. Whitehead & F. H. Weibezahn, 1981. Late Quaternary environmental history of Lake Valencia, Venezuela. Science 214: 1299–1305.

Chang, K., 1986. The archaeology of ancient China (4th ed.) Yale University Press, New Haven. 450 p.

Deevey, E. S., M. Brenner & M. W. Binford, 1983. Paleolimnology of the Peten Lake District, Guatemala, III. Late Pleistocene and Gamblian environments of the Maya area. Hydrobiologia 103: 211–216.

Deevey, E. S., M. Brenner, A. M. Moore, M. W. Binford, K. T. Dorsey, X. Song, Z. Wang & R. Long, 1988. Paleolimnology and limnology of Qilu Hu, Tonghai County, Yunnan – preliminary results. Journal of Yunnan University 10 (supplement): 57–62.

Deevey, E. S. & M. Stuiver, 1964. Distribution of natural isotopes of carbon in Linsley Pond and other New England lakes. Limnol. Oceanogr. 9: 1–11.

Håkanson, L. & M. Jansson, 1983. Principles of lake sedimentology. Springer-Verlag, N.Y. 316 p.

Huffman, E. W. D., Jr., 1977. Performance of a new automatic carbon dioxide analyzer. Microchem. 22: 567–573.

Leyden, B. W., 1984. Guatemalan forest synthesis after Pleistocene aridity. Proc. Natl. Acad. Sci. (USA) 81: 4856–4859.

Li, X. & D. Walker, 1986. The plant geography of Yunnan Province, southwest China. J. Biogeogr. 13: 367–397.

Liu, J., L. Tang, Y. Qiao, M. J. Head & D. Walker, 1986. Late Quaternary vegetation history at Menghai, Yunnan Province, southwest China. J. Biogeogr. 13: 399–418.

340

Lin, S., Y. Qiao & D. Walker, 1986. Late Pleistocene and Holocene vegetation history at Xi Hu, Er Yuan Province, southwest China. J. Biogeogr. 13: 419–440.

Moore, A. M., M. Brenner, M. W. Binford, E. S. Deevey & X. Ou, 1988. Field notes on a paleolimnological expedition to five lakes on the Yunnan Plateau. Journal of Yunnan University 10 (supplement): 28–36.

Parkinson, J. A. & S. E. Allen, 1975. A wet oxidation procedure for the determination of nitrogen and mineral nutrients in biological material. Community Soil Sci. Plant Anal. 6: 1–11.

Sun, X., Y. Wu, Y. Qiao & D. Walker, 1986. Late Pleistocene and Holocene vegetation history at Kunming, Yunnan Province, southwest China. J. Biogeogr. 13: 441–476.

Tonghai Office of Historical Records, 1985. A survey of Tong Hai County. 149 p.

Walker, D., 1986. Late Pleistocene-early Holocene vegetational and climatic changes in Yunnan Province, southwest China. J. Biogeogr. 13: 477–486.

Watts, W. A., 1975. A late Quaternary record of vegetation from Lake Annie, south-central Florida. Geology 3: 344–346.

Wright, H. E., Jr., D. H. Mann & P. H. Glaser, 1984. Piston corers for peat and lake sediments. Ecology 65: 657–659.

Yen, S., 1987. Evidence for the origin of crops in Yunnan based on lowland rice. Yunnan Cultural Relics 21: 18–25.

Hydrobiologia **214**: 341–345, 1991.
J. P. Smith, P. G. Appleby, R. W. Battarbee, J. A. Dearing, R. Flower, E. Y. Haworth, F. Oldfield & P. E. O'Sullivan (eds), 341
Environmental History and Palaeolimnology.

Sedimentary features and the evolution of lake Honghu, central China

Shuming Cai & Zhaolu Yi
Institute of Geodesy and Geophysics Academia Sinica, 54 Xu Dong Road, Wuchang, Hubei 430077 People's of Republic of China

Key words: palaeolimnology, ^{210}Pb dating, lake level changes, lake sediments, swamp deposits

Abstract

Palaeolimnological studies of the sediments of Honghu, a large shallow lake in Central China, were used to investigate the history of the origin, formation, and changing extent of the lake. The results indicate that Honghu is a naturally-dammed feature, formed about 3000 years ago by meandering of the Changjiang river. Lake level then fell in the period after 2500 BP, but during the Jin epoch (265–420 AD) it recovered, only to fall once more after the Song dynasty (969–1279 AD). The most recent episodes of reflooding are dated to ca. 400 years ago, and to the late nineteenth century.

Introduction

Honghu is the largest shallow water lake on the Jianghan plain of Hubei province, central China. Its surface area is 355 km², average depth is 35 m. There are two different opinions about the formation of Honghu. One is that Honghu is a remnant of a formerly more extensive lake (Yen and Chen, 1960; Hun, 1965); the other is that it is a lateral lake of the Changjiang River (Hutchinson, 1957; Cai & Guan, 1979). Between 1985 and 1988 we made observations in the Honghu area. These included 30 cores and 7 detailed soil profiles in the crop fields around the lake (Fig. 1), and an investigation of the distribution of ancient lake sediments exposed in a newly-dug canal. The research indicates that Honghu is a dammed lake, no more than 3000 years old and has experienced periods of contraction, expansion and swamp formation.

Landforms and geological setting

Honghu (Fig. 1) is situated between the Changjiang and Dongjin rivers. The lake is impounded by a dam, except on the southeastern side where it is contained by the natural levees of the Changjiang River.

In the Honghu area Quaternary sediments are about 200 m thick (Cai and Guan, 1982, 1986). They consist mainly of alluvial facies with pebbles and sand. The uppermost layer is silty clay which is 20–30 m deep. The modern sediments in the Honghu lake are about 0.5–2.0 m thick. Beneath the lacustrine deposits are alluvial flat facies with red silts and clays.

Processes of sedimentation

Three main processes of sedimentation operate in Honghu. These are:

1. Physical sedimentation produced by erosion, transport and deposition from the catchment.
2. Chemical sedimentation. The colloidal content of suspended sediment entering the lake is between 30 and 50%. Because the lake water is alkaline the sediment is deposited as a gel.

342

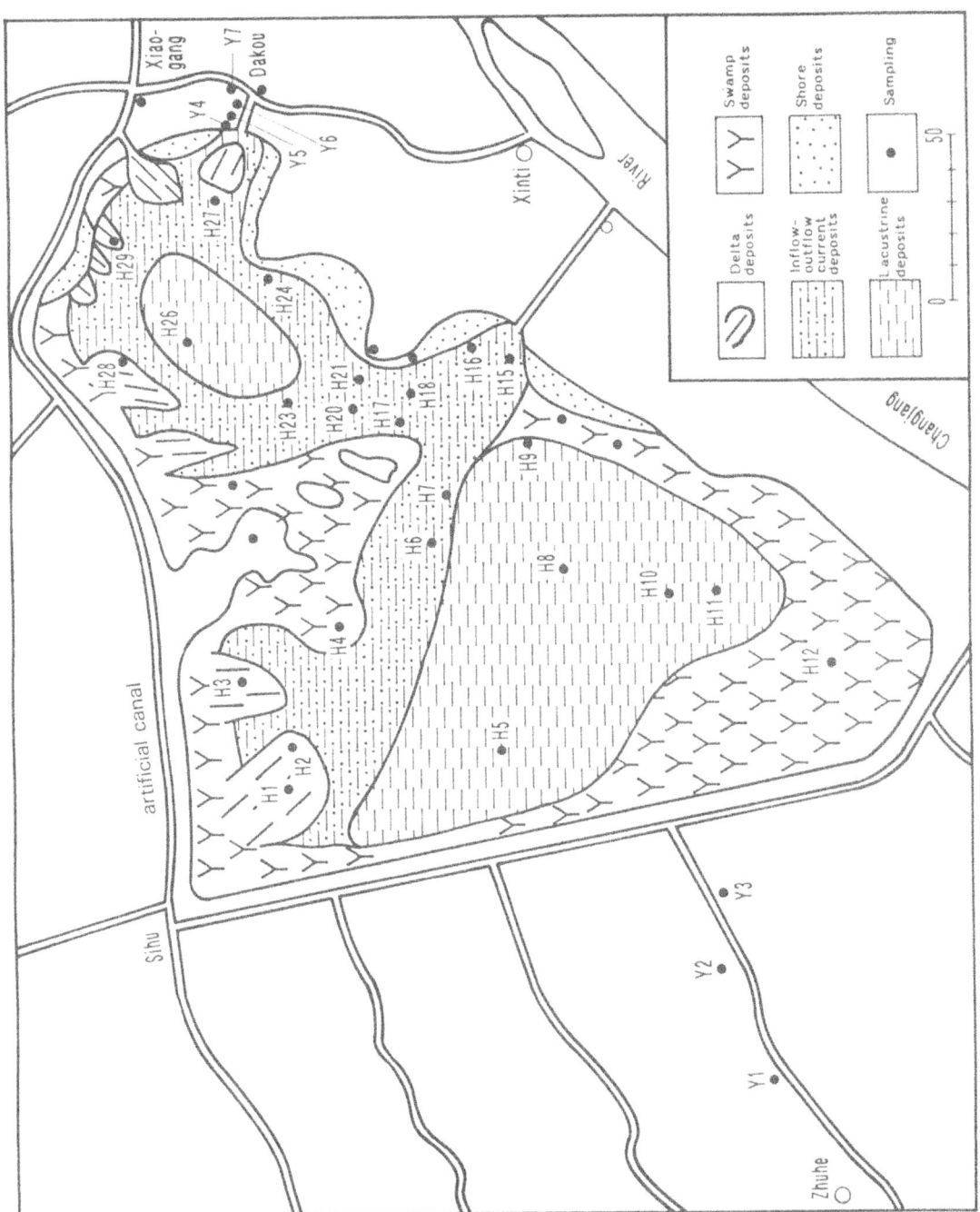

Fig. 1. Sampling location and sediments type in Honghu Lake.

343

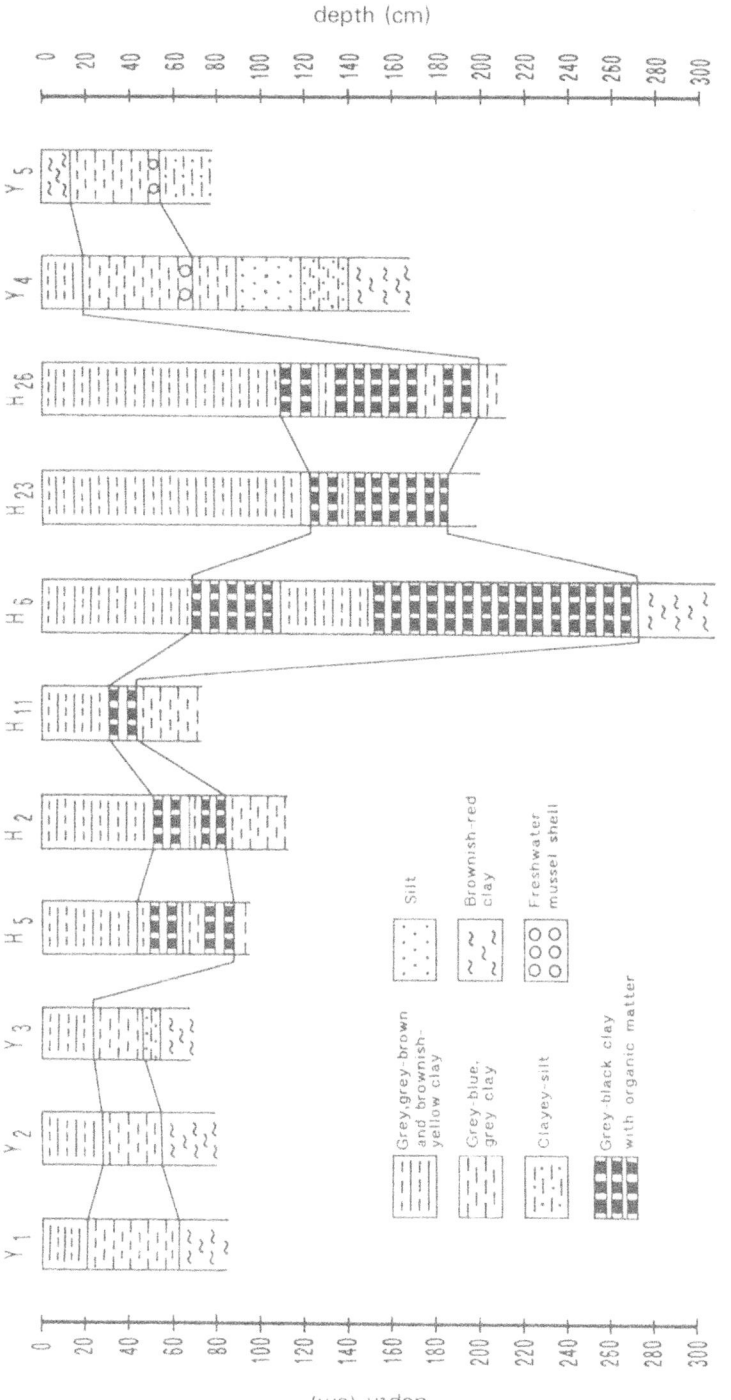

Fig. 2. Comparison of sedimentary profiles in Honghu Lake area.

3. Organic sedimentation. Honghu is a shallow water lake and aquatic plants grow across the bed. Along the shore are found extensive stands of *Zizania lat.folia* (Griseb.) Stapf and *Nelumbo nucifera* Gaertn. In the central regions of the lake grows *Potamogeton maackianus* A Bennett. The organic matter content of the sediment is comparatively high, reaching a maximum of 12%.

These three processes produce the following types of sediment (Fig. 1; Fig. 2).

Deltaic deposits

The Neijin River enters the lake from the north (Fig. 1). The sediments deposited at the mouth of this river form a delta. At present this can only be seen on the eastern bank near Dakou. The deltaic topsets, foresets and bottomsets are mainly clay. From Landsat MSS images we can see seven deltaic lobes.

Inflow-outflow current deposits

These sediments have been brought in by the Neijin and Changjiang Rivers. The river flow extends in an unconfined channel across the lake. The deposits consist mainly of clay and silt. The pattern of water flow from the rivers is complex and partly controlled by human agencies. Since 1958 the main source of sediment has been the Neijin River.

Still water lacustrine deposits

These are formed in the still waters on both sides of the main river courses. They consist of yellow-brown, grey-brown or grey-white clays.

Swamp deposits

These occur beneath the areas of active plant growth where organic detritus accumulates.

Shore deposits

The major part of the natural levee of the Changjiang River in the south-eastern part of Honghu consists of silt. By wave action, silt particles move towards the centre of the lake.

Sediment stratigraphy (Figure 2)

In vertical section the deposits vary greatly. For example, they show changes from lacustrine to swamp sediments and back again. The sequence of deposits is not the same in all parts of the lake.

Dating, sedimentation rate and sediment influx

A radiometric dating framework for Honghu Lake has been provided by ^{210}Pb and ^{14}C analyses. Lead 210 data indicates that the rate of sedimentation varies from place to place. It may reach $0.16–0.19$ cm y^{-1} with a sediment influx of between 0.13 and 0.18 g cm^{-2} y^{-1}. The sedimentation rate in still water is relatively small; between 0.07 and 0.09 cm y^{-1} with an influx of $0.05–0.07$ g cm^{-2} y^{-1}. According to ^{14}C dates the age of the bottom of the lacustrine deposits in the western part of Honghu is 2500 ± 70 years. The age of the overlying swamp deposit is 940 ± 65 years.

Discussion

The formation of Honghu Lake

Honghu Lake was formed between the natural levees of the Changjiang and Dongjin rivers in the lowest part of the basin, during the Chun Qiu epoch (770–476 BC). Before then the area was part of the flood plain of the Changjiang River with the characteristic landscape of a 'water country'. It is therefore clear that it is not a remnant lake of the so called Yunmung swamp, but is a large levee-dammed lake of recent origin.

The evolution of Honghu Lake

After the formation of Honghu, its depositional processes were complex and the area experienced great changes. In the early period, the lacustrine deposits extended over the whole basin, and consisted of black-grey and grey-white sediments. According to soil profiles collected around Honghu lacustrine sediments are distributed to the west of the modern lake. They show that from 2500 years BP, Honghu was separated into two small lakes. Western Honghu lake was larger than the east. Above the black grey or grey white lacustrine deposits are more than 50 cm of swamp deposits. The ^{14}C age of the bottom of this layer is between 890–960 years.

In the lowermost swamp deposits there is a lacustrine layer. This suggests that the process of swamp formation was interrupted for a short period when the lake emerged again. According to historical documents, during the Jin epoch (265–420 AD), Honghu turned into a great lake. After the Song epoch (960 AD–1279 AD) the lake again became a swamp with extensive beds of *Phragmites australis*. The result of our investigations are therefore in accord with the historical record.

At the time of the development of the swamp the average annual temperature fluctuated several times. Meanwhile, Chinese sea level was low, which increased the gradient of the Changjiang river and Hanshui rivers. At the same time, low temperatures meant both decreased rainfall and flooding, so that the lake became replaced by swamp.

Above the swamp deposits are grey-brown and yellow brown modern sediments. They cover the whole lake area, and some of the fields around the lake. By ^{210}Pb determination we estimate that the age of the bottom of the present deposit is 440 years. This denotes that before 400 years ago, Honghu extended rapidly, and the eastern and western part merged into one big lake.

After that, Honghu was reduced in area in the recent period. Before the 19th century owing to the influence of human activity, the area of Honghu was 20% of that of the modern lake. Rapid expansion took place at the end of the 19th century. The reason for this is that lake water was ponded by the natural levee of the Chiangjiang River. In the 1950's, 1960's and 1970's, reclamation of the lake took place and Honghu has turned into an almost-closed lake system.

Acknowledgement

The authors gratefully acknowledge the help of Drs J. P. Smith & P. E. O'Sullivan in the preparation of this paper.

References

Cai, S. M. & Z. H. Guan, 1979. A study on the geology of Lake Donghu. Oceanol. Limnol. Sinica 10: 385–394 (in Chinese).

Cai, S. M. & Z. H. Guan, 1982. The Comment about the Ancient Yun Meng swamp. Oceanol. Limnol. Sinica 13: 130–142 (in Chinese).

Cai, S. M. & Z. H. Guan, 1986. Formation and evolution of Lake Dongjin. Sediments Research 1: 31–34 (in Chinese).

Hun, D. F., 1965. The geology, formation and development of three large freshwater lake in the lower basins of the Changjiang. Oceanol. Limnol. Sinica 7: 392–422 (in Chinese).

Hutchinson, G. E., 1957. A Treatise on Limnology, I. Wiley and sons, New York, N.Y., 1015 pp.

Yen, H. R. & G. L. Chen, 1960. The Landforms and Quaternary geology of the Middle and Lower Basins of the Changjiang. National Geography Association Symposium. Science Press, Beijing: 6–44 (in Chinese).

Hydrobiologia **214**: 347–357, 1991.
J. P. Smith, P. G. Appleby, R. W. Battarbee, J. A. Dearing, R. Flower, E. Y. Haworth, F. Oldfield & P. E. O'Sullivan (eds), 347
Environmental History and Palaeolimnology.
© 1991 *Kluwer Academic Publishers.*

Palaeolakes of the south central Sahara – problems of palaeoclimatological interpretation

Roland Baumhauer
Geographisches Institut, Universität Würzburg, Am Hubland, D-8700 Würzburg, Germany

Key words: sedimentology, palaeolimnology, palaeoclimatology, central Sahara, Holocene

Abstract

The results of palaeoecological studies of Holocene swamp and lake deposits of a number of endorheic depressions of the Kawar, Djado and Great Erg of Bilma region of eastern Niger are presented, comprising analysis of their stratigraphy, sedimentology, diatom flora and macrofossils.

The investigations demonstrate that various palaeolakes have reacted differently in space and time and by type of lake to climatic conditions. Some of the lakes reacted rapidly to changes in the precipitation regime, as evidenced by changing size, level, water balance and water chemistry, while perennial fresh-water lakes nearby show changes relatively independent of short-term climatic fluctuations. These facts suggest a more complex influence of local and regional geomorphological, hydrological and hydrogeological factors on the Holocene lake evolution than a mere climatic dependence. Beyond doubt humidity considerably increased during early and mid-Holocene periods, but the stratigraphical and ecological status obtained for individual endorheic depressions seems to be mainly a reflection of differences among groundwater catchments, aquifers of different size, local topographical conditions and changes in geomorphology (e.g. dune activity) – superimposed on the major climatic tendencies – and thus of a considerable diversity of palaeoenvironmental conditions occurring within the region at any given time of the Holocene.

Introduction

In the history of the earth climate has changed on many spatial and temporal scales. One method of quantifying climate history data as well as to reconstruct palaeoenvironments in the use of temporal variations in the extent of lake levels. The changes in the size of a lake as a result of climatic fluctuations have often been demonstrated, either by using proxy data (e.g. Street & Grove, 1979; Servant, 1983; Petit-Maire & Riser, 1983; Durand, 1984; Ritchie & Hayne, 1987) or by using hydrological and energy-balance models

(e.g. Kutzbach, 1980; Adams & Tetzlaff, 1984; Adams, 1987).

This paper shows some of the problems of palaeoclimatological interpretation associated with stratigraphical, sedimentological, palaeoecological and chronological studies of Holocene swamp and lake deposits in the south central Sahara.

The Holocene records of Kawar, Seggedim and Dibella region of eastern Niger have been chosen as study sites. Kawar, Seggedim and Dibella are comparable sites representing former lakes in endorheic depressions in front of cuestas.

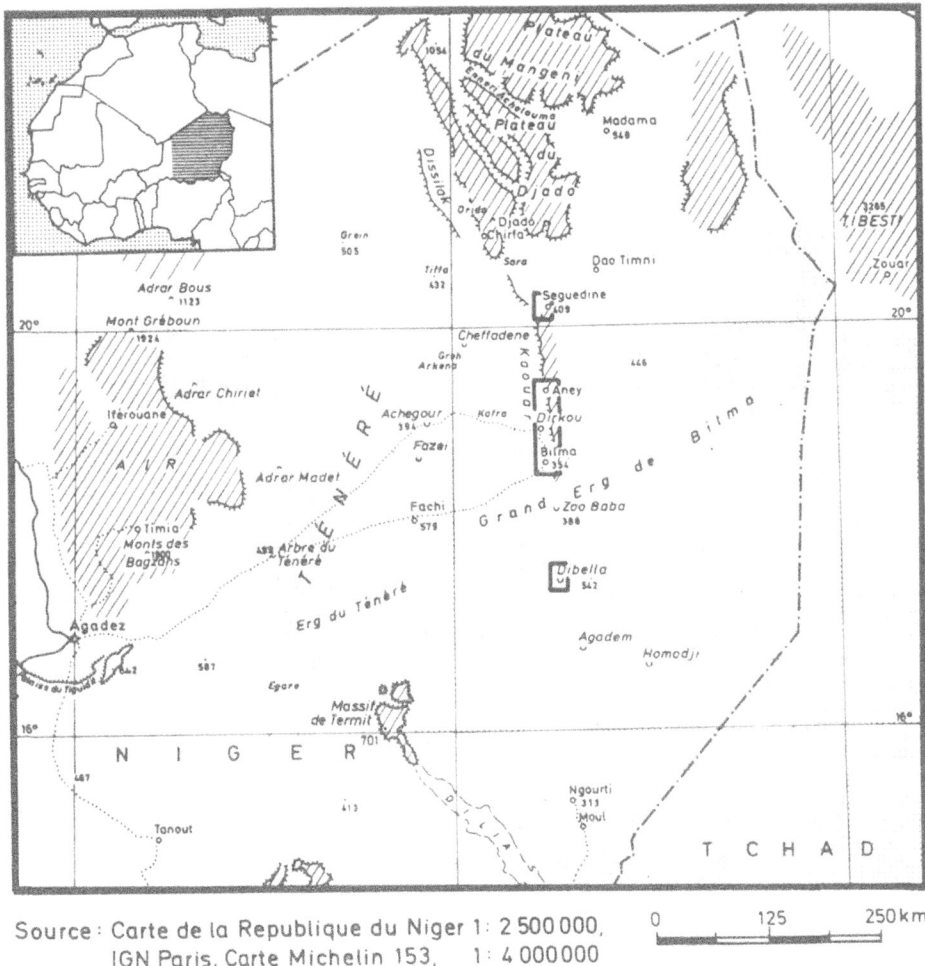

Source: Carte de la Republique du Niger 1:2 500 000,
IGN Paris, Carte Michelin 153,   1:4 000 000

Fig. 1. Map of eastern Niger. Frames indicate areas of investigation.

Study area

The endorheic depressions of Kawar, Seggedim and Dibella lie at the northern margin of the Chad-Basin in northeastern Niger aligned between Djado-Plateau and modern-day Lake Chad following a meridional tectonic structure that probably extends further north to the western margin of the Murzuk-Basin (Fig. 1). They are typically situated in the western foreland of cuestas or isolated plateaus of Nubian Sandstone. These form free-faced escarpments of an absolute height of 500–640 m and a relative height of 150–200 m.

The Kawar is situated in the southern part of the foreland depression of the cuesta of Bilma (maximum height: 548 m) at the eastern margin of Ténéré. The depression is orientated approximately north-south with a length of 80 km, a width of 6 km and an overdeepening ranging from 20 m (altitude: 380 m) in the northern part of the depression to 80 m (altitude: 350 m) in the Bilma region. Though the annual precipitation is only 18 mm and the potential evaporation rate is very high (*ca.* 3000 mm/yr) the depression of Kawar contains four lakes with a total surface area of about 0.5–200 ha according to seasonal variations. These groundwater-fed lakes are shallow, saline (Na-CO_3-SO_4-Cl) lakes surrounded by belts of salt marsh.

The Seggedim (Emi Bao) foreland depression some 150 km north of Bilma, but still part of the large cuesta of the Bilma depression zone, consists of a sebka of approximately 10 km² with some groundwater inflows in the centre and the eastern margin, stretching 2 km in an east-west direction with a maximum width of 7 km. The cuesta has a relative height of about 220 m (altitude: 639 m). The surrounding plains are mostly covered with a serir formed by the eroded gravels of some conglomeratic layers of the cuesta.

The area of Dibella, 120 km south of the Kawar depression consists of several narrow plateau relicts, some 500 m high, but projecting only slightly above the rolling dune landscape of the southern Erg of Bilma. The predominate dunes are fixed today. In contrast the few tradewind-orientated longitudinal dunes are very active. These lie in a leeward position at the western rim of the plateaus

and cross the deep foreland depression subdividing it into several units. The depression has a surface area of approximately 15 km² and, at deepest point, an altitude of about 368 m. An artificial spring is filled with slightly brackish water some 0,5 m below the surface, showing the modern-day groundwater table.

Kawar

In the Kawar depression, particularly in the Bilma region, the southern part of the foreland depression of the Bilma cuesta, the synoptic view of more than 40 investigated stratigraphical profiles shows two major phases of Holocene lacustrine sedimentation, dated from *ca.* 9300 to 7500 BP (Servant, 1983; Baumhauer, 1986), with a maximum at about 8000 BP and 6500 to 5000 BP

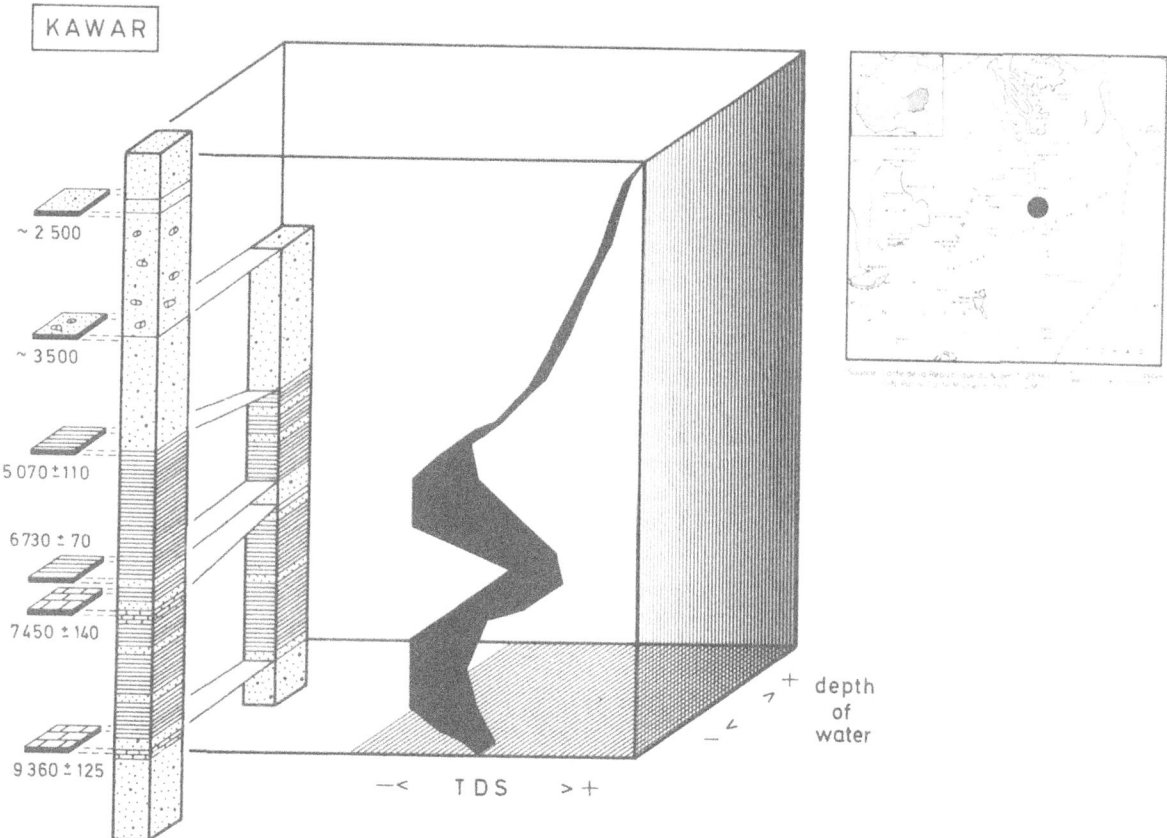

Fig. 2. Schematic reconstruction of Holocene lake phases in the depression of Kawar.

350

(Servant, 1983; Baumhauer, 1986) separated by accumulations of aeolien and illuviated sands only in the littoral zone of the palaeolake (footzone area of the cuesta), while in the central part a continuous lacustrine and swamp sequence persisted. Since *ca.* 5000 BP, after a long regression phase, characterized by the on-lap of carbonate-, gypsum-, and ferricrusts interbedded with clayey-silty sands several ephemeral or permanent shallow swampy waterbodies persisted during the late Neolithic (Fig. 2). In the northern part of the Kawar depression four groundwater-fed hypersaline ponds with seasonal and interannual fluctuations persisted up to the present.

The lacustrine and swamp sequence consists of diatomaceous deposits and sandy/silty clays with diatom concentrations varying from 0 to 10^6 valves mg^{-1}. Generally, percentage diatom analyses have been performed on one sample every 40 cm at least by counting 500 valves per level.

The taxa encountered show comparable spectra for the early/mid Holocene limnic phases. In the central part of the Kawar depression the diatom assemblage is composed of planktonic species. The most common species include *Fragilaria brevistriata*, *Cyclotella kutzingiana*, and *Cyclotella stelligera* (Fig. 3). In the foot-zone area of the cuesta (shore area of the Holocene lake) predominate littoral epiphytic and euplanktonic species with an abundance of *Cyclotella ocellata*, *Epithemia zebra* and *Melosira granulata*. Therefore the diatom spectra encountered characterize a perennial and deep freshwater lake in the early and mid Holocene. Only in the beginning of the Holocene and during a short regression phase between the early and mid-Holocene period do evaporite-crusts, ferricrusts and the diatom assemblages, predominated by *Campylodiscus clypeus* and *Melosira granulata angustissima*, indicate a shallow lake or swamp environment.

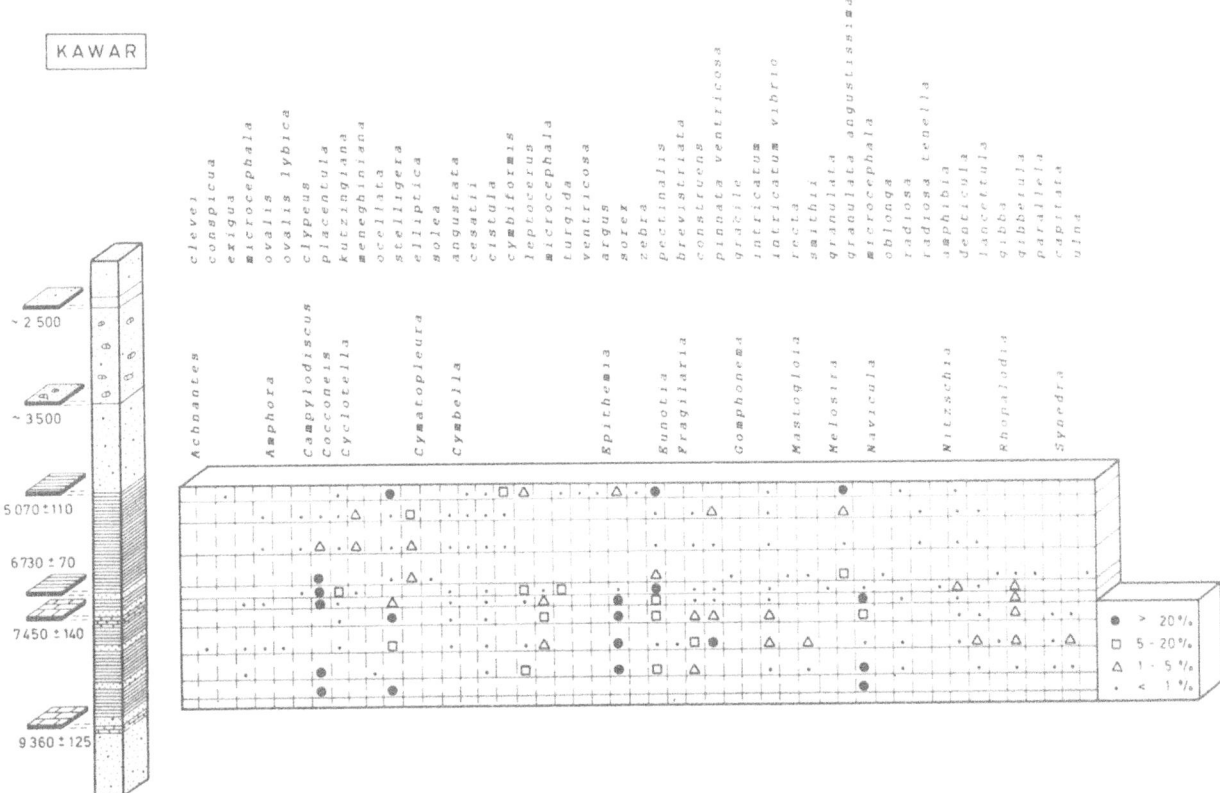

Fig. 3. Stratigraphy and percentages of the diatom taxa in the central part of Kawar depression.

351

The abundance of *Melosira granulata* and above all the domination of freshwater *Cyclotella spp.* indicate high inputs of dissolved silica. Normally an increase of local/regional precipitation leads to solute dilution, but the input of dissolved silica can only be attributed to groundwater (Hurley *et al.*, 1986, Gasse, 1987). The high content of dissolved silica may be derived from clay mineral transformation of *kaolinite* (sensu stricto) to *mica-montmorillonites* (Baumhauer, 1986). The complete absence of mollusc shells in the lacustrine sequences of Kawar as well as the very high dissolved silica content of modern-day artesian water of the Kawar oasis supports the groundwater hypothesis (Faure, 1963; Baumhauer & Hagedorn, 1989).

The palaeoecological and geomorphological evidence (Baumhauer, 1986) suggest that during the early Holocene maximum at about 8000 BP the lake level reached at least 25 m above the present ground of the Kawar depression, and *ca.* 15 m during the mid-Holocene.

Seggedim

An 8 m core from the central section of the sebka of Emi Bao/Seggedim shows a continuous accumulation of lacustrine and swamp sediments (Fig. 4). The drill log (Fig. 5) shows an upper part (0–5,3 m) of yellow to brownish sands rich in evaporites (NaCl) with a very low content of fine-

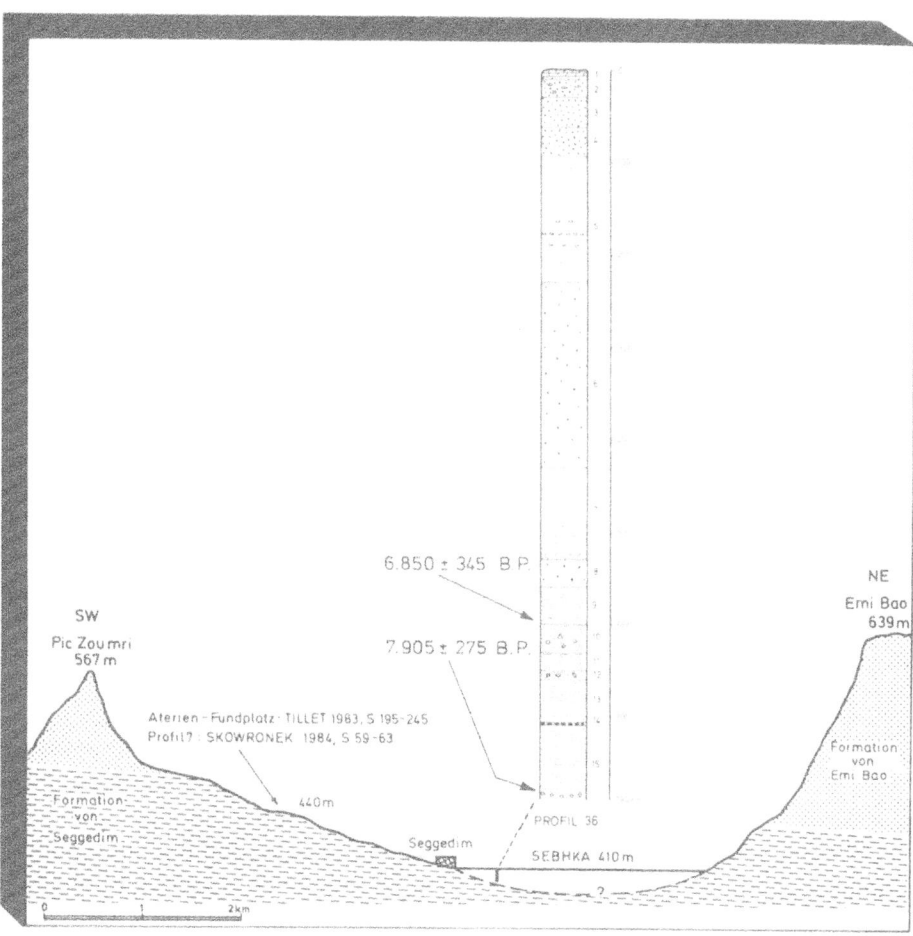

Fig. 4. Cross section of the depression of Seggedim showing the stratigraphy and location of Seggedim core (lithofacies is marked in Fig. 5).

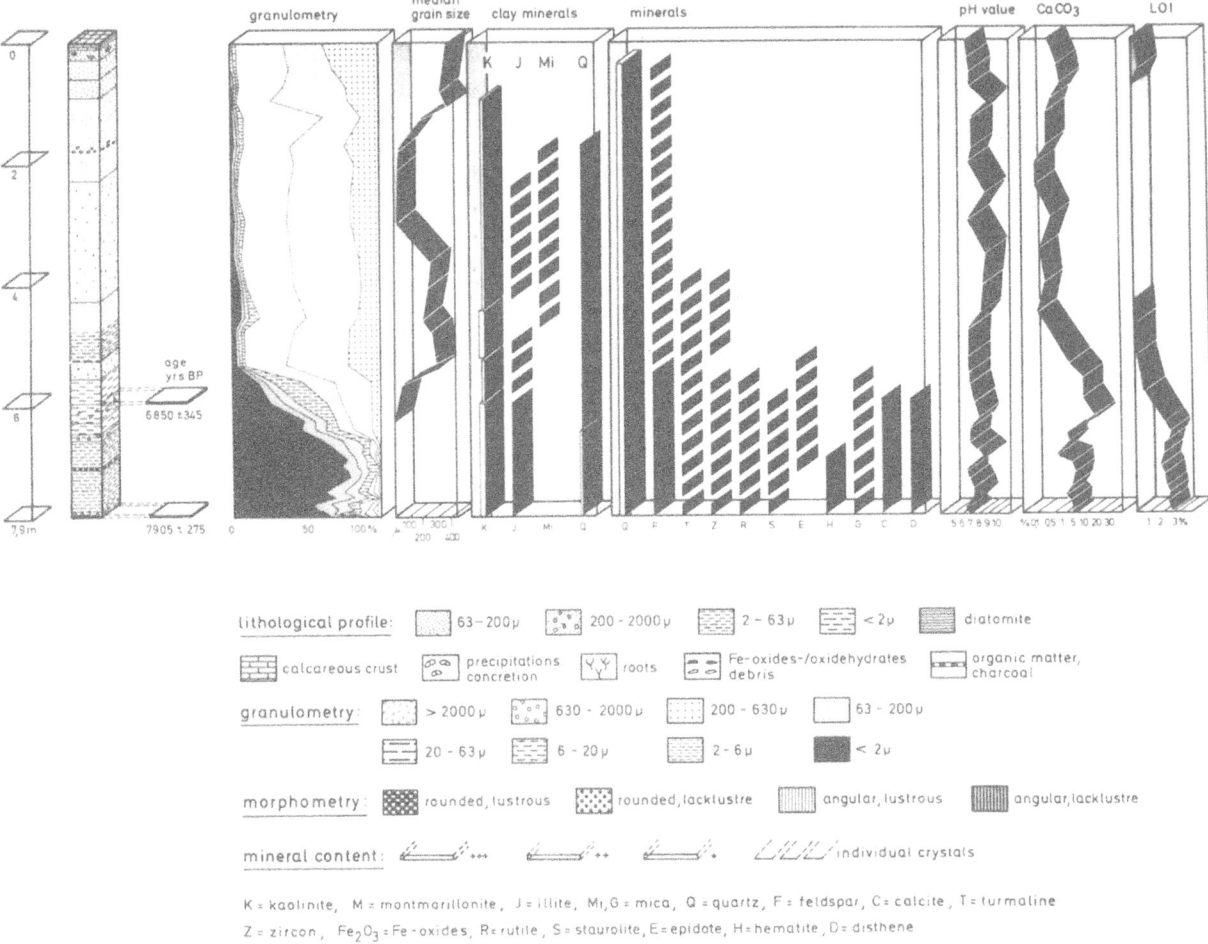

Fig. 5. Lithofacies and results of sedimentological analyses of Seggedim core.

grained material, representing the sedimentation milieu of a sebka. Between 4,2 m and 5,3 m the sand become browner. At 5,3 m below the surface a rapid change to brown clays occurs. Below 6,4 m the brown clays change to greenish and blue clays containing a large amount of charcoal and macro-remains of large grasses of the *Phragmites* type.

The sedimentological investigations and the microfossil record (phaeopigments, *Pediastrum* and diatoms) indicate the desiccation of a freshwater lake to a sebka environment. Pollen spectra from the lower part of the core confirm the environmental conditions at that time showing an open vegetation consisting of *Acacia-Maerua-Capparis* communities dominated by *Gramineae* and *Cyperaceae* (Baumhauer, & Schulz, 1984). Detailed results of the biological remains of the depression of Seggedim will be discussed elsewhere.

Dates at 7,9 m (7905 ± 275 BP) and at 5,9 m (6850 ± 345 BP) during the desiccation phase from freshwater lake to sebka, show a rapid change in environmental conditions. In contrast to Kawar, only 150 km away, the Seggedim Holocene lake dried up, whereas in the Bilma region the deposits indicate the persistence of previous conditions at least till 5000 BP (Fig. 6).

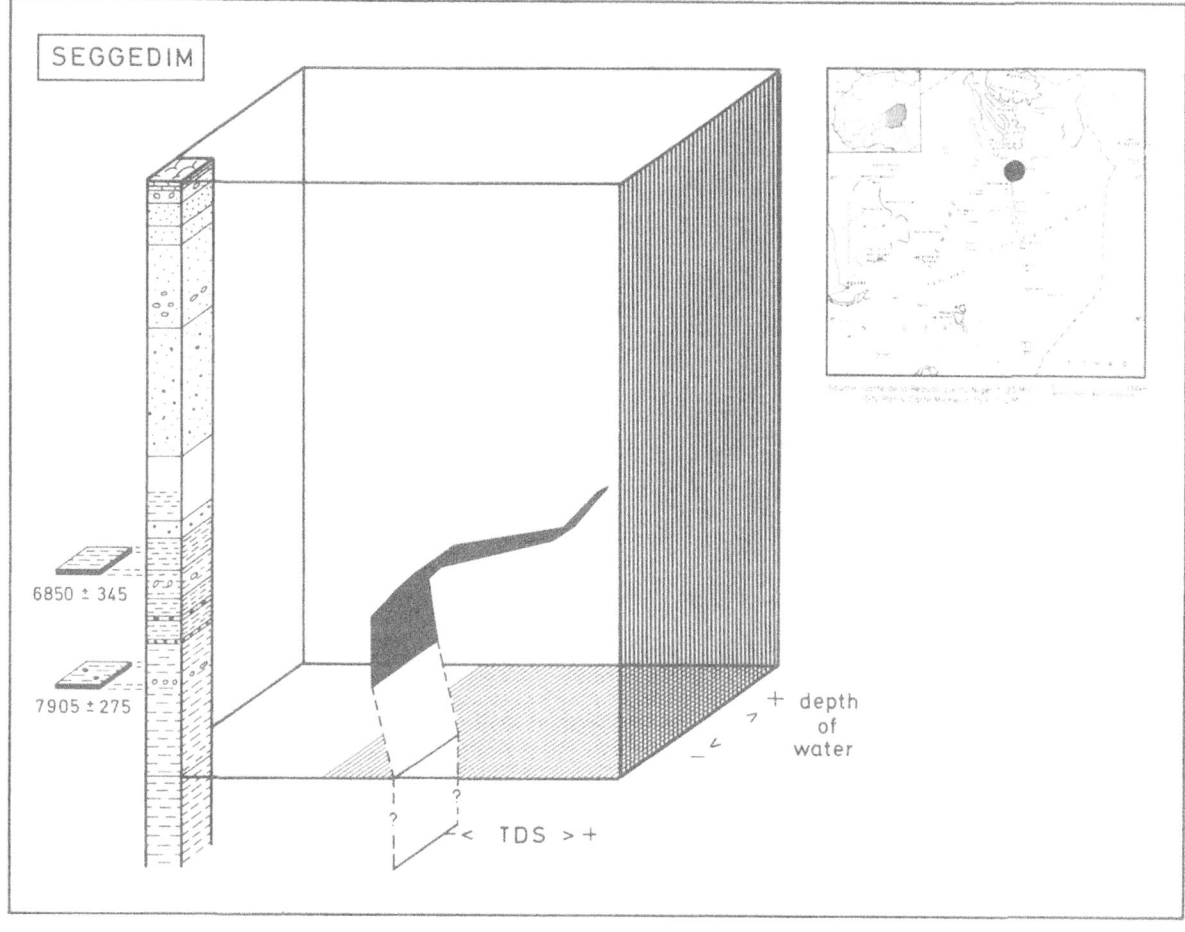

Fig. 6. Schematic reconstruction of Holocene lake phase in the depression of Seggedim.

Dibella

In the depression of Dibella, located at the southern margin of the Central Erg of Bilma, the phase of lacustrine sedimentation started after a dune-sand accumulation at about 10 000 BP (Baumhauer, 1987). Oxidation horizons (FeO) and decomposed organic matter underlying the lacustrine deposits indicate the erosion of a palaeosol by accelerated fluvial processes. It may be supposed that in situ weathering processes occurred at that time, documented by a distinct increase in the content of feldspar, mica, illite and mixed-layer clays towards the top of the thick layer of fine sand (Fig. 7). At the same time as this increase in fluvial activity and weathering the de-

pression probably divided by dune-cutting into several basins, filled with water.

In contrast with the palaeolakes of Kawar and Seggedim, the lacustrine sediments, dominated by laminated diatomites with diatom concentrations $> 10^6$ values mg^{-1}, indicate a freshwater environment typical for the littoral zone of a deeper lake only for a short period in the early Holocene. Planktonic/euplanktonic taxa (*Cyclotella stelligera, Melosira granulata, Melosira spp., Synedra ulna*) are abundant in layer 16 (Fig. 8). The taxa encountered in the major part of the diatomite sequence reflect an instability of the Dibella depression waterbody with many short-term fluctuations in environmental conditions (Fig. 8). Inter-stratified evaporite- and ferricrusts also indi-

354

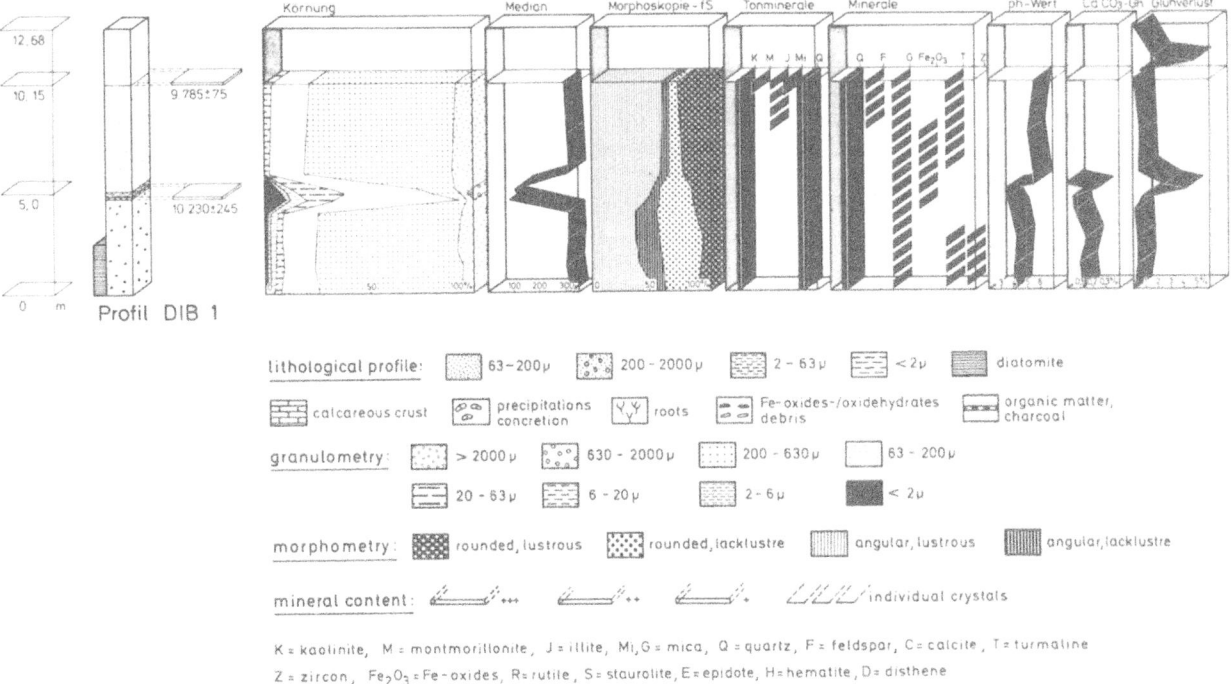

Fig. 7. Lithofacies and result of sedimentological analyses of Dibella profile DIB 1.

cate the frequent changes of water and salt balance until *ca.* 7000 BP. (Fig. 9). This is in contrast with the topographical situation if the lacustrine sediments, the top of which reaches some 20 m above the present depression. Therefore the fossil records indicate that the water level as well as the chemical and biological remains of the Holocene waterbody(ies) might have been

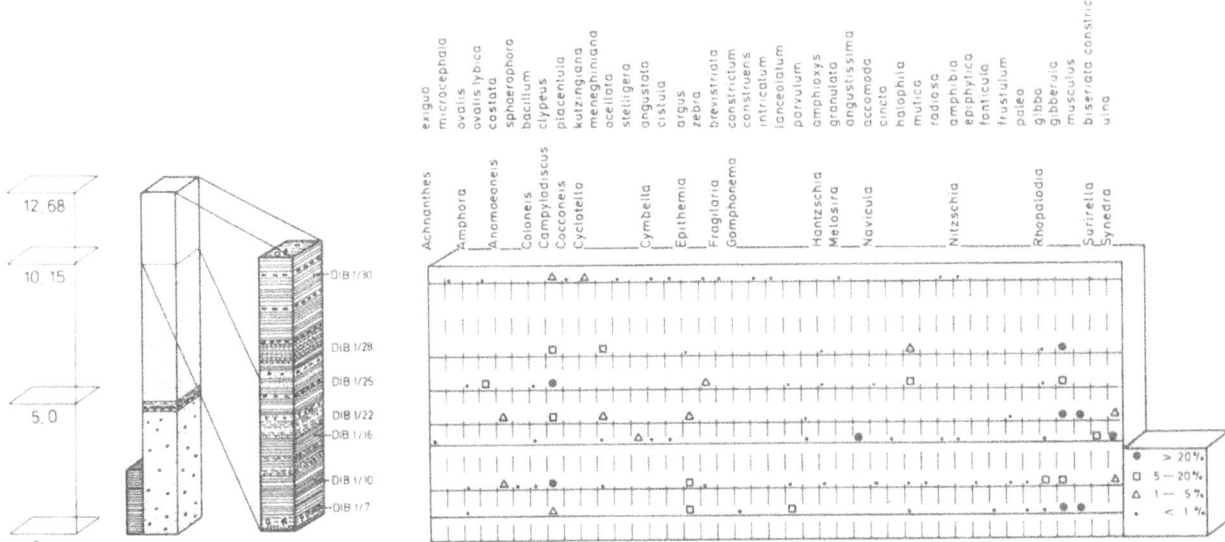

Fig. 8. Stratigraphy and percentages of the diatom taxa of Dibella profile DIB 1.

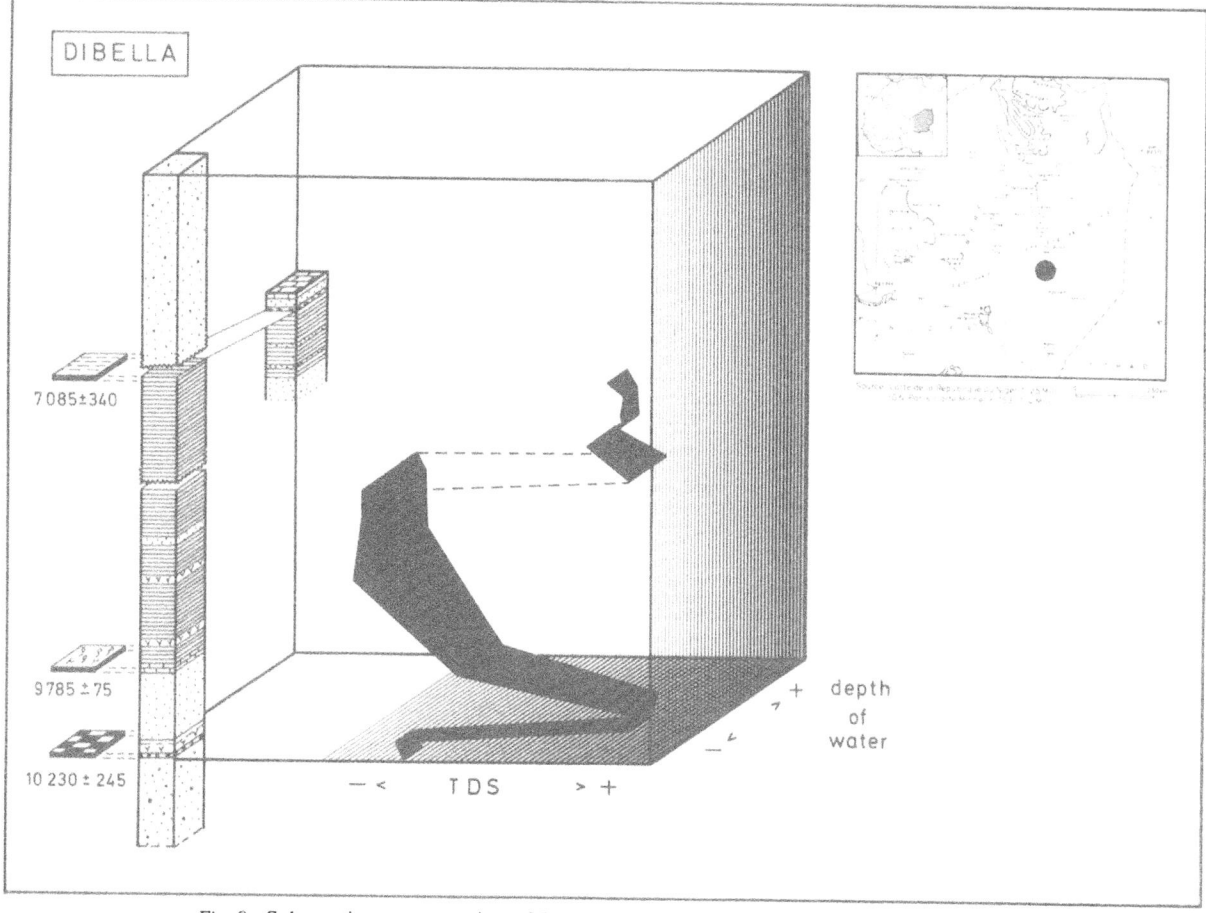

Fig. 9. Schematic reconstruction of Holocene lake phases in the depression of Dibella.

rapidly modified not only by climatic fluctuations but also by variations in geomorphological status, for example a sub-division of the depression into several units by dunes. Nevertheless several analysed palaeosols, found on the surface of flattened fossil dunes of the Ogolien/Kanemien period surrounding the depression of Dibella are a good indication of greatly increased precipitation at that time (Völkel, 1989; Grunert *et al.*, 1989).

Generally, the early Holocene Dibella lake sediment sequence suggests a history very similar to those of Holocene interdunal waterbodies in the Great Erg of Bilma and Ténéré region, not only in time but also in palaeoecological conditions (Baumhauer, in prep.).

Conclusions

Comparing the Holocene lacustrine sediments, the investigations demonstrate that various palaeolakes have reacted differently in space and time and by the type of lake to climatic conditions (Fig. 10).

Beyond doubt humidity must have been considerably increased during the early and mid-Holocene periods, and therefore the general palaeoclimatological reconstructions which have been deduced from lacustrine accumulations as well as from water and energy balance models of closed lakes for the south Sahara are comparable. But the (regional) palaeoclimatic conditions can-

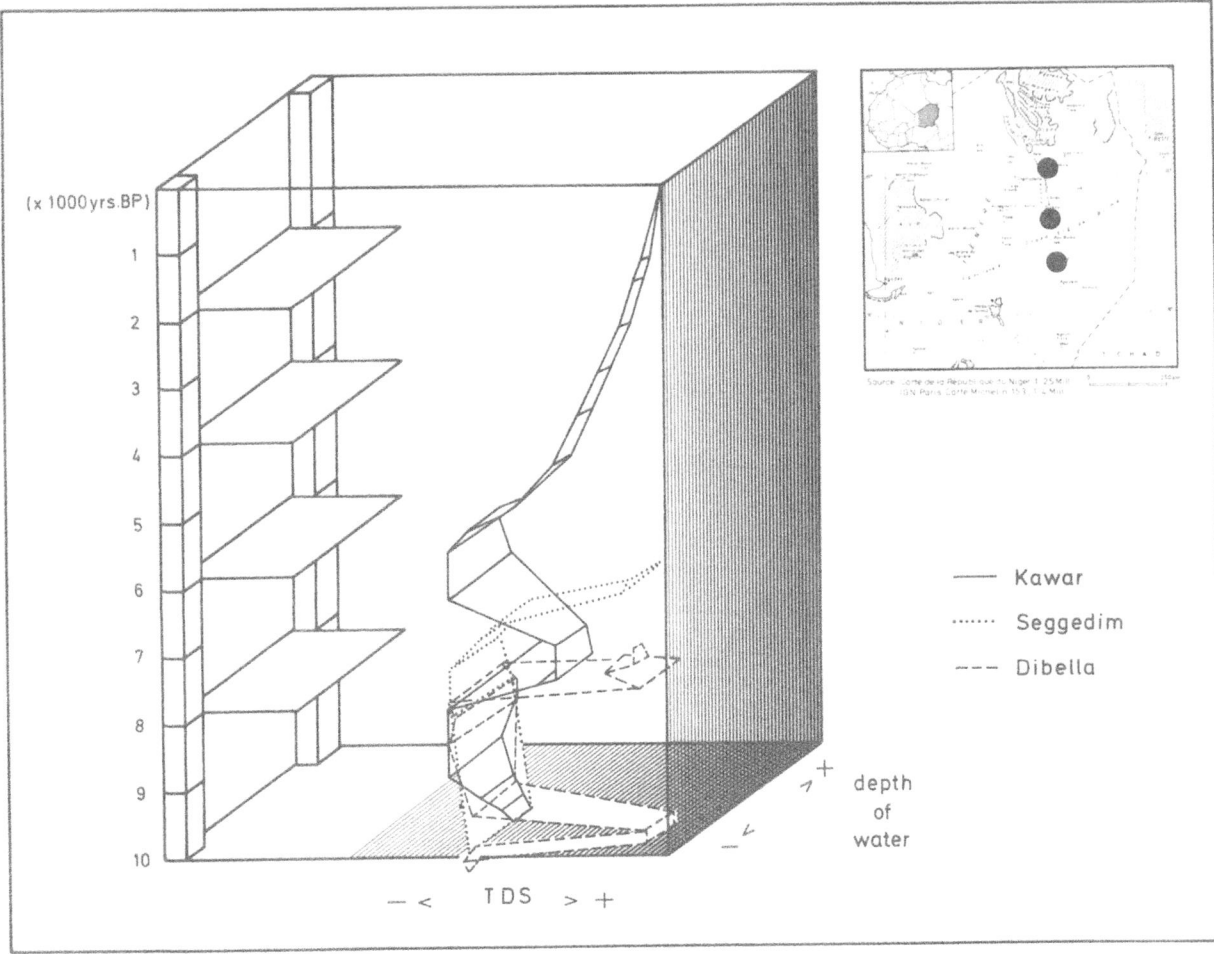

Fig. 10. Schematic reconstruction of Holocene lake phases in the depressions of Kawar, Seggedim and Dibella.

not explain the differences in the behaviour of the individual palaeolakes. Waterbodies with clear changes in water balance and water chemistry occurred, while perennial freshwater lakes persisted nearby.

This spatial and temporal diversity in lake history suggests a more complex influence of local and regional geomorphological, hydrological and hydrogeological factors on the Holocene lake evolution in the investigated area than a mere climatic dependence. The stratigraphical and ecological status obtained for individual endorheic depression seems to be mainly a reflection of differences in groundwater catchments, aquifers of different size, local topographical conditions and

changes in geomorphology (e.g. dune activity), superimposed on the major climatic tendencies.

This approach shows that caution is necessary when drawing large scale (zonal) palaeoclimatological conclusions from palaeolimnological data without applying local/regional correction factors.

References

Adams, L. J., 1987. Ein Wasser- und Energiebilanz-Modell von abflußlosen Seen und seine Anwendung in der Paläoklimatologie von Nordwest-Afrika. Ber. d. Inst. Meteo. u. Klimatologie, Hannover 29, 127 p.

Adams, L. J. & G. Tetzlaff, 1984. Did lake Chad exist around 18 000 yrs. BP?. Arch. Met. Geoph. Biocl., B34: 299–308.

Baumhauer, R., 1986. Zur jungquartären Seenentwicklung im Bereich der Stufe von Bilma (NE-Niger). Würzb. Geograph. Arbeiten 65, 235 p.

Baumhauer, R., 1987. Holozäne limnische Akkumulationen im Bereich der Stufen von Zoo Baba und Dibella (NE-Niger). Palaeoecol of Africa 18: 167–177.

Baumhauer, R. & E. Schulz, 1984. The holocene lake of Séguédine, Kaouar, Niger. Palaeoecol of Africa 16: 283–290.

Baumhauer, R. & H. Hagedorn, 1989. Probleme der Grundwassererschließung im Kawar (Niger). Die Erde 120: 11–20.

Durand, A., Fontes, J. Ch., Gasse, F., Icole, M. & J. Lang, 1984. Le Nord-Ouest du lac Tchad au quaternaire: étude de paléoenvironnements alluviaux, éoliens, palustres et lacustres. Palaeoecol. of Africa. 16: 215–243.

Faure, H., 1963. Inventaires des évaporites du Niger (Mission 1963). Min. Trav. Publ., des Mines et de l'Hydraulique, Niamey Niger, 350 p.

Gasse, F., 1987. Diatoms for reconstructing palaeoenvironments and palaeohydrology in tropical semi-arid zones. Hydrobiologica 154: 127–163.

Grunert, J., Baumhauer, R. & J. Völkel, 1989. Paleolakes between Kaouar and Lake Chad (E-Niger): The example of Zoo Baba and Dibella. In: Kogbe, C. A. (Hrsg.); African Continental Sediments (in press).

Hurley, J. P., Armstrong, D. < <E., Kenoyer, G. J. & C. J. Bowser, 1986. Ground water as a silica source for diatom production in a precipitation-dominated lake. Science 227: 1576–1578.

Kutzbach, J. E., 1980. Estimates of past climate at palaeolake Chad, North Africa, based on a hydrological and energy-balance model. Quat. Res. 14: 210–223.

Petit-Maire, N. & J. Riser, 1983. Sahara ou Sahel. CNRS, Marseille, 473 p.

Ritchie, J. C. & C. V. Haynes, 1983. Holocene vegetation zonation in the Eastern Sahara. Nature 330: 352–354.

Servant, M., 1983. Séquences continentales et variations climatiques: évolution du bassin du Tchad au Cénozoique supérieur. Trav. et Doc. Orstom 159, 573 p.

Street, F. A. & A. T. Grove, 1979. Global maps of lake level fluctuations since 30 000 y. BP. Quat. Res. 12: 83–118.

Völkel, J., 1989. Geomorphologische und pedologische Untersuchungen zum jungquartären Klimawandel in den Dünengebieten Ost-Nigers (Südsahara und Sahel. – Bonner Geograph. Abh. 79, 258 p.

Hydrobiologia **214**: 359–365, 1991.
J. P. Smith, P. G. Appleby, R. W. Battarbee, J. A. Dearing, R. Flower, E. Y. Haworth, F. Oldfield & P. E. O'Sullivan (eds), 359
Environmental History and Palaeolimnology.
© 1991 *Kluwer Academic Publishers.*

Holocene environments in the central Sahara

Erhard Schulz

Geographisches Institut, Am Hubland, D 8700 Würzburg, Germany

Key words: central Sahara, Holocene, palynology, Palaeolakes, palaeoenvironment, palaeoclimate

Abstract

Palynological investigations of corings in the sebkhas of Taoudenni (N-Mali) and Segedim (N-Niger), archaeological excavations in the Acacus Mts. (SW-Libya) and charcoal records in the central Ténéré (Niger) give evidence for a northward shift of the desert-savanna boundary to 22°–20° N during the middle Holocene. Between Niger and S-Libya there was a ecological gradient from the sudanian, sahelian and saharan savannas to a denser saharan desert vegetation. After a transition phase between 6000 and 4000 BP the saharan desert vegetation was finally established in the Taoudenni and Segedim region and this degraded from ca. 2000 BP to its present condition.

During the middle Holocene the central Sahara had a monsoonal summer rain climate with an effective rainfall of 250–300 mm per year near the desert-savanna boundary (ca. 22° N). Interaction between the monsoon and the atlantic cyclones also allowed rainfall in other periods of the year.

Introduction

The steadily growing demand for water in desert areas and the need to mine water as a non renewable resource (Thorweihe, 1988) is the main background for the investigations of palaeo-climates and of former recharge estimates of groundwater in these areas. For the central and southern Sahara a great deal of information derives from the presence and history of palaeo-lakes (Baumhauer, 1986; Gasse, 1988; Kutzbach, 1980; Petit Maire & Riser, 1983). These palaeo-lakes provide information about regional environments e.g. vegetation, catchment area as well as local environments e.g. water chemistry, algal content and different lake levels. Discrimination between local and regional scales of information is important for the reconstruction of former climates and for estimations of non-renewable resources like fossil groundwater.

This paper deals with the reconstruction of the Holocene vegetation, its regional differences and its development during the Holocene. Based on the former vegetation some palaeoclimatological inferences are made. Cores from a variety of palaeolakes in the central Sahara provided samples rich enough in pollen and macrofossils to establish a vegetation history for the middle and late Holocene. These cores were taken in Taoudenni (N-Mali) (Petit-Maire, 1986; Fabre & Petit-Maire, 1987; Petit-Maire, 1988) and Segedim (Baumhauer & Schulz, 1984; Schulz, 1987). Additional information comes from archaeological sites in Dj. Acacus (SW-Libya) (Barich, 1987) and from charcoal findings from Fachi (Ténéré) (Neumann, 1988).

Methods

The Taoudenni sequence was sampled from an open pit in the upper part and cored with a

360

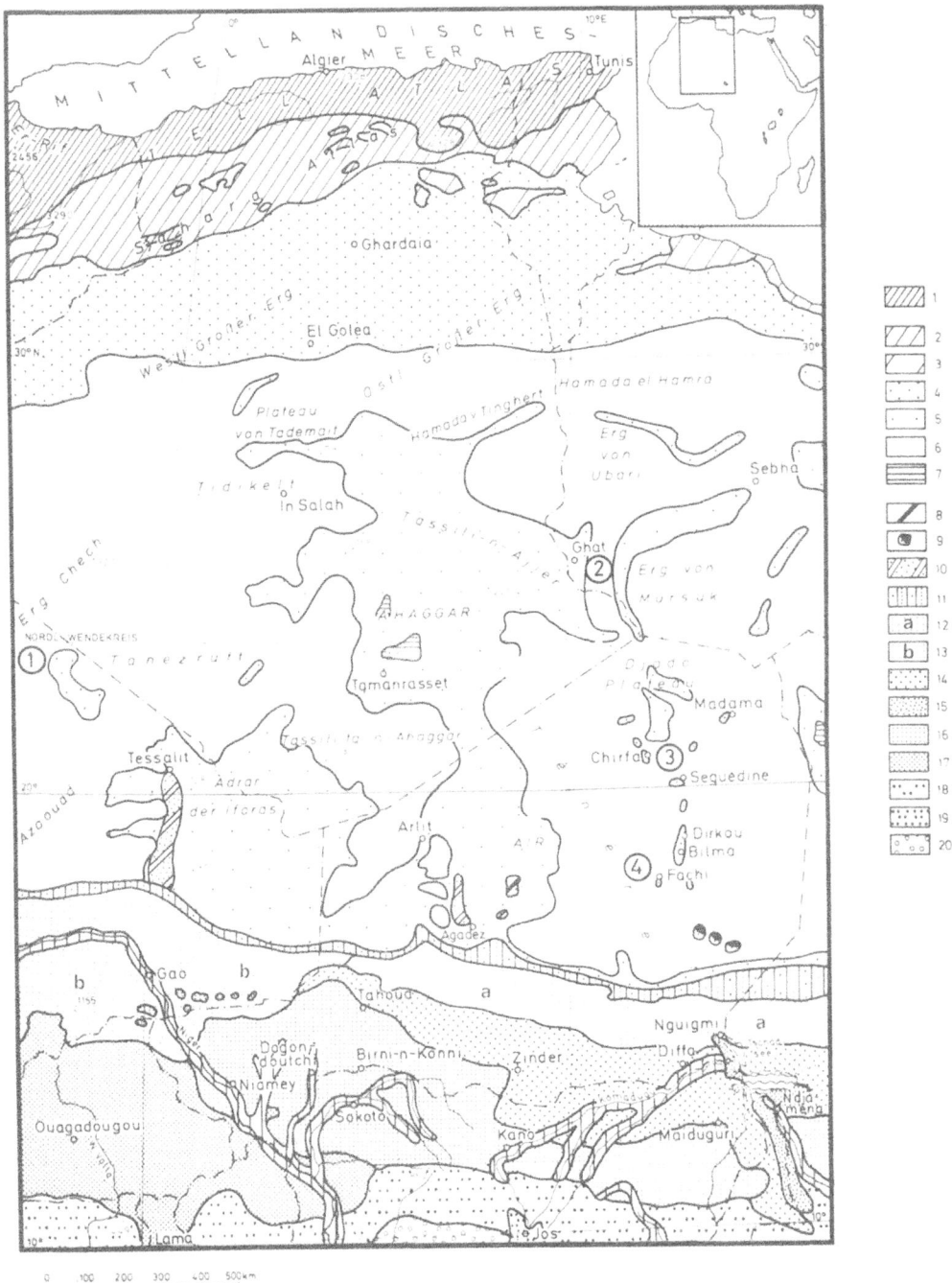

Fig. 1. Schematic vegetation map of the Sahara and the surrounding areas also shown are the sites of the holocene records. (Schulz, 1987). Inset shows the study area within Northern Africa.
1. Medit. shrub and forest (*Quercus, Pistacia, Pinus, Cedrus*). 2. Medit. steppe (*Stipa, Lygeum*). 3. Chenopodiaceae vegetation of the coast and the chotts. 4. Semidesert (*Artemisia, Ephedra*, Chenopodiaceae). 5. Contracted vegetation of the desert (*Acacia, Panicum, Tamarix, Stipagrostis*). 6. Ephemeric vegetation of achabs. 7. Diffuse *Artemisia-Ephedra* vegetation of the high mountains of the central Sahara. 8. Acacia-Commiphora-Rhus savanna of the high plateaus of S-Air. 9. Maerua-savanna of the plateaus

modified Livingstone corer for the lower part. The Segedine core was drilled and samples were taken directly from the inner part of the helix. The samples from Dj. Acacus come from archaeological excavations in the rock shelters of Uan Muhuggiag and Tin-n-Torha. All samples were prepared using the classical combination of HF-, HCL-, and acaetolysis treatments. The pollen spectra were counted to the point of consistancy of the percentages.

Present conditions

All corings and excavations were made in a desert environment where, at present, the permanent vegetation is restricted to wadis and depressions. In these localities groundwater and runoff can compensate the insufficiant annual precipitation. This kind of vegetation is mainly composed of *Acacia-Panicum* or *Tamarix-Stipagrostis* communities. The present rainfall in the area ranges between 5 and 30 mm per year. The monsoonal summer rain regime is predominant but intrusions of atlantic cyclones also occur.

The southern boundary of the desert is marked by the transition to diffuse vegetation of the *Acacia-Panicum* type (Fig. 1). This transition is relatively sharp and is clearly visible across the continent (Schulz, 1988). The region receives an annual precipitation of about 250 mm. To the South, the sahelian and sudanian savannas are characterised by the different tree layer and by a variable understorey of annual and permanent grasses and herbs. The floristic composition and density of the sahelian and sudanian savannas are for the most controlled by anthropogenic factors. They provide only a restricted model for the interpretation of former environments. The desert-savanna bondary on the other hand seems to be mainly climatically controlled.

Investigations of the present pollen rain (Cour & Duzer, 1976; Schulz, 1984, 1987) provide information about the dispersal of the different floristic elements. A certain part of the pollen, such as *Alnus*, *Pinus*, *Ephedra* or Combretaceae, is transported over long distances, but the bulk derives from local or regional vegetation. In the dune areas of southern Niger the transition zone of desert to savanna is clearly marked in the pollen spectra by the combination of high percentages of Gramineae and Cyperaceae (*Cyperus conglomeratus* as the typical dune plant). These differences enable the desert-savanna transition to be reconstructed.

Holocene conditions

Both the Taoudenni and Segedim records start in the middle Holocene. Cores and samples of the Taoudenni sequence were taken in the center of the depression of Taoudenni in the Agorgott salt mines. The depression is surrounded by cuestas of marine carboniferous clay and silt stones. The Taoudenni record is built up of alternating layers of salts and clays covered by a thick layer of reddish clays (Fabre & Petit-Maire, 1988).

The pollen diagram is divided into three parts (Fig. 2). The lower part (III) covers the period from about 8000 to 6000 BP. It is characterised by high percentages of Gramineae and Cyperaceae and the continuous presence of *Acacia*, *Capparis*, *Cassia*, *Fagonia* and especially *Grewia*. This indicates the proximity of the desert-savanna boundary to the site and the saharan *Acacia-Capparis-Panicum* savanna as the main vegetation type. The Sudanean as well as the temperate or mediterranean elements in the pollen spectra are due to long distance transport.

During a transition phase (II) between ca. 6000 and 4000 BP there was still a continuous presence

of SE-Niger. 10. Semi-diffuse *Acacia-Panicum* vegetation (enlarged wadi vegetation). 11. *Acacia-Panicum* savanna. 12. *Commiphora-Acacia* savanna. 13. *Acacia-Leptadenia-Commiphora* savanna. 14. *Piliostigma-Bauhinia-Acacia* savanna. 15. *Acacia* thornbush around Lake Chad. 16. Combretaceae savanna. 17. *Parkia-Butyrospermum-Terminalia* savanna. 18. *Isoberlinia-Daniella-Pterocarpus* savannas. 19. *Isoberlinia-Carissa-Ficus* savannas on the Jos-Plateau. 20. *Afzelia-Lophira* savannas and open forests. Sites: 1 Taoudenni 2 Tadrart Acacus 3 Segedim 4 Fachi-Dogonbolo

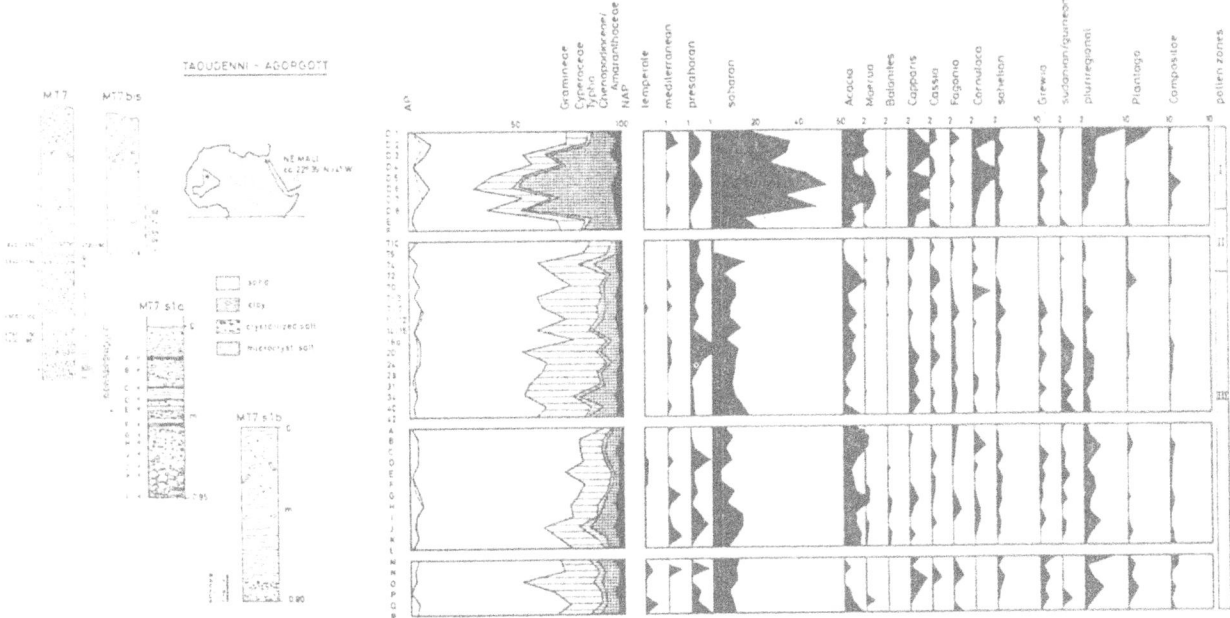

Fig. 2. Simplified pollen diagram of the Taoudenni-Agorgott record. The columns MT7 and MT7 bis represent the stratigraphy taken in the Agorgott salt mines. MT7 s1 and MT7 s2 refer to corings taken in the same pits. The correlation is based on the stratigraphy of the different salt layers. Numbers on the pollen spectra refer to those on the sediment columns.

of *Acacia*, *Capparis*, *Fagonia*, *Cassia* and in lesser extend *Grewia* but the percentages of Cyperaceae were decreasing. Also, the proportion of the long distance transport component diminished. This shows that during this period the savanna-desert boundary had moved to the South but that a remarkable *Acacia-Maerua-Capparis* vegetation remained in the depression.

At about 4000 BP there is a distinct change in the composition of the pollen spectra (I). The sharp rise of the Chenopodiaceae/Amaranthaceae together with *Balanites* and *Cornulaca* represents the establishment of the saharan desert vegetation which has degraded from ca. 2000 BP to its present condition. Elements characteristic of a very open vegetation such as *Plantago* and *Compositae* are predominant but trees like *Acacia*, *Balanites* and *Capparis* were still present in the local vegetation.

The Segedim core was taken in the center of the depression of Segedim (NE-Niger) in front of the Emi Bao cuesta made of karstified cretaceous sand and siltstones. The record consists of blueish fresh water clays in the lower part and sandy sebhka sediments in the upper part (Fig. 3). Only the lower part of the core provided pollen bearing sediments from the period from ca. 8000 to 7000 BP. The pollen spectra are characterised by the combination of high percentages of the Gramineae and Cyperaceae and the presence of *Acacia*, *Maerua*, *Capparis* and *Grewia*. Again this indicates the presence of the desert-savanna transition zone and a plant cover of a loose Acacia-Maerua-Panicum savanna. These savannas occur today on the sandstone plateaus of Se-Niger ca. 600 km to the South (Schulz, 1988). The long distance transport component is relatively low in the pollen spectra.

Archaeological excavation in the central and southern Djebel Acacus (SW-Libya) (Barich, 1987) yielded a pollen sequence covering the middle Holocene. However, samples from

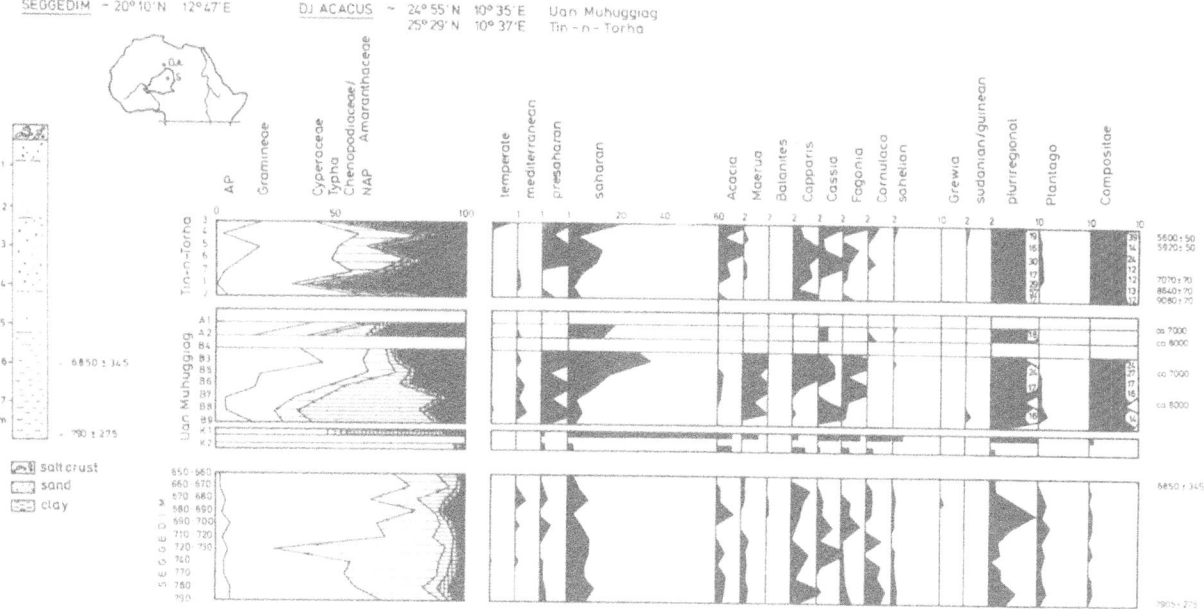

Fig. 3. Simplified pollen diagrams of the Segedim record and the Djebel Acacus excavations. The sediment column gives the stratigraphy of the Segedim core. Numbers on the pollen spectra refer to depth in the core. The Uan Muhuggiag and Tin-n-Torha sequences represent the excavations in rock shelters. The sample numbers refer to the original field numbers. K1 and K2 are coproliths from the lower layers of the Uan Muhuggiag.

excavations are not completely comparable to continuous cores but they give information about regional vegetation types and human influences. The pollen spectra (Fig. 3) of both sites show an open saharan vegetation dominated by Gramineae and pre-saharan elements like *Artemisia* or *Ephedra* as well as *Plantago* and Compositae. The high frequency of *Typha* in the Uan Muhuggiag record indicates a nearby swamp or even material collected by man. Two coprolites of sheep (K1) and goat (K2) gave pollen spectra which show the differences between grasers and browsers feeding on the same vegetation.

A charcoal flora from Fachi-Dogonbolo in the eastern Ténéré (Niger) dated to 7010 BP consists of sudanian elements like *Terminalia, Annona, Crataeva, Celtis, Ximenia, Ficus* etc. (Neumann, 1988). This gives direct proof of a sudanian savanna at 19° N during the middle Holocene.

Comparing the pollen records and the charcoal flora indicates that there was an ecological gradient along the Niger-Libya transect ranging from the sudanian, sahelian and saharan savannas to an enriched desert vegetation during the middle Holocene.

Holocene landscape and Holocene climate

The main difference between the mid-Holocene plant cover and that of present day is caused by the northward shift of the savanna-desert boundary to about 20°–22° N in Niger and Mali. This region was formerly covered by a saharan savanna composed of *Acacia, Capparis* and *Maerua* and tussock grasses as well as a denser herb cover. A large number of achab floras – short life therophyte communities – variable in time and space, were also characteristic of the region. Together with the tree elements, they stabilised the landscape with a loose root mat and made soil formation possible. Depending on relief and on increasing continentally to the East, the vegetation consisted of a complicated mosaic of different types rather than a simple northward sequence of vegetation belts.

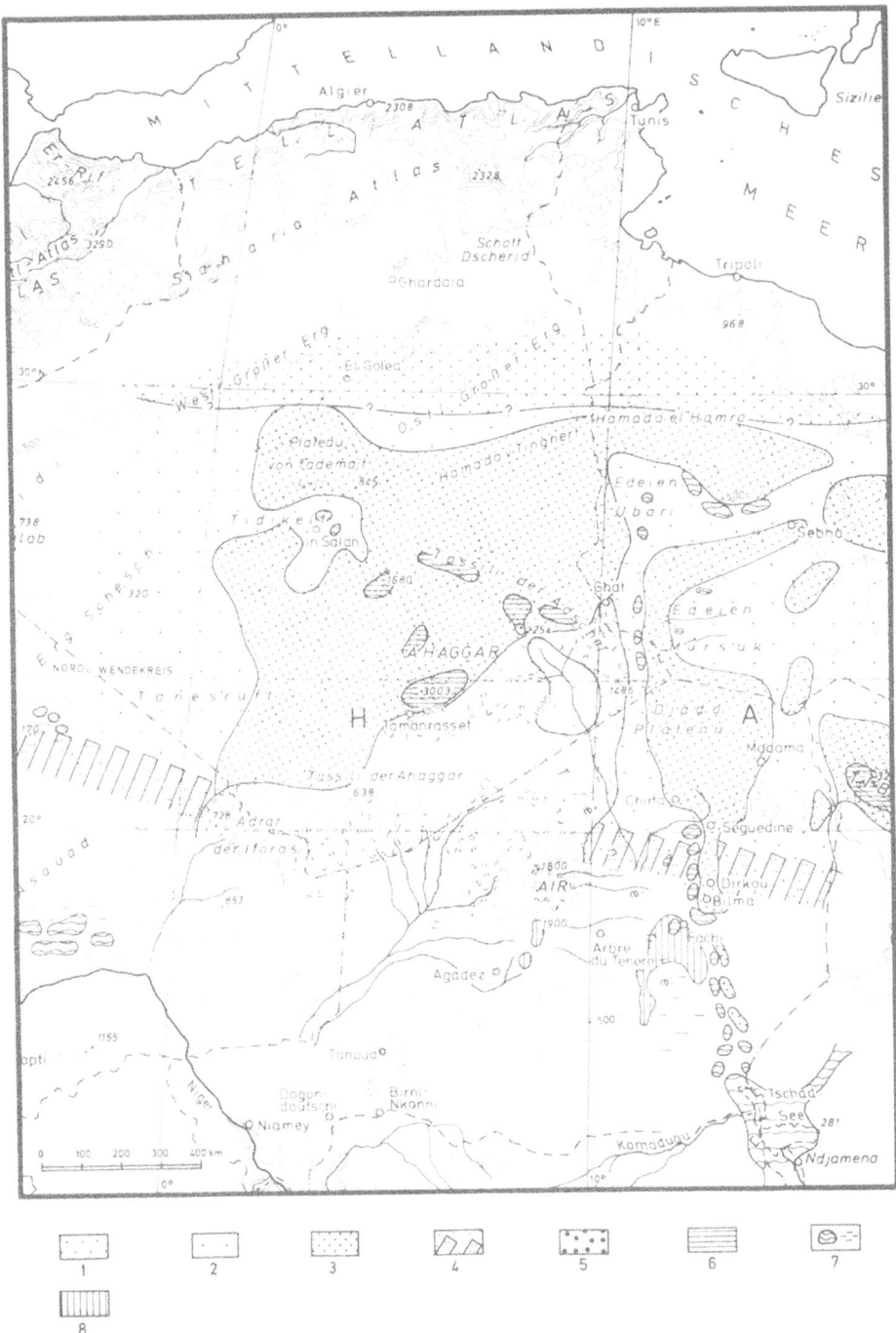

Fig. 4. Schematic map of middle holocene vegetation of the central and southern Sahara (Schulz 1987). 1. semideserts. 2. tussock grass and Chenopodiaceae vegetation, achabs. 3. Maerua-Acacia savannas and Acacia-Panicum/Tamarix-Stipagrostis communities of the central Sahara. 4. desert-savanna transition. 5. Acacia-Commiphora Celtis savanna. 6. Artemisia-Ephedra semidesert of the high mountains of the central Sahara. 7. lakes and swamps in different extensions. 8. sudanian savannas.

The second important point concerning the former landscape was the presence of an organised hydrological system with various lakes and swamps which were mostly groundwater fed. Also, the karstic systems played an important role in the formation of the aquifers (Sponholz, 1989).

From the presence of the Savanna-desert boundary at 20°–22° N one may estimate, in comparison with present conditions, a climate dominated by monsoonal summer rain with an effective annual precipitation of about 250–300 mm for the region. The elements of long distance transport in the pollen spectra also indicate a strong interaction with atlantic cyclones giving the change of additional rainfall throughout the year.

Estimates of rainfall based on the evaporation rate of the salt-layers in the Taoudenni region (Petit-Maire, 1986) may be re-evaluated following the discovery of Holocene neotectonic activity in the depression of Taoudenni (Fabre *et al.*, 1989). Carboniferous salts might have been remobilised along the different dolomite dykes in the depression and deposited in the sebkhas. This underlines again the individuality of palaeolakes.

Acknowledgements

I am gratefully indebted to the Deutsche Forschungsgemeinschaft for the financial support.

References

Barich, B., (ed.), 1987. Archaeology and Environment in the Libyan Sahara. Cambridge Monogr. in Afr. Arch. 23. BAR Int. Ser. 368. 347 p.

Baumhauer, R., 1986. Zur jungquartären Seenentwicklung im Bereich der Stufe von Bilma (NE-Niger). Würzburger Geogr. Arb. 65, 235 p.

Baumhauer, R. & Schulz, E., 1984. The holocene lake of Seguedine, Kaouar, NE-Niger. Palaeoecol. Afr. 16: 283–90.

Cour, P. & D. Duzer, 1976. Persistance d'un climat hyperaride au Sahara central et meridional au cours de l'Holocène. Rev. Geogr. Phys. Geol. Dynamique. 2, XVIII: 175–198.

Fabre, J. & N. Petit-Maire, 1988. Holocene climatic evolution at 22–23° N from two palaeolakes in the Taoudenni area (northern Mali). Palaeogeogr., Palaeoclimatol., Palaeoecol. 65: 133–148.

Fabre, J., P. Carbonel, J. Riser, I. Øxnevad, 1989. Deformations recentes au cœur du craton ouest africain (Taoudenni, Mali). C.R. Acad. Sci. Paris. 308: 1561–1566.

Gasse, F., 1988. Diatoms, palaeoenvironments and palaeohydrology in the western Sahara and the Sahel. Würzburger Geogr. Arb. 69: 233–254.

Kutzbach, J. E., 1980. Estimates of postclimate at palaolake Chad, North Africa, based on a hydrological and energy balance model. Quat. Res. 14: 210–223.

Neumann, N., 1988. Die Bedeutung von Holzkohleuntersuchungen für die Vegetationsgeschichte der Sahara – das Beispiel Fachi/Niger. Würzburger Geogr. Arb. 69: 71–86.

Petit-Maire, N., 1986. Paleoclimatologie du Sahara occidental et central pendant les deux derniers optima climatiques aux latitudes paratropicales. INQUA-ASEQUA Changements globeaux en Afrique durant le Quaternaire: 375–379.

Petit-Maire, N. & J. Riser, 1983. Sahara ou Sahel? Marseille, 473.

Schulz, E., 1984. The recent pollen rain in the eastern central Sahara. A transect between northern Libya and southern Niger. Palaeoecol. Afr. 16: 245–253.

Schulz, E., 1987. Die holozäne Vegetation der zentralen Sahara (N-Mali, N-Niger, SW-Libyen). Palaeoecology of Africa 18: 143–161.

Schulz, E., 1988. Der Südrand der Sahara. Würzburger. Geogr. Arb. 69: 167–210.

Sponholz, B., 1989. Karsterscheinungen in nichtkarbonatischen Gesteinen der östlichen Republik Niger. Würzburger Geogr. Arb. 78: 265 p.

Thorweihe, U., 1988. Das Grundwasser der Sahara. Wissenschaftsmagazin TU Berlin, 11: 98–102.

Hydrobiologia **214**: 367–372, 1991.
J. P. Smith, P. G. Appleby, R. W. Battarbee, J. A. Dearing, R. Flower, E. Y. Haworth, F. Oldfield & P. E. O'Sullivan (eds), 367
Environmental History and Palaeolimnology.

Chronology of the major palaeohydrological events in NW Africa during the late Quaternary: PALHYDAF results

Fontes, J.Ch. & F. Gasse
Laboratoire d'Hydrologie et de Géochimie Isotopique, Bât. 504, Université Paris-Sud, 91405, Orsay Cedex, France

Key words: Late Quaternary, palaeohydrology, radiometric chronology, Sahara, Sahel

Abstract

On the basis of African examples, the paper draws attention to some geochemical and sedimentological problems which commonly occur in the establishment of a reliable chronology from lacustrine sediments. New results are then provided on the ages of the successive wet climatic phases in NW Africa since 150 ka ago.

Introduction

PALHYDAF (Palaeohydrology of Africa) project aims to understand the causes and mechanisms of climatic changes in Africa north of the Equator since the last interglacial (130 ka). The principle is to reconstruct palaeoenvironmental and palaeohydrological fluctuations in closed basins by studying cores and natural exposures of lacustrine or palustral sediments. Northwest Africa is subjected to two major climatic influences, the Atlantic monsoon, responsible for summer rains up to 30 °N, and the Polar Front which brings cool air and winter rains. An attempt is made to detect the latitudinal migrations through time of the conflict zone between these air masses and the related origin of palaeoprecipitation. Therefore, sites were selected along two north-south transects (Fig. 1). The investigated basins lie in arid or semi-arid areas of general internal drainage and are mainly supplied by groundwater. Special attention is thus given to the relationships between the periods of recharge of the aquifers (groundwater datings) and the related rises in lake levels.

Techniques used and detailed results are presented elsewhere (Fontes & Gasse, 1991). This paper focusses on the age of the major palaeohydrological events.

Difficulties in establishing a reliable chronology

Significance of radiocarbon dates

Besides the limits of applicability of the radiocarbon method inherent to the ^{14}C period, radiocarbon ages may be biased by different processes.

a. Older carbon contamination from groundwater supply. Example of Wadi el Akarit (site 1) (Fontes *et al.*, 1983; Zouari, 1988; Fontes & Gasse, 1991).

Wadi el Akarit is perennial because it is supplied by an artesian aquifer. Along its course, vegetation dams create small ponds where organism-rich peat and carbonates have accumulated during the last few decades. Radiocarbon dates on modern material demonstrate the importance of older carbon contamination in poorly

368

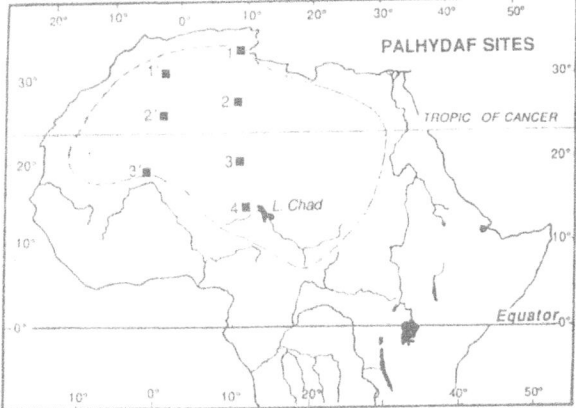

Fig. 1. Location map of the PALHYDAF sites.

mixed groundwater-fed waterbodies[1]. The ^{14}C age of the aquifer (18.5 ka B.P.) is responsible for the 'old' apparent ^{14}C ages of the living organisms (9.7–15.2 ka B.P.) measured on the shells and organic content of gastropods and on macrophytes. An exchange-mixing model (Fontes *et al.*, 1983) was used to account for the incomplete equilibrium between the Total Dissolved Inorganic Carbon (TDIC) of the water and the atmospheric CO_2. This model dates the Early and Mid-Holocene deposits from 10.5 to 5.7 ka B.P., while the apparent ^{14}C ages range from 13.8 to 11.6 ka B.P.

b. Input of detrital material. Example of Wadi el Akarit (site 1).

During the Late Glacial episode, the Wadi el Akarit deposits were contaminated by aeolian particles windblown from the surrounding Cretaceous hills, as demonstrated by the occurence of Cretaceous coccoliths and ostracods. The relative abundance of Cretaceous dolomite (2.6%) considerably biased the ^{14}C dates of bulk sediments (22.4 to > 40 ka B.P.). It is thus necessary to separate authigenic material. AMS datings of individual shells provide apparent ^{14}C ages of 13.3 and 16.7 ka B.P. for this interval (Zouari, 1988).

[1] This effect of dilution of the recent carbon by the 'old' carbon from the aquifer is often improperly named 'the hard water effect' since the dilution does not depend on the Ca^{2+} and Mg^{2+} hardness of the water.

c. Recrystallization. Example of Adrar Bous (site 3).

At Adrar Bous (Dubar, 1988), the rough stratigraphy of a 8 m core shows three major Late Quaternary sedimentary environments: 1) a flood plain (gravelly silts with carbonate concretions); 2) swamp with carbonate sedimentation; 3) shallow lake (diatomite ^{14}C dated on organic matter and on carbonates from ≈ 10 to 7.5 ka B.P.). Below the Holocene diatomite, the ^{14}C ages on carbonates increase regularly from 19.9 ka B.P. (1.2 m in depth) to 30.6 ka B.P. (2.20 m), and then decrease up to 14.7 ka B.P. (6 m). This is interpreted as a recrystallization of the carbonates during a rise of the aquifer which lies today at a depth greater than that of the core. This is deduced from results of chemical and stable isotope analyses of inorganic carbonates. During stage 2, the ^{18}O contents regularly increase (-7 to $-2‰$), reflecting increasing evaporation rate which may lead to a concentration of the water. However, Sr/Ca and Mg/Ca ratios which are salinity-indicators, show a reverse evolution. These apparent discrepancies are attributed to an *in situ* recrystallization of the carbonates which tends to remove trace elements from the calcite crystals. Therefore, the ^{14}C dates reflect a rise of the water table ≈ 30–20 ka B.P., but do not indicate the age of the original carbonates. Other chronometers are thus necessary to reconstruct the evolution of the basin.

$^{230}Th/^{234}U$ disequilibrium method

a. Principle

Dissolved uranium including ^{234}U coprecipitates with Ca in carbonates, and is also gained (through adsorption on decaying organic matter from carbonates) from pore water after the death of organisms. Then ^{230}Th builds up in the carbonate through decay of ^{234}U. This method damps the effect of recrystallization in comparison with the ^{14}C method because of the longer radioactive period of ^{230}Th (75200 years) and because ^{230}Th is insoluble and stays within the carbonate crystal. Converting the deficit of ^{230}Th into determi-

Table 1. Comparison of ^{14}C and Th/U ages obtained on the same samples or on the same stratigraphical units.

Locality	[14]C ages ka B.P.	Th/U ages ka	References
South Tunisia			
Chotts Fejej and Jerid[1]	39–25	150; 90	} Fontes, Gasse & Coll, 1987
	42.8		} Zouari, 1988; Causse *et al.*, 1989
North Algeria-Saoura			
Agdal	26.3; 27.3	51	} Gibert, unpublished
	21.4; 36.1	23	} Gibert, unpublished
South Algeria-Erg Chech			
Kadda	13.1–17.8	14.7	}
	34.9	26.8	}
	27.0–28.9	75–80	} Causse *et al.*, 1988
	19.4	111	} Gibert, 1989
	16.2	110	}
	18.8–35.5 to ≥40	150	}
Libya			
Wadi Shati		130; 80	} Gaven *et al.*, 1981;
North Niger-Aïr			
Adrar Bous	30.6–19.9	88	} Dubar, 1988

[1] The high ratios [234]U/[238]U demonstrates the continental origin of the waters of these palaeosebkhas, which were attributed to a marine invasion of Late Pleistocene age by Richards & Vita-Finzi (1982).

nations of age may be questionable because of the possible presence of allochtonous [230]Th. This may be due to the inclusion of detrital particles in the dated material. This problem is theoretically solved by using the isochron method (Schwarcz & Skoflek, 1982).

The homogeneity of the uranium content in a set of samples assumed to be synchronous on the basis of field arguments suggests that neither gain nor losses of U occurred. It is thus necessary to analyse several samples and different types of material for a given stratigraphical level. The internal consistency of a set of data is the main warranty of the Th/U ages.

b. Discrepancies between ^{14}C and Th/U ages
Table 1 compares the ages obtained by the two methods on the same samples or on thc same stratigraphical units. Although good agreement are observed in several cases, major discrepancies appear in many Late Pleistocene sections. These results lead to a reconsideration of previous syntheses and concepts on the climatic evolution

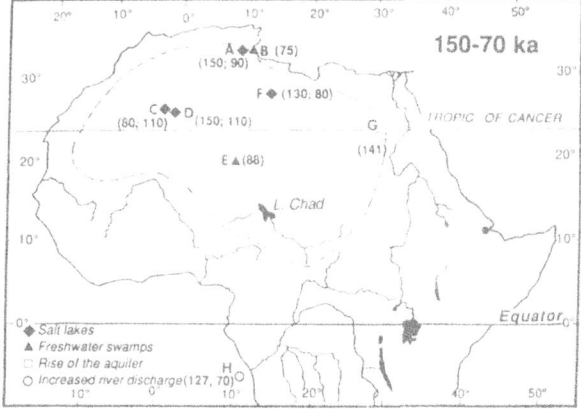

Fig. 2. The period 150–75 ka
Th/U PALHYDAF ages (ka) (except F to H)
A: Chotts Fejej and Jerid (Fontes, Gasse & coll., 1987; Zouari, 1988; Causse *et al.*, 1989).
B: Wadi el Akarit. (Causse *et al.*, 1989)
C: Erg Chech, Kadda (Causse *et al.*, 1988; Gibert, 1989)
D: Azzel Matti (Causse *et al.*, 1989; Gibert, 1989)
E: Aïr, Adrar Bous (Dubar, 1988)
F: Southern Egypt (Szabo *et al.*, 1989)
G: Wadi Shati (Gaven *et al.*, 1981)
H: Congo deep-sea fan (Gasse *et al.*, 1989)

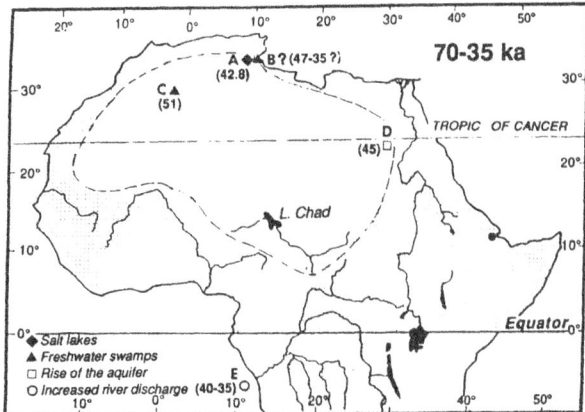

Fig. 3. The period 70–35 ka
Th/U and ^{14}C PALHYDAF ages (ka) (except D and E)
A: Chott Fejej (Causse *et al.*, 1989)
B: Wadi el Akarit (Zouari, 1988)
C: Wadi Saoura, Agdal (Gibert, unpublished)
D: Southern Egypt (Szabo *et al.*, 1989)
E: Congo deep-sea fan (Gasse *et al.*, 1989)

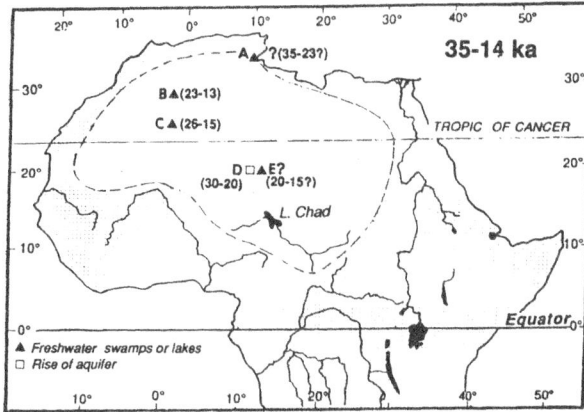

Fig. 4. The period 35–14 ka
Th/U and ^{14}C PALHYDAF ages (ka B.P.)
A: Wadi el Akarit (Fontes *et al.*, 1983; Zouari, 1988)
B: Wadi Saoura, Agdal (Gibert, 1989, and unpublished)
C: Erg Chech (Kadda (Gibert, 1989, and unpublished)
D: Aïr, Adrar Bous (Dubar, 1988)
E: Aïr, Tin Ouaffadene (Dubar, 1988)

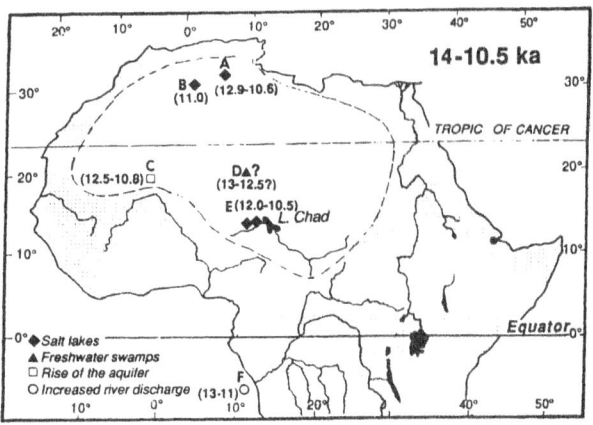

Fig. 5. The period 14–10.5 ka
^{14}C PALHYDAF ages (ka B.P.)
A: Sebkha Mellala (Gibert *et al.*, 1990)
B: Great Western Erg, Daiet el Melah (Gasse *et al.*, 1987)
C: Aïr, Adrar Bous (Dubar, 1988)
D: Mali, Arawan-Azawad ridge (Fontes *et al.*, 1991)
F: Chad basin (Ari Koukouri, Bwori, Maine Soroa, Bougdouma) (Durand *et al.*, 1984; Gasse, 1987; Fontes and Gasse, 1991, Téhet *et al.*, 1990; Gasse *et al.*, 1990)

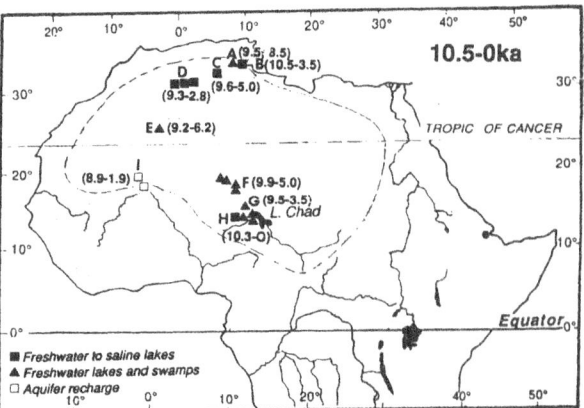

Fig. 6. The period 10.5–0 ka BP
A: Chott Fejej (Zouari, 1988; Fontes & Gasse, 1990)
B: Wadi el Akarit (Fontes *et al.*, 1983; Zouari, 1988)
C: Sebkha Mellala (Gibert *et al.*, 1990; Gasse *et al.*, 1990)
D: Great Western Erg (Daiet et Mellah, Hassi el Mejna, Hassi Cheikh) (Fontes *et al.*, 1985; Gasse *et al.*, 1987)
E: Erg Chech, Kadda (Gibert, 1989)
F: Aïr (Temet; Adrar Bous, Tin Ouaffadene, Izouzadene) (Dubar, 1988)
G: Termit (Fontes & Gasse, 1991)
H: Chad basin (N'Guigmi, Ari Koukouri, Maine Soroa, Bougdouma, Bwori) (Durand *et al.*, 1984; Gasse, 1987; Fontes & Gasse, 1991; Téhet *et al.*, 1990)
I: Mali, Arawan-Azawad ridge (Fontes *et al.*, 1991)

in Northwest Africa since the last interglacial (Fontes and Gasse, 1989).

Successive wet climatic phases in NW Africa since 150 ka

Figures 2 to 6 summarize the chronology of the

major hydrological events recorded from PALHYDAF investigations.

The period 150–70 ka B.P. (Fig. 2)

a. Large saline lakes over the north western Sahara
Sebkha (playa) deposits occur in numerous depressions of the northwestern Sahara and are commonly characterized by accumulations of shells (mainly *Cerastoderma glaucum*) reflecting shallow, saline waters. Several of these deposits were previously attributed to a 40–20 ka humid phase on the base of ^{14}C dates (Tab. 1). We show that, in the investigated basins, the last great humid period before the Holocene one is much older, since the U/Th ages fall between 150 and 80 ka. This is in good agreement with recent results from Egypt (Szabo *et al.*, 1989).

b. Southward extension of humid conditions
At Adrar Bous, ^{14}C ages of 30–20 ka B.P. obtained for pre-Holocene deposits are due to diagenesis of older carbonates. These carbonates, which are associated with freshwater organisms, exhibit a Th/U age of 88 ka, indicating that conditions wetter than today, and wetter than at 30–20 ka B.P., occurred in the Central Sahara at that time. In the equatorial zone, increases of the Congo river discharge are indicated by peaks of river transported freshwater diatoms in the Congo deep-sea fan sediments ≈ 130–125 ka and 70 ka.

At present, it is not possible to decide whether these indices of wet conditions are synchronous at the scale of the whole of west Africa. In the case of the Sahara, further investigations are needed to distinguish between the occurrence of a long humid phase centered on 130 ka, or of several humid episodes in the same time range.

The period 70–35 ka (Fig. 3)

Although Th/U ages of ≈ 51 and ≈ 43 ka were obtained at Agdal and Chott Fejej respectively, further data are needed before inferring the occurrence of humid phase(s) during this long time interval.

The period 35–14 ka (Fig. 4)

In northern Africa, the period 35–20 ka B.P. is commonly regarded as a humid phase, while the 20–13 ka interval is known as a generally dry episode (except at ≈ 18 ka B.P. around the southern margin of the Mediterranean) (Street & Grove, 1979; Street-Perrott & Roberts, 1983).

Palhydaf Th/U dates (Fontes & Gasse, 1991; Table 1) suggest that the Late Pleistocene humid phase on the Sahara was probably not as important as previously proposed on the basis of ^{14}C chronology, although consistent indices of conditions wetter than today are found. In Central Sahara, the aquifer rose at ≈ 30–20 ka B.P., but did not reach the surface at Adrar Bous. In a neighbouring depression, a diatomite is ^{14}C dated at 20–15 ka B.P. but this age needs to be confirmed. Northward, good agreement between Th/U and ^{14}C ages suggests a sub-continuous palustral sedimentation from ≈ 26.8 to 14.7 ka at Kadda (Erg Chech) and from ≈ 23 to 14–13 ka at Agdal (Wadi Saoura).

The period 14–0 ka B.P.

New insights into the Post-Glacial interval 14–10.5 ka B.P. (Fig. 5) and a better knowledge of the Holocene (10.5–0 ka, Fig. 6) are provided.

The hypothesis of a progressive re-establishment of humid conditions from the Equator to the Maghreb between 12.5 and 6 ka B.P. (Street-Perrott & Roberts, 1983), adopted by several authors (Fabre et Petit-Maire, 1988; Rognon, 1987; Lézine et Casanova, 1989), is not confirmed (Gasse *et al.*, 1990). The first rises in lake level occurred as soon as 15–14 ka B.P. in the northern Sahara (Gibert *et al.*, 1990) and before 13–12 ka B.P. in the Sahel (Téhet *et al.*, 1990)). Groundwater datings in northern Mali indicate a period of recharge of the aquifers by 12.5–11 ka (Fontes *et al.*, 1991). Lake fluctuations are abrupt and correlate well with the steps of the deglaciation as recorded through [18]O profiles in Atlantic deep-sea cores and with the European climatostratigraphy. This is not surprising if one con-

372

siders that rainfall fluctuations over the Continent are linked to changes in the global pattern of ocean currents, SST, and solar radiation. The source term i.e. the global availability of precipitable atmospheric moisture, directly linked to the temperature of tropical oceans appears as a major controlling factor for rainfall variations over Africa north of the Equator at the 10^2–10^3 a time scale (Gasse et al., 1990).

References

Causse, C., G. Conrad, J. Ch. Fontes, F. Gasse, E. Gibert & A. Kassir, 1988. Le dernier 'Humide' pléistocène du Sahara nord-occidental daterait de 80–100000 ans. – C.R. Acad. Sc. Paris, 306, II: 1459–1464.

Causse, C., R. Coque, J. Ch. Fontes, F. Gasse, E. Gibert, H. Ben Ouezdou & K. Zouari, 1989. Two high levels of continental waters in the southern Tunisian chotts at about 90 and 150 ka. Geology 17: 922–925.

Dubar, C., 1988. Eléments de paléohydrologie de l'Afrique Saharienne: les dépôts quaternaires d'origique aquatique du Nord-Est de l'Aïr (Niger). Thesis, Université Paris-Sud, Orsay, 197 pp.

Durand, A., J. Ch. Fontes, F. Gasse, M. Icole & J. Lang, 1984. Le Nord-Ouest du lac Tchad au Quaternaire: étude de paléo-environnements alluviaux, éoliens, palustres et lacustres. Palaeoecology of Africa 16: 215–243.

Fabre, J. & N. Petit-Maire, 1988. Holocene climatic evolution at 22–23 °N from two palaeolakes in the Taoudenni area (Northern Mali). Palaeogeogr., Palaeoclimatol., Palaeoecol. 65: 133–148.

Fontes, J. Ch & F. Gasse, 1989. On the ages of humid Holocene and Late Pleistocene phases in North Africa. Remarks on 'Late Quaternary climatic reconstruction for the Maghreb (North Africa)' by P. Rognon. Palaeogeogr., Palaeoclimatol., Palaeoecol. 70: 393–398.

Fontes, J. Ch. & F. Gasse, 1991. PALHYDAF Programme. Objectives, methods, major results. Palaeogeogr., Palaeoclimatol., Palaeoecol. in press.

Fontes, J. Ch., J. N. Andrews, W. M. Edmunds, A. Guerre & Y. Travi, 1991. Paleorecharge by the Niger River (Mali) deduced from groundwater geochemistry. Wat. Res. Res., in press.

Fontes, J. Ch., R. Coque, L. Dever, A. Filly & A. Mamou, 1983. Paléohydrologie isotopique de l'Oued el Akarit (Sud Tunisie) au Pléistocène supérieur et à l'Holocène. Palaeogeogr., Palaeoclimatol., Palaeoecol. 43: 41–62.

Fontes, J. Ch., F. Gasse, Y. Callot, J. C. Plaziat, P. Carbonel, P. A. Dupeuble & I. Kaczmarska, 1985. Freshwater to marine-like environments from Holocene lakes in northern Sahara. Nature 317, n° 6038: 608–610.

Fontes, J. Ch., F. Gasse & Collaborateurs, 1987. Programme PALHYDAF: Etat d'avancement, janvier 1987. Premier méridien: Sud Tunisie-Sud Niger. Géodynamique 2(2): 139–142.

Gasse, F., 1987. Diatoms for reconstructing palaeoenvironments and palaeohydrology in tropical semi-arid zones. Hydrobiologia 154: 127–163.

Gasse, F., J. Ch. Fontes, J. C. Plaziat, P. Carbonel, I. Kaczmarska, P. De Deckker, I. Soulie-Märsche, Y. Callot & P. A. Dupeuble, 1987. Biological remains, geochemistry and stable isotopes for the reconstruction of environmental and hydrological changes in the Holocene lakes from North Sahara. Palaeogeogr., Palaeoclimatol., Palaeoecol. 60: 1–46.

Gasse, F., B. Stabell, E. Fourtanier & Y. Van Iperen, 1989. Freshwater diatom influx in intertropical Atlantic: relationships with continental records from Africa. Quat. Res. 32: 229–243.

Gasse, F., A. Téhet, A. Durand, E. Gibert & J. Ch. Fontes, 1990. The arid-humid transition in the Sahara and the Sahel during the last deglaciation. Nature 346: 141–146.

Gaven, C., C. Hillaire-Marcel & N. Petit-Maire, 1981. A Pleistocene lacustrine episode in southeastern Libya. Nature 290: 131–135.

Gibert, E., 1989. Géochimie et paléohydrologie des bassins lacustres du nord-ouest saharien. Programme Palhydaf, site 2. Thesis, Université Paris-Sud, Orsay, 210 pp.

Gibert, E., M. Arnold, G. Conrad, P. De Deckker, J. Ch. Fontes, F. Gasse & A. Kassir, 1990. Retour des conditions humides au Tardiglaciaire au Sahara septentrional (Sebkha Mellala, Algérie). Bull. Soc. géol. France 6: 497–504.

Lézine, A. M. & J. Casanova, 1989. Pollen and hydrological evidence for the interpretation of past climates in tropical West Africa during the Holocene. Quat. Sci. Rev. 8: 45–55.

Richards, G. W. & C. Vita-Finzi, 1982. Marine deposit 35000–25000 years old in the Chott el Djerid, Southern Tunisia. nature 295: 54–55.

Rognon, P., 1987. Late Quaternary climatic reconstruction for the Maghreb (North Africa). Palaeogeogr., Palaeoclimatol., Palaeoecol. 58: 11–34.

Schwarcz, H. & I. Skoflek, 1982. New dates for the Hundary archaeological site. Nature 295: 590–591.

Street, F. A. & A. T. Grove, 1979. Global maps of lake-level fluctuations since 30000 B.P.. Quat. Res. 12: 83–118.

Street-Perrott, F. A. & N. Roberts, 1983. Fluctuations in closed-basin lakes as an indicator of past-atmospheric circulation patterns. In: F. A. Street-Perrott et al. (Editors). Variations in the Global Water Budget, Reidel, Dordrecht, 331–345.

Szabo, B. J., W. P. McHucgh, C. G. Schaber, J. Haynes & C. S. Breed, 1989. Uranium-Series dated authigenic carbonates and Acheulian sites in southern Egypt. Science 243: 1053–1056.

Téhet, R., F. Gasse, A. Durand, P. Schroeter & J. Ch. Fontes, 1990. Fluctuations climatiques du Tardiglaciaire à l'Actuel au Sahel (Bougdouma, Niger Méridional). C.R. Acad. Sci. Paris 311: 253–258.

Zouari, K., 1988. Géochimie et sédimentologie des dépôts continentaux d'origine aquatique du Quaternaire supérieur du Sud Tunisien: interprétations paléohydrologiques et paléoclimatiques. Thesis, Université Paris-Sud, Orsay, 246 pp.

Hydrobiologia **214**: 373–382, 1991.
J. P. Smith, P. G. Appleby, R. W. Battarbee, J. A. Dearing, R. Flower, E. Y. Haworth, F. Oldfield & P. E. O'Sullivan (eds), 373
Environmental History and Palaeolimnology.

Paleolimnology, William Morris and *The Magic Flute*[1]

P.E. O'Sullivan
Department of Environmental Sciences, Polytechnic South West, Plymouth PL4 8AA, UK

Key words: paleolimnology, William Morris, *The Magic Flute*, paradigms

Introduction

At the last congress in Ossiach (thanks to the splendid organisation), there were many interesting late-night conversations. One was between myself, Heikki Simola and Maureen Longmore, who sadly is not here this time. The conversation was about 'what paleolimnologists should do'. Maureen characteristically got straight to the point by saying something like (and I apologise if I misquote her) "Paleolimnology is fun, and paleolimnologists are lucky to have such interesting work, but there is a great deal of it which says 'Here is our lake, here is our core, here is our diagram of results – we found that the lake had been disturbed[2] by human action'. Then what? What do we do with our results?"

What I think Maureen was saying was – Is paleolimnology by itself enough? It *is* interesting and rewarding – we can all agree on that or we would not be here. But what do we *do* about our results? As we are studying pressing and urgent environmental problems, we are also asking – what do paleolimnologists do about such problems? I was, as usual, both stimulated and disturbed by Maureen's question. I went away from

the congress wondering what to do about it. I decided that at the next congress I would try to give Maureen an answer. She isn't here to receive it, but this lecture is my attempt. I don't wish to lay down rules for anybody else. This is just my view of where paleolimnologists, as scientists, could be going. I may be quite wrong.

William Morris

Why paleolimnology, William Morris and *The Magic Flute*? Well, first of all, these are three of the great interests of my life. Paleolimnology, and *The Magic Flute*, presumably need no introduction, but I might just say a little about William Morris, whom elsewhere (Coleman & O'Sullivan, 1990; O'Sullivan, 1987, 1990), I have described as my hero.

William Morris lived from 1834–1896. He was born into a wealthy family and after university he decided first to become an architect, then a painter, and finally settled on being a designer of household furnishings, fabrics, textiles, carpets, glass, wallpapers – an interior designer. He was also a famous poet, especially for *The Earthly Paradise*, published in 1868.

In the last third of his life, Morris became active in politics, owing to his interest in the relationship between art and society. He wrote many lectures and essays, and a novel called *News from Nowhere* (1890) in which he described his ideal society of the future. As far as his art was concerned, he was part of the great European literary and philosophical movement called Romanticism, and later I

---

[1] This is the text of an informal lecture given on the Saturday evening of the Ambleside conference, outside the formal programme. I submitted it for publication at the request of a number of participants. (PO'S).

[2] I once asked Ed Deevey 'Disturbance of what?' Ed lit his pipe and grinned, but, characteristically, or so I believe, he didn't answer! But Ed was a true 'green' scientist in the best sense of the word. For confirmation see *Q. R. Biol.* **17(1)**, (1942).

will be saying how I think his ideas, and *The Magic Flute*, are relevant to paleolimnology!

Paleolimnology and paradigms

Perhaps one of the most important things to have happened since the last congress in 1985 is that the world turned 'green'. By this I mean that after twenty years of desperate urgency (but bungling inactivity on the part of the great and powerful) environmental issues finally reached the political agenda. They had done so earlier in some countries such as Federal Germany, but in general 1989 was the year that 'green' issues achieved the political prominence they deserve.

One characteristic of Greens, or so they claim, is that they do not fit into the conventional left/right political spectrum. Some green ideas are very conservative, others very radical. Some social scientists see the rise of green ideas as a fundamental challenge to Western society. They call this the rise of the Alternative Environmental Paradigm.

Here we are using the terminology of the historian of science, Thomas L. Kuhn (1972). Kuhn believes that science proceeds via a series of revolutions, separated by periods of quiescence which he calls paradigms. A paradigm is a set of beliefs or principles that underlies our thoughts and actions. It is often quite unconscious, so that we believe that the ideas that make up our paradigms are 'true'. One characteristic of the current social paradigm is the belief that economic growth is fundamental and inevitable – that human beings somehow instinctively need and want growth, and that organisation and societies can only survive by growing. But this is not an established scientific 'fact', it is only one of our beliefs. It is not a universal.

We will come to the question of paradigms in more detail in just a moment, but let us finish with Kuhn. Kuhn sees the development of science in terms of a series of revolutions, separated by periods of adherence to particular paradigms. In the Middle Ages, in Europe, the ideas of Aristotle were so powerful that we could call the scientific

paradigm of that time the Aristotelian paradigm. In the C17, medieval science, in the west, was replaced by the ideas of the Scientific Revolution, which Kuhn says eventually became the Newtonian paradigm. In the early C20 another revolution replaced Newton's ideas with the Einsteinian paradigm. Some physicists, notably Fritjof Capra (1983), believe that physics at least, if not yet the rest of science, has entered the post Einsteinian paradigm.

Kuhn's ideas, as far as Kuhn is concerned, apply only to science. But some social scientists and historians believe that they can be applied to other aspects of society as well, so that we may speak of social, as well as scientific paradigms. One great period of change in the west – a paradigm shift – is said to be the Scientific Revolution/Renaissance/Reformation of the C16 and C17. Some social scientists (eg Catton & Dunlap, 1980) call this the period of the establishment of our present western society and its ideas – the establishment of what is called the Dominant Social Paradigm (DSP).

So we have the ideas of the present society, the DSP, and those of the greens – the Alternative Environmental Paradigm (AEP). What are their characteristics? These are compared in a table by Stephen Cotgrove (Fig. 1). The core values of the DSP are growth, nature valued as a resource, and domination over nature. In the DSP, nature is something which is there for us, and though an increasing number of us think we would be wise to take care of it, in the last analysis we often decide in our favour rather than that of nature.

The economy is 'risk and reward', via the market. There are social welfare programs, but many economists and many politicians believe that the only real wealth is that created by manufacturing. The society is centralised. Decisions are made at the centre and although there may be democracy, it is a representative democracy in which the expert is a very powerful figure. We as experts may applaud this, but it is not an inherently democratic decision-making process.

In the DSP, we are confident about our ability to solve problems. We see nature as something which if carefully managed, provides ample re-

	Dominant Social Paradigm	Alternative Environmental Paradigm
CORE VALUES	Material (economic growth)	Non-material (self-actualisation)
	Natural environment valued as resource	Natural environment intrinsically valued
	Domination over nature	Harmony with nature
ECONOMY	Market forces	Public interest
	Risk and reward	Safety
	Rewards for achievement	Incomes related to need
	Differentials	Egalitarian
	Individual self-help	Collective/social provision
POLITY	Authoritative structures: (experts influential)	Participative structures: (citizen/worker involvement)
	Hierarchical	Non-hierarchical
	Law and order	Liberation
SOCIETY	Centralised	Decentralised
	Large-scale	Small-scale
	Associational	Communal
	Ordered	Flexible
NATURE	Ample reserves	Earth's resources limited
	Nature hostile/neutral	Nature benign
	Environment controllable	Nature delicately balanced
KNOWLEDGE	Confidence in science and technology	Limits to science
	Rationality of means	Rationality of ends
	Separation of fact/value, thought/feeling	Integration of fact/value, thought/feeling

Fig. 1. The structure of two conflicting social paradigms (after Cotgrove & Duff, 1980).

serves of abundant resources. In this paradigm nature is not generally seen as benevolent. It is viewed either as neutral, or as something hostile to be tamed. Perhaps one of the most important characteristics of the DSP, for us as scientists, is that it believes that 'facts' and 'values' should be kept quite separate, and that whatever scientists may 'feel' about their work, we should keep those feelings to ourselves, and not allow them to influence our work, and in particular the 'objective' scientific 'facts' that we produce.

In the AEP, growth is not viewed as fundamental. Instead, the emphasis is upon self and self-realisation – realisation of the self as part of a community, and as part of nature, living with both the community and with nature, in harmony. Rather than growth, the emphasis is upon balance. Nature is not just there as a resource, but possesses value in itself, as Nature, independent of any values we may place upon it. The economy is a collective one, in which emphasis is upon need rather than demand. Production also, is for need only, rather than wants. The market mechanism with its emphasis upon competition, is seen as wasteful of scarce resources. The society is decentralised and organised on a small scale. This is felt to be necessary so that people can retain, or regain, control over their own lives, and thus the amount of resources they consume. (How many of us would like to live 'greener' lives but at present still find it very difficult to do so?)

The earth in the AEP is seen as a benign provider of resources, but these are available on a strictly limited bases. Nature is thought of as living in a balance which we humans had better not disrupt. Although in the DSP, this balance is also recognised, it is seen there as something to be learned about and then manipulated. In the AEP, knowledge of the balance is seen as something which would reinforce the idea of living by nature's laws, rather than trying to alter them.

In the AEP, there are said to be limits to scientific enterprise. Here is where perhaps we as paleolimnologists may find most difficulty. Whereas in the DSP, the important question is *How*? the question in the AEP is *Why*? And in the AEP it is thought of as quite permissible that scientists should allow their feelings to influence their work – NOT to distort the information they obtain, of course, but to recognise that what we select as a 'fact', is *in fact*, strongly affected by our ideas and beliefs.

All of us here, for example, deep inside, do not study lakes merely because we want to generate objective scientific 'facts'. We want to do that, and for many reasons we *need* to do that, but we also study lakes, and their sediments, because they are fascinating, marvellous, and *beautiful*. We all know this to be the case. And we are scientists because we are curious about the world and the universe, and because doing science is immensely satisfying. It is one of the great things in life. I personally don't understand how anyone could *not* want to be a scientist.

So, in the AEP, it is recognised that the separation between feeling and thought of the DSP is not necessarily a valid one. As scientists we are taught that to allow our feelings to influence our work leads to distortion, but *not* recognising that it is impossible ever to be completely objective is also a distortion. It is a distortion of a different kind, but it is still a distortion. A third way therefore is to recognise that feelings affect what we select as 'facts', and allow for this in our action. This should lead to enquiry with a more open mind – surely a desirable aim for any science.

Paleolimnology and the scientific revolution

What has this to do with *Die Zauberflöte*? To answer this question we must examine the origin of the DSP; where it comes from and how it developed. In order to do so, it is convenient here to begin at the Scientific Revolution of the C16–17th, but it must also be recognised that embedded in the DSP are attitudes that go far back into the development of western civilisation; for example to the Ancient Greeks (Passmore, 1978). Unfortunately this is interesting but I don't have time to deal with it here.

I must recognise that what I have to say may be rather Western, and rather British in its scope. I apologise for my ignorance, and I would welcome any information that you think might help me! Let us begin with Francis Bacon, who lived from 1561 to 1626. He was a very powerful man, at one time the Lord Chancellor (head of the justitiary) of England. It is to Frances Bacon that we owe the origin of the idea of progress via the application of science, and also the scientific or inductive method of hypothesis testing by experiment, that we all use in our work. In *The New Atlantis* of 1627 he described his idea of how science could be used to the benefit of, and to transform, society.

The New Atlantis is a community of scientists who gather knowledge about nature, and use it to the benefit of all. It is therefore what we would call today a research institute, or 'think tank', but with one important difference, which is that Bacon saw the scientists as being in charge of society – benevolent rulers who use scientific knowledge to the benefit of all. It is not a democratic system, but one in which the scientist, the expert, plays a key role. Because the scientists are wise, they govern wisely. I'm not sure I agree with Bacon about the possibility of this, and I also am not sure that all of us, as scientists, would really want this role. However...

Why did Bacon want to increase scientific knowledge? The answer is quite clear in many of his writings – he wanted to increase human power over nature. He saw this in a positive way – in that we could learn how to combat hunger, how to understand disease, how to resist the effects of climate, how to change nature for our own purposes. And there can be no doubt whatsoever that we owe to Bacon, and others who thought like him, a great deal in terms of progress and the comfort of living today. Particularly significant was that he realised the power which would ensue from the fusion of science with technology, in which he was way ahead of his time. But, in this desire to dominate nature, to overcome her, to 'wrest her secrets from her', some writers (for example Carolyn Merchant, 1979) suggest that Bacon possessed too hostile an attitude, and that he underestimated the complexity of nature, and overestimated our ability, not to understand, but to *control*.

Other important scientists and philosophers figure in this discussion, but again I know that I don't have time to dwell on them at length. By these I mean for example René Descartes (1596–1650) to whom we owe the use of mathematics as the language of the sciences, but also the idea that human mind is supremely rational and outside nature, and that nature itself is really only a mechanism that we can understand if we take it to pieces. (How to put it back together is surely now the more important question!) These ideas, and those of Bacon, and of Isaac Newton, led to the concept of the universe as a place governed by fixed, unchanging laws which had to be obeyed, but which could also be used to control nature. In the humanities, these ideas contributed to the development of what is called Classicism, which brings us at last, to *The Magic Flute*.

In *The Magic Flute*, Mozart and Schickaneder (1791) describe the adventures of a prince (Tamino) and a princess (Pamina) who meet and fall in love, but who must pass through trials in order to prove their wisdom, before they may be united and rule together. In particular, Tamino must show that he is wise and understanding and rational, before he can win Pamina and realise his destiny. His kingdom will be the Temple of Light and Wisdom which in the story is ruled by Sarastro, a wise and tolerant high-priest. The enemy of Sarastro is the Queen of the Night who represents disorder, chaos and darkness. The real catch in the story is of course that Pamina is the daughter of the Queen of the Night, and that she can only marry Tamino, and live in a realm of wisdom, joy and light, by renouncing her mother.

Pamina	Mich rufet ja die Kindespflicht denn meine Mutter...
Sarastro	... steht in meiner Macht. Du würdest um dein Glück gebracht, wenn ich dich ihren Händen liesse.
Pamina	Mir klingt der Muttername süsse! Sie ist es, sie ist es...
Sarastro	... Und ein stolzes Weib! Ein Mann muss eure Herzen leiten, denn ohne ihn pflegt jedes Weib aus ihrem Wirkungskreis zu schreiten.[3]

Leaving aside the feminist implications of all this misogyny (but see Jordanova, 1980), it is clear that Sarastro's temple is the analogue of Bacon's *New Atlantis*. The wise philosopher priest Sarastro explains to Tamino and Pamina that it is only by Reason that they will achieve wisdom and be worthy to rule, and that they must renounce Darkness and Unreason to do so. The way forward is into a world governed by wise philosopher-scientist-priests, in which the common people, represented by Papageno, who wants only a little wife to take care of him and bear his many children (another interesting feminist question), are taken care of by the wise, but do not share in the decision-making process (either because they are too ignorant, or because, like Papageno, they don't care!)

My argument anyway, and I am not sure I have supported it very well, is that Classicism and the Enlightenment led to the development in our society of an attitude towards nature that involved domination and control, and that this was rather an arrogant attitude, as we are now finding out. Men – Tamino and Sarastro – were identified with Reason and Light, and Women – the Queen

---

[3] Pamina A child's duty calls me, for my mother
  Sarastro ... stands in my power. You would be robbed of your happiness I left you in her hands.
  Pamina To me, the sound of my mother's name is sweet. It is she, it is she...
  Sarastro And a proud woman! A man must guide your heart. For without a man, a woman would not fulfill her aim in life.

of the Night and Pamina – with Unreason, Chaos and Darkness – wild nature! So, just as Nature could be dominated by the forces of Reason, so also could Woman, who was identified with Nature (Merchant, 1979). We are just beginning to realise what many people have suspected all along, that the DSP does involve a considerable amount of arrogance and is certainly over-optimistic about our ability to control nature. We are probably just as far from being able to do so as we were in Bacon's time, and it looks more and more likely that we will be fooling ourselves if we continue to try.

Some environmentalists therefore have attacked Classicism as being hostile to nature, and the root cause of many of our serious ecological problems. However, in *The Magic Flute* there are many moments of great (romantic!) beauty and perhaps we should hear one now!

William Morris

How then does Morris come into this question? Well, as stated earlier, Morris was a member of the Romantic tradition, and in England at least, Romanticism was both a political and a philosophical reaction to Classicism, as well as an artistic one. It was also a reaction against the emerging industrial society, and many of the early Romantics (in England) deplored not only the despoilation of the landscape, but the hard, 'gradgrind' economic rationality that caused it – the subservience of all other values to economic ones.

Therefore like Wordsworth, they welcomed the French Revolution, and in this, and many other things, they were influenced by Jean-Jacques Rousseau. From Rousseau, the Romantics obtained their emphasis upon freedom and personal expression ('Man is born free but is everywhere in chains'), and their belief in the importance of Nature, both as a source of artistic inspiration, and of mental relief from the dirt, noise and over-crowding of the city. Such values are also found (of course) in Transcendentalism in North America, and in European Romanticism. Nature is seen as wild, but basically benign,

and a place where a more simple, more moral, more natural existence may be lived.

In England the Romantics mostly became disillusioned with the French Revolution, as it did not lead, in their estimation, to a society as free as they believed was desirable. By the time Morris comes into this story, Romanticism had become a rather negative phenomenon, backward-looking and unworldly, rejecting the progress generated by industrialisation.

For most of his life Morris shared these views, but when he was 45 (a dangerous age!) he became interested in politics, and developed his ideas about art and society, and in particular about the nature of Work. And this is always a difficult thing to write and talk about, as many people, of course, would prefer not to think or talk about work at all, and see it as something that is a necessary evil, in order to live, and if we are lucky, to buy time not to work, and to have leisure. And this is precisely the attitude to work that Morris thought was a root cause of many problems, including what we would today call environmental problems.

For Morris, work was something that must be pleasurable. When the majority of people have work which is dirty, debilitating, boring or repetitive, this is hard to defend, but scientists are amongst the few fortunate groups in our present society for whom this is true. Why else would we go out in a gale to collect cores from a freezing lake? Why else would we stay up late counting microfossils etc etc etc? *It's fun!*

So, we are lucky – we follow Morris's model, in that most of our work is pleasurable, and is something in which we are able to express ourselves fairly freely (although here in the UK the government is working on that!). Morris realised this, and late in life he also realised something else, which is that if all work was made pleasurable (or he would say *remade* pleasurable) then production would be only for needs, not wants, and products could be made which enhance our lives without spoiling nature, or at least not nearly so much. As an artist he saw the artist's role in this very clearly, but he also thought that scientists could be involved as well

And Science – we have loved her well, and followed her diligently, what will she do? I fear she is so much in the pay of the counting-house, the counting-house and the drill-sargeant, that she is too busy, and will for the present do nothing. Yet there are matters which I should have though easy for her; say for example teaching Manchester how to consume its own smoke, or Leeds how to get rid of its superfluous black dye without turning it into the river, which would be as much worth her attention as the production of the heaviest of heavy black silks, or the biggest of useless guns. Anyhow, however it be done, unless people care about carrying on their business without making the world hideous, how can they care about Art? I know it will cost much both of time and money to better these things even a little; but I do not see how these can be better spent than in making life cheerful and honourable for others and for ourselves; and the gain of good life to the country at large that would result from men seriously setting about the bettering of the decency of our big towns would be priceless, even if nothing specially good befell the arts in consequence: I do not think matters hopeful if men turned their attention to such things, and I repeat that, unless they do so, we can scarcely even begin with any hope our endeavours for the bettering of the arts.

From *The Lesser Arts* (1877)

Morris also realised that production for needs not wants, would change the appearance of landscape. It is perhaps important to note that what he meant by Art was the production of *all* goods and services – all human *artefacts*. In *News from Nowhere* he describes the landscape of a future society in which production is only for need, and work is pleasurable. The rivers and the air are clean, and forests have sprung up where land was once used to over-produce, or to grow crops which are no longer needed. Industry is non-polluting, and what we would call today alternative

technology is in use. All this has been achieved by removal from production of what Morris called surplus value – the production of cheap, shoddy throw away goods.

'This is how we stand. England was once a country of clearings amongst the woods and wastes, with a few towns interspersed, which were fortresses for the feudal army, markets for the folk, gathering places for the craftsmen. It then became a country of huge and foul workshops and fouler gambling-dens, surrounded by an ill-kept, poverty-stricken farm, pillaged by the masters of the workshops. It is now a garden, where nothing is wasted and nothing is spoilt, with the necessary dwellings, sheds and work-shops scattered up and down the country, all trim and neat and pretty. For, indeed, we should be too much ashamed of ourselves if we allowed the making of goods, even on a large scale, to carry with it the appearance, even, of desolation and misery. Why, my friend, those housewives we were talking of just now would teach us better than that.'

'... our villages are something like the best of such places, with the church or mote-house of the neighbours for their chief building. Only note that there are no tokens of poverty about them: no tumble-down picturesque: which, to tell you the truth, the artist usually availed himself of to veil his incapacity for drawing architecture. Such things do not please us, even when they indicate no misery. Like the mediaevals, we like everything trim and clean, and orderly and bright; as people always do when they have any sense of architectural power; because then they know that they can have what they want, and they won't stand any nonsense from Nature in their dealings with her.'

'Besides the villages, are there any scattered country houses?' said I. 'Yes, plenty' said Hammond; 'in fact, except in the wastes

380

and forests and amongst the sand-hills (like Hindhead in Surrey), it is not easy to be out of sight of a house; and where the houses are thinly scattered they run large, and are more like the old colleges than ordinary houses as they used to be. That is done for the sake of society, for a good many people can dwell in such houses, as the country dwellers are not necessarily husbandmen; though they almost all help in such work at times. The life that goes on in these big dwellings in the country is very pleasant, especially as some of the most studious men of our time live in them, and altogether there is a great variety of mind and mood to be found in them which brightens and quickens the society here.'

'I am rather surprised,' said I, 'by all this, for it seems to me that after all the country must be tolerably populous.' 'Certainly,' said he; 'the population is pretty much the same as it was at the end of the nineteenth century;[4] we have spread it, that is all. Of course, also, we have helped to populate other countries – where we were wanted and were called for.'

Said I: 'One thing, it seems to me, does not go with your word of 'garden' for the country. You have spoken of wastes and forests, and I myself have been the beginning of your Middlesex and Essex forest. Why do you keep such things in a garden? and isn't it very wasteful to do so?' 'My friend,' he said, 'we like these pieces of wild nature, and can afford them, so we have them; let alone that as to the forests, we need a great deal of timber, and suppose that our sons' sons will do the like. As to the land being a garden, I have heard that they used to have shrubberies and rockeries in gardens once; and though I might not like artificial ones, I assure you that some of the natural rockeries of our garden are worth

seeing. Go north this summer and look at the Cumberland and Westmoreland ones – where, by the way, you will see some sheep feeding, so that they are not so wasteful as you think; not so wasteful as forcing-grounds for fruit out of season, I think. Go and have a look at the sheep-walks high up on the slopes between Ingleborough and Pen-y-gwent, and tell me if you think we waste the land there by not covering it with factories for making things that nobody wants, which was the chief business of the nineteenth century.' 'I will try to go there,' said I. 'I won't take much trying,' said he.
from *News From Nowhere* (1890).

Paleolimnology

How does this relate to Paleolimnology? Well it relates to us in several ways I think, the first of which is that as already stated, the distinction between work and leisure, for many of us, does not really exist. Some things we find a bore, some things are not connected with science, but mostly it's a blur, the distinction is not clear. Second, as Morris also suggested, pleasurable work would mean that the distinction between art and science would also begin to disappear. Scientists express themselves via science just as much as artists express themselves via art. Who amongst us has not considered the 'art' of paleolimnology, or thought of some particular study by a colleague as being very *beautiful* work?

So paleolimnology begins, as a way of looking at the world, to satisfy some of Morris's most important criteria. It is pleasurable, and it allows great self-expression (as does a great deal of science). Also, of course, it follows the Romantic rather than the Classical view of the world, in that it is an integrative rather than a divisive subject. Paleolimnologists may not consider themselves experts in all aspects of the field, but we all have a greater or a lesser idea how the physics, chemistry, biology and humanity of lakes and their watersheds fit together, and we have a strong sense of interaction in nature through time. In

[4] ca. 40 million

this, I am sure we follow not Descartes and his mechanistic, reductionist approach to Nature, but an holistic, integrative one. Ed Deevey (1969), and David Frey (1979, 1984, 1989) have dealt with this topic much more exhaustively than I have been able to here.

Paleolimnology and paradigm shifts

The World is changing, fast! Nowhere, of course, during 1989, was the pace of change so fast as in Eastern Europe. In the West, some feel very confident that our very way of life has somehow been vindicated by all this change elsewhere. Nothing could be further from the truth! Smug and arrogant essays like 'The End of History' (Fukuyama, 1989) are no substitute for scientific and intellectual rigour!

If paleolimnology is to respond to changes that are happening elsewhere in society, in which direction could we go? Again, in answering this question, I wish to emphasise that these are only my opinions, and I am not trying the legislate for anyone else at all.

Several writers have tried to do more than speculate about the nature of society, and of science, in the 'Post-Industrial' future. One is James Robertson (1983) who suggests that in the near future, the following paradigm shifts, which to a certain extent he thinks have already taken place, will become widely accepted.

The concept of Growth will change from the merely economic to that of a more flexible type of growth in a steady state economy in which production will be for needs not wants. There will be growth in some parts of the economy, and contraction in others; and different kinds of growth, intellectual, moral, spiritual, as well as economic, will be accepted as legitimate. Similarly, Wealth will be thought of not as 'the flow of goods and services', but in terms of stocks – how many trees we have left in the forest, rather than how many we have transformed into chipboard, copies of *The Sun*, or matchsticks. There will also be an emphasis upon what environmental economists call positional goods – the experience obtained by

being in a specific place at a particular point in time (e.g. by the shore of Längsee in early September 1985!), which is, of course, both non-quantifiable and irreplaceable.

Work will become much more like Morris's concept of 'useful' work – something which is pleasurable and rewarding, and which produces a commodity or service of lasting value to the community. Thus people in general will produce less, consume fewer resources, and have more time for self-expression. Education will need to respond to this increased personal autonomy, rather than as now, being confined to producing members of a 'work force', for a purely economic rule. It will not, however, be 'education for leisure'. We can't all end up selling each other tickets!

In such a society, power will need to be dispersed to individuals, and societies will need to become much more democratic. By this I do not mean the kind of over-centralised representative democracy we now seem to have developed in the west, but a much more participating democracy in which participation is seen not just as a right, but also as responsibility, and as an educative process. How else will we achieve the kind of control over our own lives which we will need if we are to adopt the truly 'green' lifestyle for which so many of us long?

Another person who has written about the 'post-industrial' future, and who has given rather more attention to the nature of a truly 'green' science, is Hazel Henderson (1983). She suggests that Post-Industrial science rejects domination of nature as a goal, and substitutes for it understanding, and *wisdom*, and a reverence for the beauty and complexity of Nature. (Aren't we all *fascinated* by the beauty of freshwater lakes?).

A green science would also be descriptive and explanatory, but not manipulative. Control of Nature would not be an objective, and the importance of future generations of scientists as potentially more sophisticated investigators would be recognised. This is, of course, already an important principle in archaeology, where investigators deliberately leave much of a site unexcavated, in order to preserve it for the future, and I was recently reminded by Rick Battarbee of the similar importance of not using very destructive coring methods at very small lakes.

As already suggested, 'green' science will become much less reductionist. The wider, social context in which scientists operate will be recognised, and more of 'the data' will be seen to be 'relevant'. A manipulative attitude to Nature is not easy to sustain in an holistic context. Finally, 'green' science will be much less interventionist. Paleolimnology is, of course, ideal for recording the damage done to Nature by human intervention, but Maureen's question could also be phrased – how do we put things back? Paleolimnology also lends itself to this restorative approach, in that earlier states of the lake and its catchment can be identified, some of which may represent more stable conditions. But really, 'putting things back' requires a more fundamental shift and in society at large, in which 'green' scientists could practise 'green' science in an entirely different way.

Practising paleolimnology in a more democratic society would involve most of the things we do now, and in particular, acting as advisers to the wider community as to how a more sustainable lifestyle might be developed. I can cite, for example, the role played by Tommy Edmondson in the restoration of Lake Washington (Chasan, 1971), and the many similar (perhaps smaller-scale!) causes in which many of us are already involved in our 'spare' time. 'Green' scientists would not therefore distance themselves from the rest of society, nor operate in an elitist, 'white-coat' context. Instead they would become more closely involved in a more democratic, less centralised society. Scientists are, after all, people too, and recognising this is a more objective, more holistic view.

Anyway, I have reached the end of my discourse, and I am not going to tell you how the save the world (although I think I know how!). I would welcome anyone's comments if they wish to make them.

References

Bacon, F., 1627. The New Atlantis. Reprinted in The World's Classics, Oxford University Press, London (1969).

Capra, F., 1983. The Turning Point: Science, Society and The Rising Culture. Flamingo, London.

Catton, W. L., jr. & R. E. Dunlap, 1980. A new ecological paradigm for post-Exuberant Sociology. Am. Behavioural Scientist, 24: 15–47.

Chasan, D. J., 1971. The Seattle area wouldn't allow death of its lake. Smithsonian 2: 6–12.

Coleman, S. & P. E. O'Sullivan, 1990. William Morris and News from Nowhere: a vision for our time. Green Books, Hartland, Devon, U.K.

Cotgrove, S. & A. Duff, 1980. Environmentalism, middle-class radicalism, and politics. Sociol. Rev. 28: 331–351.

Deevey, E. S., jr, 1942. A re-examination of Thoreau's 'walden'. Q. R. Biol. 17(1): 1–11.

Deevey, E. S., jr, 1969. Coaxing history to conduct experiments. BioScience 19: 40–43.

Frey, D. G., 1969. The rationale of Paleolimnology. Mitt. Int. Ver. Limnol. 17: 7–18.

Frey, D. G., 1974. Paleolimnology. Mitt. Int. Ver. Limnol. 20: 95–123.

Frey, D. G., 1988. What is paleolimnology? J. Paleolimnol. 1: 1–5.

Fukuyama, F., 1989. The End of History? National Interest (Summer 1989). Washington, DC. Referred to in The Guardian, 4 November 1989, London and Manchester, UK, p. 25.

Henderson, H., 1983. The Warp and the Weft: the coming synthesis of Ecophilosophy and Ecofeminism. In Caldecott, L. & S. Leland, (eds), Reclaim the Earth: Women speak out for Life on Earth, The Women's Press, London.

Kuhn, T. L., 1972. The Structure of Scientific Revolutions. University of Chicago Press.

Merchant, C., 1979. The Death of Nature: Women, Ecology and the Scientific Revolution. Wildwood House, London.

Jordanova, L. J., 1980. Natural facts: a historical perspective on science and sexuality. In McCormack, C. & M. Strathern (eds), Nature, Culture and Gender. Cambridge University Press.

Mozart, W. A. & E. Schickaneder, 1791. Die Zauberflote (The Magic Flute). Published by Cassell, London (1971).

O'Sullivan, P. E., 1987. Environmental Science and Environmental Philosophy 2 – Environmental Science and the coming social paradigm. Int. J. Env. Studies 28: 257–267.

O'Sullivan, P. E., 1990. Struggle for the vision fair – Morris and Ecology. J. William Morris Soc. VIII(4): 5–9.

Passmore, J., 1978. Man's Responsibility for Nature. Duckworth, London.

Robertson, J., 1983. The Sane Alternative: a choice of Futures. J. Robertson, Cholney, Oxon, UK.

The manufacturer's authorised representative in the EU is Springer
Nature Customer Service Centre GmbH, Europaplatz 3, 69115 Heidelberg,
Germany. If you have any concerns regarding our products, please
contact ProductSafety@springernature.com

Printed and bound by CPI Group (UK) Ltd, Croydon, CR0 4YY

23/04/2026

02095657-0004